*Broadband Circuits
for Optical Fiber
Communication*

Broadband Circuits for Optical Fiber Communication

Eduard Säckinger

A JOHN WILEY & SONS, INC., PUBLICATION

TK
7871.58
.B74
S23
2005

Copyright © 2005 by John Wiley & Sons, Inc. All rights reserved.

Published by John Wiley & Sons, Inc., Hoboken, New Jersey.
Published simultaneously in Canada.

No part of this publication may be reproduced, stored in a retrieval system or transmitted in any form or by any means, electronic, mechanical, photocopying, recording, scanning or otherwise, except as permitted under Section 107 or 108 of the 1976 United States Copyright Act, without either the prior written permission of the Publisher, or authorization through payment of the appropriate per-copy fee to the Copyright Clearance Center, Inc., 222 Rosewood Drive, Danvers, MA 01923, (978) 750-8400, fax (978) 646-8600, or on the web at www.copyright.com. Requests to the Publisher for permission should be addressed to the Permissions Department, John Wiley & Sons, Inc., 111 River Street, Hoboken, NJ 07030, (201) 748-6011, fax (201) 748-6008.

Limit of Liability/Disclaimer of Warranty: While the publisher and author have used their best efforts in preparing this book, they make no representation or warranties with respect to the accuracy or completeness of the contents of this book and specifically disclaim any implied warranties of merchantability or fitness for a particular purpose. No warranty may be created or extended by sales representatives or written sales materials. The advice and strategies contained herein may not be suitable for your situation. You should consult with a professional where appropriate. Neither the publisher nor author shall be liable for any loss of profit or any other commercial damages, including but not limited to special, incidental, consequential, or other damages.

For general information on our other products and services please contact our Customer Care Department within the U.S. at 877-762-2974, outside the U.S. at 317-572-3993 or fax 317-572-4002.

Wiley also publishes its books in a variety of electronic formats. Some content that appears in print, however, may not be available in electronic format.

Library of Congress Cataloging-in-Publication Data:

Säckinger, Eduard, 1959–
  Broadband circuits for optical fiber communication / Eduard Säckinger.
    p. cm.
  Includes bibliographical references and index.
  ISBN 0-471-71233-7 (cloth)
    1. Fiber optics. 2. Optical communications.—Equipment and supplies. 3. Broadband amplifiers. 4. Integrated circuits, Very large scale integration. I. Title.

TK7871.58.B74S23 2005
621.383'75—dc22   2004060617

Printed in the United States of America

10 9 8 7 6 5 4 3 2 1

To my wife, Hye-Sun

Preface

This book is the result of lecturing on "Broadband Circuits for Optical Fiber Communication" over the past several years (at Agere Systems and Lucent Technologies seminars; VLSI Symposium, June 2000; MEAD Microelectronics, 2001–2002). During this period, I experimented with various ways of presenting the material and eventually settled on the structure used for this book, which I found worked best. Compared with the lectures, which were limited to just a few hours, this book permits me to go into more detail and to provide many more examples.

Scope. We discuss five types of broadband circuits: transimpedance amplifiers, limiting amplifiers, automatic gain control (AGC) amplifiers, laser drivers, and modulator drivers. Some background information about optical fiber, photodetectors, lasers, and modulators is provided to elucidate the system environment in which these circuits operate. A summary of receiver theory is given at the outset to streamline the discussion of the receiver circuits in the later chapters.

For each of the five circuit types, I proceed as follows. First, the main specifications are explained and illustrated with example numerical values. In many IC design projects, a significant amount of time is spent determining the right specifications for the new design. Therefore, emphasis is put on how these specs relate to the system performance. Next, the circuit concepts are discussed in a general manner. At this point, we try to abstract as much as possible from specific semiconductor technologies, bit rates, and so forth. Then, these general concepts are illustrated with practical implementations taken from the literature. A broad range of circuits in

MESFET, HFET, BJT, HBT, BiCMOS, and CMOS technologies are covered. Finally, a brief overview of product examples and current research topics are given.

The focus of this book is on circuits for digital, continuous-mode transmission, which are used, for example, in SONET, SDH, and Gigabit Ethernet applications. Furthermore, we concentrate on high-speed circuits in the range of 2.5 to 40 Gb/s, typically used in long-haul and metro networks. Circuits for burst-mode transmission, which are used, for example, in passive optical networks (PON), as well as analog receiver and transmitter circuits, which are used, for example, in hybrid fiber-coax (HFC) cable-TV systems, also are discussed.

It is assumed that the reader is familiar with basic analog IC design as presented, for example, in *Analysis and Design of Analog Integrated Circuits* by P. R. Gray and R. G. Meyer [34] or a similar book [7, 57].

Style and Audience. My aim has been to present an overview of the field, with emphasis on an intuitive understanding. Many references to the literature are made throughout this book to guide the interested reader to a more complete and in-depth treatment of the various topics. In general, the mindset and notation used are those of an electrical engineer. For example, whenever possible we use voltages and currents rather than abstract variables, we use one-sided spectral densities as they would appear on a spectrum analyzer, we prefer the use of noise bandwidths over Personick integrals, and so forth. Examples are given frequently to make the material more concrete. Many problems, together with their answers, are provided for the reader who wants to practice and deepen his understanding of the learned material. The problem and answer sections also serve as a repository for additional material, such as proofs and generalizations that would be too distracting to present in the main text of this overview. I hope this book will be useful to students or professionals who may wish for some survey of this subject without becoming embroiled in too much technical detail.

Acknowledgments. I would like to thank my colleagues at the Bell Laboratories and Agere Systems, from whom I have learned much of what is presented in this book. I also would like to thank Behzad Razavi, who got this book project started by inviting me to the VLSI Symposium 2000 and asking me to present a tutorial on "Broadband Circuits for Optical Fiber Communications," which later evolved into this book with the same title. I am grateful to Vlado Valence, Ibi and Gabor Temes, and all the other people at MEAD Microelectronics who have made teaching in their course a pleasure.

I am deeply indebted to the many reviewers who have given freely of their time to read through the book, in part or in full. In particular, I am most grateful to Behnam Analui, California Institute of Technology; Prof. Hercules Avramopoulos, National Technical University of Athens; Dr. Kamran Azadet, Agere Systems; Dr. Alexandru Ciubotaru, Maxim Integrated Products; Dr. Sherif Galal, Broadcom Corp.; Dr. Yuriy M. Greshishchev, PMC-Sierra Inc.; Prof. Renuka Jindal, Universisty of Louisiana at Lafayette; Dr. Helen H. Kim, MIT Lincoln Laboratory; Dr. Herwig Kogelnik, Bell Laboratories, Lucent Technologies; Dr. Patrik Larsson, utMOST Technologies; Dr. Marc Loinaz, Aeluros Inc.; Dr. Sunderarajan Mohan, Barcelona Design Inc.;

Dr. Kwok Ng, Agere Systems; Nicolas Nodenot, National Semiconductor; Dr. Yusuke Ota, Zenko Technologies Inc.; Joe H. Othmer, Agere Systems; Prof. Sung-Min Park, Ewha Women's University, Seoul; Prof. Ken Pedrotti, University of California, Santa Cruz; Prof. Khoman Phang, University of Toronto; Hans Ransijn, Agere Systems; Prof. Behzad Razavi, University of California, Los Angeles; Prof. Hans-Martin Rein, Ruhr-Universität, Bochum, Germany; Fadi Saibi, Agere Systems; Dr. Leilei Song, Agere Systems; Prof. Sorin Voinigescu, University of Toronto; Jim Yoder, Agere Systems; and Dr. Ty Yoon, Intel Corporation.

Finally, I would like to thank my wife, who has endured more than two years of "weekend work" during which I have converted my lecture notes into this book manuscript.

Despite the effort made, there are undoubtedly some mistakes left in this book. If you have any corrections or suggestions, please e-mail them to edi@ieee.org. Thank you!

E. SÄCKINGER

Tinton Falls, New Jersey
September 9, 2004

Contents

	Preface	*vii*
1	*Introduction*	*1*
2	*Optical Fiber*	*11*
	2.1 Loss and Bandwidth	*11*
	2.2 Dispersion	*14*
	2.3 Nonlinearities	*18*
	2.4 Pulse Spreading due to Chromatic Dispersion	*19*
	2.5 Summary	*22*
	2.6 Problems	*23*
3	*Photodetectors*	*25*
	3.1 p-i-n Photodetector	*25*
	3.2 Avalanche Photodetector	*31*
	3.3 p-i-n Detector with Optical Preamplifier	*34*
	3.4 Summary	*40*
	3.5 Problems	*42*

xii CONTENTS

4 Receiver Fundamentals — 45
- 4.1 Receiver Model — 45
- 4.2 Bit-Error Rate — 47
- 4.3 Sensitivity — 54
- 4.4 Personick Integrals — 66
- 4.5 Power Penalty — 70
- 4.6 Bandwidth — 73
- 4.7 Adaptive Equalizer — 82
- 4.8 Nonlinearity — 86
- 4.9 Jitter — 90
- 4.10 Decision Threshold Control — 95
- 4.11 Forward Error Correction — 96
- 4.12 Summary — 100
- 4.13 Problems — 101

5 Transimpedance Amplifiers — 105
- 5.1 TIA Specifications — 105
  - 5.1.1 Transimpedance — 105
  - 5.1.2 Input Overload Current — 107
  - 5.1.3 Maximum Input Current for Linear Operation — 108
  - 5.1.4 Input-Referred Noise Current — 108
  - 5.1.5 Bandwidth and Group-Delay Variation — 111
- 5.2 TIA Circuit Concepts — 112
  - 5.2.1 Low- and High-Impedance Front-Ends — 112
  - 5.2.2 Shunt Feedback TIA — 113
  - 5.2.3 Noise Optimization — 121
  - 5.2.4 Adaptive Transimpedance — 130
  - 5.2.5 Post Amplifier — 132
  - 5.2.6 Common-Base/Gate Input Stage — 133
  - 5.2.7 Current-Mode TIA — 134
  - 5.2.8 Active-Feedback TIA — 135
  - 5.2.9 Inductive Input Coupling — 136
  - 5.2.10 Differential TIA and Offset Control — 137
  - 5.2.11 Burst-Mode TIA — 141
  - 5.2.12 Analog Receiver — 143
- 5.3 TIA Circuit Implementations — 145
  - 5.3.1 MESFET and HFET Technology — 145
  - 5.3.2 BJT, BiCMOS, and HBT Technology — 147
  - 5.3.3 CMOS Technology — 149

	5.4	Product Examples	151
	5.5	Research Directions	151
	5.6	Summary	154
	5.7	Problems	156
6	*Main Amplifiers*		*159*
	6.1	Limiting vs. Automatic Gain Control (AGC)	159
	6.2	MA Specifications	161
		6.2.1 Gain	161
		6.2.2 Bandwidth and Group-Delay Variation	164
		6.2.3 Noise Figure	165
		6.2.4 Input Dynamic Range	169
		6.2.5 Input Offset Voltage	171
		6.2.6 Low-Frequency Cutoff	173
		6.2.7 AM-to-PM Conversion	175
	6.3	MA Circuit Concepts	176
		6.3.1 Multistage Amplifier	176
		6.3.2 Techniques for Broadband Stages	179
		6.3.3 Offset Compensation	203
		6.3.4 Automatic Gain Control	207
		6.3.5 Loss of Signal Detection	211
		6.3.6 Burst-Mode Amplifier	212
	6.4	MA Circuit Implementations	213
		6.4.1 MESFET and HFET Technology	213
		6.4.2 BJT and HBT Technology	215
		6.4.3 CMOS Technology	221
	6.5	Product Examples	224
	6.6	Research Directions	226
	6.7	Summary	227
	6.8	Problems	228
7	*Optical Transmitters*		*233*
	7.1	Transmitter Specifications	234
	7.2	Lasers	237
	7.3	Modulators	247
	7.4	Limits in Optical Communication Systems	253
	7.5	Summary	256
	7.6	Problems	257

xiv CONTENTS

8	Laser and Modulator Drivers		259
	8.1	Driver Specifications	259
		8.1.1 Modulation and Bias Current Range (Laser Drivers)	259
		8.1.2 Output Voltage Range (Laser Drivers)	261
		8.1.3 Modulation and Bias Voltage Range (Modulator Drivers)	261
		8.1.4 Power Dissipation	263
		8.1.5 Rise and Fall Times	264
		8.1.6 Pulse-Width Distortion	265
		8.1.7 Jitter Generation	265
		8.1.8 Eye-Diagram Mask Test	267
	8.2	Driver Circuit Concepts	268
		8.2.1 Current-Steering Output Stage	268
		8.2.2 Back Termination	273
		8.2.3 Predriver	276
		8.2.4 Pulse-Width Control	279
		8.2.5 Data Retiming	280
		8.2.6 Automatic Power Control (Lasers)	282
		8.2.7 End-of-Life Detection (Lasers)	285
		8.2.8 Automatic Bias Control (MZ Modulators)	286
		8.2.9 Burst-Mode Laser Driver	287
		8.2.10 Analog Laser/Modulator Driver	290
	8.3	Driver Circuit Implementations	294
		8.3.1 MESFET and HFET Technology	294
		8.3.2 BJT and HBT Technology	297
		8.3.3 CMOS Technology	302
	8.4	Product Examples	305
	8.5	Research Directions	305
	8.6	Summary	308
	8.7	Problems	309
Appendix A	Eye Diagrams		313
Appendix B	Differential Circuits		321
	B.1	Differential Mode and Common Mode	322
	B.2	The Modes of Currents and Impedances	324
	B.3	Common-Mode and Power-Supply Rejection	325

Appendix C   S Parameters — *329*
   C.1   Definition and Simulation — *329*
   C.2   Matching Considerations — *333*
   C.3   Differential S Parameters — *339*

Appendix D   Transistors and Technologies — *343*
   D.1   MOSFET and MESFET — *343*
   D.2   Heterostructure FET (HFET) — *348*
   D.3   Bipolar Junction Transistor (BJT) — *351*
   D.4   Heterojunction Bipolar Transistor (HBT) — *355*

Appendix E   Answers to the Problems — *359*

Appendix F   Notation — *385*

Appendix G   Symbols — *387*

Appendix H   Acronyms — *399*

References — *407*

Index — *425*

1
Introduction

Optical Receiver and Transmitter. Figure 1.1 shows the block diagram of a typical optical receiver and transmitter. The optical signal from the fiber is received by a *photodetector* (PD), which produces a small output current proportional to the optical signal. This current is amplified and converted to a voltage by a *transimpedance amplifier* (TIA or TZA). The voltage signal is amplified further by either a *limiting amplifier* (LA) or an *automatic gain control amplifier* (AGC amplifier). The LA and AGC amplifier are collectively known as *main amplifiers* (MAs) or *post amplifiers*. The resulting signal, which is now several 100 mV strong, is fed into a *clock and data recovery circuit* (CDR), which extracts the clock signal and retimes the data signal. In high-speed receivers, a *demultiplexer* (DMUX) converts the fast serial data stream into n parallel lower-speed data streams that can be processed conveniently by the digital logic block. Some CDR designs (those with a parallel sampling architecture) perform the DMUX task as part of their functionality, and an explicit DMUX is not needed in this case [47]. The digital logic block descrambles or decodes the bits, performs error checks, extracts the payload data from the framing information, synchronizes to another clock domain, and so forth. The receiver just described also is known as a *3R receiver* because it performs signal *re-amplification* in the TIA (and the AGC amplifier, if present), signal *re-shaping* in the LA or CDR, and signal *re-timing* in the CDR.

On the transmitter side, the same process happens in reverse order. The parallel data from the digital logic block are merged into a single high-speed data stream using a *multiplexer* (MUX). To control the select lines of the MUX, a bit-rate (or half-rate) clock must be synthesized from the slower word clock. This task is performed by a *clock multiplication unit* (CMU). Finally, a *laser driver* or *modulator driver* drives

2 INTRODUCTION

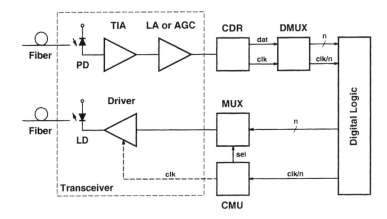

Fig. 1.1 Block diagram of an optical receiver (top) and transmitter (bottom).

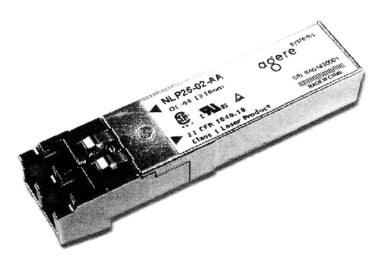

Fig. 1.2 A 2.5-Gb/s transceiver in a "small form-factor pluggable" (SFP) package (5.7 cm × 1.4 cm × 1.1 cm). The two fibers are plugged in from the left (LC connectors). Reprinted by permission from Agere Systems, Inc.

the corresponding optoelectronic device. The laser driver modulates the current of a *laser diode* (LD), whereas the modulator driver modulates the voltage across a *modulator*, which in turn modulates the light intensity from a *continuous wave* (CW) laser. Some laser/modulator drivers also perform data retiming, and thus require a bit-rate (or half-rate) clock from the CMU (dashed line in Fig. 1.1).

A module containing a PD, TIA, MA, laser driver, and LD, that is, all the blocks shown inside the dashed box of Fig. 1.1, often is referred to as a *transceiver*.[1] See Fig. 1.2 for a so-called *small form-factor* transceiver module. A module that contains all the functionality of the transceiver plus a CDR, DMUX, CMU, and MUX frequently is called a *transponder*. In this book, we discuss the PD, TIA, MA, as well as the laser/modulator and their drivers in greater detail.

Modulation Schemes. The most commonly used modulation format in optical communication is the *non-return-to-zero* (NRZ) format shown in Fig. 1.3(a). This format is a form of *on-off keying* (OOK): the signal is *on* to transmit a one bit and is *off* to transmit a zero bit. When the signal (i.e., the laser light) is on, it stays on for the entire bit period. For example, when transmitting the periodic bit pattern "010101010..." at 10 Gb/s in NRZ format, a 5-GHz square wave with 50% duty cycle is produced.[2]

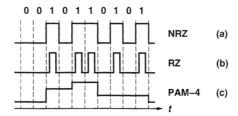

Fig. 1.3 Modulation schemes: (a) NRZ, (b) RZ, and (c) PAM-4.

In high-speed and long-haul transmission (e.g., fiber links between two continents), the *return-to-zero* (RZ) format, shown in Fig. 1.3(b), generally is preferred. In this format, the pulses, which represent the one bits, occupy only a fraction (e.g., 50%) of the bit period. Compared with the NRZ signal, the RZ signal requires less signal-to-noise ratio for reliable detection but occupies a larger bandwidth because of its shorter pulses.[3] An important advantage of this narrow-pulse format is that more pulse distortion and spreading can be tolerated without disturbing the adjacent bits.

---

[1] The term *transceiver* is a contraction of the words "*trans*mitter" and "*receiver*".
[2] In some standards, such as Fast Ethernet and FDDI, the *non-return-to-zero change-on-ones* (NRZ1 or NRZI) format is used. This format also is based on NRZ modulation, but before modulation, the bit stream is passed through a line coder that changes its (binary) output value when the bit to be transmitted is a one and leaves the output value unchanged when the bit is a zero.
[3] To receive data at a bit-error rate of 10^{-12}, we need a signal-to-noise ratio of 16.9 dB for NRZ modulation, 15.7 dB for 50%-RZ modulation, and 23.9 dB for PAM-4 modulation assuming additive Gaussian noise

Thus, this format is more immune to effects of fiber nonlinearity and polarization-mode dispersion. On the downside, faster, more expensive transceiver components (laser/modulator, photodetector, front-end electronics, etc.) are required to handle the shorter pulses. Furthermore, in optical multichannel systems, the wavelengths can be packed less densely because the RZ signal occupies a wider bandwidth than the NRZ signal for a given bit rate. Several variations of the RZ modulation, such as the *chirped return-to-zero* (CRZ) modulation, the *carrier-suppressed return-to-zero* (CS-RZ) modulation, and the *return-to-zero differential phase-shift keying* (RZ-DPSK) modulation also are used, but a discussion of these modulation schemes is beyond the scope of this book.

Since the late 1980s, the TV signals in *community-antenna television* (CATV) systems often are transported first *optically* from the distribution center to the neighborhood before they are distributed to the individual homes on conventional coaxial cable. This combination, called *hybrid fiber-coax* (HFC), has the advantage over an all-coax system in that it saves many electronic amplifiers (the loss in a fiber is much lower than the loss in a coax cable) and provides better signal quality (lower noise and distortions). In the optical part of the HFC system, the laser light is modulated with multiple radio-frequency (RF) carriers, so-called subcarriers, each one corresponding to a different TV channel. This method is known as *subcarrier multiplexing* (SCM).[4] Then, each subcarrier is modulated with a TV signal, for example, *amplitude modulation with vestigial sideband* (AM-VSB) is used for analog TV channels and *quadrature amplitude modulation* (QAM) is used for digital TV channels.

In contrast to NRZ and RZ modulation, which produce a two-level digital signal (laser light on or off), the AM-VSB and QAM modulation used in CATV applications produce continuous or multilevel analog signals. Figure 1.3(c) illustrates a multilevel signal produced by *pulse amplitude modulation* (PAM), a modulation scheme that is related to QAM. In this example, groups of two successive bits are encoded with one of four signal levels, and hence this format is known as PAM-4. Compared with the NRZ signal, the PAM-4 signal requires a higher signal-to-noise ratio for reliable detection but occupies a narrower bandwidth because its symbol rate is only half the bit rate. Similarly, the analog signals distributed over CATV/HFC systems require a high signal-to-noise ratio and transceivers with a good linearity to minimize distortions.

In the remainder of this book, we always assume that we are dealing with NRZ modulation, except if stated otherwise, as, for example, in the sections on analog receiver and transmitter circuits.

Line Codes. Before data bits are modulated onto the optical carrier, they usually are preconditioned with a so-called *line code*. The line code provides the transmitted bit stream with the following desirable properties: *DC balance*, short *run lengths*,

---

(cf. Problems 4.4 and 4.6). The signal bandwidth measured from DC to the first null is B for NRZ modulation, $2B$ for 50%-RZ modulation, and $B/2$ for PAM-4 modulation, where B is the bit rate.

[4]In contrast to *discrete multitone* (DMT) modulation (or *orthogonal frequency division multiplexing* [OFDM], the RF modulated equivalent), which uses overlapping channel spectra, SCM keeps a frequency gap between the channels.

and a high *transition density*. A DC balanced bit stream contains the same number of zeros and ones on average. This is equivalent to saying that the average *mark density* (number of one bits divided by all bits) is 50%. A DC balanced signal has the nice property that its average value (the DC component) is always centered halfway between the zero and one levels (half of the peak-to-peak value). This property often permits the use of AC coupling between circuit blocks, simplifying their design. Furthermore, it is desirable to keep the number of successive zeros and ones, the run length, to a small value. This provision reduces the low-frequency content of the transmitted signal and limits the associated *baseline wander* (a.k.a. *DC wander*) when AC coupling is used. Finally, a high transition density is desirable to simplify the clock recovery process.

In practice, line coding is implemented as either *scrambling*, *block coding*, or a combination of the two:

- *Scrambling*. In this case, a *pseudorandom bit sequence* (PRBS) is generated with a feedback shift register and xor'ed with the data bit stream (see Fig. 1.4). Note that the data can be descrambled with the same arrangement, provided the descrambling PRBS generator is synchronized with the scrambling PRBS generator. Scrambling provides DC balance without adding overhead bits to the bit stream, thus preserving the bit rate. On the down side, the maximum run length is not strictly limited, that is, there is a small chance for very long runs of zeros or ones, which can be hazardous. In practice, runs up to 72 bits usually are expected. The scrambling method is used in the United States telecommunication system described in the SONET (synchronous optical network) standard [188] and the almost identical SDH (synchronous digital hierarchy) standard used in Europe and Japan.

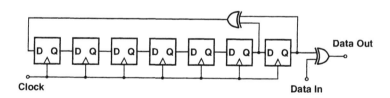

Fig. 1.4 Implementation of a SONET scrambler.

- *Block Coding*. In this case, a contiguous group of bits (a block) is replaced by another slightly larger group of bits such that the average mark density becomes 50% and DC balance is established. For example, in the 8B10B code, 8-bit groups are replaced with 10-bit patterns using a look-up table [198]. The 8B10B code increases the bit rate by 25%; however, the maximum run length is strictly limited to five zeros or ones in a row. The 8B10B code is used in the

6   INTRODUCTION

Gigabit Ethernet (GbE, 1000Base-SX, 1000Base-LX) and Fiber Channel data communication systems.[5]

- *Combination.* In the serial 10-Gigabit Ethernet (10-GbE) system, DC balance is established first by scrambling the bit stream and then by applying a block-code (64B66B code) to it. This combination features low overhead ($\approx$ 3% increase in bit rate) and a run length that is strictly limited to 66 bits.

Continuous Mode vs. Burst Mode. It is important to distinguish two types of transmission modes because they call for different circuit designs: *continuous mode* and *burst mode*. The signals corresponding to these two modes are shown schematically in Fig. 1.5.

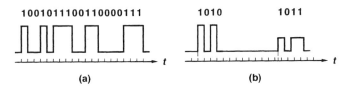

Fig. 1.5 (a) Continuous-mode vs. (b) burst-mode signals (schematically).

In continuous-mode transmission, a continuous, uninterrupted stream of bits is transmitted as shown in Fig. 1.5(a). The transmitted signal usually is DC balanced using one of the line codes described earlier. As a result, AC coupled circuits normally can be used. In burst-mode transmission, data are transmitted in short *bursts*, with the transmitter remaining silent (laser off) in between bursts. See Fig. 1.5(b) for an illustration, but note that practical bursts are much longer than those shown in the figure, typically longer than 400 bits.

Bursts can be fixed or can be variable in length. Bursts that encode ATM (asynchronous transfer mode) cells have a *fixed length*, they always contain 53 bytes plus a preamble (e.g., 3 bytes). Bursts that encode Ethernet frames have a *variable length* (70–1524 bytes). In either case, the bursts start out with a preamble (a.k.a. overhead) followed by the payload. The burst-mode receiver uses the preamble to establish the decision threshold level (slice level) and to synchronize the receiver clock. In passive optical network systems, to be discussed shortly, bursts arrive *asynchronously* and with strongly *varying power levels* (up to 30 dB); therefore, the clock signal must be synchronized and the slice level adjusted for every single burst (cf. Fig. 1.5(b)).

The average value (DC component) of a burst-mode signal varies with time, depending on the burst activity. If the activity is high, it may be close to the halfway point between the zero and one levels, as in a continuous mode system; if the activity is low, the average drifts arbitrarily close to the zero level. This means that the burst-mode signal is *not* DC balanced, and in general, AC coupling cannot be used because

---

[5]The 4B5B code used in Fast Ethernet (100Base-TX, 100Base-FX), FDDI, and so forth, does not achieve perfect DC balance; the worst-case unbalance is 10% [136].

it would lead to excessive baseline wander. (Note that the mark density *within a burst* may well be 50%, but the overall signal is still not DC balanced.) This lack of DC balance and the fact that bursts often arrive with varying amplitudes necessitate specialized amplifier and driver circuits for burst-mode applications. Furthermore, the asynchronous arrival of the bursts requires specialized fast-locking CDRs.

In the remainder of this book we always assume that we are dealing with DC-balanced, continuous-mode signals, except if stated otherwise, as, for example, in the sections on burst-mode circuits.

Optical Networks. We must distinguish two important types of optical networks: the simple *point-to-point connection* and the *point-to-multipoint network*. In the following, we discuss these networks and how continuous-mode and burst-mode transmissions are used with them.

An optical point-to-point connection between two *central offices* (CO) is illustrated schematically in Fig. 1.6(a). An example for such a connection is a SONET OC-192 link operating at 10 Gb/s (9.953 28 Gb/s to be precise), a bit rate that can carry about 130,000 voice calls. Point-to-point links are used over a wide range of bit rates and distances, from short computer-to-computer links to ultra-long-haul undersea lightwave systems.

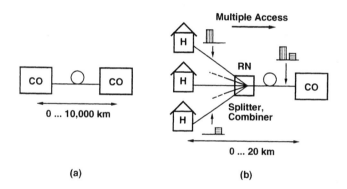

Fig. 1.6 Example of (a) a point-to-point link and (b) a point-to-multipoint network.

Point-to-point connections can be assembled into more complex structures such as *ring networks* and *active star networks*. Examples for ring networks are provided by SONET/SDH rings and FDDI token rings. An active star is formed, for example, by Gigabit Ethernet links converging into a hub. It is important to realize that each individual optical connection of the star has a transceiver on both ends and therefore forms an optical point-to-point link. This is in contrast to a *passive star network* or an optical point-to-multipoint network, where multiple optical fibers are coupled with a passive optical device. We discuss the latter network type below.

Continuous-mode transmission is used on almost all point-to-point connections. One exception occurs in half-duplex systems, in which bidirectional communication is implemented by periodically reversing the direction of traffic following a ping-pong pattern, so-called *time compression multiplexing* (TCM; a.k.a. *time division duplexing*). Such systems require burst-mode transmitters and receivers. However, for bandwidth efficiency reasons, TCM systems are limited to relatively short links (e.g., home networking applications) and are not widely used. In all other cases of bidirectional transmission, for example, with two fibers, so-called *space division multiplexing* (SDM), or two wavelengths, so-called *wavelength division multiplexing* (WDM), continuous-mode transmission is used.

Passive Optical Network. A *passive optical network* (PON) is illustrated schematically in Figure 1.6(b). A feeder fiber from the central office (CO) runs to a *remote node* (RN), which houses a passive optical power splitter/combiner. From there, around 32 fibers branch out to the subscribers. If these fibers extend all the way to the *homes* (H), as shown in Fig 1.6(b), this system is known as a *fiber-to-the-home* (FTTH) system. Alternatively, if the fibers terminate at the curb, the system is known as a *fiber-to-the-curb* (FTTC) system. The final distribution from the curb to the homes is accomplished, for example, by twisted-pair copper wires or radio. All systems that bring the fiber relatively close to the subscriber are collectively known as FTTx systems.

In a traditional telephony access network, the connection between the CO and the remote node is a digital, possibly optical line. The final distribution from the remote node to the subscribers, however, is accomplished with analog signals over twisted-pair copper wires. Thus, the remote node must be *active*; that is, it needs to be powered to perform the conversion from the high-speed digital signal to the analog signals. In contrast, a PON system is all optical and *passive*. Because a PON does not require outside power supplies, it is low in cost, easy to maintain, and reliable.

A PON is a point-to-multipoint network because the optical medium is *shared* among the subscribers. Information transmitted downstream, from the CO to the subscriber, is received by all subscribers, and information transmitted upstream, from the subscribers to the CO, is superimposed at the passive combiner before it is received at the CO. To avoid data collisions in the *upstream direction*, the subscriber data must be buffered and transmitted in short bursts. The CO must coordinate which subscriber can send a burst at which point in time. This method is known as *time division multiple access* (TDMA) and requires *burst-mode transmission*. The *downstream direction* is more straightforward: the CO tags the data with addresses and broadcasts it to all subscribers in sequential order. Each subscriber simply selects the information with the appropriate address tag. This method is known as *time division multiplexing* (TDM), and conventional continuous-mode transmission can be used. Upstream and downstream transmissions usually are separated by using two different wavelengths (WDM bidirectional transmission).

The most promising PON systems are (i) BPON (broadband passive optical network), which carries the data in ATM cells and hence also is known as ATM-PON [30, 52], and (ii) EPON (Ethernet passive optical network), which carries the data in Ethernet frames, as the name implies [48]. In general, PON FTTx networks are limited to relatively small distances (<20 km) and currently are operated at modest bit rates (50 Mb/s–1.25 Gb/s). In a typical BPON FTTH scenario, 16 to 32 homes share a bit rate of 155 Mb/s, giving each subscriber an average speed of 5 to 10 Mb/s. This is sufficient for fast Internet access, telephone service, and video on demand. Sometimes, an all-optical CATV service is provided over the PON infrastructure by means of a third wavelength.

Besides the TDM/TDMA approach outlined above, there are several other types of PON systems. For example, the WDM-PON system, where a different wavelength is assigned to each subscriber, has been studied extensively. In WDM-PON systems, data collisions are avoided without the need for burst-mode transmission. However, the optical WDM components required for such a system currently are too expensive, making WDM-PON uneconomical. For more information on PON systems see [40, 69].

Book Outline. In Chapter 2, we introduce the communication channel presented by the optical fiber. Its loss, bandwidth, and various forms of dispersion are described briefly. The relationship between pulse spreading and transmitter linewidth is given and is used later in Chapter 7.

Chapter 3 is the first of four chapters dealing with the receiver. We start by studying the responsivity and noise properties of three photodetector types: the p-i-n photodetector, the avalanche photodetector, and the optically preamplified p-i-n detector. In Chapter 4, we present the receiver at the system level. This chapter introduces terminology and concepts that simplify the discussion in later chapters. First, we analyze how noise in the receiver causes bit errors. This leads to the definition of the receiver sensitivity. Next, after introducing the concept of power penalty, we study the impact of the receiver's bandwidth and frequency response on its performance. The adaptive equalizer, used to mitigate distortions in the received signal, is covered briefly. We then turn to other receiver impairments, such as nonlinearity (in analog receivers), jitter, decision threshold offset, and sampling time offset. We conclude with a brief description of forward error correction, a technique that can improve the receiver's performance dramatically. In Chapter 5, we discuss the transimpedance amplifier. We start by introducing the main specifications. Next, we discuss circuit concepts in a general, and as much as possible, technology independent manner. This includes the shunt-feedback architecture and variations thereof, noise optimization procedures, and special techniques for burst-mode and analog TIAs. Then, we illustrate these concepts with practical implementations in a broad range of technologies. We conclude with a brief overview of product examples and current research topics. In Chapter 6, we discuss the main amplifier, that is, the limiting and AGC amplifier. As in the previous chapter, we proceed from specifications to circuit concepts to implementation examples. The circuit concepts covered include the multistage architecture, techniques for broadband stages, offset compensation, and automatic gain control.

Chapter 7 is the first of two chapters dealing with the transmitter. We start by studying various types of lasers and optical modulators. In Chapter 8, we discuss the driver circuits for directly modulated lasers as well as external modulators. As in Chapters 5 and 6, we proceed from specifications to circuit concepts to implementation examples. The circuit concepts covered include the current-steering output stage with and without back termination, the predriver with pulse-width control, data retiming, automatic power control, and special techniques for burst-mode and analog drivers.

Appendices A through D provide additional material on eye diagrams, differential circuits, S parameters, transistors, and technologies. Answers to all problems are given in Appendix E. The notation for currents, voltages, and noise quantities used in this book is explained in Appendix F. For reference, Appendix G defines all symbols and Appendix H defines all acronyms used in this book.

About Numerical Examples. In the following chapters, we make extensive use of numerical examples. Frequently when we introduce a new quantity or relationship, we illustrate it with so-called *typical values*. In my own learning experience, this approach is most helpful: it makes the subject more concrete and promotes a feeling for the numerical values. However, specialists tend to be quite critical about such values because they are never quite right. Typical values may change over time as the technology advances or they may depend on several conditions that may or may not be met in a particular case. It therefore is important to take the subsequent typical values only as an illustration and not as the basis for your next design project!

Further Reading. The book *High-Speed Circuits for Lightwave Communications* edited by K. Wang contains an interesting collection of papers discussing many insightful case studies [197]. The textbook *Design of Integrated Circuits for Optical Communications* by B. Razavi presents an overview of circuits for optical communication, including phase-locked loops and clock and data recovery circuits [141]. The latest developments in integrated circuits for optical communication are published in several magazines and at various conferences, for example, the *IEEE Journal of Solid-State Circuits* and the *IEEE International Solid-State Circuits Conference* are excellent sources. Information about optical fibers, lasers, detectors, optical amplifiers, as well as optical networks can be found in the textbooks *Fiber-Optic Communication Systems* by G. Agrawal [5] and *Optical Networks: A Practical Perspective* by R. Ramaswami et al. [136]. The continually updated book series *Optical Fiber Telecommunications* (so far Volumes I–IV have been published) covers the latest developments in optical/optoelectronic components and systems [58, 59, 60, 61, 93]. Finally, I recommend *City of Light: The Story of Fiber Optics* by Jeff Hecht for an entertaining and informative historical account on fiber optics [43].

# 2
Optical Fiber

In this chapter, we introduce the communication channel presented by the optical fiber. However, it is well beyond the scope of this book to treat this subject in depth. Therefore, we concentrate here on the main characteristics of optical fiber that will be useful later in this book. In particular, we describe the loss, the bandwidth, and the various forms of fiber dispersion. The relationship between pulse spreading and transmitter linewidth is discussed in preparation for Chapter 7. For a more complete treatment of this subject, many excellent books, such as [5, 136, 168], are available.

2.1 LOSS AND BANDWIDTH

Loss. As the optical signal propagates over a long stretch of fiber, it becomes attenuated because of scattering, absorption by material impurities, and other effects. The attenuation is measured in dBs ($10 \cdot \log$ of power ratio) and is proportional to the length of the fiber. *Fiber attenuation* or *fiber loss* is therefore specified in dB/km.

As shown in Fig. 2.1, *silica glass* has two low-loss windows, one around the wavelength $\lambda = 1.3\,\mu\text{m}$ and one around $\lambda = 1.55\,\mu\text{m}$, which both are used for optical fiber communication.[1] The popular single-mode fiber has a loss of about 0.25 dB/km at the 1.55-μm wavelength and 0.4 dB/km at the 1.3-μm wavelength. Because the loss is lower at 1.55 μm, this wavelength is preferred for long-haul communication.

---

[1]Note that these wavelengths, and all wavelengths we refer to later, are defined in the *vacuum*. Thus, an optical signal with $\lambda = 1.55\,\mu\text{m}$ has a wavelength of about 1 μm in a glass fiber, which has a refractive index of about 1.5!

A third wavelength window around $\lambda = 0.85\,\mu$m, where the loss is about 2.5 dB/km, is used for short-reach (data) communication applications, mostly because low-cost optical sources and detectors are available for this wavelength.

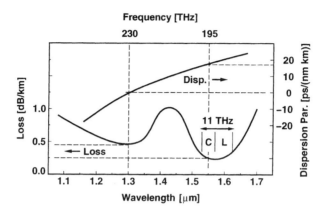

Fig. 2.1 Loss and dispersion parameter D of a standard single-mode fiber.

The loss of modern silica-glass fiber is phenomenally low compared with that of an RF coax cable at high frequencies. A high-performance RF coax cable operating at 10 GHz has an attenuation of about 500 dB/km. Compare this with 0.25 dB/km for a fiber! On a historical note, it is interesting to know that low-loss fiber was not easy to produce. In 1965, the best glass fiber had a loss of around 1,000 dB/km. It was estimated that for a fiber to be useful for optical communication, its loss must be reduced to 20 dB/km or less, that is, an improvement by *98 orders of magnitude* was required!! It is therefore understandable that in '65 most researchers thought that using glass fiber for optical communication was a hopelessly crazy idea. They spent their time working on "reasonable" approaches to optical communication such as metal pipes that contain periodically spaced lenses (so-called *confocal waveguides*) or pipes heated in such a way that the air in them formed *gas lenses*. Nevertheless, in 1970, a research team at the *Corning Glass Works* managed to reduce the fiber loss below 20 dB/km by using ultrapure silica glass rather than the ordinary compound glass [43]. So, next time your circuit parameters are off by 98 orders of magnitude, don't give up...

In comparison with silica glass fiber, *plastic optical fiber* (POF) is very cheap to manufacture and also permits the use of low-cost connectors because of its large core size of almost 1 mm. However, it has a huge loss of about 180 dB/km, even when operated in the "low-loss" window at 0.65 μm (visible red). It therefore is restricted to very-short reach applications such as home networks and consumer electronics.

Although the loss of silica glass fiber is very low, it is still not low enough for ultralong-haul communication (e.g., 10,000 km of fiber) without optoelectronic repeaters. What can we do to reduce the loss further? First, we must operate the fiber at the 1.55-μm wavelength where loss is the lowest. Then, we can use periodically spaced *optical in-line amplifiers* to boost the signal and thus effectively reduce the optical loss. Two types of fiber amplifiers are in use: (i) the *erbium-doped fiber amplifier* (EDFA), which provides gain in the 1.55-μm band, and (ii) the *Raman amplifier*, which can provide distributed gain in the transmission fiber itself at a selectable wavelength (13 THz below the pump frequency). A third type of fiber amplifier is in the research stage: the *fiber optical parametric amplifier* (FOPA), which promises gain over a very wide bandwidth (e.g., 200 nm) at a selectable wavelength.

Bandwidth. In addition to the very low loss, optical fiber also has a huge bandwidth. By *bandwidth*, we mean the range of optical frequencies or wavelengths for which the fiber is highly transparent. For example, the low-loss window around the 1.55-μm wavelength is subdivided into two bands (C band for "conventional" and L band for "long-wavelength"), which together provide a bandwidth of 95 nm, corresponding to about 11 THz (see Fig. 2.1). This means that we can transmit about 4 Tb/s of information over a fiber using the C+L bands and the non-return-to-zero (NRZ) modulation format, which achieves a spectral efficiency of about 0.4 b/s/Hz. For example, a backbone connection consisting of 100 parallel fibers could transport up to 400 Tb/s of information, or a fiber-to-the-home (FTTH) system based on a passive optical network (PON) in which each feeder fiber serves 100 subscribers could be upgraded to 40 Gb/s per user, if the demand should arise! This should be enough access bandwidth for the foreseeable future, and that's why FTTH advocates tout their system as *future proof*. [→ Problem 2.1]

Bandwidth and Dispersion. Given that we have more than 10 THz of bandwidth in the fiber, could we take a 1.55-μm laser, modulate it with a 4 Tb/s NRZ data stream, and use this arrangement for optical transmission, at least in theory? Besides the fact that the electronic circuits and the optoelectronic devices (laser, detector) would be too slow, the propagation of the optical signal over the fiber would present a major limit! The received signal would be totally distorted already after a very short distance because of *dispersive effects* in the fiber. The transmitted optical signal in our hypothetical system has a very large spectral width, filling all of the C and L band. Although each spectral component is in the low-loss window and arrives intact at the other end of the fiber, each component is delayed by a *different* amount, and the superposition of all components, the received signal, is severely distorted. The dependence of delay on wavelength is known as *chromatic dispersion*, and we discuss it in more detail in the next section.

It therefore is important to distinguish between two types of *fiber bandwidths*: the *bandwidth for the optical carrier*, which is very wide (>10 THz), and the *bandwidth for the modulation signal*, which is limited by dispersion and is much, much smaller. For example, the modulation-signal bandwidth for 1 km of standard single-mode fiber is just a few 10 GHz, given a 1.55-μm source with a 1-nm linewidth. We discuss this bandwidth and its dependence on linewidth further in Section 2.4. For 1 km of graded-index multimode fiber, the modulation-signal bandwidth is only 300 MHz to 3 GHz, and for a step-index multimode fiber, it is even lower at 6 MHz to 50 MHz [168].

Does this mean that we cannot really use the huge bandwidth that the fiber offers? Yes we can, if we use multiple optical carriers at different wavelengths and modulate each one at a modest bit rate, then the transmission becomes much more resilient to the various forms of dispersion. For example, instead of one wavelength modulated at 4 Tb/s, we could use 400 wavelengths, each one modulated at 10 Gb/s. This approach is known as *dense wavelength division multiplexing* (DWDM).

2.2 DISPERSION

Modal Dispersion. An optical fiber consists of a core surrounded by a cladding that has a slightly lower refractive index than the core such that the light beam is guided by total internal reflection, as shown in Fig. 2.2. In principle, air, which has a lower refractive index than glass, could act as the cladding. However, the fiber surface then would be extremely sensitive to dirt and scratches, and two fibers touching each other would leak light. The invention of the *clad fiber* was a major breakthrough on the way to a practical optical fiber [43].

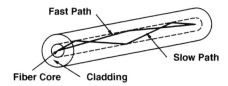

Fig. 2.2 Modal dispersion in a multimode fiber.

Depending on the size of the fiber core, there is only a single or several pathways (so-called modes) for the light beam to propagate through the fiber. The core of a *multimode fiber* (MMF) is large enough (50–100 μm) for the light to take multiple pathways from the transmitter to the receiver, as shown in Fig. 2.2 (typically, several hundred modes exist). Each path has a slightly different propagation delay, thus producing a distorted (spread out) pulse at the receiver end. This effect is known as *modal dispersion*. The time difference between the longest and shortest path ΔT for a so-called *graded-index multimode fiber* (GRIN-MMF) can be approximated by [5]

$$\Delta T = \frac{(n_{\text{cor}} - n_{\text{clad}})^2}{8c \cdot n_{\text{cor}}} \cdot L, \tag{2.1}$$

where L is the fiber length, c is the speed of light in vacuum, and n_{cor} and n_{clad} are the refractive indices of the core and cladding, respectively. With the typical values $n_{\text{cor}} = 1.48$ and $n_{\text{cor}} - n_{\text{clad}} = 0.02$ and a fiber length of 1 km, we find a propagation-delay variation of about 113 ps. Thus, modal dispersion is significant even for short fiber links. For example, at 10 Gb/s, the fiber length is limited to about 100 to 300 m. The core of a *single-mode fiber* (SMF) is much smaller (8–10 μm) and permits only one pathway (a single mode) of light propagation from the transmitter to the receiver, and thus distortions due to modal dispersion are suppressed.[2] Note that the word 'mode' in SMF, MMF, and modal dispersion refers to *pathway* modes only. We see later that the single pathway mode in an SMF can be decomposed into two *polarization* modes, both of which propagate through the fiber and may cause polarization-mode dispersion.

SMF is preferred in telecommunication applications where distance matters (ultra-long-haul, long-haul, metro, and access networks). MMF is mostly used within buildings for data communication (computer interconnects) and in consumer electronics. Its reach can be extended somewhat by the use of an adaptive equalizer in the receiver (cf. Section 4.7). Because the MMF has a larger core size, alignment of the fiber with another fiber or a laser chip is less critical. A transverse alignment error between a laser and an SMF of just 0.5 μm causes a power penalty of about 1 dB, whereas the laser-to-MMF alignment is about 5× less critical [174]. Thus, components interfacing to MMF generally are lower in cost.

Chromatic Dispersion. Chromatic dispersion, also called *group-velocity dispersion* (GVD), is another source of signal distortions and is caused by different wavelengths (colors) traveling at different speeds through the fiber. Figure 2.3 illustrates how the group delay varies with wavelength for 1 km of standard SMF. We recognize that the change in group delay is large around 1.55 μm, whereas it is nearly zero at 1.3 μm. In practice, chromatic dispersion is specified by the *change* in group delay per nm wavelength and km length:

$$D = \frac{1}{L} \cdot \frac{\partial \tau}{\partial \lambda}, \qquad (2.2)$$

where D is known as the *dispersion parameter*, L is the fiber length, τ is the group delay, and λ is the wavelength. A standard SMF operated at 1.55 μm has $D = 17\,\text{ps}/(\text{nm} \cdot \text{km})$, which means that a change in wavelength of 1 nm will change the group delay by 17 ps in a 1-km piece of fiber (cf. Fig. 2.3). The dependence of D on wavelength is plotted together with the fiber loss in Fig. 2.1. [→ Problem 2.2]

How much pulse distortion is caused by chromatic dispersion depends on the spectral linewidth of the transmitter. If the transmitter operates at precisely a single wavelength, which implies an ideal, zero-linewidth laser without modulation, chromatic dispersion doesn't matter. However, if the transmitter operates over a range of wavelengths, as it does when transmitting information using a real laser, chromatic dispersion causes pulse distortions. The propagation time difference between the slowest and fastest wavelength for a transmitter that emits light over the range $\Delta\lambda$

---

[2]The reader may wonder why a 8- to 10-μm core is small enough to ensure single-mode propagation of light with a wavelength of 1.3 to 1.55 μm. It turns out that the condition for single-mode propagation is that the core diameter must be $d < \lambda \cdot 0.766/\sqrt{n_{\text{cor}}^2 - n_{\text{clad}}^2}$ [5]. Because the difference between n_{cor} and n_{clad} is small (less than 1% for an SMF), the core can be made quite a bit larger than the wavelength λ, simplifying the light coupling into the fiber. Another advantage of the clad fiber!

16 OPTICAL FIBER

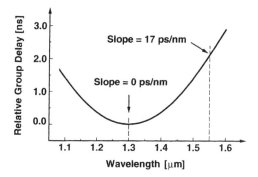

Fig. 2.3 Relative group delay as a function of wavelength for 1 km of standard SMF.

can be derived from Eq. (2.2) as

$$\Delta T = |D| \cdot \Delta\lambda \cdot L. \tag{2.3}$$

The effects of chromatic dispersion on pulse spreading are rather important when designing a transmission system, and we discuss them in more detail in Section 2.4.

What can we do to reduce the chromatic dispersion $|D| \cdot L$? Because dispersion is a linear phenomenon, it can be reversed by applying an equal amount of *negative* dispersion. This method is known as *dispersion compensation*, and so-called *dispersion compensating fiber* (DCF) with a large negative value of D such as $-100 \, \text{ps}/(\text{nm} \cdot \text{km})$ is available for this purpose. For example, to compensate for the positive dispersion of 100 km of standard SMF, we can append 17 km of DCF, resulting in an overall dispersion of zero: $100 \, \text{km} \cdot 17 \, \text{ps}/(\text{nm} \cdot \text{km}) + 17 \, \text{km} \cdot (-100 \, \text{ps}/(\text{nm} \cdot \text{km})) = 0.$[3] Note that in such a system, at first each bit spreads out over many adjacent bit slots, resulting in a total mess, but then the DCF pulls all the bits back together again, producing a crisp optical signal. Alternatively, we can transmit at the $1.3\text{-}\mu\text{m}$ wavelength, where the dispersion parameter D of a standard SMF is much smaller than at $1.55 \, \mu\text{m}$ (see Fig. 2.1). But, as we know, at the $1.3\text{-}\mu\text{m}$ wavelength the fiber loss is higher. To resolve this dilemma, fiber manufacturers have come up with the so-called *dispersion-shifted fiber* (DSF), which has a value of D close to zero at the $1.55\text{-}\mu\text{m}$ wavelength while preserving the low loss of an SMF. This fiber, however, has a disadvantage in DWDM systems that we discuss in Section 2.3 on nonlinearities. The effects of chromatic dispersion also can be mitigated in the electrical domain by using a receiver with an equalizer (cf. Section 4.7). Because the optical phase information becomes lost in the detection process, this method is less capable than optical dispersion compensation; however, it may have a cost advantage.

---

[3] Typically, perfect dispersion compensation can be achieved only at a single wavelength. In a DWDM system, the middle channel can be compensated perfectly, whereas the outer channels retain some residual dispersion.

Polarization-Mode Dispersion. Another source of distortions is *polarization-mode dispersion* (PMD), which is caused by different polarization modes traveling at different speeds. This effect occurs in fibers with a slightly elliptic core or asymmetrical mechanical stress. Figure 2.4 illustrates how, in such a fiber, horizontally and vertically polarized light propagates at slightly different speeds. The difference in arrival time, ΔT, is known as *differential group delay* (DGD). Because, in general, the transmitter is exciting both polarization modes (horizontal and vertical), the receiver sees two time-shifted copies of the transmitted sequence superimposed on top of each other. The strength of each sequence depends on the alignment of the transmitter's linearly polarized light with the axes of the elliptic core.

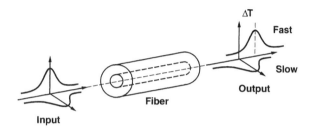

Fig. 2.4 Polarization-mode dispersion in a short fiber segment.

In a long stretch of fiber, the situation is complicated by the fact that the fiber's effect on the polarization state (birefringence) changes randomly along its *length*. As a result, we have the following modifications. (i) The input polarization states that correspond to the fast and slow propagation are no longer linear states (i.e., horizontal and vertical) but general elliptic polarization states, which differ from fiber to fiber. These states are known as the input *principle states of polarization* (PSP). (ii) The differential group delay, averaged over many fibers, is proportional to the *square root* of the fiber length L (rather than the length itself) and can be written

$$\overline{\Delta T} = D_{PMD} \cdot \sqrt{L}, \tag{2.4}$$

where D_{PMD} is the *polarization-mode dispersion parameter*. In addition to the statistical uncertainty, PMD also varies slowly over *time*, that is, the PSPs and D_{PMD} are a function of time. As a rule of thumb, we must keep $\overline{\Delta T}$ below 10% of the bit interval ($0.1/B$) to keep the power penalty due to PMD almost permanently below 1 dB.[4] Another complication in real systems is the wavelength dependence of both the input PSPs and the DGD, which makes the PMD distortions look more complex than just a simple superposition of time-shifted copies of the transmitted signal (so-called *higher-order PMD* distortions).

---

[4]More precisely, to ensure an outage probability of $<10^{-7}$, that is, <3 seconds/year during which the power penalty exceeds 1 dB, we need $\overline{\Delta T} < 0.1/B$ for RZ modulation and $\overline{\Delta T} < 0.07/B$ for NRZ modulation [74].

What can we do against PMD? Recently manufactured fiber has a very low PMD parameter around $D_{PMD} = 0.1\,\text{ps}/\sqrt{\text{km}}$. This means that in a 100-km fiber, the average delay is only 1 ps; no problem, even for 40 Gb/s. Older fiber, which is widely deployed and has a slightly elliptic cross section of the fiber core as a result of manufacturing tolerances, has a much larger PMD parameter around $D_{PMD} = 2\,\text{ps}/\sqrt{\text{km}}$. In the latter case, we can use optical or electrical PMD mitigation techniques. In a simple optical PMD compensator, a fiber segment with strong intentional PMD, a so-called *polarization maintaining fiber* (PMF), is placed in front of the receiver to undo the PMD accumulated during transmission. In between the transmission fiber and the compensation fiber, a polarization controller must be inserted to make sure that the fast output PSP is fed into the slow input PSP of the compensation fiber and vice versa. Because the PMD is time varying, the polarization controller must be adjusted continuously with a feedback signal derived from the quality of the received signal. In a simple electrical PMD compensator, the distorted optical signal is converted first to an electrical signal with a photodetector (losing the polarization information). Then, an adaptive equalizer as part of the receiver removes as much distortion as possible (see Section 4.7). Again, because of the time-varying nature of PMD, it is necessary to adjust the equalizer characteristics continuously as a function of the received signal quality.

2.3 NONLINEARITIES

Attenuation and dispersion are known as *linear* fiber effects because they can be described by a linear relationship between the fields of the lightwaves at the input and output of the fiber. Apart from these linear effects, the fiber suffers from a number of *nonlinear* effects that may distort and attenuate the optical signal or may produce crosstalk between optical channels. The most important of these effects are *self-phase modulation* (SPM), *cross-phase modulation* (CPM or XPM), *stimulated Raman scattering* (SRS), *stimulated Brillouin scattering* (SBS), and *four-wave mixing* (FWM). These effects become important in long optical links operated at high optical power levels. In single-wavelength systems, SPM, SRS, and SBS can cause pulse distortion and attenuation. In addition to that, in DWDM systems, CPM, SRS, SBS, and FWM can cause crosstalk between optical channels.

What can we do against these nonlinear effects? In general, nonlinear effects can be suppressed if the transmitted optical power is kept sufficiently low. One way to lower the transmit power without impacting the bit-error rate performance is to use forward error correction (cf. Section 4.11). As we have said, in a DWDM system, the bits in different channels may interact with each other through nonlinear effects, resulting in a change of pulse shape and amplitude (cf. Section 6.2.7). The longer the interacting bits stay together, the stronger the crosstalk distortions become. For this reason, it is advantageous if the different wavelength channels propagate at slightly different speeds, that is, if there is a small amount of chromatic dispersion. A special fiber, called *nonzero dispersion-shifted fiber* (NZ-DSF), has been developed that has a small value for $|D|$ around 1 to 6 ps/(nm · km), large enough to create a "walk-off" between the bit streams, thus reducing nonlinear interactions, but small enough to limit the amount of dispersion compensation needed or to avoid it altogether.

2.4 PULSE SPREADING DUE TO CHROMATIC DISPERSION

Modal dispersion can be suppressed with the use of SMF, and PMD is only of concern in long-haul and high-speed systems, but almost every optical transmission system is affected by chromatic dispersion. In the following, we investigate the pulse distortions caused by chromatic dispersion in greater detail.

Nonlinear Character of Optical Fiber Communication. To transmit an optical signal over a fiber, we modulate the *intensity* of a light source, and to receive the signal, we detect the *intensity* of the light.[5] Let's assume that the fiber in between the transmitter and the receiver exhibits dispersion. We said earlier that dispersion is a linear phenomenon, but linear in the *fields* not the *intensity*! Thus, the complete communication channel can be modeled as shown in Fig. 2.5. The signal is converted to light with a proportional intensity; the light is carried by an electromagnetic field, which is proportional to the square-root of the intensity; the field disperses linearly in the fiber; the field at the end of the fiber is characterized by an intensity, which is proportional to the square of the field; finally, the light intensity is detected and a signal proportional to it is generated. This is a *nonlinear* system! Note that the nonlinearity here is due to the square law relating the intensity and the field and has nothing to do with the nonlinearity in fibers discussed in the previous section.

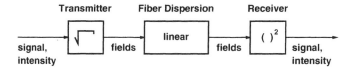

Fig. 2.5 Communication channel with intensity modulation, fiber dispersion, and intensity detection.

Now we understand that, in general, we cannot apply linear system theory to analyze the pulse distortions caused by fiber dispersion, making this a rather nasty problem. (This same nonlinearity will bother us again when discussing optical and electrical dBs, and then again when we talk about optical noise.) However, there is an approximation that we can use under certain conditions. If we use a light source with a bandwidth much greater than the signal bandwidth, we can approximately describe the channel with a linear response [11]. With such a wide-linewidth source, the transmitter linewidth $\Delta\lambda$ is approximately equal to the source linewidth $\Delta\lambda_S$,

---

[5]This method is known as *direct detection.* An alternative is *coherent detection,* but this subject is beyond the scope of this book.

and the effects of the modulating signal on $\Delta\lambda$ can be neglected. In practice, this is the case for transmitters with a light-emitting diode (LED) or a Fabry-Perot (FP) laser source, as we see in Chapter 7. If we further assume that the source spectrum is Gaussian and we are operating at a wavelength far from zero dispersion, the impulse response of the channel turns out to be

$$h(t) = h(0) \cdot \exp\left(-\frac{t^2}{2 \cdot (\Delta T/2)^2}\right), \quad (2.5)$$

where

$$\Delta T = |D| \cdot \Delta\lambda \cdot L \quad (2.6)$$

and $\Delta\lambda$ is the 2σ-linewidth of the transmitter (or source). In other words, a Dirac impulse injected into one end of the fiber will spread out slowly into a Gaussian pulse as it propagates along the fiber. The 2σ-width of the spreading pulse is the ΔT given in Eq. (2.6). For example, a very narrow pulse launched into a standard SMF will spread out to 17 ps after 1 km given a 1.55-μm source with a linewidth of 1 nm. Thus, the dispersion parameter D tells us how rapidly a narrow pulse is spreading out.

Time-Domain Analysis. Now that we have a linear model, we are on familiar territory and we can calculate how a regular (non-Dirac) data pulse spreads out. The math is easiest if we assume that the transmitted pulse is Gaussian. The convolution of the (Gaussian) input pulse with the (Gaussian) impulse response produces a Gaussian output pulse! We find the relationship between the 2σ-width of the input pulse, T_{in}, and the 2σ-width of the output pulse T_{out} to be

$$T_{out} = \sqrt{T_{in}^2 + \Delta T^2}. \quad (2.7)$$

This situation is illustrated in Fig. 2.6. Although in practice we are not likely to use Gaussian pulses, this simple calculation helps us to understand under which circumstances chromatic dispersion becomes important. For example, on 1 km of standard SMF with a source linewidth of 1 nm at 1.55 μm, a 100-ps pulse will broaden to $\sqrt{(100\,\text{ps})^2 + (17\,\text{ps})^2} = 101.4\,\text{ps}$. [$\rightarrow$ Problem 2.3]

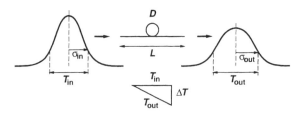

Fig. 2.6 Pulse spreading due to chromatic dispersion.

The maximum amount of spreading, ΔT, that normally can be tolerated in an NRZ-modulated system is equal to half a bit period [46]:[6]

$$\Delta T \leq \frac{1}{2B}. \qquad (2.8)$$

This amount of spreading increases the pulse width by $\sqrt{1^2 + 0.5^2} = 1.12$ or about 12% and causes a power penalty of approximately 1 dB [46].

As we can see from Eqs. (2.6) and (2.8), the linewidth of the source is of critical importance in determining the amount of pulse spreading in a dispersive fiber and thus the maximum bit rate and distance over which we can transmit.

Frequency-Domain Analysis. Given the expression for the impulse response of a dispersive fiber, we can easily transform it into the frequency domain and discuss the corresponding channel bandwidth. Transforming the Gaussian impulse response in Eq. (2.5) yields the Gaussian frequency response

$$H(f) = H(0) \cdot \exp\left(-\frac{(2\pi f)^2 (\Delta T/2)^2}{2}\right). \qquad (2.9)$$

The 3-dB bandwidth of this optical intensity response can be found by setting the equation equal to $1/2 \cdot H(0)$ and solving for f. Together with Eq. (2.6), we find [5][7]

$$BW_{3dB} = \frac{0.375}{\Delta T} = \frac{0.375}{|D| \cdot \Delta \lambda \cdot L}. \qquad (2.10)$$

This is the modulation-signal bandwidth due to chromatic dispersion first introduced in Section 2.1. Its value decreases as the fiber length L, the linewidth $\Delta \lambda$, or the dispersion parameter D increases. For example, 1 km of standard SMF with a source linewidth of 1 nm at 1.55 μm has a bandwidth of just 22 GHz. If we replace the SMF with an NZ-DSF that has a dispersion parameter of only $D = 5\,\text{ps}/(\text{nm} \cdot \text{km})$, the bandwidth increases to 75 GHz. [→ Problem 2.4]

Finally, what is the interpretation of the spreading limit, Eq. (2.8), in the frequency domain? Inserting Eq. (2.8) into Eq. (2.10), we find

$$BW_{3dB} \geq 0.75 \cdot B. \qquad (2.11)$$

This means the fiber bandwidth must be made larger than 3/4 of the bit rate to avoid excessive distortions. [→ Problem 2.5]

Narrow-Linewidth Source. What happens if we use a light source with a bandwidth much smaller than the signal bandwidth? In practice, this situation occurs for

---
[6] Note that in terms of the mean-square impulse spread $\sigma_T^2 = (\Delta T/2)^2$, this limit is $\sigma_T \leq 1/(4B)$.
[7] In the electrical domain, this bandwidth is the 6-dB bandwidth, because three optical dBs convert to six electrical dBs (cf. Section 3.1)!

22 OPTICAL FIBER

transmitters with a DFB laser, as we see in Chapter 7. Under these circumstances, the source linewidth $\Delta\lambda_S \ll \Delta\lambda$ can be neglected and the transmitter linewidth $\Delta\lambda$ is determined by the modulation signal, that is, the modulation format, the bit rate, and possibly spurious frequency modulation of the optical carrier, so-called *chirp*.

As we said earlier, for a narrow-linewidth source, the communication channel becomes nonlinear and cannot be described by linear system theory. In particular, the concept of modulation-signal bandwidth cannot be applied strictly. We can still interpret Eqs. (2.5) and (2.6) loosely to mean that the pulse spreading, ΔT, tends to increase with increasing transmitter linewidth, $\Delta\lambda$, that is, for higher bit rates and stronger chirp. However, there are important exceptions to this rule, such as for optical pulses with *negative* chirp and for so-called *solitons*. Pulses with negative chirp[8] are characterized by a temporary decrease in optical frequency (red shift) during the leading edge and an increase in frequency (blue shift) during the trailing edge. Although this chirp broadens the transmitter linewidth, these pulses become *compressed* up to a certain distance in a dispersive medium with $D > 0$. Solitons are short ($\approx$ 10 ps) and powerful optical pulses of a certain shape. Although the shortness of these pulses increases the transmitter linewidth, they do not broaden at all. This is so because chromatic dispersion is counterbalanced by self-phase modulation, one of the nonlinear fiber effects. [$\rightarrow$ Problem 2.6]

2.5 SUMMARY

Optical silica-glass fiber is characterized by a very low loss of about 0.25 dB/km and a huge bandwidth of more than 10 THz when operated in the 1.55-μm wavelength band.

On the down side, various types of dispersion cause the optical pulses to spread out in time and to interfere with each other. The following types of dispersion can be distinguished:

- Modal dispersion, which only occurs in multimode fibers.

- Chromatic dispersion, which is small at 1.3 μm but presents a significant limitation at the 1.55-μm wavelength in standard single-mode fibers. The impact of chromatic dispersion on pulse spreading can be ameliorated by using narrow-linewidth transmitters.

- Polarization-mode dispersion, which is significant in high-speed, long-haul transmission over older types of fiber and is slowly varying with time.

Furthermore, at elevated power levels, nonlinear effects can cause attenuation and pulse distortions as well as crosstalk in dense wavelength division multiplexing (DWDM) systems.

---

[8]In the literature, the polarity of chirp is not consistently defined. In this book, we use the term *negative chirp* to describe a leading edge with red shift.

2.6 PROBLEMS

2.1 Wavelength and Frequency. (a) At what frequency oscillates the electromagnetic field of a 1.55-μm optical signal? (b) An optical filter has a bandwidth of 0.1 nm at 1.55 μm. What is its bandwidth in Hertz?

2.2 Group-Delay Variation. Assume that the dispersion parameter D depends linearly on the wavelength. It has a value of zero at 1.3 μm and 17 ps/(nm·km) at 1.55 μm. Calculate the dependence of the group delay on wavelength.

2.3 Pulse Spreading. Derive the pulse-spreading rule for Gaussian pulses given in Eq. (2.7) from the impulse response given in Eq. (2.5).

2.4 Fiber Response. Derive the frequency response for a fiber given in Eq. (2.9) from the impulse response given in Eq. (2.5).

2.5 1-dB Dispersion Penalty. What is the highest bit rate, B, at which we can transmit an NRZ signal while incurring less than 1 dB of attenuation due to dispersion? Use Eq. (2.9) to estimate the attenuation and assume that most of the data signal's energy is located at the frequency $B/2$ (as for the "101010..." sequence).

2.6 Pulse Compression. Explain qualitatively why a pulse with negative chirp initially becomes compressed for a fiber with $D > 0$.

2.7 Transmission System at 1,310 nm. A 1.31-μm transmitter with a 3-nm linewidth launches a 2.5-Gb/s NRZ signal with 1 mW into a standard SMF. (a) How long can we make the fiber before the power is attenuated to -24.3 dBm? (b) How long can we make the fiber before chromatic dispersion causes too much pulse spreading? Assume $D = 0.5$ ps/(nm·km).

2.8 Transmission System at 1,550 nm. Now we use a 1.55-μm transmitter with the same linewidth, bit rate, and launch power as in Problem 2.7. How does the situation change?

2.9 Transmitter Linewidth. (a) In which system, Problem 2.7 or 2.8, would it make sense to use a narrow-linewidth transmitter? (b) How far could we go, if we reduce the linewidth to 0.02 nm?

2.10 Fiber PMD. We are using "old" fiber with $D_{PMD} = 2\,\text{ps}/\sqrt{\text{km}}$. Do we have to be concerned about PMD in one of the above transmission systems?

3
Photodetectors

The first element in an optical receiver is the *photodetector*. The main characteristics of this device, in particular its responsivity and noise properties, have a significant impact on the receiver's performance. For example, the receiver's sensitivity is determined largely by the characteristics of the photodetector and the transimpedance amplifier. There are three types of photodetectors that commonly are used: the p-i-n photodetector, the avalanche photodetector (APD), and the optically preamplified p-i-n detector, which we discuss in this order. More information on photodetectors can be found in [5, 136, 183].

3.1 P-I-N PHOTODETECTOR

The simplest detector is the *p-i-n photodetector*, also called *p-i-n photodiode*, shown in Fig. 3.1. A p-i-n photodetector consists of a p-n junction with a layer of intrinsic (undoped or lightly doped) semiconductor material sandwiched in between the p- and the n-doped material. The junction must be reverse biased to create a strong electric field in the intrinsic material. Photons incident on the i-layer create electron-hole pairs, which become separated by the electric drift field. As a result, a photocurrent appears at the terminals.

The width W of the i-layer, which is approximately equal to the depletion-layer width under reverse-bias conditions, controls the trade-off between efficiency and speed of the detector. The fraction of photons that create electron-hole pairs is called the *quantum efficiency* and is designated by η. The wider W is made, the better the chances that a photon is absorbed in the detector and thus the higher the quantum effi-

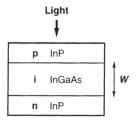

Fig. 3.1 p-i-n photodetector (schematically).

ciency. However, the wider W is made, the longer it takes for the electrons and holes to traverse the i-layer, making the photodetector response slower (cf. Eq. (3.4)). To avoid this dilemma, the i-layer in very fast photodetectors is illuminated from the side. Now, the quantum efficiency is controlled by the horizontal dimension, whereas the speed (bandwidth) is controlled by the vertical W dimension. The *waveguide photodetector* and the *traveling-wave photodetector* belong to this class of horizontally illuminated detectors. For horizontally illuminated photodetectors, a bandwidth-efficiency product $(BW \cdot \eta)$ in the range of 55 to 76 GHz is possible, whereas for vertically illuminated p-i-n detectors, 20 to 35 GHz has been achieved [64].

Most semiconductor materials are transparent, that is, they don't absorb photons, at the 1.3-μm and 1.55-μm wavelengths commonly used in telecommunication applications. For example, silicon only absorbs photons with $\lambda < 1.06\,\mu$m, GaAs only with $\lambda < 0.87\,\mu$m, and InP only with $\lambda < 0.92\,\mu$m. Therefore, a semiconductor compound with a narrow bandgap must be used for the i-layer. For example InGaAs, which is sensitive in the 1.0- to 1.65-μm range, is a popular choice. The p- and n-layers of the InGaAs photodetector shown in Fig. 3.1 are made from InP material, which is transparent at the wavelength of interest, and thus the i-layer can be illuminated from the top or bottom as shown. Detectors sensitive at the 0.85-μm wavelength, used in data communication, typically are based on GaAs or silicon.

Electron-hole pairs created outside of the drift field diffuse very slowly (e.g., at 4 ns/μm) to the drift-field region, where they make a delayed contribution to the photocurrent. As a result, the intended current pulse corresponding to the optical signal is followed by an undesired "tail current." In burst-mode receivers, this tail current can cause problems when a very strong burst signal is followed by a very weak one [120]. Tail currents can be minimized by using transparent materials for the p- and n-layers and precisely aligning the fiber to the active part of the i-layer.

Responsivity. The light-current relationship for a p-i-n photodetector can be derived easily: of all incident photons, the fraction η creates electron-hole pairs. Each photon has the energy hc/λ, where h is the Planck constant, and each electron carries the elementary charge q. Remember that current is "electron charge per time" and optical power is "photon energy per time." Thus, the electrical current (I_{PIN})

produced for a given amount of optical power (P) illuminating the photodetector is

$$I_{PIN} = \eta \cdot \frac{\lambda q}{hc} \cdot P. \tag{3.1}$$

The constant relating I_{PIN} and P is known as the *responsivity* of the photodetector and is designated by the symbol $\mathcal{R}$:

$$I_{PIN} = \mathcal{R} \cdot P \quad \text{with} \quad \mathcal{R} = \eta \cdot \frac{\lambda q}{hc}. \tag{3.2}$$

For example, for the commonly used wavelength $\lambda = 1.55\,\mu\text{m}$ and the quantum efficiency $\eta = 0.64$, we obtain a responsivity $\mathcal{R} = 0.8\,\text{A/W}$. This means that for every milli-Watt of optical power incident onto the photodetector, we obtain 0.8 mA of current. The responsivity of a typical InGaAs p-i-n photodetector is in the range 0.6 to 0.9 A/W [5].

A Two-for-One Special. Let's examine the relationship in Eq. (3.2) in more detail. It implies that if we double the power we obtain twice as much current. Now this is very odd! Usually, the power grows with the *square* of the current and not linearly with the current. For instance in a wireless receiver, if we double the RF power, we obtain $\sqrt{2}$ more current from the antenna. Or if we double the current flowing into a resistor, 4× as much power is dissipated into heat. This square relationship between power and current is the reason why we use "10 log" to calculate power dBs and "20 log" to calculate current or voltage dBs. When using this convention, a 3-dB increase in RF power translates into a 3-dB increase in current from the antenna, or a 3-dB increase in current results in a 3-dB increase in power dissipation in the resistor. For a photodetector, however, a 3-dB increase of optical power translates into a 6-dB increase in current. What a bargain! [→ Problems 3.1 and 3.2]

Wireless Receiver with a Photodetector? In contrast to optical receivers, wireless receivers are using antennas to detect electromagnetic waves. The rms current that is produced by an antenna under matched conditions is

$$i_{ANT}^{rms} = \sqrt{P/R_{ANT}}, \tag{3.3}$$

where P is the received power[1] and R_{ANT} is the antenna resistance. For example, for a $-50\,\text{dBm}$ signal (10^{-8} W), we obtain approximately $14\,\mu\text{A}$ rms from an antenna with $R_{ANT} = 50\,\Omega$.

What if we would use a hypothetical hyperinfrared photodetector instead of the antenna to detect a 1-GHz RF signal? This photodetector is made sensitive to low-energy RF photons ($4\,\mu\text{eV}$) through advanced bandgap engineering. It is cooled down to a few milli-Kelvins to prevent dark currents caused by thermal electron-hole generation. With Eq. (3.2), we can calculate the responsivity of this detector

---

[1]More precisely, P is the power incident on the effective aperture of the antenna [75].

to be an impressive 120 kA/W assuming that $c/\lambda = 1$ GHz and $\eta = 0.5$. So at the same received power level of -50 dBm, we obtain a current of 1.2 mA: about $86\times$ more than with the old fashioned antenna! The reason for this is, of course, that the photodetector produces a current proportional to the square of the electromagnetic field, that is, the intensity (cf. Fig. 2.5), whereas the antenna produces a current directly proportional to the field.

But don't launch your start-up company to market this idea just yet! What happens if we reduce the received power? After all, this is where the detector's responsivity matters the most. The signal from the photodetector decreases *linearly*, whereas the signal from the antenna decreases more slowly following a *square-root* law. After we are down to -90 dBm (10^{-12} W), we obtain approximately $0.14\,\mu$A from the antenna and $0.12\,\mu$A from the photodetector (see Fig. 3.2): about the same! Unfortunately, our invention turns out to be a disappointment ... [$\rightarrow$ Problem 3.3]

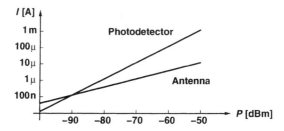

Fig. 3.2 Responsivity of an antenna and a photodetector at 1 GHz.

Bandwidth. Let's get back to more practical work! Figure 3.3 shows two equivalent AC circuits for 10-Gb/s p-i-n photodetectors, one for a bare detector and one for a detector with packaging parasitics. The current source in both models represents the current generated by the light and has the value $i_{PIN}(t) = \mathcal{R} \cdot P(t)$. The main parasitics are the photodiode junction capacitance C_{PD} and the combination of contact and spreading resistance R_{PD}. The packaged detector has additional L-C parasitics caused by wire bonds and so forth. At the frequencies of interest, the impedance of the p-i-n detector is mostly capacitive, and therefore the detector impedance is sometimes modeled by a single capacitance.

The bandwidth of the bare photodiode is determined by two time constants: (i) the transit time, that is, the time it takes the carriers to travel through the drift-field region, and (ii) the R-C time constant given by the parasitics R_{PD} and C_{PD}. The intrinsic bandwidth of a p-i-n photodiode turns out to be [5]

$$BW = \frac{1}{2\pi} \cdot \frac{1}{W/v_n + R_{PD}C_{PD}}, \quad (3.4)$$

where W is the width of the depletion region and v_n is the carrier velocity. For high-speed applications, the reverse voltage should be chosen large enough such that the carrier velocity v_n saturates at its maximum value and the transit time (W/v_n)

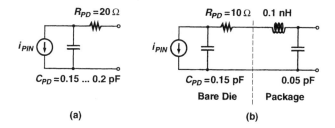

Fig. 3.3 Equivalent AC circuits for 10-Gb/s p-i-n photodetectors: (a) bare photodiode [137] and (b) photodiode with packaging parasitics [37].

is minimized. Typically, a reverse voltage of about 5 to 10 V is required for that. Photodiodes with bandwidths well in excess of 100 GHz have been demonstrated.

Shot Noise. The p-i-n photodetector not only produces the signal current I_{PIN}, but also a noise current known as *shot noise*. This noise current is the result of the photocurrent being composed of a large number of short pulses that are distributed randomly in time. Each pulse corresponds to an electron-hole pair created by a photon. If we approximate these pulses with Dirac pulses, the shot-noise spectrum is white and its mean-square value turns out to be [5]

$$\overline{i_{n,PIN}^2} = 2q I_{PIN} \cdot BW_n, \qquad (3.5)$$

where I_{PIN} is the signal current and BW_n is the bandwidth in which we measure the noise current.[2] For example, a received optical power of 1 mW generates an average current of 0.8 mA (assuming $\mathcal{R} = 0.8$ A/W) and a shot-noise current of about 1.6 μA rms in a 10-GHz bandwidth. The signal-to-noise ratio can be calculated as $10 \log(0.8 \text{ mA}/1.6 \mu\text{A})^2 = 54$ dB. [→ Problem 3.4]

We can see from Eq. (3.5) that the shot-noise current is *signal dependent*; that is, it is a function of I_{PIN}. If the received optical power increases, the noise increases, too. But fortunately, the rms noise grows only with the *square root* of the signal amplitude, so we still gain in terms of signal-to-noise ratio. If we double the power in our previous example to 2 mW, we obtain an average current of 1.6 mA and a shot-noise current of 2.26 μA; thus, the signal-to-noise ratio improves by 3 dB to 57 dB. Conversely, if the received optical power reduces, the noise reduces, too. For example, if we reduce the optical power by 3 dB, the signal current is reduced by 6 dB, but the signal-to-noise ratio degrades only by 3 dB.

If we receive a non-return-to-zero (NRZ) signal with a p-i-n photodetector, the noise on the one bits is much larger than that on the zero bits. In fact, if we turn the transmitter light source completely off during the transmission of zeros (infinite extinction ratio) and the photodetector is free of dark current (see below), then there

---

[2]More precisely, BW_n is the *noise bandwidth* of the measurement equipment (i.e., the subsequent receiver); cf. Section 4.4.

30 PHOTODETECTORS

will be no signal and therefore no noise for the zeros. Let's suppose that the received optical signal is DC balanced, has a high extinction ratio, and has the average power $\overline{P}$. It then follows that the optical power for the ones is $P_1 = 2\overline{P}$ and that for the zeros is $P_0 \approx 0$. Thus with Eq. (3.5), we find the noise currents for zeros and ones to be

$$\overline{i_{n,PIN,0}^2} \approx 0 \quad \text{and} \tag{3.6}$$

$$\overline{i_{n,PIN,1}^2} = 4q\mathcal{R}\overline{P} \cdot BW_n. \tag{3.7}$$

The precise value of $\overline{i_{n,PIN,0}^2}$ depends on the extinction ratio and dark current. Figure 3.4 illustrates the signal and noise currents produced by a p-i-n photodetector in response to an optical NRZ signal with DC balance and high extinction. Signal and noise magnitudes are expressed in terms of the average received power $\overline{P}$.

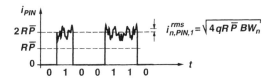

Fig. 3.4 Signal and noise currents from a p-i-n photodetector.

Dark Current. The p-i-n photodetector produces a small amount of current even when it is in total darkness. This so-called *dark current*, I_{DK}, depends on the junction area, temperature, and processing, but usually is less than 5 nA for a high-speed InGaAs photodetector. The dark current and its associated shot-noise current interfere with the received signal. Fortunately, in high-speed p-i-n receivers (2.5–40 Gb/s), this effect usually is negligible. To demonstrate this, let's calculate the optical power for which the worst-case dark current amounts to 10% of the signal current. As long as our received optical power is larger than this, we are fine:

$$\overline{P} > 10 \cdot \frac{I_{DK}(\max)}{\mathcal{R}}. \tag{3.8}$$

With the values $\mathcal{R} = 0.8$ A/W and $I_{DK}(\max) = 5$ nA, we find $\overline{P} > -42$ dBm. We see later that high-speed p-i-n receivers require much more signal power than this to work at an acceptable bit-error rate, and therefore we don't need to worry about dark current in such receivers. However, in high-sensitivity receivers (at low speeds and/or with APDs), the dark current can be an important limitation. In Section 4.5, we formulate the impact of dark current on the receiver performance in a more precise way.

Saturation Current. Whereas the shot noise and the dark current define the lower end of the p-i-n detector's dynamic range, the *saturation current* defines the upper end. At very high optical power levels, a correspondingly high density of electron-hole pairs is produced, which generates a space charge that counteracts the bias-induced drift field. The consequences are a decreased responsivity (gain compression) and reduced bandwidth. This effect is particularly important in receivers with optical preamplifiers, such as, erbium-doped fiber amplifiers (EDFAs), which readily can produce several 10 mW of optical power at the p-i-n detector. Typical values for the saturation current are in the 10 to 76 mA range [64].

3.2 AVALANCHE PHOTODETECTOR

The basic structure of the *avalanche photodetector* (APD) is shown in Fig. 3.5. Like the p-i-n detector, the avalanche photodetector is a reverse biased diode. However, in contrast to the p-i-n photodetector, it features an additional layer, the *multiplication region*. This layer provides gain through avalanche multiplication of the electron-hole pairs generated in the i-layer, also called the *absorption region*. For the avalanche process to set in, the APD must be operated at a fairly high reverse bias of about 40 to 60 V. As we said earlier, a p-i-n photodetector can be operated at a voltage of about 5 to 10 V.

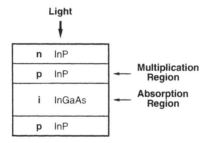

Fig. 3.5 Avalanche photodetector (schematically).

Similar to the p-i-n detector, InGaAs commonly is used for the absorption region to make the APD sensitive at long wavelengths (1.3 and 1.55 μm). The multiplication region, however, typically is made from the wider bandgap InP material, which can sustain a higher electric field.

Responsivity. The gain of the APD is called *avalanche gain* or *multiplication factor* and is designated by the letter M. A typical value for an InGaAs APD is $M = 10$. The light power P therefore is converted to electrical current I_{APD} as

$$I_{APD} = M \cdot \mathcal{R} P, \tag{3.9}$$

where $\mathcal{R}$ is the responsivity of the APD without avalanche gain, which is similar to the responsivity of a p-i-n detector. Assuming that $\mathcal{R} = 0.8$ A/W, as in our example for the p-i-n detector, and that $M = 10$, the APD generates 8 A/W. Therefore, we also can say that the APD has an effective responsivity $\mathcal{R}_{APD} = 8$ A/W, but we have to be careful to avoid confusion with the responsivity $\mathcal{R}$ in Eq. (3.9), which does not include the avalanche gain.

As we can see from Fig. 3.6, the avalanche gain M is a sensitive function of the reverse bias voltage. Furthermore, the avalanche gain also is a function of temperature and a well-controlled bias voltage source with the appropriate temperature dependence is required to keep the gain constant. The circuit in Fig. 3.7 uses a thermistor (ThR) to measure the APD temperature and a control loop to adjust the reverse bias voltage V_{APD} at a rate of 0.2%/°C to compensate for the temperature coefficient of the APD [2]. Sometimes, the dependence of the avalanche gain on the bias voltage is exploited to implement an *automatic gain control* (AGC) mechanism that acts right at the detector. Such an AGC mechanism can increase the dynamic range of the receiver.

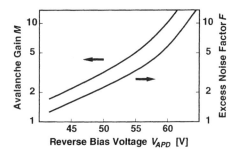

Fig. 3.6 Avalanche gain and excess noise factor as a function of reverse voltage for a typical InGaAs APD.

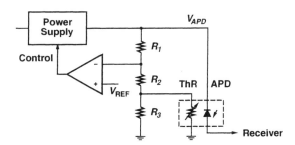

Fig. 3.7 Temperature-compensated APD bias circuit.

Avalanche Noise. Unfortunately, the APD not only provides more signal but also more noise than the p-i-n detector, in fact, *more* noise than simply the amplified shot noise that we are already familiar with. On a microscopic level, each primary carrier created by a photon is multiplied by a random gain factor: for example, the first photon ends up producing nine electron-hole pairs, the next one 13, and so on. The avalanche gain M, introduced earlier, is really just the *average* gain value. When taking the random nature of the gain process into account, the mean-square noise current from the APD can be written as [5]

$$\overline{i_{n.APD}^2} = F \cdot M^2 \cdot 2q I_{PIN} \cdot BW_n, \qquad (3.10)$$

where F is the so-called *excess noise factor* and I_{PIN} is the primary photodetector current, that is, the current before avalanche multiplication ($I_{PIN} = I_{APD}/M$). Equivalently, I_{PIN} can be understood as the current produced in a p-i-n photodetector with responsivity $\mathcal{R}$ that receives the same amount of light as the APD under discussion. In the ideal case, the excess noise factor is one ($F = 1$), which corresponds to the situation where we have a deterministic amplification of the shot noise. For a conventional InGaAs APD, this excess noise factor is more typically around $F = 6$. [$\rightarrow$ Problem 3.5]

As we can see from Fig. 3.6, the excess noise factor F increases with increasing reverse bias roughly tracking the avalanche gain M. In fact, it turns out that F and M are related as follows [5]:

$$F = k_A \cdot M + (1 - k_A) \cdot \left(2 - \frac{1}{M}\right), \qquad (3.11)$$

where k_A is the so-called *ionization-coefficient ratio*. If only one type of carrier, say electrons, participates in the avalanche process, then $k_A = 0$ and the excess noise factor is minimized. However, if electrons and holes both are participating, then $k_A > 0$ and more excess noise is produced. For an InGaAs APD, $k_A = 0.5$ to 0.7 and the excess noise factor increases almost proportional to M, as can be seen in Fig. 3.6; for a silicon APD, $k_A = 0.02$ to 0.05 and the excess noise factor increases much more slowly with M [5]. Thus from a noise point of view, the silicon APD is preferable, but as we know, silicon is not sensitive at the long wavelengths commonly used in telecommunication applications. Researchers are working on long-wavelength APDs with better noise performance than the conventional InGaAs APD. They do so by using materials with a lower k_A (e.g., InAlAs) and structures that reduce the randomness in the avalanche process.

Because the avalanche gain can be increased only at the expense of producing more noise in the detector (Eq. (3.11)), there is an optimum APD gain at which the receiver becomes most sensitive. As we see in Section 4.3, the value of this optimum gain depends, among other things, on the APD material (k_A).

From what has been said, it should be clear that the APD noise is signal dependent, just like the p-i-n detector noise. The noise currents for zeros and ones, given a DC-balanced NRZ signal with average power $\overline{P}$ and high extinction, can be found with Eq. (3.10):

$$\overline{i_{n.APD,0}^2} \approx 0 \quad \text{and} \qquad (3.12)$$
$$\overline{i_{n.APD,1}^2} = F \cdot M^2 \cdot 4q\mathcal{R}\overline{P} \cdot BW_n. \qquad (3.13)$$

The precise value of $\overline{i_{n.APD,0}^2}$ depends on the extinction ratio and dark current.

Dark Current. Similar to the p-i-n detector, the APD also suffers from a dark current. The so-called *primary dark current*, I_{DK}, is amplified, just like a signal current, to $M \cdot I_{DK}$ and produces the avalanche noise $F \cdot M^2 \cdot 2q I_{DK} \cdot BW_n$. Typically, I_{DK} is less than 5 nA for a high-speed InGaAs APD [5]. We again can use Eq. (3.8) to judge if this amount of dark current is harmful. With the values $\mathcal{R} = 0.8$ A/W and $I_{DK}(\max) = 5$ nA, we find that we are fine as long as $\overline{P} > -42$ dBm. Most high-speed APD receivers require more signal power than this to work at an acceptable bit-error rate, and dark current is not a big worry.

Bandwidth. Increasing the reverse bias not only increases the gain and the excess noise factor, but also reduces the signal *bandwidth*. Similar to a single-stage amplifier, the product of gain and bandwidth remains approximately constant and therefore can be used to quantify the speed of an APD. The gain-bandwidth product of a typical high-speed APD is in the range of 100 to 150 GHz. The equivalent AC circuit for an APD is similar to those shown for the p-i-n detector in Fig. 3.3, except that the current source is now given by $i_{APD}(t) = M \cdot \mathcal{R} P(t)$ and the parasitic capacitances typically are somewhat larger.

APDs are in widespread use for receivers up to and including 2.5 Gb/s. However, it is challenging to fabricate APDs with a high enough gain-bandwidth product to be useful at 10 Gb/s and above. For this reason, high sensitivity 10-Gb/s+ receivers often use optically preamplified p-i-n detectors. These detectors are more expensive than APDs but feature superior speed and noise performance.

3.3 P-I-N DETECTOR WITH OPTICAL PREAMPLIFIER

A higher performance alternative to the APD is the p-i-n detector with optical preamplifier or simply the *optically preamplified p-i-n detector*. The p-i-n detector operates at high speed, whereas the optical preamplifier provides high gain over a huge bandwidth (e.g., more than 10 nm corresponding to more than 1,250 GHz), eliminating the gain-bandwidth trade-off known from APDs. Furthermore, the optically preamplified p-i-n detector has superior noise characteristics when compared with an APD. However, the cost of a high-performance optical preamplifier, such as an EDFA, is substantial.

The optical preamplifier can be implemented with a so-called *semiconductor optical amplifier* (SOA), which is small and can be integrated together with the p-i-n detector on the same InP substrate. However, for best performance, the *erbium-doped fiber amplifier* (EDFA), which operates in the important 1.55-μm band and features high gain and low noise, is a popular choice. See Fig. 3.8 for the operating principle of an EDFA-preamplified p-i-n detector. An optical coupler combines the received optical signal (input) with the light from a continuous-wave pump laser. The pump laser typically provides a power of a few 10 mW at either the 0.98-μm or 1.48-μm wavelength, where the 0.98-μm wavelength is preferred for low-noise preamplifiers. The signal and the pump light are sent through an erbium-doped fiber of about 10 m, where the amplification takes place by means of stimulated emission. An optical isolator prevents reflections of the optical signal from entering back into the amplifier, which would cause instability. An optical filter with (noise) bandwidth BW_O reduces the noise of the amplified optical signal before it is converted to an electrical signal with a p-i-n photodetector. Optical noise is generated in the EDFA because of a process called *amplified spontaneous emission* (ASE). The power spectral density of this ASE noise, S_{ASE}, is nearly white.[3] Thus, we can calculate the optical noise power that reaches the photodetector as $P_{ASE} = S_{ASE} \cdot BW_O$. To keep P_{ASE} low, we want to use a narrow optical filter.

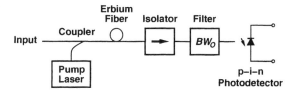

Fig. 3.8 A p-i-n photodetector with erbium-doped fiber preamplifier (schematically).

Responsivity. One of the main characteristics of the optical amplifier is its power gain, G. The gain value of an EDFA depends on the length of the erbium-doped fiber and increases with increasing pump power, as shown in Fig. 3.9.[4] A typical value is $G = 100$, corresponding to a 20-dB gain. The current produced by the p-i-n photodetector, I_{OA}, expressed as a function of the optical power at the *input* of the preamplifier, P, is

$$I_{OA} = G \cdot \mathcal{R} P, \tag{3.14}$$

where $\mathcal{R}$ is the responsivity of the p-i-n photodetector.

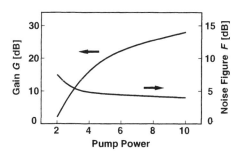

Fig. 3.9 EDFA gain and noise figure as a function of the pump power.

---

[3] In the following, S_{ASE} always refers to the noise spectral density in *both* polarization modes, that is, $S_{ASE} = 2 \cdot S'_{ASE}$, where S'_{ASE} is the noise spectral density in a single polarization mode.
[4] The pump power in Fig. 3.9 is given in multiples of the pump saturation power [5].

Because the gain depends sensitively on the pump power, EDFA modules frequently incorporate a microcontroller, which adjusts the power of the pump laser. An automatic gain control (AGC) mechanism can be implemented by controlling the pump power in response to a small light sample split off from the amplified output signal [29]. Such an AGC mechanism can increase the dynamic range of the receiver.

Whereas the APD improved the responsivity by about one order of magnitude ($M = 10$), the optically preamplified p-i-n detector can improve the responsivity by about two orders of magnitude ($G = 100$) relative to a regular p-i-n detector. So, given that $\mathcal{R} = 0.8$ A/W and $G = 100$, the effective responsivity of the combined preamplifier and p-i-n detector is 80 A/W.

ASE Noise. As we said earlier, the EDFA not only amplifies the input signal as desired, but also produces an optical noise known as ASE noise. How is this optical noise converted to an electrical noise in the photodetector? If you thought that it was odd that optical signal *power* is converted to a proportional electrical signal *current*, wait until you hear this: because the photodetector responds to the intensity, which is proportional to the *square* of the fields (cf. Fig. 2.5), the optical noise gets converted to *two* electrical beat-noise components. Roughly speaking, we get the terms corresponding to (signal + noise)^2 = (signal)^2 + 2 · (signal · noise) + (noise)^2. The first term is the desired electrical signal, the second term is the so-called *signal-spontaneous beat noise*, and the third term is known as the *spontaneous-spontaneous beat noise*. A detailed analysis reveals that the two electrical noise terms are [5]

$$\overline{i_{n.ASE}^2} = \mathcal{R}^2 \cdot \left(2P_S S_{ASE} + S_{ASE}^2 \cdot BW_O\right) \cdot BW_n. \quad (3.15)$$

The first term in Eq. (3.15), the signal-spontaneous beat noise, usually is the dominant term. This noise component is proportional to the signal power P_S at the output of the EDFA ($P_S = GP$). So, a signal-independent optical noise density S_{ASE} generates a *signal-dependent* noise term in the electrical domain! Furthermore, this noise term is *not* affected by the optical filter bandwidth BW_O, but the electrical bandwidth BW_n does have an effect. The second term in Eq. (3.15), the spontaneous-spontaneous beat noise, may be closer to your expectations.[5] Similar to the signal component, this noise current component is proportional to the optical noise power. Moreover, the optical filter bandwidth does have an effect on the spontaneous-spontaneous beat noise component. In addition to the ASE noise terms in Eq (3.15), the p-i-n photodetector also produces shot noise terms. However, the latter noise contributions are so small that they usually can be neglected. [→ Problem 3.6]

---

[5]In the literature, spontaneous-spontaneous beat noise is sometimes given as $4\mathcal{R}^2 S'^2_{ASE} BW_O BW_n$ [5] and sometimes as $2\mathcal{R}^2 S'^2_{ASE} BW_O BW_n$ [116] ($S'_{ASE} = S_{ASE}/2$, the ASE spectral density in a single polarization mode), which may be quite confusing. It seems that the first equation applies to EDFA/p-i-n systems *without* a polarizer in between the amplifier and the p-i-n detector, whereas the second equation applies to EDFA/p-i-n systems *with* a polarizer. In practice, polarizers are not usually used in EDFA/p-i-n systems because this would require a polarization controlled signal. We thus are using the $4\mathcal{R}^2 S'^2_{ASE} BW_O BW_n$ expression here.

By now you have probably developed a healthy respect for the unexpected ways optical quantities translate to the electrical domain. Now let's see what happens to the *signal-to-noise ratio* (SNR). For a continuous-wave signal with the optical power P_S incident on the photodetector, the signal power in the electrical domain is $\overline{i_S^2} = \mathcal{R}^2 P_S^2$. The electrical noise power, $\overline{i_{n,ASE}^2}$, for the same optical signal is given by Eq. (3.15). With $P_{ASE} = S_{ASE} \cdot BW_O$, the ratio of these two expressions ($\overline{i_S^2}/\overline{i_{n,ASE}^2}$) becomes

$$SNR = \frac{(P_S/P_{ASE})^2}{P_S/P_{ASE} + 1/2} \cdot \frac{BW_O}{2BW_n}. \qquad (3.16)$$

Now P_S/P_{ASE} also is known as the *optical signal-to-noise ratio* (OSNR) at the output of the EDFA measured in the optical bandwidth BW_O. If the OSNR is much larger than $1/2$ ($-3\,\text{dB}$), we can neglect the contribution from spontaneous-spontaneous beat noise (this is where the $1/2$ in the denominator comes from) and we end up with the surprisingly simple result

$$SNR = \frac{OSNR^2}{OSNR + 1/2} \cdot \frac{BW_O}{2BW_n} \approx OSNR \cdot \frac{BW_O}{2BW_n}. \qquad (3.17)$$

This means that the electrical SNR can be obtained simply by scaling the OSNR with the ratio of the optical and $2\times$ the electrical bandwidth. For example, for a receiver with $BW_n = 7.5\,\text{GHz}$, an OSNR of $14.7\,\text{dB}$ measured in a 0.1-nm bandwidth ($12.5\,\text{GHz}$ at $\lambda = 1.55\,\mu\text{m}$) translates into an electrical SNR of $13.9\,\text{dB}$. In Section 4.3, we use Eq. (3.17) to analyze optically amplified transmission systems. [→ Problem 3.7]

Noise Figure of an Optical Amplifier. Just like electrical amplifiers, optical amplifiers are characterized by a noise figure F. A typical value for an EDFA noise figure is $F = 5\,\text{dB}$, and the theoretical lower limit turns out to be $3\,\text{dB}$, as we see later. But what is the meaning of noise figure for an *optical* amplifier?

In an electrical system, the noise figure is defined as the ratio of the "total output noise power" to the "fraction of the output noise power due to the thermal noise of the source resistance." Usually, this source resistance is $50\,\Omega$. (We discuss the electrical noise figure in more detail in Section 6.2.3.) Now, an optical amplifier doesn't get its signal from a 50-Ω source, and so the definition of its noise figure cannot be based on thermal 50-Ω noise. What fundamental noise is it based on? The quantum (shot) noise of the optical source!

The noise figure of an optical amplifier is defined as the ratio of the "total output noise power" to the "fraction of the output noise power due to the quantum (shot) noise of the optical source." The output noise power is measured with a p-i-n photodetector that has a perfect quantum efficiency ($\eta = 1$) and is quantified as the detector's mean-square noise current.[6] If we write the total output noise power as $\overline{i_{n,OA}^2}$ and the fraction that is due to the source as $\overline{i_{n,OA,S}^2}$, then the noise figure is $F = \overline{i_{n,OA}^2}/\overline{i_{n,OA,S}^2}$.

Figure 3.10 illustrates the various noise quantities introduced above. At the top, an ideal photodetector is illuminated directly by the optical source and produces the DC current I_{PIN} and the mean-square shot-noise current $\overline{i_{n,PIN}^2} = 2qI_{PIN} \cdot BW_n$. In the middle, the signal from the optical source is amplified with a noiseless, deterministic amplifier with gain G. This amplifier multiplies every photon from the source into exactly G photons. The ideal photodetector now produces the DC current $I_{OA} = GI_{PIN}$ and the mean-square shot-noise current $\overline{i_{n,OA,S}^2} = G^2 \cdot 2qI_{PIN} \cdot BW_n$ (cf. Problem 3.5). Note that this quantity represents the "fraction of the output noise power due to the source." At the bottom, we replaced the noiseless amplifier with a real amplifier with gain G and noise figure F, which produces the "total output noise power." According to the noise figure definition, the ideal photodetector now produces a mean-square noise current that is F times larger than before:

$$\overline{i_{n,OA}^2} = F \cdot G^2 \cdot 2qI_{PIN} \cdot BW_n, \tag{3.18}$$

where I_{PIN} is the current produced by an ideal p-i-n photodetector receiving the same amount of light as the optical preamplifier. Note that this noise current is still based on an ideal photodetector. How large is the output noise current of an optical amplifier followed by a *real* p-i-n detector with $\eta < 1$? We have to reduce $\overline{i_{n,OA}^2}$ by the factor η^2 while taking into account that I_{PIN} also reduces by η; thus, we obtain the output noise current

$$\overline{i_{n,OA}^2} = \eta F \cdot G^2 \cdot 2qI_{PIN} \cdot BW_n. \tag{3.19}$$

As usual, the noise current on zeros and ones is different and, given a DC-balanced NRZ signal with average power $\overline{P}$ and high extinction, we find with Eq. (3.19)

$$\overline{i_{n,OA,0}^2} \approx 0 \quad \text{and} \tag{3.20}$$

$$\overline{i_{n,OA,1}^2} = \eta F \cdot G^2 \cdot 4q\mathcal{R}\overline{P} \cdot BW_n. \tag{3.21}$$

The precise value of $\overline{i_{n,OA,0}^2}$ depends on the extinction ratio, dark current, and spontaneous-spontaneous beat noise.

It is instructive to compare the noise expression Eq. (3.10) for the APD with Eq. (3.19) for the optically preamplified p-i-n detector. We discover that the excess noise factor F of the APD plays the same role as the product ηF of the optical preamplifier!

Noise Figure and ASE Noise. In Eq. (3.15), we expressed the electrical noise in terms of the optical ASE noise, and in Eq. (3.18), we expressed the electrical noise

---

[6] An equivalent definition for the noise figure of an optical amplifier is the ratio of the "input SNR" to the "output SNR," where both SNRs are measured in the electrical domain with ideal photodetectors ($\eta = 1$) and where the input SNR is based on shot noise only.

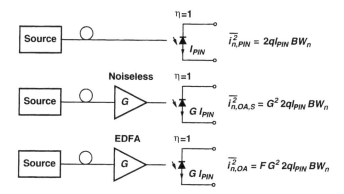

Fig. 3.10 Definition of the noise figure for an optical amplifier.

in terms of the amplifier's noise figure. Now let's combine the two equations and find out how the noise figure is related to the ASE noise spectral density. With the assumption that all electrical noise at the output of the optically preamplified p-i-n detector is described by the terms in Eq. (3.15), $\overline{i_{n,OA}^2} = \overline{i_{n,ASE}^2}$, that is, ignoring shot noise contributions, we find

$$F = \frac{\lambda}{hc} \cdot \left(\frac{S_{ASE}}{G} + \frac{S_{ASE}^2}{2 \cdot G^2 P} \cdot BW_O \right). \quad (3.22)$$

The first term is caused by signal-spontaneous beat noise, whereas the second term is caused by spontaneous-spontaneous beat noise. Note that this noise figure depends on the input power P and becomes infinite for $P \to 0$. The reason for this is that when the signal power goes to zero, we are still left with the spontaneous-spontaneous beat noise, whereas the noise due to the source does go to zero. [$\to$ Problem 3.8]

Sometimes a restrictive type of noise figure $\tilde{F}$ is defined that corresponds to just the first term of Eq. (3.22):

$$\tilde{F} = \frac{\lambda}{hc} \cdot \frac{S_{ASE}}{G}. \quad (3.23)$$

This noise figure is known as *signal-spontaneous beat noise limited noise figure* or *optical noise figure* and is independent of the input power. For sufficiently large input power levels P and small optical bandwidths BW_O, it is approximately equal to the noise figure F in Eq. (3.22). (The fact that there are two similar but not identical noise figure definitions can be confusing at times.)

Let's go one step further. A physical analysis of the ASE noise process reveals the following expression for its power spectral density [5]:

$$S_{ASE} = 2(G-1) \cdot \frac{N_2}{N_2 - N_1} \cdot \frac{hc}{\lambda}, \quad (3.24)$$

where N_1 is the number of erbium atoms in the ground state and N_2 is the number of erbium atoms in the excited state. The stronger the amplifier is "pumped," the more atoms will be in the excited state, and thus for a strongly pumped amplifier, we have $N_2 \gg N_1$. Combining Eq. (3.23) for the optical noise figure with Eq. (3.24) and taking $G \gg 1$, we find the following simple approximation for the EDFA noise figure(s):

$$F \approx \tilde{F} \approx 2 \cdot \frac{N_2}{N_2 - N_1}. \tag{3.25}$$

This equation means that increasing the pump power will decrease the noise figure until it reaches the theoretical limit of 3 dB (cf. Fig. 3.9).

Negative Noise Figure? What would an optical amplifier with a *negative* noise figure ($10 \log F < 0$ dB, $F < 1$) do? Placing such an amplifier in front of a p-i-n detector would *improve* the signal-to-noise ratio over that of an unamplified p-i-n detector. This sounds like a tricky thing to do. Now you may be surprised to learn that you can actually *buy* optical amplifiers with negative noise figures. You can buy a Raman amplifier with $F = -2$ dB or even less, if you are willing to pay more!

Consider the following: a fiber span with loss $1/G$ has a noise figure of G. The same fiber span followed by an EDFA with noise figure F has a combined noise figure of $G \cdot F$. You can prove both facts easily with the noise figure definition given earlier. For example, a 100-km fiber span with 25-dB loss followed by an EDFA with a noise figure of 5 dB has a total noise figure of 30 dB. [→ Problem 3.9]

Now, there is a type of optical amplifier, the *Raman amplifier*, that can provide distributed gain *in* the fiber span itself. The fiber span is "pumped" from the receive end with a strong laser (1 W or so) and *stimulated Raman scattering* (SRS), one of the nonlinear fiber effects, provides the gain. For example, by pumping the 100-km fiber span from above the loss may reduce from 25 dB to 15 dB and the noise figure may improve from 25 dB to 23 dB. How do you sell such an amplifier? Right, you compare it with a lumped amplifier such as an EDFA and say it has a gain of 10 dB and a noise figure of -2 dB. O.K., I'll order one but please ship it without the fiber span ...

3.4 SUMMARY

Three types of photodetectors commonly are used for optical receivers:

- The p-i-n photodetector with a typical responsivity in the range of 0.6 to 0.9 A/W (for an InGaAs detector) is used mostly in short-haul applications.

- The avalanche photodetector (APD) with a typical responsivity in the range of 5 to 20 A/W (for an InGaAs detector) is used mostly in long-haul applications up to 10 Gb/s.

- The optically preamplified p-i-n detector with a responsivity in the range of 6 to 900 A/W is used mostly in ultra-long-haul applications and for speeds at or more than 10 Gb/s.

All three detectors generate a *current* that is proportional to the received *optical power*, that is, a 3-dB change in optical power results in a 6-dB change in current.

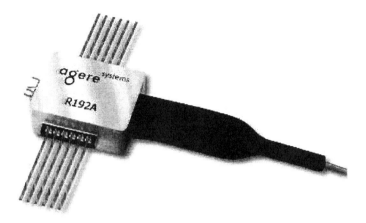

Fig. 3.11 A 10-Gb/s photodetector and TIA in a 16-pin surface-mount package with a single-mode fiber pigtail (1.6 cm × 1.3 cm × 0.7 cm). Reprinted by permission from Agere Systems, Inc.

Fig. 3.12 A packaged two-stage erbium-doped fiber amplifier with single-mode fiber pigtails for the input, output, interstage access, and tap monitor ports (12 cm × 10 cm × 2 cm). Reprinted by permission from Agere Systems, Inc.

All three detectors produce a *signal-dependent* noise current, specifically, the noise power $\overline{i_{n.PD}^2}$ grows proportional to the signal current I_{PD} (neglecting the spontaneous-spontaneous beat noise of the optically preamplified p-i-n detector). As a result, received one bits contain more noise than zero bits. The p-i-n detector produces shot noise, which often is negligible in digital transmission systems. The APD produces avalanche noise, quantified by the excess noise factor F. The optical preamplifier produces amplified spontaneous emission (ASE) noise, which is converted into two electrical noise components by the p-i-n detector. The noise characteristics of the optical preamplifier are specified by a noise figure F.

3.5 PROBLEMS

3.1 Optical vs. Electrical dBs. A p-i-n photodetector in a 1.55-μm transmission system converts the received optical signal to an electrical signal. By how many dBs is the latter signal attenuated if we splice an additional 40 km of standard SMF into the system?

3.2 Power Conservation in the Photodiode. The p-i-n photodetector produces a current that is proportional to the received optical power P. When this current runs through a resistor, it produces a voltage drop that also is proportional to the received optical power. Thus, the electrical power dissipated in the resistor is proportional to P^2. We conclude that for large values of P, the electrical power will exceed the received optical power! (a) Is this a violation of energy conservation? (b) What can you say about the maximum forward-voltage drop, V_F, of a photodiode?

3.3 Photodetector vs. Antenna. An ideal photodetector ($\eta = 1$) and antenna both are exposed to the same continuous-wave electromagnetic radiation at power level P. (a) Calculate the power level P at which the signal from the photodetector becomes equal to the rms value of the shot noise. (b) Calculate the power level P at which the rms signal level from the antenna becomes equal to the rms value of the antenna's thermal noise. (c) How do these power levels (sensitivities) for the photodetector and the antenna compare?

3.4 Shot Noise. The current produced by a p-i-n photodetector contains shot noise because the current consists of a stream of randomly generated, point-like, charged particles (electrons). (a) Does a battery loaded by a resistor also produce shot noise? (b) Explain the answer!

3.5 Amplified Shot Noise. An APD with deterministic amplification (every primary carrier generates precisely M secondary carriers) produces the mean-square noise $\overline{i_{n.APD}^2} = M^2 \cdot 2q I_{PIN} \cdot BW_n$ (Eq. (3.10)). Now, we could argue that the DC current produced by the APD is $M I_{PIN}$ and thus the associated shot noise should be $\overline{i_{n.APD}^2} = 2q \cdot (M I_{PIN}) \cdot BW_n$. What is wrong with the latter argument?

3.6 Optically Preamplified p-i-n Detector. The following equation for the noise produced by an optically preamplified p-i-n photodetector receiving the continuous-wave input power P is given in [5]:

$$\overline{i^2_{n,OA}} = 2q\mathcal{R} \cdot GP \cdot BW_n + 2q\mathcal{R} \cdot S'_{ASE} \cdot BW_O \cdot BW_n$$
$$+ 2q I_{DK} \cdot BW_n + 4\mathcal{R}^2 \cdot S'^2_{ASE} \cdot BW_O \cdot BW_n \qquad (3.26)$$
$$+ 4\mathcal{R}^2 \cdot GP \cdot S'_{ASE} \cdot BW_n + 4q\mathcal{R} \cdot S'_{ASE} \cdot BW_O \cdot BW_n.$$

Explain the origin of each term in this equation.

3.7 Optical Signal-to-Noise Ratio. Equations (3.16) and (3.17) state the relationship between SNR and OSNR for a continuous-wave signal with power P_S. How does this expression change for a DC-balanced ideal NRZ-modulated signal with high extinction and an average power $\overline{P_S}$?

3.8 Noise Figure of an Optical Amplifier. (a) Derive the equation for the noise figure of an optical amplifier, Eq. (3.22), but also include the effect of the shot noise caused by the signal current (cf. Problem 3.6). (b) What would that noise figure be, if we could build an optical amplifier free of ASE noise?

3.9 Noise Figure of a Fiber. (a) Calculate the noise figure F of an optical fiber with loss $1/G$. (b) Calculate the noise figure F of an optical system consisting of an optical fiber with loss $1/G_1$ followed by an EDFA with gain G_2 and noise figure F_2. (c) Calculate the noise figure F of an optical system with n segments, where each segment consists of an optical fiber with loss $1/G$ followed by an EDFA with gain G and noise figure F_2.

4
Receiver Fundamentals

In this chapter, we present the optical receiver at the system level. The terminology and concepts introduced here will simplify the discussion in later chapters. In the following, we analyze how noise in the receiver causes bit errors. This leads to the definition of the receiver sensitivity. After introducing the concept of power penalty, we study the impact of the receiver's bandwidth and frequency response on its performance. The adaptive equalizer, used to mitigate distortions in the received signal, is covered briefly. We then turn to other receiver impairments such as nonlinearity (in analog receivers), jitter, decision threshold offset, and sampling time offset. We conclude with a brief description of forward error correction, a technique that can improve the receiver performance dramatically. More information on receiver theory can be found in [6, 42, 83].

4.1 RECEIVER MODEL

The basic *receiver model* used in this chapter is shown in Fig. 4.1. It consists of (i) a photodetector model, (ii) a linear channel model that comprises the transimpedance amplifier (TIA), the main amplifier (MA), and optionally a low-pass filter, and (iii) a binary decision circuit with a fixed threshold (V_{DTH}). Later in the Sections 4.7, 4.10, and 4.11 we extend this basic model to include an adaptive equalizer, an adaptive decision threshold, and a multilevel decision circuit, respectively.

The *detector model* consist of a signal current source i_{PD} and a noise current source $i_{n.PD}$. The characteristics of these two current sources were discussed in Chapter 3 for the p-i-n photodetector, the avalanche photodetector (APD), and the optically

46 RECEIVER FUNDAMENTALS

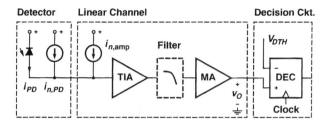

Fig. 4.1 Basic receiver model.

preamplified p-i-n detector. In all cases, we have found that the signal current is linearly related to the received optical power and that the noise current spectrum is approximately white and signal dependent.

The *linear channel* can be modeled with a complex transfer function $H(f)$ that relates the amplitude and phase of the output voltage v_O to those of the input current i_{PD}. This transfer function can be decomposed into a product of three transfer functions: one for the TIA, one for the filter, and one for the MA. But for now, we are concerned with the receiver as a whole. The noise characteristics of the linear channel are modeled by a single noise current source $i_{n.\text{amp}}$ at the input of the channel.[1] The noise spectrum of this source is chosen such that after passing through the noiseless channel $H(f)$, it produces the output noise spectrum of the actual noisy channel. In practice, the linear-channel noise $i_{n.\text{amp}}$ is determined almost completely by the input-referred noise of the TIA, which is the first element of the linear channel. Therefore, we also call this noise the *amplifier noise*. It is important to distinguish the different characteristics of the detector and amplifier noise:

- The detector noise, $i_{n.PD}$, is *nonstationary* (the rms value is varying with the bit value) and *white* (frequency independent) to a good approximation. Thus, the power spectral density (or power spectrum for short) of the detector noise must be written as a function of *time*:

$$I_{n.PD}^2(f, t) \sim \text{bit\_value}(t). \qquad (4.1)$$

- The amplifier noise, $i_{n.\text{amp}}$, is *stationary* (the rms value is independent of time) and usually is *not white*. In Section 5.2.3, we calculate the spectrum of this noise source (Eqs. (5.37), (5.40), and (5.41)) and we see that its two main components are a constant part (white noise) and a part increasing with frequency like f^2. This is the case no matter if the receiver is built with an FET or BJT front-end. The power spectrum of the amplifier noise therefore can be written in the general form

$$I_{n.\text{amp}}^2(f) = \alpha_0 + \alpha_2 f^2 + \ldots . \qquad (4.2)$$

---

[1] Note that as a result of modeling the amplifier noise with only a single noise current source, rather than a noise current and noise voltage source, the value of $i_{n.\text{amp}}$ becomes dependent on the photodetector impedance, in particular its capacitance.

The last block in our receiver model, the *decision circuit*, compares the output voltage from the linear channel, v_O, with a fixed threshold voltage, V_{DTH}. If the output voltage is larger than the threshold, a one bit is detected; if it is smaller, a zero bit is detected. Note that in contrast to the linear channel, this block is *nonlinear*. The comparison in the decision circuit is triggered by a clock signal, which typically is provided by a clock-recovery circuit.

At this point, you may wonder how appropriate a *linear* model for the TIA and MA really is, in particular if the MA is implemented as a limiting amplifier, which becomes strongly nonlinear for large input signals. Fortunately, the receiver's own noise as well as the signal levels at the sensitivity limit usually are so small that we don't have to worry about nonlinearity and limiting. Thus, for the subsequent noise and sensitivity calculations, a linear model is appropriate.

4.2 BIT-ERROR RATE

The voltage v_O at the output of the linear channel is a superposition of the desired *signal* voltage v_S and the undesired *noise* voltage v_n ($v_O = v_S + v_n$). The noise voltage v_n, of course, is caused by the detector noise and the amplifier noise. Occasionally, the instantaneous noise voltage $v_n(t)$ may become so large that it corrupts the received signal $v_S(t)$, leading to a decision error or *bit error*. In this section, we first calculate the rms value of the output noise voltage, v_n^{rms}, and then derive the bit-error rate, *BER*, caused by this noise.

Output Noise. The output noise power can be written as a sum of two components, one caused by the detector and one caused by the linear channel (amplifiers). Let's start with the amplifier noise, which is stationary and therefore easier to deal with. Given the input-referred power spectrum $I_{n,\mathrm{amp}}^2(f)$ for the amplifier noise and the transfer function of the linear channel $H(f)$, we can easily calculate the power spectrum at the output:

$$V_{n,\mathrm{amp}}^2(f) = |H(f)|^2 \cdot I_{n,\mathrm{amp}}^2(f). \tag{4.3}$$

Note that to avoid cluttered equations, we omit indices distinguishing input and output quantities. This can be done without ambiguity because we know from our model that a current indicates an input signal to the linear channel and a voltage indicates an output signal from the linear channel. Integrating the noise spectrum in Eq. (4.3) over the bandwidth of the decision circuit, BW_D, gives us the total noise power due to the amplifier experienced by the decision circuit:

$$\overline{v_{n,\mathrm{amp}}^2} = \int_0^{BW_D} |H(f)|^2 \cdot I_{n,\mathrm{amp}}^2(f)\, df. \tag{4.4}$$

This equation is illustrated by Fig. 4.2. The input noise spectrum, I_n^2, which increases with frequency as a result of the f^2 component, is shaped by the $|H(f)|^2$ frequency response, producing an output spectrum, V_n^2, which rolls off rapidly at high frequencies. Because of the rapid rolloff, the precise value of the upper integration bound (BW_D) is uncritical and sometimes is set to infinity.

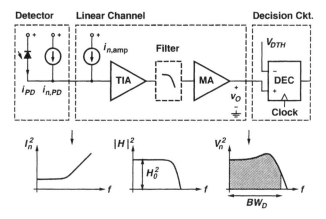

Fig. 4.2 Calculation of the total output-referred noise.

Next, we have to deal with the nonstationary detector noise. Visualize the input noise spectrum, $I_{n,PD}^2(f, t)$, as a two-dimensional surface located above the time and frequency coordinates. It can be shown [83] that this two-dimensional spectrum is mapped to the output of the linear channel as follows:

$$V_{n,PD}^2(f, t) = H(f) \cdot \int_{-\infty}^{\infty} I_{n,PD}^2(f, t - t') \cdot h(t') \cdot e^{j 2\pi f t'} \, dt', \quad (4.5)$$

where $h(t)$ is the impulse response of the linear channel. This means that the spectrum not only gets "shaped" along the frequency axis, but it also gets "smeared out" along the time axis! Potentially, this is a complex situation, because the output noise during the nth bit period depends not only on the input noise during this same period, but also depends on the input noise during all the previous bits. In some texts, this complex noise analysis is carried out to the full extent [6, 127, 177]. However, here we take the easy way out and assume that the input noise varies slowly compared with the duration of the impulse response $h(t)$. Under these circumstances, Eq. (4.5) can be simplified to the form of Eq. (4.3), with the difference that the spectra are now time dependent. Thus, the total output noise power due to the photodetector is

$$\overline{v_{n,PD}^2}(t) = \int_0^{BW_D} |H(f)|^2 \cdot I_{n,PD}^2(f, t) \, df. \quad (4.6)$$

For systems using on-off keying (OOK), this time-dependent output noise power can be described by just two values, one during the reception of zeros and one during the reception of ones. [→ Problem 4.1]

The rms noise at the output of the linear channel due to both noise sources is obtained by adding the (uncorrelated) noise powers given in Eqs. (4.4) and (4.6) under the square root:

$$v_n^{rms}(t) = \sqrt{\overline{v_{n,PD}^2}(t) + \overline{v_{n,\mathrm{amp}}^2}} \qquad (4.7)$$
$$= \sqrt{\int_0^{BW_D} |H(f)|^2 \cdot [I_{n,PD}^2(f,t) + I_{n,\mathrm{amp}}^2(f)]\,df}.$$

Again, for OOK systems, this time-dependent noise can be described by two values: $v_{n,0}^{rms}$ for the zeros and $v_{n,1}^{rms}$ for the ones.

Signal, Noise, and Bit-Error Rate. Now that we have derived the value of the output rms noise, how is it related to the bit-error rate? Figure 4.3 illustrates the situation at the input of the decision circuit, where we have the non-return-to-zero (NRZ) signal $v_S(t)$ with a peak-to-peak value v_S^{pp} and the noise $v_n(t)$ with an rms value v_n^{rms}. For now, we assume that the NRZ signal is free of distortions (intersymbol interference) and that the noise is Gaussian and signal independent (later we will generalize). The noisy signal is sampled at the center of each bit period (vertical dashed lines), producing the statistical distributions shown on the right-hand side. Both distributions are Gaussian and have a standard deviation that is equal to the rms value of the noise voltage, v_n^{rms}, which we calculated in Eq. (4.7).

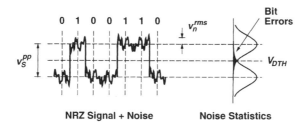

Fig. 4.3 Relationship between signal, noise, and bit-error rate.

The decision circuit determines whether a bit is a zero or a one by comparing the sampled output voltage v_O with the threshold voltage V_{DTH}, which is located at the midpoint between the zero and one levels. Note that aligning the threshold voltage with the crossover point of the two distributions produces the fewest bit errors (assuming equal probability for zeros and ones). Now we can define the *bit-error rate* (BER) as the probability that a zero is misinterpreted as a one or that a one is misinterpreted as a zero.[2]

Given the above model, we can now derive a mathematical expression for the BER. The error probabilities are given by the shaded areas under the Gaussian tails. The

---

[2] In fact, the term *bit-error rate* is misleading because it suggests a measurement of bit errors per time interval. A more accurate term would be *bit-error probability* or *bit-error ratio*, however, because of the widespread use of the term *bit-error rate*, we stick with it here.

area of each tail has to be summed with a weight of 1/2 because zeros and ones are assumed to occur with probability 1/2. Because the two tails are equal in area, we can calculate just one of them:

$$BER = \int_Q^\infty \text{Gauss}(x)\, dx \quad \text{with} \quad Q = \frac{v_S^{pp}}{2 \cdot v_n^{rms}}, \quad (4.8)$$

where Gauss(x) is the normalized Gaussian distribution (average = 0, standard deviation = 1). The lower bound of the integral, Q, is the difference between the one (or zero) level and the decision threshold, $v_S^{pp}/2$, normalized to the standard deviation v_n^{rms} of the Gaussian distribution. Note that this value is the starting point of the shaded tail in normalized coordinates. The Q parameter, also called the *Personick* Q,[3] is a measure of the ratio between signal and noise (but there are some subtle differences between Q and the signal-to-noise ratio [SNR], as we will discuss later). The integral in the above equation can be expanded and approximated as follows:

$$\int_Q^\infty \text{Gauss}(x)\, dx = \frac{1}{\sqrt{2\pi}} \int_Q^\infty e^{-\frac{x^2}{2}}\, dx$$
$$= \frac{1}{2}\,\text{erfc}\left(\frac{Q}{\sqrt{2}}\right) \approx \frac{1}{\sqrt{2\pi}} \cdot \frac{\exp(-Q^2/2)}{Q}. \quad (4.9)$$

The approximation on the far right is correct within 10% for $Q > 3$. The precise numerical values for the integral are listed in Table 4.1 for some commonly used BER values.

Table 4.1 Numerical relationship between Q and bit-error rate.

Q	BER	Q	BER
0.0	1/2	5.998	10^{-9}
3.090	10^{-3}	6.361	10^{-10}
3.719	10^{-4}	6.706	10^{-11}
4.265	10^{-5}	7.035	10^{-12}
4.753	10^{-6}	7.349	10^{-13}
5.199	10^{-7}	7.651	10^{-14}
5.612	10^{-8}	7.942	10^{-15}

A Generalization: Unequal Noise Distributions. We now drop the assumption that the noise is signal independent. We know that the noise on the ones is larger than the noise on the zeros in applications where the detector noise is significant compared

---

[3] Note that the Personick Q is different from the Q-function, $Q(x)$, used in some texts [136]. In fact, the Personick Q corresponds to the argument, x, of the Q-function.

with the amplifier noise, that is, for receivers with an optically preamplified p-i-n detector or an APD (and also in optically amplified lightwave systems, as we will see later). Given the simplified noise model introduced earlier, the rms noise alternates between the values $v_{n,0}^{rms}$ and $v_{n,1}^{rms}$, depending on whether the received bit is a zero or a one. In terms of the noise statistics, we now have two different Gaussians, one for the zeros with the standard deviation $v_{n,0}^{rms}$ and a wider, lower one for the ones with the standard deviation $v_{n,1}^{rms}$. Calculating the crossover point for the optimum threshold voltage and integrating the error tails yields [5]

$$BER = \int_Q^\infty \text{Gauss}(x)\,dx \quad \text{with} \quad Q = \frac{v_S^{pp}}{v_{n,0}^{rms} + v_{n,1}^{rms}}. \quad (4.10)$$

Of course, this equation simplifies to Eq. (4.8) for the case of equal noise distributions, $v_n^{rms} = v_{n,0}^{rms} = v_{n,1}^{rms}$. [→ Problem 4.2]

Signal-to-Noise Ratio. The term *signal-to-noise ratio* (SNR) often is used in a sloppy way; any measure of signal strength divided by any measure of noise may be called SNR. In this sense, the Q parameter is an SNR, but in this book we use the term SNR in a precisely defined way. We define SNR as the *mean-free average signal power* divided by the *average noise power*.[4] The SNR can be calculated in the continuous-time domain, before the signal is sampled by the decision circuit, or in the sampled domain (cf. Fig. 4.4). Note that in general these two SNR values are not equal. Here we calculate the continuous-time SNR: the mean-free average signal power is calculated as $\overline{v_S^2(t)} - \overline{v_S(t)}^2$, which is $(v_S^{pp}/2)^2$ for a DC-balanced ideal NRZ signal. The noise power is calculated as $\overline{v_n^2(t)}$, which can be written $1/2 \cdot (v_{n,0}^2 + v_{n,1}^2)$, given equal probabilities for zeros and ones. Thus, the SNR follows as

$$SNR = \frac{\left(v_S^{pp}\right)^2}{2 \cdot \left(\overline{v_{n,0}^2} + \overline{v_{n,1}^2}\right)}. \quad (4.11)$$

Comparing Eqs. (4.10) and (4.11), we realize that we cannot simply convert Q into SNR, or vice versa, without additional knowledge of the noise ratio $v_{n,0}^{rms}/v_{n,1}^{rms}$. However, there are two important special cases: (i) if the noise on the zeros and ones is equal (additive noise, i.e., noise dominated by the amplifier) and (ii) if the noise on the ones is much larger than on the zeros (multiplicative noise, i.e., noise dominated

---

[4]In some books on optical communication [5, 168], SNR is defined as the *peak* signal power divided by the average noise power. Here we define SNR based on the *average* power to be consistent with the theory of communication systems. Furthermore, the signal power is defined as *mean-free*, that is, the power of the mean signal $\overline{v_S(t)}^2$ is subtracted from the total power $\overline{v_S^2(t)}$ when computing the signal power to avoid a dependence of the signal power on biasing conditions. However, there is one important exception: if the signal is constant (unmodulated, continuous wave), the mean power is *not* subtracted, or else the signal power would vanish. Cf. the SNR calculations in Sections 3.1 and 3.3 where the signal was a continuous wave. The noise voltage $v_n(t)$ is mean free by definition and thus the noise power is automatically mean free.

by the detector or optical amplifiers):

$$SNR = Q^2, \quad \text{if} \quad v_{n,1}^{rms} = v_{n,0}^{rms} \tag{4.12}$$

$$SNR = 1/2 \cdot Q^2, \quad \text{if} \quad v_{n,1}^{rms} \gg v_{n,0}^{rms}. \tag{4.13}$$

For example, to achieve a BER of 10^{-12} ($Q = 7.0$), we need an SNR of 16.9 dB in the first case and 13.9 dB in the second case. [→ Problems 4.3, 4.4, 4.5, and 4.6]

At this point, you may wonder if you should use $10 \log Q$ or $20 \log Q$ to express Q values in dB. The above SNR discussion suggests $20 \log Q$ ($= 10 \log Q^2$). But an equally strong argument can be made for $10 \log Q$ (for example, look at Eq. (4.20) in the next section). So, my advice is to use Q on a linear scale whenever possible. If you must express Q in dBs, *always* clarify whether you used $10 \log Q$ or $20 \log Q$ as the conversion rule.

SNR for TV Signals. Although our focus here is on digital transmission systems based on OOK, it is instructive to compare them with analog transmission systems. An example of such an analog system is the CATV/HFC system, where multiple analog or digital TV signals or both are combined by means of subcarrier multiplexing (SCM) into a single analog signal, which is then transmitted over an optical fiber (cf. Chapter 1). To provide a good picture quality, this analog signal must have a much higher SNR than the 14 to 17 dB typical for an NRZ signal. To be more precise, we should use the term *carrier-to-noise ratio* (CNR) rather than SNR: cable-television engineers use the term CNR for RF-modulated signals such as the TV signals in an SCM system and reserve the term SNR for baseband signals such as the NRZ signal [23]. For an analog TV channel with AM-VSB modulation, the National Association of Broadcasters recommends $CNR > 46$ dB. For a digital TV channel with QAM-256 modulation and forward error correction (FEC), typically $CNR > 30$ dB is required.

And then there is E_b/N_0. There is yet another SNR-like quantity called E_b/N_0, often pronounced "ebno." Sometimes this quantity also is referred to as *SNR per bit*. E_b/N_0 is mostly used in wireless applications, but occasionally, it appears in the optical communication literature, especially when error-correcting codes are discussed. It therefore is useful to understand what it means and how it relates to Q, SNR, and BER. E_b is the energy per information bit and N_0 is the (one-sided) noise power spectral density. The E_b/N_0 concept applies to signals with white noise where the noise spectral density can be characterized by the single number N_0. This situation is most closely approximated at the *input* of the receiver as shown in Fig. 4.4, before any filtering is performed, and the noise can be assumed to be approximately white (not necessarily a good assumption for optical receivers, as we have seen). Obviously, the SNR at this point is zero because the white noise has an infinite power; however, E_b/N_0 has a finite value. As we know, after the band-limiting linear channel, we can calculate a meaningful SNR and Q value as indicated in Fig. 4.4.

The energy per bit is the average signal power times the bit interval. Let's assume that the midband gain of the linear channel is normalized to one and that the linear channel only limits the noise but does not attenuate the signal power. We thus can

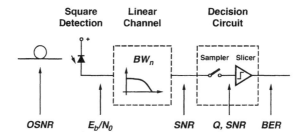

Fig. 4.4 Various performance measures in an optical receiver.

relate the average energy per information bit to the signal voltage at the decision circuit as $E_b = (\overline{v_S^2(t)} - \overline{v_S(t)}^2) \cdot T'$, where T' is the duration of the information bit. For a DC-balanced ideal NRZ signal, this can be shown to be $E_b = (v_S^{pp}/2)^2 \cdot T'$. Why this emphasis on *information* bit? Because the transmission system may use a coding scheme such as 8B10B, where groups of 8 information bits are coded into 10 bits before they are transmitted over the fiber. In this case, the period of an information bit is somewhat longer than period of a channel bit. Mathematically, we can write $T' = T/r = 1/(r \cdot B)$, where B is the channel bit rate and r is the so-called *code rate*; for example, the code rate for the 8B10B code is $r = 0.8$. Next, how is N_0 related to the rms noise at the decision circuit? Assuming additive white noise with the power spectral density N_0 at the input, we can calculate the noise voltage at the decision circuit as $(v_n^{rms})^2 = N_0 \cdot BW_n$, where BW_n is the noise bandwidth of the linear channel. Dividing E_b and N_0 reveals the following relationship with Q:

$$\frac{E_b}{N_0} = \left(\frac{v_S^{pp}}{2 \cdot v_n^{rms}}\right)^2 \cdot \frac{BW_n}{r \cdot B} = Q^2 \cdot \frac{BW_n}{r \cdot B}. \quad (4.14)$$

Thus, E_b/N_0 is equal to Q^2 scaled by the ratio of the noise bandwidth and the information bit rate. The latter ratio is related to the spectral efficiency of the modulation scheme. Thus, the main difference between E_b/N_0 and Q (or SNR) is that E_b/N_0 takes the spectral efficiency of the modulation scheme into account. In texts on communication systems and forward error correction, it usually is assumed that a matched filter receiver is used. For NRZ modulation, this means that the noise bandwidth is half the bit rate, $BW_n = B/2$ (cf. Section 4.6), leading to the simpler relationship

$$\frac{E_b}{N_0} = \frac{Q^2}{2r}. \quad (4.15)$$

For example, to achieve a BER of 10^{-12} ($Q = 7.0$) without coding ($r = 1$), we need $E_b/N_0 = 13.9$ dB, whereas with 8B10B coding ($r = 0.8$), we would need $E_b/N_0 = 14.9$ dB. [$\rightarrow$ Problem 4.7]

We conclude that if we assume DC-balanced NRZ modulation with no coding, signal-independent white noise, and a matched filter receiver, then E_b/N_0 is always

54 RECEIVER FUNDAMENTALS

3 dB lower than the Q parameter, where the dB value of the latter is calculated as $20 \log Q$.

BER, Q, SNR, OSNR, E_b/N_0, and All the Rest of It. By now you have realized that there is a bewildering variety of SNR-like quantities out there and it is time to put them in perspective (see Fig. 4.4). The primary performance measure in a digital communication system is the BER, which is measured at the output of the decision circuit. To design and test the system, we would like to relate the BER to quantities that can be measured in other parts of the system. The first and most direct predictor of BER is the Q parameter, which is determined from the sampled values at the input of the decision circuit. To calculate the BER from Q, we have to assume only that the noise is Gaussian, that the signal has two levels that are used with equal probability, and that the decision threshold is set to its optimum value. In particular, we do *not* need to make any assumptions about the spectral distribution and additiveness of the noise, neither does the shape and duration of the data pulses matter (e.g., NRZ vs. RZ modulation). This makes Q an excellent performance measure and should be used whenever possible. The sampled SNR measured at the input of the decision circuit is another, less direct predictor of the BER. To calculate the BER from the sampled SNR, we need additional information about the relative strength of the noise on the zeros and ones. To calculate the BER from the continuous-time SNR, we need additional knowledge of the shape of the data pulses. To calculate the BER from the E_b/N_0, which is measured at the input of the receiver, we need additional knowledge of the receiver's noise bandwidth and code rate.[5] Finally, to calculate the BER from the optical signal-to-noise ratio (OSNR), which is measured in the optical domain, we also need additional knowledge of the receiver's noise bandwidth. We discuss the effect of OSNR on the BER in the next section.

4.3 SENSITIVITY

Rather than asking "What is the bit-error rate given a certain signal level?," we could ask the other way round, "What is the minimum signal level needed to achieve a given bit-error rate?" This minimum signal, when referred back to the input of the receiver, is known as the *sensitivity*. The sensitivity is one of the key characteristics of an optical receiver. It tells us to what level the transmitted signal can become attenuated by the fiber and still be detected reliably by the receiver. Sensitivity can be defined in the electrical as well as the optical domain, as we will see next.

Definitions. The *electrical receiver sensitivity*, i_{sens}^{pp}, is defined as the minimum peak-to-peak signal current at the input of the receiver necessary to achieve a specified

---

[5] In the literature on communication systems, it often is said that "E_b/N_0 uniquely determines the BER for a particular modulation scheme," in apparent contradiction to what we are saying here. However, note that the above statement rests on many implicit assumptions such as additive white Gaussian noise, a signal without intersymbol interference (ISI), a matched filter receiver with optimum threshold, and no coding.

BER ($i_{\text{sens}}^{pp} = i_S^{pp}$ @ BER). The current swing i_S^{pp} at the input of the linear channel causes the output voltage swing $v_S^{pp} = H_0 \cdot i_S^{pp}$, where H_0 is the midband value of $H(f)$ (see Fig. 4.2). We thus can derive the electrical sensitivity by solving Eq. (4.8) for v_S^{pp} and dividing by H_0:

$$i_{\text{sens}}^{pp} = \frac{2Q \cdot v_n^{rms}}{H_0}. \tag{4.16}$$

Before continuing, it is useful to define the *input-referred* rms noise as the rms noise at the output of the linear channel divided by H_0:

$$i_n^{rms} = \frac{v_n^{rms}}{H_0}. \tag{4.17}$$

Now with this definition in hand, we can rewrite the electrical sensitivity Eq. (4.16) in the more compact form

$$i_{\text{sens}}^{pp} = 2Q \cdot i_n^{rms}. \tag{4.18}$$

For example, given an input-referred rms noise of 380 nA and a required BER of 10^{-12}, the electrical sensitivity is about $14.07 \cdot 380\,\text{nA} = 5.3\,\mu\text{A}$. In a situation with different amounts of noise on the zeros and ones, we can use Eq. (4.10) to obtain the more general electrical sensitivity expression

$$i_{\text{sens}}^{pp} = Q \cdot \left(i_{n,0}^{rms} + i_{n,1}^{rms} \right), \tag{4.19}$$

where $i_{n,0}^{rms} = v_{n,0}^{rms}/H_0$ is the input-referred rms noise for zero bits and $i_{n,1}^{rms} = v_{n,1}^{rms}/H_0$ is the input-referred rms noise for one bits. Finally, it should be noted that the above sensitivity expressions assume an ideal slicer with an optimally set decision threshold (same assumptions as for Eqs. (4.8) and (4.10)). A nonideal slicer will degrade the sensitivity.

The *optical receiver sensitivity*, $\overline{P}_{\text{sens}}$, is defined as the minimum optical power, averaged over time, necessary to achieve a specified BER ($\overline{P}_{\text{sens}} = \overline{P}_S$ @ BER). For a DC-balanced signal with high extinction,[6] we have $\bar{i}_S = i_S^{pp}/2$, and with Eq. (3.2), we find $\overline{P}_S = i_S^{pp}/(2\mathcal{R})$, where $\mathcal{R}$ is the responsivity of the photodetector. Thus, the optical sensitivity is

$$\overline{P}_{\text{sens}} = \frac{Q \cdot i_n^{rms}}{\mathcal{R}}, \tag{4.20}$$

or more generally,

$$\overline{P}_{\text{sens}} = \frac{Q \cdot \left(i_{n,0}^{rms} + i_{n,1}^{rms} \right)}{2\mathcal{R}}. \tag{4.21}$$

Again, an ideal slicer is assumed. For example, given an input-referred rms noise of 380 nA, a responsivity of 0.8 A/W, and a required BER of 10^{-12}, the optical sensitivity is about $7.035 \cdot 380\,\text{nA}/(0.8\,\text{A/W}) = 3.33\,\mu\text{W}$, corresponding to $-24.8\,\text{dBm}$.

---

[6]The optical receiver sensitivity defined in regulatory standards usually assumes the worst permissible extinction ratio. This leads to a lower sensitivity than derived here. More on this in Section 7.1.

Note that the optical sensitivity is based on the *average* signal value, whereas the electrical sensitivity is based on the *peak-to-peak* signal value. As a result, the optical sensitivity depends on the pulse width of the optical signal. As we have said, the average current for a DC-balanced NRZ signal is $\bar{i}_S = i_S^{pp}/2$, but for a return-to-zero (RZ) signal with 50% duty cycle, the average current reduces to $\bar{i}_S = i_S^{pp}/4$. This means that given an RZ and NRZ receiver with identical electrical sensitivities, the optical sensitivity of the RZ receiver will be 3 dB better than that of the NRZ receiver. [→ Problem 4.8]

Sometimes, the optical sensitivity for a receiver with an *ideal photodetector* is given. This sensitivity is designated by $\eta \overline{P}_{sens}$ and is useful to compare the electrical performance of different receivers while excluding the quantum efficiency η of the photodetector. With Eqs. (4.20) and (3.2), we can express this sensitivity as

$$\eta \overline{P}_{sens} = \frac{hc}{\lambda q} \cdot Q \cdot i_n^{rms}, \qquad (4.22)$$

or more generally as

$$\eta \overline{P}_{sens} = \frac{hc}{\lambda q} \cdot \frac{Q \cdot \left(i_{n,0}^{rms} + i_{n,1}^{rms}\right)}{2}. \qquad (4.23)$$

For example, given an input-referred rms noise of 380 nA, a wavelength of 1.55 μm, and a required BER of 10^{-12}, the optical sensitivity for a receiver with an ideal photodetector is about $7.035 \cdot 380\,\text{nA}/(1.25\,\text{A/W}) = 2.13\,\mu\text{W}$, corresponding to -26.7 dBm.

Dynamic Range. As we have seen, for very weak signals, random noise at the receiver causes bit errors. For very strong signals, effects such as pulse-width distortion and data-dependent jitter cause bit errors as well. Thus, besides the lower signal level, known as the sensitivity or the *sensitivity limit*, there is an upper signal level, known as the *overload limit*, beyond which the required BER cannot be met. In analogy to the sensitivity, we define the *input overload current*, i_{ovl}^{pp}, as the maximum peak-to-peak signal current for which a specified BER can be achieved. Similarly, we define the *optical overload power*, $\overline{P}_{ovl}$, as the maximum time-averaged optical power for which a specified BER can be achieved.

The *dynamic range* of a receiver is defined at its lower end by the sensitivity limit and at its upper end by the overload limit. Thus, the electrical dynamic range is $i_{sens}^{pp} \ldots i_{ovl}^{pp}$, whereas the optical dynamic range is $\overline{P}_{sens} \ldots \overline{P}_{ovl}$.

Reference Bit-Error Rates. When specifying a (electrical or optical) receiver sensitivity, we must do so with respect to a reference BER. The following BER values are commonly used for this purpose: the SONET OC-48 standard (2.5 Gb/s) requires a system bit-error rate of $\leq 10^{-10}$, which corresponds to $Q \geq 6.4$. The faster SONET OC-192 standard (10 Gb/s) is tougher and requires a system bit-error rate of $\leq 10^{-12}$ corresponding to $Q \geq 7.0$. Component manufacturers usually aim at even lower BERs, such as 10^{-15} ($Q \geq 7.9$), to meet the system BERs quoted above.

At very low BERs, such as 10^{-15}, it becomes very time consuming to perform an accurate BER measurement. For example, to collect 10 errors at a BER of 10^{-15} and a bit rate of 10 Gb/s, we have to wait on average for 10^6 seconds, which amounts to 12 days! Then there is the question: "Are 10 errors enough?" If we assume that the error statistics follow a Poisson distribution, then for n collected errors, the standard deviation is $\sqrt{n}$. Thus, for 10 collected errors, the standard deviation is 3.2, meaning that we have a 32% uncertainty; for 100 errors, this uncertainty reduces to 10%, and so forth.

Sensitivity Analysis Based on Amplifier Noise Only. Let's carry out some numerical calculations to get a feeling for optical sensitivity values. For these first calculations, we ignore the detector noise and use Eq. (4.20) with $i_n^{rms} = i_{n.\mathrm{amp}}^{rms}$ to estimate the sensitivity based on the amplifier noise only. The input-referred amplifier noise $i_{n.\mathrm{amp}}^{rms}$ is defined, as we have seen, in terms of the output noise: $i_{n.\mathrm{amp}}^{rms} = v_{n.\mathrm{amp}}^{rms}/H_0$. Alternatively, the input-referred noise can be obtained directly from the input-referred noise spectrum, but these calculations are not trivial and we spend all of Section 4.4 discussing how to do that. Now, assuming $i_{n.\mathrm{amp}}^{rms}$ is given, the sensitivity for a receiver with a p-i-n photodetector is simply

$$\overline{P}_{\mathrm{sens}.PIN} = \frac{Q \cdot i_{n.\mathrm{amp}}^{rms}}{\mathcal{R}}. \tag{4.24}$$

The APD has an M times higher responsivity, which leads to an M-fold improvement in the receiver sensitivity:

$$\overline{P}_{\mathrm{sens}.APD} = \frac{1}{M} \cdot \frac{Q \cdot i_{n.\mathrm{amp}}^{rms}}{\mathcal{R}}. \tag{4.25}$$

Similarly, the optically preamplified p-i-n detector (OA + p-i-n) has a G times higher responsivity, which leads to a G-fold improvement in the receiver sensitivity:

$$\overline{P}_{\mathrm{sens}.OA} = \frac{1}{G} \cdot \frac{Q \cdot i_{n.\mathrm{amp}}^{rms}}{\mathcal{R}}. \tag{4.26}$$

For our numerical calculations we use the typical input-referred rms noise values $i_{n.\mathrm{amp}}^{rms} = 380\,\mathrm{nA}$ at 2.5 Gb/s and $1.4\,\mu\mathrm{A}$ at 10 Gb/s, which are based on the TIA noise data from Section 5.1.4. We continue to use the typical detector values ($\mathcal{R} = 0.8\,\mathrm{A/W}$, $M = 10$, $G = 100$) introduced in Chapter 3 and we choose the reference BER to be 10^{-12}. The resulting approximate sensitivity values for all three detector types are listed in Table 4.2. We see how the sensitivities improve in proportion to the detector responsivities as we go from the p-i-n detector to the APD and finally to the optically preamplified p-i-n detector. [→ Problem 4.9]

How important is a 1-dB difference in sensitivity? Is it worth a lot of trouble to improve the sensitivity by a single dB? In a system without optical in-line amplifiers (or regenerators), a 1-dB improvement in sensitivity means that the reach is extended by 4 km. This is so simply because the attenuation of a fiber at 1.55 μm is about 0.25 dB/km. A 4-km extension is quite significant and this is why system designers usually care about small sensitivity improvements (or degradations) such as 0.05 dB.

58 RECEIVER FUNDAMENTALS

Table 4.2 Approximate receiver sensitivities at $BER = 10^{-12}$ for various photodetectors. Only the amplifier noise is considered.

Parameter	Symbol	2.5 Gb/s	10 Gb/s
Input rms noise due to amplifier	$i_{n,\text{amp}}^{rms}$	380 nA	1.4 µA
Input signal swing for $BER = 10^{-12}$	i_{sens}^{pp}	5.3 µA	19.7 µA
Sensitivity of p-i-n receiver	$\overline{P}_{\text{sens},PIN}$	−24.8 dBm	−19.1 dBm
Sensitivity of APD receiver	$\overline{P}_{\text{sens},APD}$	−34.8 dBm	−29.1 dBm
Sensitivity of OA + p-i-n receiver	$\overline{P}_{\text{sens},OA}$	−44.8 dBm	−39.1 dBm

Sensitivity Analysis Including Detector Noise. In the following, we repeat the sensitivity calculations from before, but this time taking the detector noise into account. This exercise shows us the relative significance of the detector noise. Now, because of the detector's signal-dependent noise, we have to consider two different noise values, one for the noise on the zeros and one for the noise on the ones: $\overline{i_{n,0}^2} = \overline{i_{n,PIN,0}^2} + \overline{i_{n,\text{amp}}^2}$ and $\overline{i_{n,1}^2} = \overline{i_{n,PIN,1}^2} + \overline{i_{n,\text{amp}}^2}$. The input-referred detector noise $\overline{i_{n,PIN,0/1}^2}$ is defined, as we have seen, as the output noise power due to the detector divided by H_0^2. Alternatively, this input-referred noise can be obtained from the input-referred power spectrum. Because the latter is approximately white, this can be accomplished easily by multiplying the spectrum with the noise bandwidth of the linear channel, BW_n (we show why in Section 4.4). This is exactly what we did to obtain $\overline{i_{n,PIN,0}^2}$ and $\overline{i_{n,PIN,1}^2}$ in Eqs. (3.6) and (3.7), respectively. By inserting the noise expressions $i_{n,0}^{rms} = i_{n,\text{amp}}^{rms}$ and $i_{n,1}^{rms} = \sqrt{4qR\overline{P}_{\text{sens}} \cdot BW_n + (i_{n,\text{amp}}^{rms})^2}$ into Eq. (4.21) and solving for $\overline{P}_{\text{sens}}$, we can derive the sensitivity of a p-i-n receiver:

$$\overline{P}_{\text{sens},PIN} = \frac{Q \cdot i_{n,\text{amp}}^{rms}}{\mathcal{R}} + \frac{Q^2 \cdot q \cdot BW_n}{\mathcal{R}}. \quad (4.27)$$

The first term of this equation is caused by the amplifier noise and is identical to Eq. (4.24); the second term is caused by the shot noise of the p-i-n photodetector. Using the APD noise expressions Eqs. (3.12) and (3.13), we can derive the sensitivity of an APD receiver:

$$\overline{P}_{\text{sens},APD} = \frac{1}{M} \cdot \frac{Q \cdot i_{n,\text{amp}}^{rms}}{\mathcal{R}} + F \cdot \frac{Q^2 \cdot q \cdot BW_n}{\mathcal{R}}. \quad (4.28)$$

Finally, the sensitivity of a receiver with an optically preamplified p-i-n detector (OA + p-i-n) can be derived from Eqs. (3.20) and (3.21):

$$\overline{P}_{\text{sens},OA} = \frac{1}{G} \cdot \frac{Q \cdot i_{n,\text{amp}}^{rms}}{\mathcal{R}} + \eta F \cdot \frac{Q^2 \cdot q \cdot BW_n}{\mathcal{R}}. \quad (4.29)$$

Comparing these three equations, we observe that the first term, which contains the amplifier noise, is suppressed with increasing detector gain (p-i-n → APD → OA +

p-i-n). The second term, caused by the detector noise, however, grows proportional to the noise figure (or excess noise factor) of the detector. [→ Problems 4.10 and 4.11]

Now let's evaluate these equations for the typical detector values introduced in Chapter 3 (APD: $F = 6$ corresponding to 7.8 dB; OA + p-i-n: $\eta = 0.64$, $F = 3.16$ corresponding to 5 dB). With these numbers and assuming that the noise bandwidth is 75% of the bit rate, we obtain the sensitivity numbers shown in Table 4.3. When comparing these numbers with the approximations in Table 4.2, we note the following: the sensitivities for the p-i-n receiver hardly changed at all, which means that the shot noise contributed by the p-i-n photodiode is negligible compared with the amplifier noise. The sensitivities for the APD receiver degraded by a little more than 1 dB, which means that neglecting the noise of the APD gets us only a rough sensitivity estimate. The sensitivities for the receiver with optically preamplified p-i-n detector degraded by about 3 to 4 dB, that is, we definitely need to include the noise of the optical amplifier when calculating the sensitivity of such a receiver.

Table 4.3 Receiver sensitivities at $BER = 10^{-12}$ for various photodetectors. Amplifier and detector noise is considered.

Parameter	Symbol	2.5 Gb/s	10 Gb/s
Input rms noise due to amplifier	$i_{n.\mathrm{amp}}^{rms}$	380 nA	1.4 µA
Sensitivity of p-i-n receiver	$\overline{P}_{sens.PIN}$	−24.7 dBm	−19.1 dBm
Sensitivity of APD receiver	$\overline{P}_{sens.APD}$	−33.5 dBm	−27.8 dBm
Sensitivity of OA + p-i-n receiver	$\overline{P}_{sens.OA}$	−41.5 dBm	−35.6 dBm

BER Plots. The performance of an optical receiver is best characterized by measuring the BER as a function of the received power and then plotting one against the other. These plots, known as *BER plots*, are in widespread use in the optical communication literature.

What do we expect these plots to look like? For a p-i-n receiver, operated well below its overload limit, we expect that the curves should follow the theoretical result in Eq. (4.27). Furthermore, neglecting the shot noise term, we expect that the Q parameter, corresponding to the measured BER, should be linearly related to the received optical power $\overline{P}_S$. It thus is convenient to use coordinates (or graph paper) in which this relationship is represented by a straight line. One possibility, illustrated by Fig. 4.5, is to choose $-Q$ for the y-axis and $\overline{P}_S$ for the x-axis (note that choosing $-Q$ for y makes *BER* increase with y). To make the graph more legible, we may still label the y-axis in BER units, but note that these labels are distorted by the erfc($\cdot$) function. Alternatively, if we prefer to represent the power in dBm rather than in mW, we can choose $-10 \log Q$ for the y-axis and $10 \log \overline{P}_S$ (the power in dB) for the x-axis. The linear relationship between Q and $\overline{P}_S$ is still represented by a straight line in these logarithmic coordinates. Figure 4.6 shows an example of this type of BER plot. Note that both BER plots describe the same receiver with the same sensitivity.

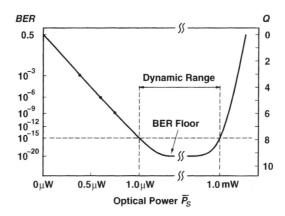

Fig. 4.5 BER plot in the linear ($\overline{P_S}$, $-Q$) coordinate system. For small power values, the plot follows a straight line down to $BER = 10^{-15}$, where the sensitivity is $1\,\mu W$.

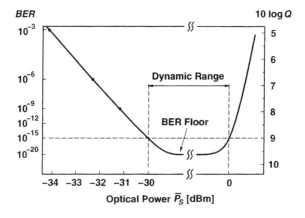

Fig. 4.6 BER plot in the ($10\log \overline{P_S}$, $-10\log Q$) coordinate system. For small power values, the plot follows a straight line down to $BER = 10^{-15}$, where the sensitivity is $-30\,\text{dBm}$.

Data points plotted on these graph papers can easily be *extrapolated* down to very low BERs. For example, Figs. 4.5 and 4.6 show how measured data points at $BER = 10^{-3}$, 10^{-6}, and 10^{-9} can be extrapolated down to 10^{-15}, revealing a sensitivity of $1\,\mu\text{W}$ (or $-30\,\text{dBm}$) at this reference BER. Such extrapolations are very convenient because they save a lot of time taking measurements. However, the result is only correct if the receiver noise closely follows a Gaussian distribution over a range of many sigmas and if there is no significant increase in noise at high received power levels. But we already know that the signal-dependent noise in APD receivers and optically preamplified systems produces a Q^2 term (cf. Eqs. (4.28) and (4.29)), which effectively makes the BER curve flatten out at high power levels. In the extreme case, the BER curve may become completely flat, which means that even for a very high received power, the BER never goes below a certain value, the so-called *BER floor*. Note that an extrapolation would go right past this floor and predict an overly optimistic BER! Moreover, at very high power levels, the receiver front-end tends to overload, which means that the signal is distorted so severely that the BER *increases* rapidly with $\overline{P_S}$. Figures 4.5 and 4.6 show BER plots that have a BER floor at 10^{-20} and an optical overload power of 1 mW (or 0 dBm); the dynamic range, defined by the sensitivity and overload limits, also is shown.

BER plots, like the ones in Figs. 4.5 and 4.6, also are a very useful diagnostic tool. For example, from a few measured data points, we can see if the data follow a straight line or if there is a bend, which may indicate a BER floor. From the slope of the curve in the linear BER plot, we can infer the amplifier noise $i_{n,\text{amp}}^{rms}$ (assuming $\mathcal{R}$ is known), and from a horizontal shift away from the point ($\overline{P_S} = 0$, $BER = 0.5$), we can infer an offset problem, and so forth. Note that BER plots don't necessarily have to be a function of received optical power, as in our examples, but also can be a function of the input current (for TIAs) or the input voltage (for MAs).

Optimum APD Gain. We saw in Section 3.2 that there must be an optimum APD gain, because the avalanche gain M can be increased only at the expense of a higher excess noise factor F, as given by Eq. (3.11). Now with Eq. (4.28) for the APD receiver sensitivity, we can derive a mathematical expression for the *optimum APD gain*, M_{opt}, that yields the best sensitivity. Intuitively, if M is chosen too low, the first term in Eq. (4.28) containing $1/M$ limits the sensitivity; however, if M is chosen too high, the second term containing F limits the sensitivity. Combining Eqs. (4.28) and (3.11) and solving for the M that minimizes $\overline{P}_{\text{sens,APD}}$ yields

$$M_{\text{opt}} = \sqrt{\frac{i_{n,\text{amp}}^{rms}}{Q \cdot k_A \cdot q \cdot BW_n} - \frac{1 - k_A}{k_A}}. \quad (4.30)$$

From this equation, we can see that M_{opt} increases with increasing amplifier noise, $i_{n,\text{amp}}^{rms}$, which makes sense because the APD gain helps to suppress this noise. Furthermore, M_{opt} decreases with increasing k_A, which means that the optimum gain for an InGaAs APD with $k_A \approx 0.6$ is smaller than that for a silicon APD with $k_A \approx 0.03$.

Note that in the case of a receiver with optically preamplified p-i-n detector, the noise figure F *decreases* with increasing gain G and therefore there is no optimum gain value, that is, a higher gain always improves the sensitivity.

Cascade of Optical Amplifiers. Ultra-long-haul fiber links without electrical regenerators became possible with the development of reliable optical in-line amplifiers. These amplifiers are inserted periodically into the fiber link to boost the optical signal. For example, let's assume an 8,000-km long fiber link connecting two continents. To compensate the fiber loss of 0.25 dB/km, we insert an optical amplifier with gain $G = 20$ dB every 80 km. This makes a total of 100 amplifiers.

With so many optical amplifiers, the noise accumulated in the fiber link becomes dominant over the noise of the receiver. In the following, we analyze this situation in more detail. We take two approaches: in a first (and unconventional) approach, we regard the whole fiber link including all the amplifiers as part of the *detector*, which leads to an analysis very similar to our previous sensitivity calculations. In a second approach, we give up the notion of sensitivity and do the analysis in terms of OSNR. Eventually, both approaches lead to the same results.

First, let's consider the whole 8,000-km link, including all the optical amplifiers to be part of the detector (see Fig. 4.7). In other words, the input to our receiver is right at the transmitter end. If we assume that the fiber loss is exactly balanced by the optical amplifier gain, then we can easily derive the "sensitivity" of this system. We regard the whole amplified link as a single optical amplifier with gain $G' = 1$ and noise figure $F' = nGF$, where n is the number of amplifiers, G is the gain of each amplifier, and F is the noise figure of each amplifier (cf. Problem 3.9). Substituting G' and F' into Eq. (4.29) for the sensitivity of an optically preamplified p-i-n receiver yields

$$\overline{P}_{\text{sens.}OAC} = \frac{Q \cdot i_{n.\text{amp}}^{rms}}{\mathcal{R}} + \eta n G F \cdot \frac{Q^2 \cdot q \cdot BW_n}{\mathcal{R}}. \qquad (4.31)$$

For example, with 100 amplifiers and the typical values introduced earlier ($G = 20$ dB, $F = 5$ dB), we get the horribly bad compound noise figure $F' = nGF = 45$ dB. The "sensitivity" $\overline{P}_{\text{sens.}OAC}$ for a 10-Gb/s system ($BW_n = 7.5$ GHz, $i_{n.\text{amp}}^{rms} = 1.4\,\mu$A, $\mathcal{R} = 0.8$ A/W, $\eta = 0.64$) comes out as $+1.7$ dBm, essentially regardless of the electrical amplifier noise $i_{n.\text{amp}}^{rms}$. This "sensitivity" is very low, but recall that it refers to the fiber link input that is at the transmitter end! So all this means is that the transmitter must launch a power of at least $\overline{P}_{\text{out}} = 1.7$ dBm into the fiber and we are fine.

We can draw two conclusions from this last example. First, it is in fact possible to send an optical signal over an 8,000-km fiber link containing 100 erbium-doped fiber amplifiers (EDFAs) and receive it at a low BER of 10^{-12}! This fact also is demonstrated by practical systems such as the commercial transpacific cable TPC-5 segment J, which is 8,620 km long and contains about 260 EDFA-type repeaters spaced 33 km apart. Second, the concept of receiver sensitivity loses its meaning in a situation where many in-line amplifiers contribute most of the system noise. Remember, receiver sensitivity is the minimum power required to achieve a certain BER based on the *receiver* noise. In optically amplified long-haul systems, the concept of

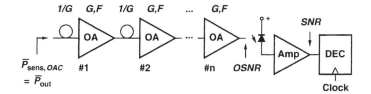

Fig. 4.7 A cascade of optical amplifiers followed by a p-i-n receiver.

sensitivity is replaced by the concept of optical signal-to-noise ratio (OSNR), which we discuss next.

Optical Signal-to-Noise Ratio. In optically amplified transmission systems, the system designer is interested in the minimum OSNR required at the receiver rather than the sensitivity of the receiver. To illustrate this, let's repeat the previous calculation but now thinking in terms of OSNR. First, we want to know how much OSNR is needed at the receiver to meet the required BER. After a chain of many EDFAs, the noise on the ones will be much larger than the noise on the zeros; therefore, with Eq. (4.13) we find the required SNR as approximately $1/2 \cdot Q^2$. With Eq. (3.17), we can convert this SNR into the required OSNR (cf. Problems 3.7 and 4.10):

$$OSNR \approx Q^2 \cdot \frac{BW_n}{BW_O}. \qquad (4.32)$$

For example, given a 7.5-GHz receiver noise bandwidth, we need an OSNR of 14.7 dB measured in a 0.1-nm optical bandwidth (12.5 GHz at $\lambda = 1.55\,\mu\text{m}$) to achieve a BER of 10^{-12}. [→ Problem 4.12]

Next, we want to know how much OSNR we have got at the end of a chain of n amplifiers. The optical signal power there is $\overline{P}_{out}$, the same as the power launched by the transmitter because the amplifier gain and the span loss are balanced. The optical noise power at the end of the chain is the sum of the noise from all n amplifiers, $n \cdot S_{ASE} \cdot BW_O$, because the gain from each EDFA output to the link output is unity. Thus, we have

$$OSNR = \frac{\overline{P}_{out}}{n \cdot S_{ASE} \cdot BW_O} \approx \frac{\overline{P}_{out}}{nGF \cdot hc/\lambda \cdot BW_O}, \qquad (4.33)$$

where, on the right-hand side, we expressed S_{ASE} in terms of the amplifier noise figure, $F \approx \tilde{F}$, using Eq. (3.23). If we transform Eq. (4.33) into the log domain and specialize it for $\lambda = 1.55\,\mu\text{m}$ and a BW_O corresponding to 0.1 nm, we obtain the useful engineering rule [207]

$$OSNR[\text{dB}] \approx 58\,\text{dB} + \overline{P}_{out}[\text{dBm}] - G[\text{dB}] - F[\text{dB}] - 10\log n. \qquad (4.34)$$

For example, with the familiar values $n = 100$, $F = 5\,\text{dB}$, and $G = 20\,\text{dB}$, we find that we need $\overline{P}_{out} = 1.7\,\text{dBm}$ to achieve the required OSNR of 14.7 dB in a 0.1-nm

optical bandwidth. Well that's the same transmit power we found with Eq. (4.31)!
[→ Problem 4.13]

When designing a practical long-haul transmission system, the above idealized OSNR calculations must be refined in a number of ways. First, OSNR penalties due to fiber dispersion, polarization effects, nonlinear pulse-shape distortions, nonlinear signal/noise mixing, and crosstalk in WDM systems must be included. Second, OSNR margins for system aging, repairs, and so forth must be allocated. Third, if forward error correction (FEC) is used, the required Q values are lower than those given in Table 4.1, leading to correspondingly lower OSNR requirements (cf. Section 4.11).

We are now finished with the main part of this section. We understand the concept of receiver sensitivity and how it is affected by the detector noise and the amplifier noise. We have briefly looked at optically amplified systems and their requirements in terms of OSNR. In the remainder of this section, we explore the theoretical sensitivity limits of optical receivers. If you are tired of this subject, you can skip ahead to Section 4.4.

Sensitivity Analysis for a Noiseless Amplifier. To study how sensitive we can make a receiver in theory, we repeat our sensitivity calculations again, but this time assuming a noiseless amplifier. We can easily derive these sensitivities, which are based on detector noise only, from Eqs. (4.27), (4.28), and (4.29) by setting $i_{n.\text{amp}}^{rms} = 0$. Numerical sensitivity values for all three receiver types, assuming the usual typical values, are listed in Table 4.4.

Table 4.4 Maximum receiver sensitivities at $BER = 10^{-12}$ for various photodetectors. A noiseless amplifier is assumed.

Parameter	Symbol	2.5 Gb/s	10 Gb/s
Sensitivity of p-i-n receiver	$\overline{P}_{\text{sens}.PIN}$	−47.3 dBm	−41.3 dBm
Sensitivity of APD receiver	$\overline{P}_{\text{sens}.APD}$	−39.5 dBm	−33.5 dBm
Sensitivity of OA + p-i-n receiver	$\overline{P}_{\text{sens}.OA}$	−44.2 dBm	−38.2 dBm

We observe that, in theory, the p-i-n receiver gives us the best sensitivity. This is so because the noise of the p-i-n detector is the lowest. But as we have seen, in a real system, the sensitivity of the p-i-n receiver is degraded by about 22 dB because of the amplifier noise (cf. Table 4.3). Regarding a receiver with optically preamplified p-i-n detector, we note that although the noise from the optical preamplifier dominated the amplifier noise in Table 4.3, we could still squeeze out another 2 to 3 dB of sensitivity with an ultra-low-noise transimpedance amplifier.

The above calculations also give us an interesting new interpretation for the excess noise factor of the APD and the noise figure of the optical amplifier. After we have calculated the maximum sensitivity for a p-i-n receiver, the maximum sensitivity for an APD receiver can be obtained simply by adding the excess noise factor F in dB: at 2.5 Gb/s, we get $-47.3\,\text{dBm} + 7.8\,\text{dB} = -39.5\,\text{dBm}$ for an APD with $F = 7.8\,\text{dB}$. Similarly, the maximum sensitivity for an OA + p-i-n receiver can be obtained by adding ηF in dB: at 2.5 Gb/s, we get $-47.3\,\text{dBm} - 1.9\,\text{dB} + 5\,\text{dB} = -44.2\,\text{dBm}$ for $\eta = 0.64$ and $F = 5$ dB. These rules can be readily explained with Eqs. (4.27), (4.28), and (4.29). In conclusion, F (for an APD receiver) and ηF (for an OA + p-i-n receiver) tell us how much less sensitive the respective receivers are compared with a p-i-n receiver in the limit of zero amplifier noise.

Before leaving this subject, we must point out a limitation of our analysis and results. At the outset of this section, we made the assumption that the noise follows a Gaussian statistics. Although this is a good assumption for the electronic amplifier noise, the detector noise typically has non-Gaussian statistics, especially when the optical signal is weak. For example, the p-i-n photodetector noise has Poisson statistics. Thus, the sensitivity results in Eqs. (4.27), (4.28), and (4.29), and derived results, such as the optimum APD gain in Eq. (4.30), lose in accuracy when the detector noise becomes a significant fraction of the amplifier noise. In particular, the results in Table 4.4, which are based on detector noise only, are not exact. Note that the Gaussian assumption tends to be conservative, that is, it underestimates the actual sensitivity.

Quantum Limit. Can we build a receiver with arbitrarily high sensitivity, at least in theory? Maybe we can combine some fancy detector with a noiseless amplifier? No, it turns out there is a *quantum limit* that cannot be surpassed.

The quantum limit is obtained from the observation that at least one photon must be detected for each transmitted one bit to have error-free reception. The number of photons, n, contained in a one bit is an integer random variable that follows the Poisson distribution:

$$\text{Poisson}(n) = e^{-M} \cdot \frac{M^n}{n!}, \qquad (4.35)$$

where M is the mean of the distribution. For equally probable zeros and ones, the total error probability equals half the error probability for zeros plus half the error probability for ones. Because nothing is transmitted for zeros (we assume high extinction), the probability of error for zeros is zero. The probability of error for ones is Poisson(0), corresponding to the situation when zero photons are received for a one bit. Thus, we conclude that $BER = 1/2 \cdot \text{Poisson}(0) = 1/2 \cdot e^{-M}$. For example, given a bit-error rate of 10^{-12}, we find that an average number of $M = -\ln(2 \cdot BER) = 27$ photons are required per one bit. An average of $M/2$ photons are required per bit (zero or one), and hence the quantum limit for the sensitivity follows as

$$\overline{P}_{\text{sens,quant}} = \frac{-\ln(2 \cdot BER)}{2} \cdot \frac{hc}{\lambda} \cdot B, \qquad (4.36)$$

where B is the bit rate.[7] Numerical values for this sensitivity assuming $\lambda = 1.55\,\mu\text{m}$ and $BER = 10^{-12}$ are listed in Table 4.5. [$\rightarrow$ Problem 4.14]

Comparing these values with those in Table 4.3, we realize that a receiver with optically preamplified p-i-n detector comes within about 12 dB of the quantum limit.

---

[7]The quantum limit derived here is for on-off keying (OOK), also called (binary) *amplitude-shift keying* (ASK), if the light is regarded as a carrier. Other modulation formats, such as *phase-shift keying* (PSK) or *frequency-shift keying* (FSK), can be detected with a somewhat better sensitivity [46].

Table 4.5 Quantum limit for the sensitivity at $BER = 10^{-12}$.

Parameter	Symbol	2.5 Gb/s	10 Gb/s
Quantum limit	$\overline{P}_{\text{sens,quant}}$	$-53.6\,\text{dBm}$	$-47.6\,\text{dBm}$

This means that such a receiver has the ability to detect correctly 16 or more photons as a one bit!

4.4 PERSONICK INTEGRALS

Total Input-Referred Noise. We saw in the previous section that the input-referred rms noise, i_n^{rms}, plays a key role in determining the receiver sensitivity. Equation (4.17) defines this noise quantity as the output noise divided by the midband value of $H(f)$: $i_n^{rms} = v_n^{rms}/H_0$. Now, by inserting Eq. (4.7) for the output noise and squaring the result, we can write this noise quantity in terms of the input-referred power spectrum:

$$\overline{i_n^2} = \frac{1}{H_0^2} \int_0^{BW_D} |H(f)|^2 \cdot I_n^2(f)\, df, \qquad (4.37)$$

where $I_n^2(f) = I_{n,PD}^2(f) + I_{n,\text{amp}}^2(f)$ is the input-referred noise power spectrum of the combined detector and amplifier noise.

Equation (4.37) is easy to use in numerical computation (simulations), but it looks quite cumbersome for analytical hand calculations. Is there an easier way to calculate the total input-referred noise from the input-referred spectrum? It is tempting just to integrate the input-referred noise spectrum over all frequencies:

$$\overline{i_n^2} \stackrel{?}{=} \int_0^\infty I_n^2(f)\, df. \qquad (4.38)$$

But this can't be right, because the integral does not converge if $I_n^2(f)$ contains the usual white and/or f^2-noise components, leading to an unbounded noise current. Maybe, we surmise, we should integrate only up to the 3-dB bandwidth of the receiver's frequency response:

$$\overline{i_n^2} \stackrel{?}{=} \int_0^{BW_{3dB}} I_n^2(f)\, df. \qquad (4.39)$$

This quantity is shown as the hatched area under the input-referred noise spectrum in Fig. 4.8. At least, now we get a finite noise current. But note that the result is very sensitive to the upper bound of the integration (the 3-dB bandwidth) because it lies in the rising part of the spectrum. So if we were to use the 1-dB bandwidth instead of the 3-dB bandwidth, the noise current would come out quite a bit different. This doesn't sound right, either.

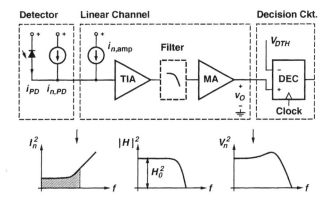

Fig. 4.8 How *not* to calculate total input-referred noise.

Noise Bandwidths. Actually, there is a simple way to calculate the total input-referred noise from the associated noise spectrum. We start out by writing the input-noise spectrum in the general form introduced in Section 4.1:

$$I_n^2(f) = \alpha_0 + \alpha_2 f^2. \tag{4.40}$$

Parameter α_0 describes the white part of the spectrum, and parameter α_2 describes the f^2-noise part (we are neglecting possible $1/f$ and f-noise terms here). Note that in contrast to Eq. (4.2), parameter α_0 includes the white noise from the amplifier *and* the detector. Now we plug this spectrum into Eq. (4.37):

$$\overline{i_n^2} = \frac{1}{H_0^2} \int_0^{BW_D} |H(f)|^2 \cdot (\alpha_0 + \alpha_2 f^2)\, df, \tag{4.41}$$

expand it as

$$\overline{i_n^2} = \alpha_0 \cdot \frac{1}{H_0^2} \int_0^{BW_D} |H(f)|^2\, df + \alpha_2 \cdot \frac{1}{H_0^2} \int_0^{BW_D} |H(f)|^2 \cdot f^2\, df, \tag{4.42}$$

and rewrite it in the form (we'll see why in a moment)

$$\overline{i_n^2} = \alpha_0 \cdot BW_n + \alpha_2/3 \cdot BW_{n2}^3, \tag{4.43}$$

where

$$BW_n = \frac{1}{H_0^2} \int_0^{BW_D} |H(f)|^2\, df, \tag{4.44}$$

$$BW_{n2}^3 = \frac{3}{H_0^2} \int_0^{BW_D} |H(f)|^2 \cdot f^2\, df. \tag{4.45}$$

The bandwidths BW_n and BW_{n2} depend *only* on the receiver's frequency response $|H(f)|$ and the decision circuit's bandwidth BW_D. The latter is uncritical as long as it is larger than the receiver bandwidth and the receiver has a steep rolloff. For simplicity, we assume in the following that the decision-circuit bandwidth is infinite. Numerical values for the bandwidths BW_n and BW_{n2} of some simple receiver responses are listed in Table 4.6. As soon as these bandwidths and the noise parameters α_0 and α_2 are known, we easily can calculate the total input-referred noise with Eq. (4.43).

Table 4.6 Numerical values for BW_n and BW_{n2}.

$H(f)$	BW_n	BW_{n2}
1st-order low pass	$1.57 \cdot BW_{3dB}$	∞
2nd-order low pass, crit. damped ($Q = 0.500$)	$1.22 \cdot BW_{3dB}$	$2.07 \cdot BW_{3dB}$
2nd-order low pass, Bessel ($Q = 0.577$)	$1.15 \cdot BW_{3dB}$	$1.78 \cdot BW_{3dB}$
2nd-order low pass, Butterworth ($Q = 0.707$)	$1.11 \cdot BW_{3dB}$	$1.49 \cdot BW_{3dB}$
Brick wall low pass	$1.00 \cdot BW_{3dB}$	$1.00 \cdot BW_{3dB}$
Rectangular (impulse response) filter	$0.500 \cdot B$	∞
NRZ to full raised-cosine filter	$0.564 \cdot B$	$0.639 \cdot B$

Why did we choose the peculiar form of Eq. (4.43)? Because it leads to a neat interpretation of the bandwidths BW_n and BW_{n2}. If we were to integrate the input spectrum Eq. (4.40) up to the 3-dB point, as suggested by the incorrect Eq. (4.39), we would get

$$\overline{i_n^2} = \alpha_0 \cdot BW_{3dB} + \alpha_2/3 \cdot BW_{3dB}^3. \tag{4.46}$$

By comparing this result with Eq. (4.43), we now understand that the latter equation can be interpreted as the result of integrating the white-noise component of the input-referred spectrum up to BW_n and the f^2-noise component up to BW_{n2}. This interpretation is illustrated graphically in Fig. 4.9. You may already have noted it: BW_n is identical to the *noise bandwidth* (a.k.a., *noise equivalent bandwidth*) of the receiver's frequency response; BW_{n2} could be called the second-order noise bandwidth, because it plays the same role as the (zeroth-order) noise bandwidth BW_n, but with the white noise replaced by f^2 noise.

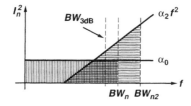

Fig. 4.9 Interpretation of BW_n and BW_{n2} as integration bounds.

Now let's go back and see how different this is from integrating the input-referred noise spectrum up to the 3-dB point. Integrating up to the 3-dB point means that we set $BW_n = BW_{n2} = BW_{3dB}$. By consulting Table 4.6, we see that this is the right thing to do in the case of the brick-wall low-pass response, but in all other cases, we incur an error. For example, in the case of a second-order Butterworth response, we underestimate the white-noise power by a factor $1.11\times$ and the f^2-noise power by a factor $3.33\times$, quite a significant difference!

We now understand that to calculate the total input-referred detector noise, $\overline{i_{n.PD}^2}$, we must multiply the spectral power density of the detector noise, which is white to a good approximation, with the noise bandwidth, $\overline{i_{n.PD}^2} = I_{n.PD}^2 \cdot BW_n$, as we did repeatedly in Chapter 3. Furthermore, to calculate the total input-referred amplifier noise, $\overline{i_{n.amp}^2}$, we must separate the white-noise and f^2-noise components and then apply Eq. (4.43); in Chapter 5, we see examples of this procedure.

Personick Integrals. For an electrical engineer, the noise bandwidths BW_n and BW_{n2} have an intuitive meaning, and this is why we have introduced them here first.[8] In the optical receiver literature (e.g., [6, 19, 62, 127, 177]), however, you find the so-called *Personick integrals* instead of the noise bandwidths. These integrals usually are designated with I_1, I_2, and I_3 and are defined such that the input-referred noise power can be written as

$$\overline{i_n^2} = \alpha_0 \cdot I_2 B + \alpha_2 \cdot I_3 B^3, \qquad (4.47)$$

where B is the bit rate. (The first Personick integral, I_1, relates to the nonstationary detector noise, which we wiped under the rug in Section 4.2.) Thus, by comparing the above equation with Eq. (4.43), the second and third Personick integrals can be identified as

$$I_2 = BW_n/B, \qquad (4.48)$$

$$I_3 = BW_{n2}^3/(3B^3). \qquad (4.49)$$

In other words, the Personick integrals I_2 and I_3 are normalized noise bandwidths. For example, if the receiver has a second-order Butterworth transfer function with a 3-dB bandwidth equal to $2/3$ of the bit rate (we will justify this choice in Section 4.6), we find with the help of Table 4.6 the values $I_2 = 0.740$ and $I_3 = 0.329$ for the Personick integrals. [$\rightarrow$ Problem 4.15]

In the theoretical receiver literature, it often is assumed that the receiver transforms ideal NRZ pulses at the input into pulses with a full raised-cosine spectrum at the output (we discuss this transfer function in Section 4.6). In this case, the Personick integrals assume the values $I_2 = 0.564$ and $I_3 = 0.087$ (cf. last entry of Table 4.6). Unfortunately, the latter values often have been used inappropriately for noise calculations of practical receivers that do *not* have the above mentioned transfer function, typically resulting in overly optimistic noise numbers [99].

---

[8]Whereas BW_n is commonly used in the engineering literature, BW_{n2} has been introduced here for didactical purposes.

4.5 POWER PENALTY

In this section, we introduce the important concept of *power penalty* (PP). This concept permits us to quantify many types of impairments in the receiver, the transmitter, and the fiber. Furthermore, it enables us to derive specification for the building blocks of the communication system. In the subsequent chapters, we frequently carry out power-penalty calculations to obtain typical IC specifications.

The basic idea is as follows. We saw in the previous section that the receiver noise determines the *basic* receiver sensitivity. The *actual* sensitivity is lower than that because of a variety of impairments such as signal distortions in the receiver, offset errors in the decision circuit, a finite extinction ratio in the transmitter, and so forth. Now the power penalty is the *loss* in optical sensitivity due to such an impairment.

More precisely, we define the power penalty PP for a particular impairment as the increase in average transmit power necessary to achieve the same BER as in the absence of the impairment. Power penalties usually are expressed in dBs using the conversion rule $10 \log PP$. Table 4.7 gives examples of impairments in various parts of an optical communication system. For many of these impairments, we later calculate the associated power penalty.

Table 4.7 Examples of impairments leading to power penalties.

Transmitter:	Extinction ratio
	Relative intensity noise
	Output power variations
Fiber:	Dispersion
	Nonlinear effects
Detector:	Dark current
TIA:	Distortions (ISI)
	Offset
MA:	Distortions (ISI)
	Offset
	Noise figure
	Low-frequency cutoff
CDR:	Decision-threshold offset
	Decision-threshold ambiguity
	Sampling-time offset
	Sampling-time jitter

Example 1: Decision-Threshold Offset. To illustrate the power-penalty concept, let's make an example and calculate the power penalty for the case that V_{DTH} is not at its optimum value, that is, for the case of a decision-threshold offset. Figure 4.10(a) shows the situation where the threshold is at its optimum value, V_{DTH}, that is, the threshold is exactly centered in between the zero and one levels. For this example we

assume an equal amount of noise on the zeros and the ones. Figure 4.10(b) shows the situation where the threshold voltage is too high. We can write this incorrect threshold voltage as

$$V'_{DTH} = V_{DTH} + \delta \cdot v_S^{pp}, \quad (4.50)$$

where δ is the threshold offset relative to the signal swing. The threshold offset shown in Fig 4.10(b) will cause many ones to be misinterpreted as zeros, thus significantly increasing the BER.

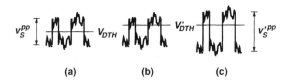

Fig. 4.10 Power penalty due to decision-threshold offset: (a) without offset, (b) with offset, and (c) with offset and increased signal swing to restore the original BER.

To restore the original BER, we need to increase the signal swing from v_S^{pp} to $v_S^{'pp}$ as shown in Fig. 4.10(c). We increase the signal swing until the difference between the one level and the decision threshold becomes the same as in Fig. 4.10(a):

$$v_S^{'pp} = v_S^{pp} + 2\delta \cdot v_S^{pp} = v_S^{pp} \cdot (1 + 2\delta). \quad (4.51)$$

As a result, the probability of misinterpreting a one as a zero is now the same as in Fig. 4.10(a). The probability of misinterpreting a zero as a one is even lower than before, so that the overall BER is a little bit lower than in Fig. 4.10(a). However, we ignore this small difference in BER because the impact on the resulting power penalty is small. Thus, to restore the original BER, we need to increase the signal swing by $v_S^{'pp}/v_S^{pp} = 1 + 2\delta$. Knowing that the signal voltage swing is proportional to the received optical power, the power penalty for a decision-threshold offset δ becomes

$$PP = 1 + 2\delta. \quad (4.52)$$

For example, a 10% decision-threshold offset causes a power penalty of 0.79 dB.

Note that in the above reasoning, we assumed that the act of increasing the transmit power does not introduce impairments of its own, such as an increase in system noise or nonlinear distortions. This is a good assumption for unamplified transmission systems with p-i-n detectors, and we continue to make this assumption unless stated otherwise. However, in systems with optical amplifiers, APDs, or both, the noise level does increase with increasing transmit power, leading to a larger power penalty for the same impairment (cf. Section 7.1).

Now let's see how we can use the power-penalty concept to derive receiver specifications. If we solve Eq. (4.52) for δ, we find

$$\delta = \frac{PP - 1}{2}. \quad (4.53)$$

72 RECEIVER FUNDAMENTALS

This means that if the largest acceptable power penalty is PP, we must control the decision threshold to a precision better than the δ given in Eq. (4.53). For example, given a worst-case power penalty of 0.05 dB ($PP = 1.0116$), the decision-threshold offset must be less than 0.58%.

Example 2: Dark Current. In Chapter 3, we mentioned the detector dark current and how it interferes with the received signal. Now we have the necessary tools to quantify this effect. The dark current by itself does not negatively impact the received signal, it just adds an offset but leaves the signal swing unchanged. As long as the receiver is able to ignore this offset, for example, because AC coupling is used, there is no power penalty for this. However, the *noise* associated with the dark current will enhance the receiver noise and cause a power penalty. Let's calculate it!

According to Eq. (3.5), the dark current I_{DK} of a p-i-n detector contributes the following shot noise:

$$\overline{i_{n,DK}^2} = 2q I_{DK} \cdot BW_n. \tag{4.54}$$

This noise power adds to the receiver noise, which we assume is dominated by the amplifier noise $\overline{i_{n,amp}^2}$. (Neglecting the detector noise overestimates the power penalty somewhat.) So the dark-current noise increases the noise power by

$$\frac{\overline{i_{n,amp}^2} + \overline{i_{n,DK}^2}}{\overline{i_{n,amp}^2}} = 1 + \frac{2q I_{DK} \cdot BW_n}{\overline{i_{n,amp}^2}}. \tag{4.55}$$

We know from Eq. (4.20) that the receiver sensitivity is proportional to the rms noise current, and thus we have found the power penalty:

$$PP = \sqrt{1 + \frac{2q I_{DK} \cdot BW_n}{\overline{i_{n,amp}^2}}}. \tag{4.56}$$

With the typical numbers for our 2.5-Gb/s receiver ($i_{n,amp}^{rms} = 380\,\text{nA}$, $BW_n = 1.9\,\text{GHz}$) and a worst-case dark current of 5 nA, we find the power penalty to be 0.000 046 dB ($PP = 1.000\,0105$). As expected, this is very, very small. For an APD detector, we had to replace I_{DK} with $F \cdot M^2 \cdot I_{DK}$, where I_{DK} is now the *primary* dark current (cf. Eq. (3.10)). In this case, the power penalty would be larger.

Now we can turn this game around and ask: "What is the maximum allowable dark current for a given maximum power penalty?" A little bit of algebra reveals

$$I_{DK} < \left(PP^2 - 1\right) \cdot \frac{\overline{i_{n,amp}^2}}{2q \cdot BW_n}. \tag{4.57}$$

With the same typical numbers as before, we find that the dark current must be less than 5.5 µA to keep the power penalty below 0.05 dB ($PP = 1.0116$). This poses no problem at all!

With these two examples, we illustrated how to compute power penalties and how to derive specifications from them. [→ Problem 4.16]

4.6 BANDWIDTH

In the following, we address the question of how large we should make the receiver bandwidth and, more generally, what frequency response we should choose. To get a feeling for the answer, consider the following dilemma: if we make the receiver bandwidth wide, the receiver preserves the signal waveform without distortions, but at the same time it picks up a lot of noise, which may corrupt the signal. We know from Section 4.2 that a lot of noise translates into a low receiver sensitivity. Alternatively, if we make the receiver bandwidth narrow, the noise is reduced and thus the sensitivity is improved, but now we are faced with signal distortions known as *intersymbol interference* (ISI). Like noise, ISI also reduces the sensitivity because the output signal swing (at the decision circuit) is reduced for certain bit sequences, that is, the signal swing for a "01010101..." sequence will be *lower* than that for a "00110011..." sequence. We can conclude from this line of reasoning that there must be an *optimum receiver bandwidth* for which the sensitivity is best. A rule of thumb for NRZ receivers says that this optimum 3-dB bandwidth is about

$$BW_{3dB} \approx \frac{2}{3} \cdot B, \tag{4.58}$$

where B is the bit rate. Similarly, we could say that the optimum bandwidth is around 60% to 70% of the bit rate.

Example: Butterworth Receiver. Figure 4.11 illustrates the trade-off between ISI and noise for the example of a 10-Gb/s receiver with a second-order Butterworth response (i.e., a maximally flat amplitude response). For this example, we further assume that the received input signal is an ideal NRZ waveform and that the input-referred noise is white. The output waveforms for three different receiver bandwidths are shown from top to bottom in the form of eye diagrams (see Appendix A for an explanation of eye diagrams).

For the moment, let's ignore the gray stripes in the eye diagrams that symbolize the noise. Figure 4.11(a) shows the eye diagram of a wideband receiver with a 3-dB bandwidth of 4/3 the bit rate, that is, twice the optimum bandwidth. As expected, we get a clean eye with almost no ISI. Figure 4.11(b) shows the eye diagram of a receiver with the optimum bandwidth, 2/3 of the bit rate. Finally, Fig. 4.11(c) shows the eye diagram of a narrowband receiver with only half the optimum bandwidth. In the latter case, we observe severe ISI resulting in a partially closed eye. In particular, we can see a trace with the full swing corresponding to the bit sequence "00110011..." and another trace with only about half the swing corresponding to the bit sequence "01010101...."

Now let's add some noise at the input of the receiver. In the case of white noise, the received noise power is proportional to the receiver bandwidth. Therefore, the rms noise voltage at the output of the receiver is proportional to the square root of the bandwidth. The gray stripes in Fig. 4.11 represent the peak value of the noise voltage, which reduces in steps of $\sqrt{2}$ as we go from the wideband, to the optimum-bandwidth, and to the narrowband receiver. For clarity, only the noise inside the eye

74 RECEIVER FUNDAMENTALS

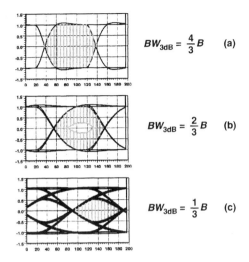

Fig. 4.11 Trade-off between ISI and noise in a receiver.

is shown; in reality, of course, noise is present on both sides of the signal trace. But wait a minute, what exactly is the peak value of the noise voltage? Doesn't Gaussian noise assume potentially unlimited values? Yes, but the trick here is to define the peak value of the noise voltage based on the required BER. For example, if we are ready to accept a BER of 10^{-12}, then we don't care about noise voltages exceeding $7.035 \cdot v_n^{rms}$ (because this happens only with a probability of 10^{-12}), and so we can take the peak value of the noise voltage as $7.035 \cdot v_n^{rms}$ or more generally $Q \cdot v_n^{rms}$ (cf. Appendix A).

Going back to Fig. 4.11, we observe that whereas the eyes for the wide and narrow-band receivers are completely closed by the noise, the eye for the optimum-bandwidth receiver with $BW_{3dB} = 2/3 \cdot B$ is open at the center. To recover the received data at the desired BER, we must make a decision in the open part of the eye, that is, the decision threshold and sampling instant must define a point in the open part of the eye. In our example, this is only possible for the optimum-bandwidth receiver.

Power Penalty due to ISI. In the following, we analyze the optimum bandwidth in a more quantitative way. The ISI caused by the finite receiver bandwidth can be quantified by a power penalty. This power penalty stems from the fact that ISI reduces the output swing for certain bit sequences such as "01010101...." The worst-case output swing can be determined from the vertical opening in the eye diagram (without noise). In Fig. 4.12(a), the signal without ISI has a vertical eye opening of V_E. In Fig. 4.12(b), the eye opening is reduced to V_E' because of ISI. To restore the original BER, we must increase the full signal swing of the distorted signal to $PP \cdot V_E$ such that its vertical eye opening becomes V_E, as indicated in Fig. 4.12(c). Hence the power penalty is

$$PP = \frac{V_E}{V'_E}. \qquad (4.59)$$

Inspecting the eye diagrams in Fig. 4.11, we see that, when sampling at the point of maximum eye opening, there is no significant vertical eye closure for the wideband and optimum-bandwidth receivers; however, for the narrowband receiver, the eye is about halfway closed (50% vertical eye closure). Therefore, the power penalties due to ISI for Figs. 4.11(a), 4.11(b), and 4.11(c) are $PP = 0\,\text{dB}$, $PP = 0\,\text{dB}$, and $PP = 3\,\text{dB}$, respectively. Note that the power penalty as given in Eq. (4.59) is somewhat pessimistic because it is based on the worst-case bit sequence. We see from Fig. 4.11(c) that some bits have a higher signal swing than V_E and thus are detected at a somewhat lower BER than the original one. [→ Problem 4.17]

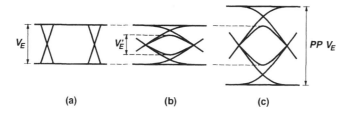

Fig. 4.12 Eye diagram (a) without ISI, (b) with ISI, and (c) with ISI and increased signal swing to restore the original BER.

Now let's combine the basic noise-based sensitivity and the power penalty due to ISI to obtain the actual receiver sensitivity. A graphical representation of this calculation is shown in Fig. 4.13. The basic noise-based sensitivity decreases with receiver bandwidth. Specifically, if we double the bandwidth, the rms-noise increases by a factor $\sqrt{2}$ (for white input noise), which reduces the sensitivity by 1.5 dB according to Eq. (4.20). The basic sensitivity is represented by the dashed line moving up from $\overline{P}_{sens0} - 3.0\,\text{dB}$ to $\overline{P}_{sens0} - 1.5\,\text{dB}$ and to $\overline{P}_{sens0}$, where $\overline{P}_{sens0}$ is an arbitrary reference value. Next, we correct the basic sensitivity with the power penalty due to ISI, which brings us to the solid line. As expected, the best sensitivity is reached near the bandwidth $2/3 \cdot B$.

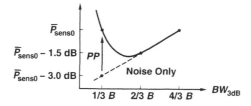

Fig. 4.13 Sensitivity as a function of receiver bandwidth.

Note that although the optimum bandwidth is around $2/3 \cdot B$, it is possible to build a practical receiver with a bandwidth of only $1/3 \cdot B$, if we are willing to accept a loss in sensitivity of about 1.5 dB (see Fig. 4.13). However, in doing so we have to be careful about the *horizontal* eye opening, which depends on the phase linearity of the receiver. Figure 4.11(c) shows that in our Butterworth receiver example the horizontal eye opening is still nearly 100% even with a bandwidth of only $1/3 \cdot B$. This narrow-band approach is attractive for 40-Gb/s systems with optical preamplifiers where a small loss in sensitivity is acceptable, if in return the receiver can be built from 13 GHz ($= 1/3 \cdot 40$ Gb/s) electronic components [149].

Bandwidth Allocation. So far we have been talking about the bandwidth of the complete receiver. As we know, the receiver consists of a cascade of building blocks: photodetector (p-i-n or APD), TIA, filter (optional), MA, and decision circuit. It is the *combination* of all these blocks that should have a bandwidth of about $2/3 \cdot B$. The combined bandwidth can be approximated by adding the inverse-square bandwidths of the individual blocks: $1/BW^2 \approx 1/BW_1^2 + 1/BW_2^2 + \ldots$. Thus, each individual block must have a bandwidth that is *larger* than $2/3 \cdot B$. There are several strategies of assigning bandwidths to the individual blocks to achieve the desired overall bandwidth. Here are three practical bandwidth allocation strategies:

- All receiver blocks (p-i-n/APD, TIA, MA, CDR) are designed for a bandwidth much larger than the desired receiver bandwidth. Then a *precise filter* is inserted, typically after the TIA, to control the bandwidth and frequency response of the receiver. Often a fourth-order Bessel-Thomson filter, which exhibits good phase linearity, is used. This method typically is used for lower-speed receivers (2.5 Gb/s and below).

- The TIA is designed to have the desired receiver bandwidth and all other blocks (p-i-n/APD, MA, CDR) are built with a much larger bandwidth. No filter is used. This approach has the advantage that the TIA bandwidth specification is relaxed, permitting a higher transimpedance and better noise performance (we study this trade-off in Section 5.2.2). But the receiver's frequency response is less well controlled compared with when a filter is used.

- All blocks together (p-i-n/APD, TIA, MA, CDR) provide the desired receiver bandwidth. No single block is controlling the frequency response and, again, no filter is used. This approach typically is used for high-speed receivers (10 Gb/s and above). At these speeds it is challenging to design electronic circuits and APDs and we cannot afford the luxury of overdesigning them.

Optimum Receiver Response. In the remainder of this section, we go beyond the empirical rule given in Eq. (4.58) and explore the questions regarding the optimum receiver response and the optimum bandwidth. Is there an optimum frequency response for NRZ receivers? Yes, but the answer depends on many factors, such as the shape of the received pulses (i.e., the amount of ISI in the received signal), the spectrum of the input-referred noise, the sampling jitter in the decision circuit, the bit estimation technique used, and so forth. Figure 4.14 shows a decision tree distinguishing the most important cases. For the following discussion, we assume that each bit is estimated independently by comparing the sampled output voltage to a threshold voltage as indicated in Fig. 4.3.[9]

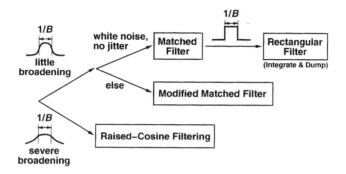

Fig. 4.14 Decision tree to determine the optimum receiver response.

If the NRZ pulses at the input of the receiver are well shaped, in particular if the pulses are broadened by less than 14% of the bit interval ($1/B$), a matched-filter response or a modified matched-filter response is the best choice [83]. In the case of white input-referred noise and the absence of sampling jitter, the *matched-filter response* gives the best results. The matched filter is defined by its impulse response $h(t)$, which must be proportional to (or matched to) a time-reversed copy of the received pulses $x(t)$, more precisely, $h(t) \sim x(T-t)$, where T is the duration of the received pulses. This definition implies that the matched-filter frequency response matches the spectral shape of the input pulses (but not generally the phase). It can be shown that in the absence of ISI, the matched-filter response maximizes the the sampled signal-to-noise ratio and thus results in the lowest BER [42, 83].

In the case of an undistorted NRZ signal, the matched filter is given by $h(t) \sim x(T-t) = x(t)$, where $x(t)$ is a rectangular pulse starting at $t = 0$ and ending at $T = 1/B$, and hence this filter is known as the *rectangular filter*. We discuss this case as well as a possible implementation (integrate and dump) in a moment. If the input-referred noise spectrum is not white or if the decision circuit exhibits sampling jitter, the concept of matched filtering can be generalized to take these effect into account [19]. This case is indicated by the box labeled "Modified Matched Filter" in Fig. 4.14.

In long-haul transmission systems, the NRZ pulses at the input of the receiver usually are severely broadened, for example, as a result of fiber dispersion (lower arrow in Fig. 4.14). If the received pulses are broader than the bit interval they overlap,

---

[9]It also is possible, and in fact better, to make a *joint* decision on a sequence of bits, for example, by using a Viterbi decoder. In this case, the optimum receiver response is the matched-filter (or modified matched-filter) response regardless of the pulse broadening (cf. Section 4.7).

78   RECEIVER FUNDAMENTALS

in other words, we have ISI. The matched-filter response discussed earlier would exacerbate the ISI problem by further broadening the pulses leading to a significant power penalty. For severely broadened pulses (more than 20% of the bit interval), *raised-cosine filtering* theoretically gives the best results [83]. Raised-cosine filtering is defined as the transformation of the (broadened) input pulses into pulses with a raised-cosine spectrum. Note that this does *not* mean that the receiver itself has a raised-cosine response. We give an example to clarify this in a moment. Now, pulses with a raised-cosine spectrum have a shape similar to $y(t) = \sin(\pi Bt)/(\pi Bt)$ and thus are free of ISI,[10] that is, they are zero at $t = nT$ for all n except 0 with $T = 1/B$ [42, 83]. Thus, the ISI problem is solved. At this point, you may wonder why we don't always use raised-cosine filtering. The answer is that the noise bandwidth of a raised-cosine receiver is wider than that of a matched filter receiver; hence if ISI in the received signal is weak, matched filtering gives the better results.

Although raised-cosine filtering is quite popular in the theoretical receiver literature, it is rarely used in practical optical receivers. For starters, the output pulses from a raised-cosine receiver extend backwards through time indefinitely (they are symmetric around $t = 0$ with a shape similar to $y(t) = \sin(\pi Bt)/(\pi Bt)$), which means that such a receiver can only be realized as an approximation. Furthermore, the shape of the received pulses must be known exactly to design the receiver's transfer function. In practice, the receiver response often is chosen to have a bandwidth of about $2/3 \cdot B$ and a good phase linearity. In case the received pulses are severely broadened or otherwise distorted, an adaptive equalizer is placed after the linear channel to reduce the ISI. We discuss this approach further in Section 4.7.

Rectangular Filter. To illustrate the concept of matched filtering, let's make a simple example. Consider that we receive an undistorted NRZ signal embedded in white noise. As we have already mentioned, the rectangular filter provides the optimum receiver response for this case.

In the time domain, this filter convolves the received ideal NRZ signal with a pulse of duration $T = 1/B$. In Fig. 4.15(a), this convolution has been carried out graphically resulting in a triangular output signal. Note that despite of the slow edges, the output signal is free of ISI when sampled at the instant of maximum eye opening (dashed line in the eye diagram). In the frequency domain, the filter has a low-pass characteristics that can be calculated by taking the Fourier transform of a pulse of duration $T = 1/B$. The normalized transfer function turns out to be

$$H(f) = \frac{\sin(\pi f/B)}{\pi f/B} \cdot e^{-j\pi f/B}. \quad (4.60)$$

The squared frequency response $|H(f)|^2$ is plotted in Fig. 4.15(b) on a lin-lin scale. The noise bandwidth of this response turns out to be $BW_n = B/2$. (The 3-dB bandwidth is slightly less than this: $BW_{3dB} = 0.443B$.) The combination of a small noise

---

[10]The particular pulse $y(t) = \sin(\pi Bt)/(\pi Bt)$ has a raised-cosine spectrum with 0% excess bandwidth. For the general case of pulses with a raised-cosine spectrum, see [42].

bandwidth and the absence of ISI are the characteristics of an ideal receiver response. However, the triangular eye shape implies that to avoid ISI, we have to sample *exactly* at the center of the eye. In other words, any sampling offset or sampling jitter will translate into a power penalty. [→ Problem 4.18]

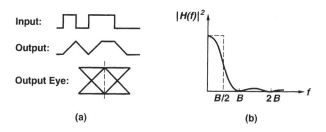

Fig. 4.15 Rectangular-filter receiver: (a) waveforms and (b) frequency response.

Integrate and Dump. As we have already pointed out, the rectangular filter convolves the received NRZ signal, $x(t)$, with a pulse of duration $T = 1/B$. This pulse is the filter's impulse response, $h(t)$, which is one in the interval from $t = 0$ to T and zero everywhere else. We thus can write the output signal $y(t)$ from the filter as

$$y(t) = \int_{-\infty}^{\infty} h(t - t') \cdot x(t') \, dt' = \int_{t-T}^{t} x(t') \, dt'. \tag{4.61}$$

The expression on the right-hand side can be interpreted as the moving average of $x(t)$ computed over the interval T. In a receiver the output signal, $y(t)$ will get sampled periodically by the decision circuit at the instant of maximum eye opening. The maximum eye opening occurs at the end of each bit period, $t = nT$, and thus the sampled signal for the nth bit is

$$y(nT) = \int_{(n-1)T}^{nT} x(t') \, dt'. \tag{4.62}$$

This expression suggests that the rectangular filter can be replaced by a circuit that integrates the received signal $x(t)$ over the bit period T. The resulting output samples $y(nT)$ are the same as those given by Eq. (4.62). Note that we need to start the integration at the beginning of each bit period, and thus the integrator must be reset quickly at the end of each bit period (alternatively, two integrators can be operated in a ping-pong fashion). For this reason, this method is called *integrate and dump* [42, 83].

The integrate-and-dump arrangement has the advantage that it lends itself well to monolithic integration. Its frequency response is well controlled and a decision circuit with "instantaneous" sampling can be avoided. Also, sampling occurs at the end of the bit period rather than in the middle, simplifying the clock-recovery circuit (data and clock edges are aligned). For CMOS implementations, see [164, 175]. However, just like the rectangular-filter receiver, the integrate-and-dump receiver is optimum

80 RECEIVER FUNDAMENTALS

only when receiving undistorted rectangular pulses with white noise, which is rarely the case in practice.

A related issue is the implementation of the clock-recovery circuit for an integrate-and-dump receiver. If the integrate-and-dump mechanism is part of the decision circuit, standard techniques can be used. However, if the integrate-and-dump mechanism is part of the TIA, as proposed in [56], it is less obvious how to obtain the phase information for the clock-recovery circuit. Note that in this case, a received signal independent of the clock signal is not available. One solution is to sample the analog output from the integrator three times per bit period: at the beginning, middle, and end. Then we compute the expression $[y(nT+1) - y(nT+0.5)] - [y(nT+0.5) - y(nT)]$, which becomes zero if the clock phase is adjusted correctly [176].

Raised-Cosine Filtering Example. To illustrate the concept of raised-cosine filtering, let's make a simple example. We want to calculate the transfer function that transforms undistorted NRZ pulses into pulses with a full raised-cosine spectrum (this transfer function is called "NRZ to full raised-cosine filter" in Table 4.6). The full raised-cosine spectrum (a.k.a., raised-cosine spectrum with 100% excess bandwidth) is defined as [83]

$$H_{FRC}(f) = \frac{1 + \cos(\pi f/B)}{2} \cdot e^{-j 2\pi f/B} \quad \text{for} \quad f < B \quad (4.63)$$

and $H_{FRC}(f) = 0$ for $f \geq B$. This spectrum guarantees that the (sinc-like) output pulses are free of ISI. The spectrum of the incoming undistorted NRZ signal is

$$H_{NRZ}(f) = \frac{\sin(\pi f/B)}{\pi f/B} \cdot e^{-j \pi f/B}. \quad (4.64)$$

The transfer function of the desired receiver response is obtained by dividing these two spectra:

$$H(f) = \frac{H_{FRC}(f)}{H_{NRZ}(f)}$$
$$= \frac{1 + \cos(\pi f/B)}{2} \cdot \frac{\pi f/B}{\sin(\pi f/B)} \cdot e^{-j \pi f/B} \quad \text{for} \quad f < B. \quad (4.65)$$

The noise bandwidth of this response turns out to be $BW_n = 0.564B$, which is about 13% larger than that of the rectangular filter. (The 3-dB bandwidth is $BW_{3dB} = 0.580B$.) Because both receiver responses produce an ISI-free output signal but the raised-cosine filtering response has a larger noise bandwidth, it is suboptimal in this case of ideal received NRZ pulses. As we pointed out earlier, raised-cosine filtering is most attractive when the received pulses are significantly broadened. Nevertheless, this NRZ to full raised-cosine filtering response and its associated Personick integrals are frequently encountered in the theoretical receiver literature.

Bandwidth of a Receiver for RZ Signals. So far we have been talking about NRZ signals, but what about the optimum bandwidth of a receiver for a 50%-RZ signal?

One way to approach this question is to observe that an RZ signal at bit rate B is like an NRZ signal at bit rate $2B$, where every second bit is a zero. Thus, we would expect that the optimum bandwidth is about twice that for an NRZ signal, that is, $BW_{3dB} \approx 4/3 \cdot B$. Another way to approach this question is the matched filter view: because the spectral width of the RZ pulse is twice that of the NRZ pulse, we would expect again that we have to double the receiver bandwidth (from $0.443B$ to $0.886B$). Finally, what does the raised-cosine approach recommend? Going through the math we find that we have to *reduce* the bandwidth from $0.58B$ for NRZ to $0.39B$ for RZ [62]! How can we explain this? Recall that the raised-cosine approach forces the same output pulses (namely sinc-like pulses) no matter whether the input consists of NRZ or RZ pulses. Therefore, the RZ receiver has to broaden the pulses more than the NRZ receiver, which explains the narrower bandwidth for the RZ receiver.

In practice, we have the following options. One, use a wide-bandwidth receiver ($\approx 1.33B$), which results in a good sensitivity but requires a clock and data recovery (CDR) circuit that can deal with an RZ signal. In particular, the sampling instant must be well timed to sample the narrow RZ pulse at its maximum value. Two, use a narrow-bandwidth receiver that converts the received RZ signal into an NRZ signal permitting the use of a standard CDR; however, the RZ to NRZ conversion lowers the signal amplitude significantly, leading to a suboptimal receiver sensitivity. Gaussian-like filters with a bandwidth of $0.375B$ are offered as low-cost RZ to NRZ converters.

Minimum Bandwidth. When talking to a communication systems specialist, he may tell you that you need at least a bandwidth of $B/2$, the so-called *Nyquist bandwidth*, for ISI free communication. What does that mean and how does it affect our receiver design?

Let's assume that our received signal is not the usual NRZ signal, but a superposition of sinc pulses of the form $x(t) = \sin(\pi Bt)/(\pi Bt)$, known as *Nyquist pulses*. One such pulse is sent for each one bit and no pulse is sent for the zero bits at the bit rate B. This communication signal has some very desirable properties [42]: its spectrum is rectangular, that is, it is flat up to $B/2$ and then drops to zero immediately. In fact, this spectrum belongs to the raised-cosine family (raised-cosine spectrum with 0% excess bandwidth) and thus the signal is free of ISI. Note that this signal can be transmitted through a channel with a brick-wall low-pass response of bandwidth $B/2$ (and linear phase) without incurring any distortion. This is so because the rectangular spectrum multiplied by the brick-wall low-pass response yields the same (scaled) rectangular spectrum. Also note that this signal is strictly bandlimited to $B/2$. A communication signal that has spectral components above $B/2$ is said to have an *excess bandwidth* (usually specified in percents relative to $B/2$). For example, the Nyquist-pulse signal has a 0% excess bandwidth, whereas the ideal NRZ and RZ signals have an infinite excess bandwidth. Now, to receive this Nyquist-pulse signal optimally, we choose the frequency response of the receiver to match the signal spectrum: again, this is a brick-wall low-pass response with bandwidth $B/2$ and, again, no ISI is incurred as a result of this response. Note that, as in the case of the matched receiver for NRZ pulses, the noise bandwidth of this receiver is $BW_n = B/2$.

82 RECEIVER FUNDAMENTALS

Does this mean that we need a 3-dB bandwidth of at least $B/2$ to receive a bit stream at the rate B? No, the Nyquist bandwidth does not refer to the 3-dB bandwidth but to the *absolute bandwidth*, which is the bandwidth where the signal is completely suppressed, that is, we also could call it the ∞-dB bandwidth. Even if the absolute bandwidth is less than $B/2$, we can still receive an error-free bit steam, but the received signal will no longer be free of ISI.

4.7 ADAPTIVE EQUALIZER

The signal at the output of the receiver's linear channel invariably contains some ISI. This ISI is caused, among other things, by dispersion in the optical fiber (modal, chromatic, and polarization mode dispersion) as well as the frequency response of the linear channel. In principle, it would be possible to remove this ISI by making the linear channel perform a raised-cosine filtering operation as we discussed in Section 4.6, but in practice it usually is impossible to predict the precise input pulse shape on which raised-cosine filtering depends. The pulse shape varies with the length of the fiber link, the quality of the fiber, chirp of the laser, and so forth, and it may even change over time. For example, polarization-mode dispersion (PMD), which is significant in long-haul transmission at high speeds (10 Gb/s or more) over older (already installed) fiber, changes slowly with time. For these reasons, it is preferred to use a linear channel that has a bandwidth of about $2/3 \cdot B$, eliminating much of the noise, followed by an adaptive *ISI canceler*.

Decision-Feedback Equalizer. The optimum realization of the ISI canceler is the *Viterbi decoder*, which performs a maximum-likelihood sequence detection of the sampled received signal based on a channel model. However, the implementation of such a decoder usually is too complex, and an *equalizer* is used instead. A popular equalizer type is the adaptive *decision-feedback equalizer* (DFE), which consists of two adaptive *finite impulse response* (FIR) filters, one feeding the received signal to the decision circuit and one providing feedback from the output of the decision circuit, as shown in Fig. 4.16.[11] The DFE is a *nonlinear* equalizer because the decision circuit is part of the equalizer structure. In contrast to the simpler linear *feed-forward equalizer* (FFE), which consists of only the first FIR filter, the DFE produces less amplified noise. For a full treatment of the DFE and the FFE, see [33, 81, 200]. Note that the MA in Fig. 4.16 must be linear (usually implemented as an automatic gain control [AGC] amplifier) to prevent nonlinear signal distortions at the input of the equalizer.

How does a DFE cancel ISI? From the input-signal waveforms in Fig. 4.17(a), we see how the bit *before* the bit currently under decision influences the signal value of the current bit. This disturbance is called *postcursor ISI*. If the preceding bit, that is, the decided bit, is a one, the signal levels of the current bit are slightly shifted *upward*

---

[11] Here we use the term *DFE* for the combination of a precursor and postcursor equalizer. Note that other authors use the term *DFE* for the postcursor equalizer *only* and use the term *FFE* for the precursor equalizer.

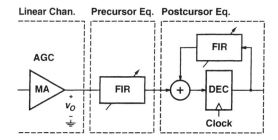

Fig. 4.16 The linear channel of Fig. 4.1 followed by an adaptive decision-feedback equalizer.

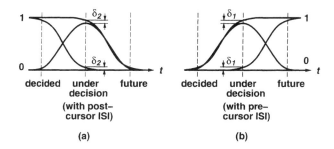

Fig. 4.17 (a) Postcursor and (b) precursor ISI in the signal before the equalizer (v_O).

compared with when the bit is a zero. This shift is marked in Fig. 4.17 with δ_2; for the later analysis, the value of δ_2 is assumed to be normalized to the signal swing. So if we know the value of the decided bit and δ_2, we can remove the postcursor ISI (at the instant of sampling) by making the necessary correction to the current signal level. This is exactly what the one-tap *postcursor equalizer* in Fig. 4.18 does. The decided bit y is available at the output of the decision circuit and is represented by the values $\{-1, 1\}$. This bit is used to compute the correction $c_2 \cdot y$, where $c_2 = -\delta_2$, which is then added to the current signal, thus compensating the postcursor ISI. Here we also can recognize a weakness of the postcursor equalizer: if a decision happens to be incorrect, it adds *more* ISI to the signal at the input of the slicer, possibly causing further decision errors. This effect is known as *error propagation*.

An alternative view of the the postcursor equalizer is that it acts as a slicer that adapts its threshold to the "situation." From Fig. 4.17(a), we see that the optimum threshold level of the current (unequalized) bit is slightly above or below the centerline, depending on whether the previous bit was a one or a zero. This suggests that we should use feedback from the decided bit at the output of the decision circuit to control the threshold level. Of course, this "adaptive threshold" view and the "ISI canceler" view are equivalent.

Now there also is some influence from the bit *after* the bit currently under decision. This disturbance is called *precursor ISI*. This at first may sound like a violation of causality, but because a typical transmission system has a latency of many bits,

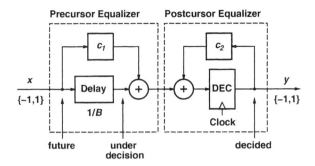

Fig. 4.18 Simple DFE to illustrate the operating principle.

precursor ISI is possible. The influence of the future bit on the current signal levels is shown in Fig. 4.17(b). If the future bit is a one, the signal levels of the current bit are slightly shifted upward (by δ_1). So if we know the value of the future bit and δ_1, we can remove the precursor ISI by making the necessary correction to the current signal level. The 2-tap *precursor equalizer* shown in Fig. 4.18 does this by delaying the input signal by one bit period so it can look into the "future" of the decision circuit. The future bit x is available at the input of the equalizer and is represented by the values $\{-1, 1\}$. This bit is used to compute the correction $c_1 \cdot x$, where $c_1 = -\delta_1$, which is then added to the current signal thus compensating the precursor ISI. [$\rightarrow$ Problem 4.19]

In practical equalizers, more taps than those shown in Fig. 4.18 are used to take the effects of additional bits before and after the current bit into account.

Weight Adaptation. How can we find the tap weights (or filter coefficients) $c_1, c_2, \ldots$ that result in the least ISI at the input of the decision circuit and how can we make them adapt to changing ISI conditions in the input signal? First, we need to define a *cost function* that tells us how close we are to the desired optimum. One possibility is to use a so-called *eye monitor* at the input of the decision circuit that measures the vertical eye opening. Its complement, the eye closure, can serve as a cost function. An arrangement similar to that in Fig. 4.24 can be used to measure this eye closure. Another possibility is to run the signal at the input of the decision circuit through a slicer and to look at the difference between the slicer input and output. This difference is a measure of the error (or ISI) in the signal. The mean-square value of this difference, which is always positive, is a popular cost function. Yet another possibility is to look at the spectral density of the signal at the input of the decision circuit and to compare it with the desired density. In the case of an NRZ receiver, the desired spectrum is the function $H_{NRZ}(f)$ given in Eq. (4.64).

Given the cost function, we now have to find a way to optimize all the tap weights (c_i) based on the information contained in it. In the case of two weights, we can visualize the cost function as a two-dimensional hilly surface where the height z is the cost function and x, y are the weights. Our job is to find the point (x, y) where

z is the smallest. A popular solution is to start at a convenient point on this surface and follow the gradient downhill. If we don't get stuck in a local minimum, we will eventually find the optimum for (x, y). This method is known as *gradient descent*. The gradient can be estimated, for example, by a procedure called *weight perturbation* (a.k.a. *dithering*) where, one after the other, each tap weight is perturbed slightly (in our two-weight example by Δx and Δy) and its effect on the cost function (Δz) is registered. At the end of this somewhat tedious procedure, we have an estimate of the gradient (in our two-weight example, the gradient is $[\Delta z/\Delta x, \Delta z/\Delta y]$). Now we can update the weights by taking a small step in the direction of the negative gradient. In case we choose as our cost function the mean-square difference between the slicer input and output, there is an elegant and efficient optimization procedure known as the *least-mean-square algorithm* (LMS), which performs a stochastic gradient descent. In this algorithm, the difference between the slicer input and output, the error, is correlated to intermediate signals in the equalizer and the correlations subsequently are used to adjust the tap weights.

Implementation Issues. The equalizer for an optical receiver can be implemented in the digital or analog domain. For a digital implementation, the biggest challenge is the A/D converter, which samples and digitizes the signal from the linear channel. For a 10-Gb/s NRZ receiver, we need a converter with about 6 bits of resolution sampling at 10 GHz. For an analog realization, the delays can be implemented with cascades of buffers and the taps with analog multipliers and current summation nodes. The challenge here is to achieve enough bandwidth and precision over process, supply voltage, and temperature. An analog 10-Gb/s DFE with 8 precursor taps and one postcursor tap is described in [20]. This equalizer can reduce the PMD induced power penalty at a differential group delay (DGD) of 70 ps from 8.5 dB to 2.5 dB. An analog 10-Gb/s FFE with 5 taps, which can compensate up to 50 ps of DGD is described in [9].

When implementing a high-speed postcursor equalizer, the feedback loop that goes through the FIR filter, the summation node, and the decision circuit often presents a speed bottleneck. Fortunately, this loop can be removed by using parallelism, as shown in Fig. 4.19 for the example of a one-tap equalizer. Two parallel decision circuits are used: one is slicing for the case that the previous bit was a zero and the other one for the case that the previous bit was a one. Then a multiplexer in the digital domain selects which result to use. This and other speed optimization methods are described in [65].

Another issue relates to the implementation of the clock-recovery circuit. Ideally, we would like to extract the clock signal from the equalized waveform, but note that in a DFE, this signal *depends* on the clock signal and its phase. If the clock phase is offset from the center of the eye, some decisions are likely to be incorrect. These errors introduce distortions in the equalized waveform, hampering an accurate clock extraction. Error propagation through the feedback path of the DFE exacerbate the problem further. Thus, clock recovery has to be performed either before the equalization (if there is little ISI) or else clock recovery and equalization must be combined carefully [179].

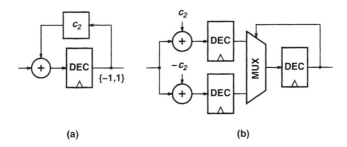

Fig. 4.19 One-tap postcursor equalizer: (a) conceptual representation and (b) high-speed implementation.

4.8 NONLINEARITY

In Section 4.1, we introduced the *linear channel* as an abstraction for the TIA followed by an optional filter, followed by the main amplifier. How linear does this channel have to be? If the linear channel is followed directly by a decision circuit, as shown in Fig. 4.1, linearity is of little concern and we may even use a limiting amplifier for the MA. In this case, amplitude distortions do no harm as long as the crossover points of the signal with the decision threshold are preserved. Nevertheless, we have to make sure that the nonlinearity doesn't introduce pulse-width distortions and jitter, which would reduce the horizontal eye opening. If the linear channel is followed by some type of signal processor, such as the equalizer shown in Fig. 4.16, linearity becomes important. In this case, we want to design the linear channel such that gain compression and other nonlinear distortions remain small. If the linear channel is part of a receiver for analog AM-VSB or QAM signals, as, for example, in a CATV/HFC application, then linearity is of foremost importance. In this case, we must design the linear channel such that the combined effects of harmonic distortions, intermodulation distortions, cross-modulation distortions, and so forth remain below the threshold of perception.

In the following, we discuss how to characterize and quantify nonlinearity. A straightforward way to describe a nonlinear DC transfer curve $y = f(x)$ is by expanding it into a power series:

$$y = A \cdot \left(a_0 + x + a_2 \cdot x^2 + a_3 \cdot x^3 + a_4 \cdot x^4 + a_5 \cdot x^5 + \ldots \right), \quad (4.66)$$

where A is the small-signal gain and a_i are the normalized power-series coefficients characterizing the nonlinearity. Note that $A \cdot a_0$ represents the output-referred offset and that $a_1 = 1$ because, for convenience, all coefficients have been normalized to the small-signal gain. The nonlinear AC characteristics could be described by writing Eq. (4.66) as a Volterra series in which the small-signal gain becomes frequency dependent, $A(f)$, and the coefficients a_i become kernels in the frequency domain: $a_2(f_1, f_2), a_3(f_1, f_2, f_3)$, and so forth, but in the following analysis, we assume that the frequency dependence is weak and that the coefficients can be taken to be constant.

Gain Compression. A simple measure of nonlinearity is the loss of gain experienced by large signals relative to the small-signal gain. For an input signal swinging from $-X$ to X, we find the broadband large-signal gain with Eq. (4.66) to be $[y(X) - y(-X)]/[X - (-X)] = A \cdot (1 + a_3 \cdot X^2 + a_5 \cdot X^4 + \ldots)$. When normalized to the small-signal gain A, we obtain the expression

$$1 + a_3 \cdot X^2 + a_5 \cdot X^4 + \ldots, \qquad (4.67)$$

which describes how the broadband large-signal gain varies with signal strength X. For practical amplifiers, a_3 usually is negative, meaning that the gain reduces for large signals.

To discuss the narrowband large-signal gain, we take the input signal as a sine wave $x(t) = X \cdot \sin(2\pi f t)$ with amplitude X and frequency f. With Eq. (4.66) we find that the output signal contains harmonic components at f, $2f$, $3f$, and so forth. Specifically, the output-signal components at the frequency f are $y(t) = A \cdot (1 + 3/4 \cdot a_3 \cdot X^3 + 5/8 \cdot a_5 \cdot X^5 + \ldots) \cdot \sin(2\pi f t)$; thus, the narrowband large-signal gain is $A \cdot (1 + 3/4 \cdot a_3 \cdot X^2 + 5/8 \cdot a_5 \cdot X^4 + \ldots)$. When normalized to the small-signal gain A, we obtain the so-called *gain compression* (GC) as a function of the input amplitude X:

$$GC = 1 + 3/4 \cdot a_3 \cdot X^2 + 5/8 \cdot a_5 \cdot X^4 + \ldots. \qquad (4.68)$$

Note that for negative values of a_3 and a_5, the narrowband gain is somewhat larger than the broadband gain, that is, filtering out the harmonic distortion products increases the swing of the output signal. Frequently, the input amplitude X for which $GC = -1$ dB ($0.89\times$) is specified, this amplitude is known as the 1-dB gain compression point.

Harmonic Distortions. A more sophisticated measure of nonlinearity in broadband circuits is the *harmonic distortion*. Again, we take the input signal as a sine wave $x(t) = X \cdot \sin(2\pi f t)$ with amplitude X and frequency f. As we know, the output signal contains harmonic components at f, $2f$, $3f$, and so forth. For small signals X, the most significant output-signal components at the frequencies $2f$ and $3f$ are $y(t) = A \cdot [-1/2 \cdot a_2 \cdot X^2 \cdot \cos(4\pi f t) - 1/4 \cdot a_3 \cdot X^3 \cdot \sin(6\pi f) + \ldots]$. Now, the nth-order harmonic distortion HDn is defined as the ratio of the output-signal component (distortion product) at frequency nf to the fundamental at f. Thus, for small signals X, we find the following expressions [34, 82]:

$$HD2 \approx 1/2 \cdot |a_2| \cdot X, \qquad (4.69)$$

$$HD3 \approx 1/4 \cdot |a_3| \cdot X^2. \qquad (4.70)$$

From these expressions, we note that a 1-dB increase in the input signal X causes a 1-dB increase in $HD2$ and a 2-dB increase in $HD3$. In general, higher-order harmonics depend more strongly on the input signal amplitude: the nth-order harmonic distortion product is proportional to X^n, or equivalently, the nth-order harmonic distortion, HDn, is proportional to X^{n-1}. In practice, often only $HD2$ and $HD3$ are considered because the higher-order harmonics drop off very rapidly for small signals. Also

88 RECEIVER FUNDAMENTALS

note that the nth-order harmonic distortion originates from the nth-order coefficient in the power series. This means that for a differential circuit, which has small even-order coefficients, *HD2* usually is small compared with *HD3*. Often, *total harmonic distortion* (THD) is used to describe the nonlinearity with a single number:

$$THD = \sqrt{HD2^2 + HD3^2 + \ldots}. \tag{4.71}$$

The THD can be expressed as a percentage value (distortion products as a fraction of the fundamental amplitude) or in dB using the conversion rule $20 \log THD$. The input dynamic range of an amplifier can be specified, for example, as the maximum value of X for which $THD \leq 1\%$.

Intermodulation Distortions. In CATV/HFC applications, the input signal to the linear channel is not a single sine wave but contains many frequency components (carriers). This means that we also have to be concerned about *intermodulation distortions* in addition to the harmonic distortions. Let's start with the simple two-tone case, that is, we apply a superposition of two equally strong sine waves at frequencies f_1 and f_2 to the input of the channel: $x(t) = X \cdot [\sin(2\pi f_1 t) + \sin(2\pi f_2 t)]$. With Eq. (4.66), we find that the output signal contains two second-order intermodulation products at $f_1 + f_2$ and $|f_1 - f_2|$ and four third-order intermodulation products at $2f_1 + f_2$, $2f_1 - f_2$, $2f_2 + f_1$, and $2f_2 - f_1$. Interestingly, the two second-order products have the same amplitude, and all four third-order products have equal amplitudes (among themselves) as well. In analogy to the harmonic distortion, we define the intermodulation distortion *IMDn* as one of the (equally strong) nth-order distortion products in the output signal normalized to one of the two (equally strong) fundamental tones. It turns out that the second- and third-order intermodulation distortions for the two-tone case are [82]

$$IMD2 \approx |a_2| \cdot X, \tag{4.72}$$

$$IMD3 \approx 3/4 \cdot |a_3| \cdot X^2, \tag{4.73}$$

where X is the amplitude of one of the two (equally strong) tones at the input; as usual, for the approximations to be valid, we assume that this amplitude is small. Compared with the harmonic distortions in Eqs. (4.69) and (4.70), we find the same dependence on the amplitude X and power-series coefficients a_i. However, the *IMD2* is twice (6 dB) as strong as *HD2*, and the *IMD3* is three times (9.5 dB) as strong as *HD3*. In addition to the intermodulation products, of course, we still have the harmonic distortion products corresponding to each tone. Figure 4.20 summarizes all the second- and third-order distortion products for the two-tone case. [→ Problem 4.20]

RF engineers, who design narrow-band systems, typically worry only about the third-order intermodulation products $2f_1 - f_2$ and $2f_2 - f_1$, which fall back into the band of interest (see Fig. 4.20). The other intermodulation and harmonic distortion products are "out of band" and can be ignored. In this situation, the value X for which $IMD3 = 1$ (extrapolated from $IMD3(X)$ where $IMD3 \ll 1$) is commonly used as a measure of the input dynamic range and is known as the *input-referred 3rd-order*

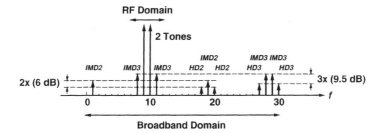

Fig. 4.20 Second- and third-order distortion products caused by two tones with frequencies $f_1 = 9$ and $f_2 = 10$ in conjunction with a nonlinearity.

intercept point (IIP3). Unfortunately, life is more difficult for the broadband engineer and we have to worry about all those distortion products. Actually, so far we looked at only the two-tone case, and things become more complicated as we add more tones.

Now let's add a third tone at the frequency f_3. Again, we get n harmonic distortion products for each one of the three tones at the frequencies nf_1, nf_2, and nf_3. Then, we get second-order intermodulation products at all permutations of $|f_i \pm f_j|$ (6 products in total). Then, we get third-order intermodulation products at all permutations of $2f_i \pm f_j$ (12 products in total). But we are not finished; we also get third-order intermodulation products at all combinations of $|f_1 \pm f_2 \pm f_3|$ (4 products in total). These products are the so-called *triple beats* and they are twice as strong as the two-tone third-order intermodulation products. For small signals, the triple-beat distortion can be written as

$$TBD3 \approx 3/2 \cdot |a_3| \cdot X^2. \tag{4.74}$$

Note that this triple-beat distortion is twice (6.0 dB) as strong as the *IMD*3 distortion and six times (15.6 dB) as strong as the third-order harmonic distortion, *HD*3.

Composite Distortions. In a CATV/HFC system with, say, 80 TV channels, the broadband signal contains 80 carriers, each one playing the role of a tone in the above analysis. All these carriers produce a huge number of harmonic and intermodulation products in the presence of a nonlinearity. To measure the effect of these products on a particular channel, this channel is turned off while all the other channels are operating. Then, the composite distortion products falling into the turned-off channel are measured. Usually, all channels are tested in this way to find the worst-case channel with the most distortion products. In the North American Standard channel plan, the carriers are spaced 6 MHz apart and are offset 1.25 MHz upward from multiples of 6 MHz. As a result of this offset, all even-order products fall 1.25 MHz above or below the carriers, whereas all odd-order products fall on the carriers or 2.5 MHz above or below the carriers. Thus, the composite even- and odd-order products have different effects on the picture quality and can be measured separately with the appropriate bandpass filters [23].

The composite even-order products usually are dominated by second-order intermodulation products. When normalized to the carrier amplitude, they are called *composite second order* (CSO) distortion. The composite odd-order products usually are dominated by triple-beat products. When normalized to the carrier amplitude, they are called *composite triple beat* (CTB) distortion. These composite distortions can be calculated by summing the power of the individual distortions (assuming phase-incoherent carriers). In the case of equal-power carriers, we can write

$$CSO = \sqrt{N_{CSO}} \cdot IMD2 \approx \sqrt{N_{CSO}} \cdot |a_2| \cdot X, \qquad (4.75)$$

$$CTB = \sqrt{N_{CTB}} \cdot TBD3 \approx \sqrt{N_{CTB}} \cdot 3/2 \cdot |a_3| \cdot X^2, \qquad (4.76)$$

where N_{CSO} and N_{CTB} are the number of second-order intermodulation products and triple-beat products, respectively, falling into the turned-off channel. These beat counts can be fairly high[12]; for example, in an 80-channel system, the maximum N_{CSO} is 69 and occurs for channel 2, whereas the maximum N_{CTB} is 2,170 and occurs for channel 40 [132]. CSO and CTB usually are expressed in dBc, that is, dB relative to the carrier amplitude, using the 20 log CSO and 20 log CTB conversion rules, respectively. The National Association of Broadcasters recommends that both CSO and CTB should be less than -53 dBc for analog TV [132]. CATV amplifiers usually are designed for a CSO and CTB of less than -70 dBc.

4.9 JITTER

So far we talked about how noise and ISI affect the signal voltage at the decision circuit and how we have to set the decision threshold to minimize bit errors. However, the decision process not only involves the signal voltage, but also the signal *timing*. In Fig. 4.21, we see how the decision process is controlled by a decision threshold voltage V_{DTH} as well as a sampling instant t_S. The decision threshold voltage slices the eye diagram horizontally, whereas the sampling instant slices the eye diagram vertically. The two slicing lines intersect at the so-called *decision point*. ISI and noise not only occur in the signal voltage domain, but also in the time domain. ISI in the time domain is known as *data-dependent jitter* (DDJ), and noise in the time domain is known as *random jitter* (RJ). We can characterize ISI and noise with a histogram of the voltage values at the sampling instant (Fig. 4.21, right), and similarly, we can characterize (data-dependent and random) jitter with a histogram of the zero-crossings relative to the decision threshold voltage (Fig. 4.21, bottom).

Data-Dependent and Deterministic Jitter. Data-dependent jitter is produced when the signal edge moves slightly in time, depending on the values of the surrounding bits. For example, the sequence "...110" may have a falling edge that is a little bit retarded relative to the sequence "...010." As a result, the eye diagram contains double edges. Data-dependent jitter is caused for example by (i) an insufficient bandwidth, (ii) an insufficient phase linearity, (iii) baseline wander due to an

---

[12] A rough estimate is $N_{CSO}(\max) \approx N/2$ and $N_{CTB}(\max) \approx 3/8 \cdot N^2$, where N is the number of channels.

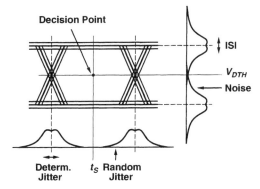

Fig. 4.21 Eye diagram at the input of the decision circuit with ISI, noise, deterministic, and random jitter.

insufficient low-frequency cutoff, (iv) reflections on cables due to an impedance mismatch, and (v) a TIA or MA operated beyond its overload limit. The histogram of (pure) data-dependent jitter is bounded and usually is discrete (non Gaussian), similar to the histogram of ISI in the signal voltage domain. Because of its boundedness, data-dependent jitter can be specified by its peak-to-peak value t_{DDJ}^{pp}.

Data-dependent jitter belongs to a larger class of jitter known as *deterministic jitter* (DJ) [107]. This class also includes *periodic jitter* (PJ) and *bounded uncorrelated jitter*, which can arise as a result of crosstalk from other signal lines or disturbances from the power and ground lines. Furthermore, it includes *duty-cycle distortion jitter*, which occurs if the rising and falling edges do not cross each other (on average) at the decision threshold voltage. Note that we similarly could extend the concept of ISI by defining a class of "deterministic signal disturbances" that includes ISI, crosstalk, power/ground bounce, and so forth. Like data-dependent jitter, deterministic jitter is bounded and can be specified by its peak-to-peak value t_{DJ}^{pp}.

Random Jitter. Random jitter, as the name implies, is random, that is, it is not related to the data pattern or any other deterministic cause. Random jitter is produced, for example, by noise on signal edges with a finite slew rate. The finite slew rate translates the signal voltage uncertainty into a timing (zero-crossing) uncertainty. Random jitter also is caused by carrier mobility variations due to instantaneous temperature fluctuations. The histogram of (pure) random jitter can be well approximated by a Gaussian distribution similar to the histogram of noise in the signal voltage domain.[13] Because of that, random jitter usually is specified by its rms value t_{RJ}^{rms}, which corresponds to the standard deviation of its Gaussian distribution.

---

[13] Note, however, that the distribution of random jitter originating from the noise voltage on the signal edges is limited by the rise and fall times, that is, the tails of the Gaussian are clipped.

Total Jitter. A typical jitter histogram, as shown in Fig. 4.21, contains both types of jitter. The inner part of the histogram is the result of deterministic jitter, the Gaussian tails are the result of random jitter. Mathematically, the total histogram is the convolution of the histogram due to deterministic jitter with the histogram due to random jitter. In this case of composite jitter, it is less obvious how to specify the amount of jitter. Some commonly used methods to quantify *total jitter* (TJ) are as follows:

- Specify the peak-to-peak value as it appears in the histogram of the total jitter, t_{TJ}^{pp}, *and* also specify how many samples were taken. The latter is important because the peak-to-peak value increases with the number of samples taken (due to the random jitter component).

- Perform a so-called *BERT scan* using a *bit-error rate test set* (BERT). In this measurement, the BERT is used to observe the BER while scanning the sampling instant t_S horizontally across the eye. The BER is low when sampling at the center of the eye and goes up when approaching the eye crossings to the left and right; hence, this curve is known as the *bathtub curve* (cf. Appendix A). The total jitter amount, t_{TJ}^{pp}, is defined as the separation of the two points, to the left and the right of the eye crossing, where the bathtub curve assumes a specified BER such as 10^{-12}.

- Decompose the total jitter into a random and a deterministic part, then specify the random jitter as an rms value, t_{RJ}^{rms}, and the deterministic jitter as a peak-to-peak value, t_{DJ}^{pp}. Unfortunately, decomposing the two types of jitter is tricky and often is not very accurate. The following two methods can be used. First, test the system with a clock-like pattern to determine the random jitter. Then, test the system with a repetitive data pattern to determine the data-dependent jitter. The averaging feature of the oscilloscope can be used to suppress random jitter in the latter measurement. A second method is to obtain the total jitter histogram and to fit two Gaussians to its tails. The standard deviations of the Gaussians provide an estimate for the random jitter. The separation of the means of the Gaussians provide an estimate for the deterministic jitter.

The peak-to-peak value of the total jitter for a given BER can be related to its deterministic and random jitter components as follows:

$$t_{TJ}^{pp} < t_{DJ}^{pp} + 2Q \cdot t_{RJ}^{rms}. \tag{4.77}$$

Note that because the worst-case deterministic and random jitter normally do not coincide, the above equation only provides an upper bound for the total jitter. For example, given a deterministic jitter of 0.3 UI peak-to-peak and a random jitter of 0.02 UI rms, we can conclude that the total jitter is less than 0.58 UI peak-to-peak, if we refer it to $BER = 10^{-12}$ (or if we collect about 10^{12} samples for the histogram).

[→ Problem 4.21]

Jitter Bandwidth. The jitter we discussed so far is so-called *wideband jitter*. It also is possible, and required by some standards, to measure jitter in a specified *jitter bandwidth*. For example, SONET OC-48 defines that the jitter must be measured in the 12 kHz to 20 MHz bandwidth. How do we do that? Should we pass the data signal through a filter with the specified bandwidth and then measure the jitter? No, we are not supposed to filter the signal itself, but the *jitter* of the signal! Figure 4.22 illustrates the difference between the data signal (v_O) and its jitter (t_J) with an example. The jitter t_J of a given data edge is defined as its deviation from the ideal location. In the upper graph, these deviations are indicated with bold lines; in the lower graph, the same deviations are represented by dots together with an interpolation (dashed curve). It is the frequency content of that lower signal that we are interested in.

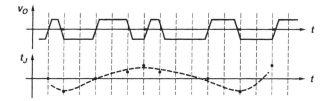

Fig. 4.22 Data signal with jitter (upper curve). Dependence of the edge jitter on time (lower curve).

Conceptually, jitter can be filtered with a *phase-locked loop* (PLL) with well-defined jitter transfer characteristics and very low jitter generation, called *golden PLL*. For example, to remove high-frequency jitter we can run the data signal into a golden PLL with a jitter transfer characteristics of 0 dB and a bandwidth equal to the desired jitter bandwidth. Now, the recovered clock from the PLL contains only the low-frequency jitter and we can determine its value by displaying the jitter histogram of the clock signal on a scope. To determine jitter in a passband, we can use a second golden PLL with a bandwidth equal to the lower corner of the desired jitter bandwidth. This PLL also is fed by the data signal, but its output is used to trigger the scope. Because the scope input and the trigger input both get the same amount of low-frequency jitter, it is suppressed (a common-mode signal in the time domain) and only the desired high-frequency jitter appears in the jitter histogram.

In practice, a so-called *jitter analyzer* can be used to measure the jitter in the desired bandwidth. Alternatively, a *time interval analyzer* can be used to measure the time intervals between zero crossings of the data signal. From the statistics of these time intervals, it is possible to calculate the power-density spectrum of the jitter [107]. Finally, a *spectrum analyzer* can be used to measure the phase noise of a clock-like data pattern. From the spectral noise data, it is possible to calculate the time-domain rms jitter in the desired bandwidth [107].

Data vs. Clock Jitter. So far we have been talking about jitter in data signals, but jitter also affects clock signals such as the sampling clock at the decision circuit. This so-called *sampling jitter* can be visualized as a random deviation of the sampling instant t_S from its ideal location. If the sampling jitter is uncorrelated with the data jitter, it causes an increase in BER, which can be described by a power penalty. However, if the sampling jitter tracks the data jitter, the receiver performance is improved. The latter occurs, for example, if the sampling clock is recovered from the data signal with a PLL and the data jitter is within the jitter transfer bandwidth of the PLL.

How do we define and specify clock jitter? First, we have to realize that we defined the data jitter t_J with respect to a reference clock signal. For the eye diagram in Fig. 4.21, we implicitly assumed a reference clock as trigger source, and in Fig. 4.22 the reference clock appears in the form of the ideal edge locations shown as vertical dashed lines. In practice, the reference clock often is provided by the pulse pattern generator used to generate the test data signal (cf. Appendix A). Now, clock jitter can be defined in the same way, that is, relative to a reference clock signal. This type of clock jitter is known as *absolute jitter* and we use $t_J(n)$ to indicate the time deviation of the nth rising (or falling) clock edge relative to the reference clock. Just as discussed for the data jitter, we can represent $t_J(n)$ as a histogram, separate it into random and deterministic components, specify it by rms or peak-to-peak values, and filter it in the frequency domain.

However, in practice a reference clock may not be available, and therefore it is more convenient, in the case of clock signals, to use the *period jitter* and *cycle-to-cycle jitter* measures. Period jitter is the deviation of the clock period relative to the period of the reference clock and can be expressed in terms of absolute jitter as $t_J(n+1) - t_J(n)$. Note that this jitter can be measured by triggering an oscilloscope on one clock edge and by taking the histogram of the subsequent edge of the same type (rising or falling). Cycle-to-cycle jitter is the deviation of the clock period relative to the previous clock period and can be expressed in terms of absolute jitter as $[t_J(n+1) - t_J(n)] - [t_J(n) - t_J(n-1)]$. For example, if the absolute jitter on the clock edges is uncorrelated (white jitter), then the rms period jitter is $\sqrt{2}\times$ the absolute jitter and the rms cycle-to-cycle jitter is $\sqrt{6}\times$ the absolute jitter.

Jitter and BER. When we discussed the BER of a receiver in Section 4.2, we considered only *noise* in the signal voltage at the input of the decision circuit as a source of bit errors. After the above discussion, it should be clear that the BER also is affected by *jitter*. Just like noise may cause the sampled voltage to be on the "wrong side" of the decision threshold voltage, random jitter may cause the edge to move past the sampling instant such that one bit is sampled twice while the adjacent bit is ignored. In general, we have to consider noise, ISI, and the different types of jitter jointly to calculate the BER accurately. Fortunately, in practice, the BER usually is *mostly* determined by the signal voltage noise and our discussion in Section 4.2 remains valid. However, it is important to know how low the jitter has to be, such that it does not significantly impact the BER of the receiver.

To answer the above question, we now calculate the BER assuming that there is deterministic and random jitter but *no* voltage noise at all. The CDR in the receiver is characterized by a jitter tolerance, that is, the maximum amount of peak-to-peak jitter, t_{JTOL}^{pp}, it can tolerate without making errors (or for which the BER is very low).

An ideal CDR would be able to tolerate at least one bit period of jitter ($t_{JTOL}^{pp} = 1/B$), but in practice the setup and hold time of the decision circuit, the sampling jitter, and so forth limit the jitter tolerance (at high jitter frequencies) to a lower value. After subtracting from this jitter tolerance, t_{JTOL}^{pp}, the deterministic jitter, t_{DJ}^{pp}, we are left with the margin for the random jitter. In analogy to the discussion in Section 4.2, the BER due to jitter becomes

$$BER = \int_Q^\infty \text{Gauss}(x)\, dx \quad \text{with} \quad Q > \frac{t_{JTOL}^{pp} - t_{DJ}^{pp}}{2 \cdot t_{RJ}^{rms}}. \tag{4.78}$$

Note that because the worst-case deterministic and random jitter normally do not coincide, the above equation only provides an upper bound for the BER (or a lower bound for Q). Furthermore, the accuracy of the relationship depends on how well the random jitter follows a Gaussian distribution. With Eq. (4.78) we find, for example, that a jitter tolerance of 0.7 UI combined with a deterministic jitter of 0.3 UI peak-to peak and a random jitter of 0.02 UI rms results in a BER of less than 10^{-23} ($Q > 10$). Thus, for this jitter scenario, the BER is almost exclusively determined by the signal voltage noise, in agreement with our previous assumption.

Finally, note that data jitter observed in the receiver also may originate in the transmitter or the regenerators along the way. Limiting the jitter generation in the transmitter is important, and we return to this issue in Section 8.1.7.

4.10 DECISION THRESHOLD CONTROL

The optimum decision threshold (slice level) is at the point where the probability distributions of the zero and one bits intersect (cf. Fig. 4.3). In the case that the noise distributions have equal widths, the optimum slice level is centered halfway between the zero and one levels. This slice level is attained automatically in an AC-coupled receiver, given a DC-balanced signal and no circuit offsets. If there is more noise on the ones than the zeros, for example because of optical amplifiers or an APD, the optimum slice level is below the center (see Fig. 4.23) and a simple AC-coupled receiver produces more errors than necessary. [→ Problem 4.22]

One way to address the latter issue is to make the slice level variable and manually to adjust it until optimum performance is reached. An intentional offset voltage in the decision circuit or the preceding MA can be used for this purpose. This method is called *slice-level adjust* and we discuss its implementation in Section 6.3.3.

Slice-level adjust is fine, but it would be more convenient to have a system that *automatically* adjusts the slice level for the lowest possible bit-error rate. Moreover, an automatic mechanism has the advantage that it can track variations in the signal swing and noise statistics over time, making the system more robust. This automatic method is called *slice-level steering*. It can be extended to find not only the optimum *slice level* but also the optimum *sampling instant* and is then known as *decision-point steering* [67, 170].

The difficulty in finding the optimum slice level automatically is that we normally can't determine the BER, which we seek to minimize, from the received bit sequence

96 RECEIVER FUNDAMENTALS

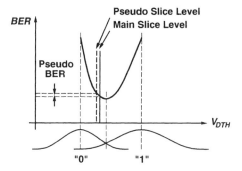

Fig. 4.23 Optimum decision threshold for unequal noise distributions.

only.[14] However, there is a trick: we can use an eye monitor to measure the so-called *pseudo bit-error rate* and can minimize the latter. This scheme can be implemented by simultaneously slicing the received signal at two slightly different levels: a main slice level and a pseudo slice level (see Figs. 4.23 and 4.24). The output from the main slicer (the decision circuit) is the regular data output; the output from the pseudo slicer is identical to that of the main slicer, except for the pseudo errors. Any discrepancy between the two bit streams is detected with an XOR gate and is counted as a pseudo error. Figure 4.23 illustrates the resulting pseudo BER. It can be seen that the pseudo BER and the actual BER have almost identical minima, thus minimizing the pseudo BER also minimizes the actual BER. Now, the controller in Fig. 4.24 performs the following steps: (i) put the pseudo slice level a small amount above the main level and measure the pseudo BER; (ii) put the pseudo slice level below the main level by the same small amount and measure the pseudo BER again; and (iii) adjust the main slice level into whichever direction that gave the smaller pseudo BER. Iterate these steps until both pseudo BERs are the same and you have found a close approximation to the optimum slice level. (The smaller the difference between main- and pseudo-slice level, the better the approximation.)

4.11 FORWARD ERROR CORRECTION

We found in Section 4.2 that we need an SNR of about 17 dB to receive an NRZ bit stream at a BER of 10^{-12}. Can we do better than that? Yes we can, if we use error-correcting codes. A simple (but impractical) example of such a code is to send each bit three times. At the receiver, we can analyze the (corrupted) 3-bit codewords and correct single-bit errors. For example, if we receive the code word '101,' we know

---

[14] An important exception occurs for systems with forward error correction (cf. Section 4.11). In such cases, we not only have a good estimate of the BER at the decision circuit, but we also know if errors occurred mostly on zeros or ones. This information can be used directly to control the slice level.

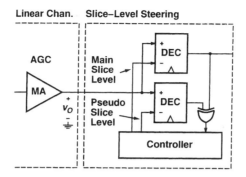

Fig. 4.24 The linear channel of Fig. 4.1 followed by a circuit for slice-level steering.

that an error occurred and that the correct code word is most likely '111.' This method of adding redundancy at the transmitter and correcting errors at the receiver (without repeat requests) is known as *forward error correction* (FEC). In practice, sophisticated codes such as the *Reed-Solomon* (RS) code and the *Bose-Chaudhuri-Hocquenghem* (BCH) code are used.

By how much can we lower the SNR requirements if we use FEC? Shannon's *channel capacity theorem* (a.k.a. *information capacity theorem*) asserts that with sufficient coding, error-free transmission over a channel with additive white Gaussian noise is possible if

$$B \leq BW \cdot (1 + \log_2 SNR), \tag{4.79}$$

where B is the information bit rate and BW is the channel bandwidth [42]. To get a rough estimate for the potential of FEC, let's assume that we use Nyquist signaling, which requires a channel bandwidth of $BW = B/(2r)$, where B/r, is the channel bit rate, that is, the bit rate after coding with the code rate r. Solving Eq. (4.79) for SNR, we find $SNR = 2^{2r} - 1$, thus with $r = 0.8$, the necessary SNR is just 3.1 dB. That's about 14 dB less SNR than what we need without coding! Therefore, FEC is a powerful technique, which is commonly used in ultra-long-haul optical transmission systems such as undersea lightwave systems. It permits more noise accumulation and eye closure for a given transmit power while maintaining a low BER.

FEC Based on Reed-Solomon Code. A typical FEC system, often used in SONET systems [53] and based on the Reed-Solomon code RS(255,239), operates as follows: the data stream into the encoder, the so-called payload, is cut up (framed) into blocks of 238 data bytes. A framing byte is appended to each data block, making it a 239-byte block. This block is then encoded with the RS(255,239) code, which adds 16 bytes of redundancy, producing a 255-byte block. Before transmitting the encoded block, it is run through a so-called 16× *interleaver*. This means that rather than transmitting a complete 255-byte block at a time, one byte is transmitted from the first block, then one byte from the second block, and so forth, until block 16 is reached; then the process continues with the next byte from the first block, and so forth. Interleaving spreads burst errors, which may occur during transmission, into multiple received blocks, thus increasing the error-correcting capacity for bursts.

The encoder described above increases the transmitted bit rate by 7% ($255/238 = 15/14 = 1.071$), which is equivalent to saying that the code rate r is $14/15 = 0.933$. This means that slightly faster hardware is needed in the transceiver front-end. However, the benefit of the code is that up to 8 byte errors can be corrected per block,[15] thus significantly lowering the BER. Furthermore, thanks to the interleaving, burst errors up to 1024 bits in length ($16 \times 8 \times 8$) can be corrected. The precise improvement in BER depends on the incoming BER and the distribution of the bit errors in the received signal. In a typical transmission system with FEC based on RS(255,239), an incoming BER of 10^{-4} (i.e., BER at the output of the decision circuit) can be boosted to 10^{-12} after error correction. This is an improvement of eight orders of magnitude! [$\rightarrow$ Problem 4.23]

Coding Gain. The performance of an FEC code can be discussed graphically with the so-called *waterfall curves* shown in Fig. 4.25. The x-axis shows the SNR at the input of the decision circuit, which is equal to Q^2 for the case of additive noise (cf. Eq. (4.12)), and the y-axis shows the BER. One curve, labeled "Uncorrected," shows the BER at the input of the decoder and the other curve, labeled "Corrected," shows the BER at the output of the decoder. The first curve ("Uncorrected") corresponds directly to the integral in Eq. (4.8), which also is tabulated in Table 4.1. At the output of the decoder, the BER is lower, as can be seen from the second curve ("Corrected"). The more this curve is pushed to the lower left relative to the uncorrected curve, the better the FEC code works. For example, let's assume that the incoming BER is $BER_{in} = 10^{-4}$, whereas the outgoing BER is $BER_{out} = 10^{-12}$. According to Table 4.1, the incoming BER corresponds to $Q_{in} = 3.719$. Now, we could say that to get $BER_{out} = 10^{-12}$ *without* coding we would need $Q_{out} = 7.035$. Thus, the addition of FEC relaxed the SNR requirement by Q_{out}^2/Q_{in}^2, which in our example is 5.5 dB. This quantity is known as the *coding gain*.[16] Note that the coding gain depends on the desired BER_{out}, as can be seen from Fig. 4.25.

It could be argued that the coding gain as calculated above is not quite fair. This is so because without coding, the bit rate would be somewhat lower, which would permit us to reduce the bandwidth of the receiver and thus the rms noise. To be more specific, we could reduce the receiver bandwidth by the code rate r and, assuming white noise, the rms noise by $\sqrt{r}$; thus, Q_{in} would improve to $Q_{in}/\sqrt{r}$. It is common to distinguish between the *gross coding gain* as defined above (Q_{out}^2/Q_{in}^2) and the *net*

---

[15] In general, an RS(n,k) code with a block length of n symbols and a message length of k symbols can correct $(n-k)/2$ symbol errors.

[16] It is most common to define coding gain as the reduction in SNR requirement at the input of the decision circuit, as we did it here. However, sometimes coding gain is defined as the improvement in *optical sensitivity* of the receiver, which is Q_{out}/Q_{in} (cf. Eq. (4.20)), corresponding to 2.8 dB in our example.

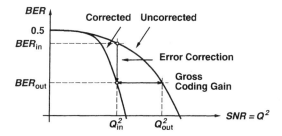

Fig. 4.25 Waterfall curves on a log-log scale describing the performance of an FEC code.

electrical coding gain (NECG), which is defined as

$$NECG = r \cdot \frac{Q_{\text{out}}^2}{Q_{\text{in}}^2} \qquad (4.80)$$

and takes the bit rate increase into account. In the above example, the gross coding gain is 5.5 dB, whereas the NECG is 5.2 dB assuming $r = 14/15$. In Section 4.2, we have introduced E_b/N_0, an SNR-type measure that includes a normalization with respect to the code rate: $E_b/N_0 = Q^2/(2r)$ (cf. Eq. (4.15)). Thus, we can express NECG conveniently as the ratio of two 'ebno's

$$NECG = \frac{(E_b/N_0)_{\text{out}}}{(E_b/N_0)_{\text{in}}}, \qquad (4.81)$$

where $(E_b/N_0)_{\text{in}} = Q_{\text{in}}^2/(2r)$ is needed to achieve BER_{out} *with* FEC and $(E_b/N_0)_{\text{out}} = Q_{\text{out}}^2/2$ is needed to achieve BER_{out} *without* FEC. Thus, if we plot the waterfall curves as a function of E_b/N_0 rather than Q or SNR, the horizontal displacement directly represents the NECG.

Soft-Decision Decoding. An FEC system that corrects errors based on the binary values from the decision circuit, as we discussed it so far, is known as a *hard-decision decoder*. Although many transmission errors can be corrected in this way, more errors can be corrected if the *analog* values of the received samples are known. This is so because the confidence with which the bit was received can be taken into account when correcting errors. The latter system is known as a *soft-decision decoder* [33, 81].

Typically, a soft-decision decoder achieves about 2 dB more coding gain than a hard-decision decoder. Soft-decision decoding is particularly attractive when used with a family of codes known as *turbo codes* or the so-called *low-density parity-check codes* (LDPC) [12, 42].

Front-End Implementation Issues. In general, the use of FEC in a transmission system has little impact on the front-end circuits, which are the focus of this book. However, the following considerations must be kept in mind. First, the bit rate increases from a few percents up to about 25% for high-performance codes. Thus,

100 RECEIVER FUNDAMENTALS

more bandwidth and faster hardware is required. Second, the CDR must be able to recover the clock from a very noisy signal corresponding to a BER of the order of 10^{-4}. Third, if soft-decision decoding is used, a multilevel slicer is required.

Figure 4.26 shows a receiver in which the linear channel of Fig. 4.1 is followed by a slicer with four different output states (similar to a 2-bit flash A/D converter) to permit some degree of soft-decision decoding. The four states correspond to "hard zero," "soft zero," "soft one," and "hard one." They can be encoded into two bits such that one bit represents the likely bit value and the other bit the confidence level. Of course, a soft-decision decoder with more than two bits can be built, but most of the benefit is realized with the simple 2-bit implementation (about 1.5 dB improvement in coding gain). The slicer outputs are fed into the decoder logic, which detects and corrects the errors. Note that the MA in Fig. 4.26 must be linear to preserve the analog sample values, and thus usually is realized as an AGC amplifier.

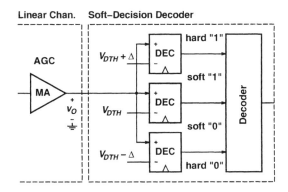

Fig. 4.26 The linear channel of Fig. 4.1 followed by a 2-bit soft-decision decoder.

4.12 SUMMARY

The noise of the receiver (and other sources) disturbs the data signal to be received; occasionally, the noise's instantaneous value becomes so large that a bit error occurs. The two main contributions to the receiver noise are (i) the amplifier noise (mostly from the transimpedance amplifier [TIA]) and (ii) the detector noise. For the case in which the receiver noise can be modeled as additive Gaussian noise, there exists a simple mathematical expression that relates the bit-error rate (BER) to the peak-to-peak signal swing and the rms value of the noise.

The electrical sensitivity is the minimum input signal (peak-to-peak) needed to achieve a specified BER. The optical receiver sensitivity is the minimum optical input power (averaged over time) needed to achieve a specified BER. The electrical sensitivity depends on the total input-referred receiver noise. The optical sensitivity depends on the total input-referred receiver noise, detector noise, and detector responsivity. Whereas the noise of a p-i-n photodetector has little impact on the optical sensitivity, the noise of an avalanche photodetector (APD) or optically preamplified p-i-n detector is significant. In optically amplified lightwave systems for ultra-long-haul transmission, the concept of sensitivity is replaced by the concept of optical signal-to-noise ratio (OSNR). The total input-referred noise of a receiver is calculated by integrating the output-referred noise spectrum at the decision circuit and then referring it back to the input. Alternatively, Personick integrals can be used to compute the total input-referred noise from the input-referred noise spectrum.

A power penalty describes the reduction in optical sensitivity as a result of a system impairment such as a decision-threshold offset at the receiver, dispersion in the fiber, or a finite extinction ratio at the transmitter. The optimum receiver bandwidth for non-return-to-zero (NRZ) modulation is about 2/3 of the bit rate. The optimum frequency response of the receiver depends on the shape of the received pulses. An adaptive equalizer can be used to cancel intersymbol interference (ISI) and automatically respond to varying received pulse shapes.

For digital modulation schemes, such as NRZ or return-to-zero (RZ), linearity of the receiver and transmitter are of secondary importance; however, for analog modulation, as used in CATV/HFC applications, linearity is critical. ISI and noise appear not only in the signal domain but also in the time domain, where they are known as data-dependent jitter and random jitter, respectively. In systems with APDs or optical amplifiers, the optimum decision threshold is somewhat below the halfway point between the zero and one levels. In this situation, either a manual slice-level adjustment or an automatic decision-point steering scheme can be used to optimize the performance. Some transmission systems use forward error correction (FEC) to boost the BER performance. One advanced type of FEC decoder (the soft-decision decoder) requires the receiver to have a multilevel slicer instead of the regular binary decision circuit.

4.13 PROBLEMS

4.1 Filtered Detector Noise. Assume that the impulse response of the linear channel, $h(t)$, is always zero except during the time interval $[0\ldots\xi]$. Under what conditions can Eq. (4.5) be simplified to $V_{n.PD}^2(f,t) = |H(f)|^2 \cdot I_{n.PD}^2(f,t)$?

4.2 BER for the Case of Unequal Noise. Derive the BER for the case of unequal Gaussian noise distributions on the zeros and ones as given by Eq. (4.10). Approximate the optimum threshold voltage assuming that it is strongly determined by the exponential factors of the Gaussians.

4.3 SNR Requirement for the Case of Unequal Noise. Let ξ be the ratio between the rms noise on the zeros and the ones of an ideal NRZ signal. Now, derive a more general relationship between SNR and Q than the one given by Eqs. (4.12) and (4.13). Assume Gaussian noise distributions and equal probabilities for zeros and ones.

4.4 SNR Requirement for RZ Modulation. (a) Derive the (continuous-time) SNR requirement for an ideal RZ signal with duty cycle ξ ($\xi = 1$ corresponds to NRZ) such that it can be detected at a specified BER. Assume additive Gaussian noise and equal probabilities for zeros and ones. (b) What is the SNR value for $\xi = 0.5$ and $BER = 10^{-12}$? (c) What is the requirement for the *sampled* SNR given an RZ signal?

4.5 SNR Requirement for Finite-Slope NRZ Modulation. (a) Derive the (continuous-time) SNR requirement for an NRZ signal with linear slopes where each slope occupies the fraction $\xi < 1$ of a bit period such that it can be detected at a specified BER. Assume additive Gaussian noise and equal probabilities for zeros and ones. (b) What is the SNR value for $\xi = 0.3$ and $BER = 10^{-12}$? (c) What is the requirement for the *sampled* SNR given a finite-slope NRZ signal?

4.6 SNR Requirement for PAM-4. (a) Derive the SNR requirement for an ideal PAM-4 signal such that it can be detected at a specified BER. Assume additive Gaussian noise and that all symbols are equally likely. (b) What is the SNR value for $BER = 10^{-12}$?

4.7 E_b/N_0 and SNR. (a) Derive the relationship between the E_b/N_0 at the input of the linear channel and the SNR at the output of the linear channel. Assume that the linear channel only limits the noise but does not attenuate the signal. (b) Under which circumstances become E_b/N_0 and SNR identical?

4.8 Sensitivity for Finite Extinction Ratio. Equation (4.21) specifies the receiver sensitivity assuming an optical signal with high extinction ratio. Generalize this equation assuming that the power for the zeros is not 0 but $P_0 = P_1/ER$, where P_1 is the power for the ones and ER is the extinction ratio.

4.9 Sensitivity of p-i-n Receiver. Engineers use the following rule to estimate the sensitivity of a p-i-n receiver:

$$\overline{P}_{\text{sens.}PIN} \, [\text{dBm}] \approx -21.53 \, \text{dBm} + 10 \log \left(i_{n,\text{amp}}^{rms} \, [\mu A] \right) \\ - 10 \log \left(\mathcal{R} \, [A/W] \right) \quad (4.82)$$

Explain the origin of this equation. What is the meaning of -21.53 dBm?

4.10 Sensitivity of p-i-n Receiver with Finite OSNR. Calculate the sensitivity of a p-i-n receiver assuming that the received optical signal contains noise from in-line amplifiers. The latter noise is specified by the OSNR. Neglect shot and spontaneous-spontaneous beat noise. (a) What is the sensitivity in the limit $OSNR \to \infty$? (b) What minimum $OSNR$ is required given a high received power?

4.11 Sensitivity of Optically Preamplified Receiver. The sensitivity in Eq. (4.29) takes only amplifier and signal-spontaneous noise into account. Derive a more precise expression that includes the effect of spontaneous-spontaneous noise. (a) Write $\overline{P}_{sens,OA}$ as a function of S_{ASE}. (b) Write $\overline{P}_{sens,OA}$ as a function of the optical noise figure $\tilde{F}$, which is defined in Eq. (3.23).

4.12 OSNR Requirement for NRZ Receiver. The approximation in Eq. (4.32) takes only signal-spontaneous noise into account. Derive a more precise expression that includes the effect of spontaneous-spontaneous noise. Write Q as a function of $OSNR$.

4.13 Sensitivity vs. OSNR. Show that the sensitivity analysis leading up to Eq. (4.31) is equivalent to the OSNR analysis based on Eqs. (4.32) and (4.33).

4.14 Quantum Limit. Consider the following "alternative derivation" of the quantum limit in Eq. (4.36). We start with the sensitivity of a p-i-n receiver in Eq. (4.27), set the amplifier noise to zero ($i_{n.amp}^{rms} = 0$), insert the responsivity of an ideal p-i-n detector (Eq. (3.2) with $\eta = 1$), and use the optimum noise bandwidth of a receiver matched to NRZ pulses, $BW_n = B/2$, (cf. Section 4.6), which leads us to

$$\overline{P}_{sens,quant} = \frac{Q^2}{2} \cdot \frac{hc}{\lambda} \cdot B. \qquad (4.83)$$

Compared with Eq. (4.36), we have a Q^2 term instead of the $-\ln(2 \cdot BER)$ term. Given $BER = 10^{-12}$, we find $Q^2 = 49$, whereas $-\ln(2 \cdot BER) = 27$. Explain the discrepancy!

4.15 Personick Integral. Assume that the input-referred noise spectrum contains a term that increases linearly with frequency: $I_n^2(f) = \alpha_0 + \alpha_1 f + \alpha_2 f^2$. To describe the effect of this term on the total input-referred noise, a Personick integral I_f is introduced such that $\overline{i_n^2} = \alpha_0 \cdot I_2 B + \alpha_1 \cdot I_f B^2 + \alpha_2 \cdot I_3 B^3$. How is I_f related to the frequency response $|H(f)|$?

4.16 Power Penalty due to Finite Extinction Ratio. Calculate the power penalty incurred when transmitting a nonzero power level for the zero bits. Assume that the power for the zeros is $P_0 = P_1/ER$, where P_1 is the power for the ones and ER is the extinction ratio.

4.17 Power Penalty due to ISI. Calculate the power penalty incurred when passing an ideal NRZ waveform through (a) a first-order low-pass filter and (b) a second-order Butterworth low-pass filter. Assume that the filtered signal is sampled at the instant of maximum eye opening. (c) How large are the power penalties for $BW_{3dB}/B = 1/3, 2/3$, and $4/3$ for each filter?

4.18 Rectangular-Filter Response. Calculate the transfer function of the rectangular filter (magnitude and phase). This filter has an impulse response $h(t)$, which is one in the interval from 0 to $1/B$ and zero everywhere else.

4.19 Feed-Forward Equalizer. Calculate the transfer function of the 2-tap precursor equalizer shown in Fig. 4.18. How large is the magnitude of this transfer function at DC and $f = B/2$?

4.20 Second-Order Distortions. Given the nonlinear transfer function in Eq. (4.66) with $a_i = 0$ for $i \geq 3$ and the two-tone input signal $x(t) = X \cdot [\sin(\omega_1 t) + \sin(\omega_2 t)]$, calculate the output signal. How large are the distortion products and the output offset?

4.21 Jitter Histogram. An NRZ signal has transitions with ideal timing except for data patterns ending in "...010," in which case the falling edge is advanced by ΔT, and for data patterns ending in "...101," in which case the rising edge is advanced by ΔT. (a) How does the jitter histogram look like and how large is the peak-to-peak jitter, t_{DJ}^{pp}? (b) Now let's add Gaussian random jitter with the rms value t_{RJ}^{rms} to the NRZ signal. How does the jitter histogram look now?

4.22 Optimum Slice Level. Assume that the sampled voltage values v_O for the zeros and ones follow the arbitrary probability distributions Zero(v_O) and One(v_O), respectively, and that the zeros and ones occur with equal probability. What slice level V_{DTH} leads to the lowest BER?

4.23 Forward Error Correction. At the input of the FEC decoder we have $BER = 10^{-4}$; the bit errors are random and independent (no bursts). For simplicity, assume that the probability for two or more errors in the same byte is very small. What is the BER at the output of the RS(255,239) decoder?

5
Transimpedance Amplifiers

In this chapter, we focus on the *transimpedance amplifier* (TIA). We start by introducing the main specifications of this amplifier. Then, we discuss TIA circuit concepts in a general and, as much as possible, technology-independent manner. Subsequently, we illustrate these concepts with practical implementations in a broad range of technologies. We conclude with a brief overview of product examples and current research topics.

5.1 TIA SPECIFICATIONS

In the following, we discuss the main specifications of the TIA: the transimpedance, the input overload current, the maximum input current for linear operation, the input-referred noise current, the bandwidth, and the group-delay variation. For a discussion of S parameters, see Appendix C.

5.1.1 Transimpedance

Figure 5.1 shows the input current i_I and the output voltage v_O of a TIA with single-ended and differential outputs. Figure 5.2 illustrates the relationship between v_O and i_I with an example.

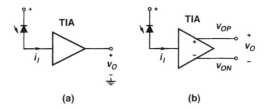

Fig. 5.1 Input and output signals of (a) single-ended and (b) differential TIA.

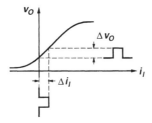

Fig. 5.2 DC transfer function of a (single-ended) TIA.

Definition. The *transimpedance*, Z_T, is defined as the output voltage change, Δv_O, per input current change, Δi_I:[1]

$$Z_T = \frac{\Delta v_O}{\Delta i_I}. \tag{5.1}$$

Thus, the higher the transimpedance, the more output signal is produced for a given input signal. The transimpedance is specified either in units of Ω or dBΩ. In the latter case, the value in dBΩ is calculated as $20\log(Z_T/\Omega)$; for example, 1 kΩ corresponds to 60 dBΩ.

When operated with a small AC signal, the TIA's transfer function is fully described by the frequency-dependent transimpedance $|Z_T(f)|$ and the frequency-dependent phase shift $\Phi(f)$ between the input and the output signal. A generalization of the transimpedance defined in Eq. (5.1) to the complex quantity $Z_T = |Z_T| \cdot \exp(j\Phi)$ captures both aspects. Note that this complex quantity can also be expressed as $Z_T = V_o/I_i$, where V_o is the output voltage phasor and I_i is the input current phasor. In the case where Z_T is real, usually at low frequencies, the transimpedance also is known as the *transresistance*, R_T.

The transimpedance usually is specified for an input signal that is small enough such that the transfer function can be regarded as linear. For larger input signals, the transfer function becomes nonlinear (compressive), as illustrated in Fig. 5.2, which

---

[1] The term *transimpedance* derives from the older term *transfer impedance*, which indicates that the voltage and current defining the impedance are measured at two different ports. In contrast, the term *driving point impedance* indicates that the defining voltage and current are measured at the same port.

causes the transimpedance to drop. Finally, for very large input signals, pulse-width distortion and jitter may set in (cf. Section 5.1.2).

Some TIAs have differential outputs, as shown in Fig. 5.1(b), and therefore the output voltage can be measured single endedly, $v_O = v_{OP}$ or $v_O = v_{ON}$, or differentially, $v_O = v_{OP} - v_{ON}$. It is important to specify which way the transimpedance was measured because the differential transimpedance is twice as large as the single-ended one.

Most high-speed TIAs have 50-Ω outputs. These outputs must be properly terminated with 50-Ω resistors when measuring the transimpedance or else the value will come out too high.

Typical Values. In general, it is desirable to make the transimpedance of a TIA as high as possible because a high transimpedance relaxes the gain and noise requirements for the subsequent main amplifier (cf. Section 6.2). However, we see in Section 5.2.2 that there is an upper limit to the transimpedance that can be achieved with the basic shunt-feedback topology. This limit depends on many factors, including the bit rate and the technology used. For higher bit rates, it is harder to obtain a high transimpedance while meeting the required bandwidth. Therefore, we typically see lower transimpedance values for 10-Gb/s parts than for 2.5-Gb/s parts. Typical values for the differential transimpedance are

$$2.5\,\text{Gb/s TIA:} \qquad R_T = 2.0\,\text{k}\Omega \ldots 4.0\,\text{k}\Omega, \tag{5.2}$$

$$10\,\text{Gb/s TIA:} \qquad R_T = 500\,\Omega \ldots 2.0\,\text{k}\Omega. \tag{5.3}$$

Much higher transimpedance values than these can be found in some commercial TIA chips (cf. Table 5.2). These chips contain a post amplifier that follows the basic shunt-feedback TIA and boosts the overall transimpedance. We discuss this topology in Section 5.2.5.

5.1.2 Input Overload Current

The input signal current i_I into the TIA (supplied by either a p-i-n photodetector or an avalanche photodetector [APD]) is shown schematically in Fig. 5.3. It is important to distinguish between the signal's peak-to-peak value i_I^{pp} and its average value $\bar{i}_I$. In the case of a DC-balanced non-return-to-zero (NRZ) signal with high extinction, as illustrated in Fig. 5.3, the peak-to-peak value is twice the average value: $i_I^{pp} = 2\bar{i}_I$.

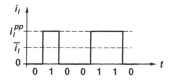

Fig. 5.3 TIA input signal current: peak-to-peak value and average value.

Definition. Many TIAs produce severe pulse-width distortion and jitter for large input signals. These undesirable distortions cause the bit-error rate (BER) of the receiver to increase rapidly with received power (cf. Figs. 4.5 and 4.6). For some critical input signal, the BER will go above the specified value such as 10^{-12}. This critical signal current, measured peak-to-peak, is called the *input overload current*, i_{ovl}^{pp}, of the TIA. For $i_I^{pp} > i_{ovl}^{pp}$, the TIA is said to be overloaded. The overload current also defines the upper end of the TIA's dynamic range.

Typical Value. The input overload current required for a TIA can be derived from the largest expected optical signal, $\overline{P}_{ovl}$, at the receiver input: $i_{ovl}^{pp} > 2\mathcal{R}\overline{P}_{ovl}$, where $\mathcal{R}$ is the photodetector responsivity and a DC-balanced signal with high extinction is assumed. For example, assuming the largest optical signal at the receiver is 0 dBm (e.g., given by the SONET OC-48 intermediate reach specification [188]) and a responsivity of 0.8 A/W, then the minimum required overload current is $i_{ovl}^{pp} = 1.6$ mA peak-to-peak.

5.1.3 Maximum Input Current for Linear Operation

Another specification for the input signal is the *maximum input current for linear operation*, i_{lin}^{pp}. This current is always smaller than the overload current i_{ovl}^{pp}.

Definition. Several definitions for i_{lin}^{pp} are in use. For example, we can define i_{lin}^{pp} as the peak-to-peak input current for which the transimpedance drops by 1 dB (about 11%) below its small-signal value. Alternatively, i_{lin}^{pp} can be defined as the peak-to-peak input current for which the output swing reaches 80% of its fully limited value. In analog CATV/HFC applications, harmonic and intermodulation distortions are of great significance, and hence i_{lin}^{pp} is defined such that these distortions remain small (cf. Section 4.8).

A TIA with a fixed transimpedance, R_T, must have a large output voltage swing to accommodate a high i_{lin}^{pp}. In particular, the output swing must be larger than $R_T \cdot i_{\text{lin}}^{pp}$ to avoid signal compression. Alternatively, the transimpedance can be made adaptive such that its value reduces for large input signals (cf. Section 5.2.4).

Typical Values. For digital (e.g., SONET) receivers that perform linear signal processing, such as equalization, on the TIA's output signal, the requirements for i_{lin}^{pp} are similar to those for i_{ovl}^{pp} discussed before, that is, around 1 mA peak-to-peak. In other applications, where the TIA's output signal is processed by a (nonlinear) limiting amplifier, i_{lin}^{pp} is uncritical and can be as small as 10 μA.

5.1.4 Input-Referred Noise Current

The input-referred noise current is one of the most critical TIA parameters. Often the noise of the TIA dominates all other noise sources (e.g., the noise from the photodetector, main amplifier, etc.) and therefore determines the sensitivity of the receiver (i.e., the lower end of the receiver's dynamic range). However, in optically amplified long-haul transmission systems, the noise contributed by the optical amplifier(s) often becomes so large that the TIA noise becomes less critical.

Definition. Figure 5.4(a) shows a noiseless TIA with an *equivalent noise current source*, $i_{n.TIA}$, at the input. This current source is chosen such that, together with the noiseless TIA, it reproduces the output noise of the actual noisy TIA. The current provided by the equivalent noise current source also is known as the *input-referred noise current*.

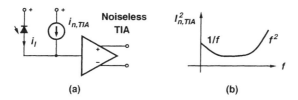

Fig. 5.4 (a) Input-referred noise current and (b) typical power spectrum.

Two points must be made about this definition. First, because the noise model in Fig. 5.4(a) consists of only a noise current source (rather than a noise current and a noise voltage source), the value of the input-referred noise current depends on the source impedance. The source impedance is determined mostly by the photodetector capacitance, and hence this capacitance must be specified when quoting the input-referred noise current. Second, in the case of a TIA with differential outputs, we have to decide if we should choose the input-referred noise current such that it reproduces the single-ended output noise or the differential output noise. These two noise currents are not necessarily the same. They are equal only if all the output noise is contained in the differential mode (and the common mode is noise free), that is, if the two single-ended noise voltages are fully anticorrelated. For example, in simulations with a particular differential TIA design, the input-referred rms noise current was found to be 30% higher when reproducing the single-ended rather than the differential output noise. For a differential-output TIA, it is reasonable to define the input-referred noise current such that it reproduces the differential output noise, because the TIA presumably will drive the next block differentially. [→ Problem 5.1]

The input-referred noise current in Fig. 5.4 can be quantified in a number of different ways.

- **Input-Referred Noise Current Spectrum.** The power spectral density of the input-referred noise current, $I_{n.TIA}^2(f)$, or the *input-referred noise current spectrum* for short, is illustrated schematically in Fig. 5.4(b). Note that this power spectrum is measured in pA^2/Hz (the square root of this spectrum, $I_{n.TIA}(f)$, is measured in $pA/\sqrt{Hz}$) and typically consists of a white part, an f^2 part at high frequencies, and possibly a $1/f$ part at low frequencies. In Section 5.2.3, we analyze the shape of this spectrum in more details. Because this spectrum is *not white*, it cannot be characterized by a single number, but instead a graph must be provided. To compare the noise performance of different TIAs, it is necessary to look at the whole spectrum up to about 2× the TIA's bandwidth. (Remember, in Section 4.4 we showed that the relevant noise bandwidths of a TIA can be about twice the 3-dB bandwidth.) Furthermore, note that the sensitivity of the TIA does not depend solely on this noise spectrum, but also is affected by the TIA's frequency response $|Z_T(f)|$.

- **Input-Referred RMS Noise Current.** Another measure for the input-referred noise current that relates directly to the sensitivity and can be expressed by a single number (in nA) is its rms noise value, $i_{n.TIA}^{rms}$. As we discussed at length in Section 4.4, this *input-referred rms noise current*, or *total input-referred noise current*, is determined by dividing the rms output noise voltage by the TIA's midband transimpedance value. The rms output noise voltage, in turn, is obtained by integrating the output-referred noise spectrum and taking the square root. Thus, we have

$$i_{n.TIA}^{rms} = \frac{1}{R_T}\sqrt{\int_0^{>2BW} |Z_T(f)|^2 \cdot I_{n.TIA}^2(f)\, df}, \quad (5.4)$$

where $|Z_T(f)|$ is the frequency response of the transimpedance and R_T is its midband value (cf. Eq. (4.37)). For analytical calculations, the integration can be carried out to infinity; for simulations (and measurements), it usually is enough to integrate up to about 2× the TIA's bandwidth, after which the contributions to the rms output noise become negligible. Note that the input-referred rms noise current $i_{n.TIA}^{rms}$ directly determines the (electrical) sensitivity of the TIA (cf. Eq. (4.18)):

$$i_{sens}^{pp} = 2\mathcal{Q} \cdot i_{n.TIA}^{rms}, \quad (5.5)$$

and therefore this noise measure is a good metric to compare different TIAs (designed for the same bit rate). Furthermore, $i_{n.TIA}^{rms}$ usually is the main contribution to the linear-channel noise, $i_{n.amp}^{rms}$, and thus has a major impact on the electrical sensitivity of the receiver.

- **Averaged Input-Referred Noise Current Density.** Sometimes the input-referred noise current is described by the *averaged input-referred noise current density*. This quantity is defined as the input-referred rms noise current, $i_{n.TIA}^{rms}$, divided by the square-root of the TIA's 3-dB bandwidth. It is important to realize that this is different from averaging the input-referred noise spectrum, $I_{n.TIA}^2(f)$, over the TIA's bandwidth. For example, in simulations with a particular 10-Gb/s TIA design, the input-referred noise current spectrum, $I_{n.TIA}$, averaged over the 3-dB bandwidth of 7.8 GHz was found to be 15.4 pA/$\sqrt{\text{Hz}}$ (incorrect way to determine the averaged input-referred noise), whereas the input-referred rms noise divided by the square-root of the bandwidth turned out to be 18.0 pA/$\sqrt{\text{Hz}}$ (correct way to determine the averaged input-referred noise). For the same design, the input-referred noise current spectrum at 100 MHz was as low as 8.8 pA/$\sqrt{\text{Hz}}$.

[→ Problem 5.2]

Typical Values. Typical values for the input-referred rms noise current seen in commercial parts are

$$2.5\,\text{Gb/s TIA:} \quad i_{n.TIA}^{rms} = 380\,\text{nA}, \tag{5.6}$$

$$10\,\text{Gb/s TIA:} \quad i_{n.TIA}^{rms} = 1{,}400\,\text{nA}. \tag{5.7}$$

The 10-Gb/s parts are almost 4× as noisy as the 2.5-Gb/s parts. In Section 5.2.3, we discuss the dependence of $i_{n.TIA}^{rms}$ on the bit rate in more detail.

5.1.5 Bandwidth and Group-Delay Variation

Definition. The TIA *bandwidth*, BW_{3dB}, is defined as the (upper) frequency at which the transimpedance $|Z_T(f)|$ dropped by 3 dB below its midband value. This bandwidth also is called the 3-dB bandwidth to distinguish it from the noise bandwidth.

The bandwidth specification alone does not say anything about the phase of $Z_T(f)$. We know that even if the frequency response $|Z_T(f)|$ is flat up to a sufficiently high frequency, distortions in the form of data-dependent jitter may occur if the phase linearity of $Z_T(f)$ is insufficient. A common measure for phase linearity is the variation of the group delay with frequency. The *group delay*, τ, is related to the phase, Φ, as $\tau(\omega) = -d\Phi/d\omega$.

As we know from Sections 4.6 and 4.9, the bandwidth and group-delay variation are important parameters determining the amount of ISI and jitter introduced by the TIA. Moreover, the bandwidth also affects the amount of noise picked up by the receiver. A well-controlled bandwidth (e.g., independence of the received power) is particularly important in optically amplified transmission systems. In these systems, the BER and the signal-to-noise ratio (SNR) strongly depend on the optical signal-to-noise ratio (OSNR) and the electrical receiver bandwidth (cf. Eq. (4.32)).

Typical Values. The desired 3-dB bandwidth of a TIA depends on which bandwidth allocation strategy is chosen for the receiver (cf. Section 4.6). If the TIA sets the receiver bandwidth, its bandwidth is chosen around $0.6B$ to $0.7B$. If the receiver bandwidth is controlled in another way (e.g., with a filter), a wider TIA bandwidth around $0.9B$ to $1.2B$ is chosen:

$$2.5\,\text{Gb/s TIA:} \quad BW_{3dB} = 1.5\,\text{GHz} \ldots 3\,\text{GHz}, \tag{5.8}$$

$$10\,\text{Gb/s TIA:} \quad BW_{3dB} = 6.0\,\text{GHz} \ldots 12\,\text{GHz}. \tag{5.9}$$

Typically, a group delay variation, $\Delta\tau$, of less than $\pm 10\%$ of the bit period (± 0.1 UI) over the specified bandwidth is required to limit the generation of data-dependent jitter. This corresponds to

$$2.5\,\text{Gb/s TIA:} \quad |\Delta\tau| < 40\,\text{ps}, \tag{5.10}$$

$$10\,\text{Gb/s TIA:} \quad |\Delta\tau| < 10\,\text{ps}. \tag{5.11}$$

5.2 TIA CIRCUIT CONCEPTS

In the following, we discuss TIA circuit concepts in a general and, as much as possible, technology-independent manner. This includes the shunt-feedback architecture and variations thereof, noise optimization procedures, offset control, and special techniques for burst-mode and analog TIAs. Additional information on TIAs can be found in [6].

5.2.1 Low- and High-Impedance Front-Ends

The shunt-feedback TIA is by far the most popular circuit for converting a small photodetector current into a voltage signal. But before turning to this circuit, we want to mention the two alternative circuits shown in Fig. 5.5, which are known as the *low-impedance front-end* and the *high-impedance front-end*. Both front-ends essentially consist of just a resistor from the photodetector to ground. In the case of the low-impedance front-end, this resistor typically is 50 Ω, whereas for the high-impedance front-end, it is much larger than 50 Ω. Note that the low-impedance front-end inherently features a 50-Ω output, whereas the high-impedance front-end must be followed by a buffer to provide a 50-Ω output.

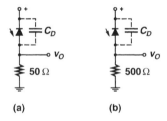

Fig. 5.5 (a) Low-impedance and (b) high-impedance front-end.

Let's consider the low-impedance front-end first. The 50-Ω resistor converts the photodetector current i_I into a voltage v_O following Ohm's law: $v_O = 50\,\Omega \cdot i_I$. Note that, although completely passive, this simple circuit qualifies as a transimpedance amplifier with the transimpedance $R_T = 50\,\Omega$. (One may argue, however, that the prefix "trans" is not appropriate because the input and output are represented by the same node.) A first drawback that we recognize here is the low transimpedance value and thus the small output voltage produced by this front-end for small detector currents.[2] Another drawback is the significant noise current associated with the small 50-Ω resistor. The input-referred noise current spectrum is white and is given by the thermal resistor noise $I_{n,\text{res}} = \sqrt{4kT/R}$, which is about $18\,\text{pA}/\sqrt{\text{Hz}}$ for $R = 50\,\Omega$ (neglecting important noise contributions from the subsequent buffer or amplifier).

---

[2]If the low-impedance front-end is loaded by an amplifier with 50-Ω inputs, its transimpedance reduces to 25 Ω. To keep the discussion simple, we assume here that both front-ends are driving capacitive loads only.

On the plus side, the low-impedance front-end has a large bandwidth. For example, assuming a photodetector capacitance of $C_D = 0.15$ pF and an external load capacitance of the same value, the bandwidth is a respectable 11 GHz. [→ Problem 5.3]

To get around the problems of the low-impedance front-end, we may consider increasing the value of the resistor, which brings us to the high-impedance front-end. Assuming a resistor of 500 Ω as shown in Fig. 5.5(b), the transimpedance increases to $R_T = 500$ Ω, a reasonable value for a 10-Gb/s TIA. The noise improves, too, and is now down to $5.8\,\text{pA}/\sqrt{\text{Hz}}$ (neglecting important noise contributions from the subsequent buffer or amplifier). But unfortunately, the bandwidth reduces to a mere 1.1 GHz, way too little for a 10-Gb/s TIA. It is possible to enhance the bandwidth of the high-impedance front-end by following it with an equalizer that boosts the high frequencies. However, there is another problem with this front-end concerning the input overload current. Assuming a peak-to-peak current of 2 mA from the photodetector and a supply voltage of 5 V, the reverse bias voltage of the photodetector drops from 5 V to 4 V whenever a one bit is received (or, in the case of insufficient bandwidth, whenever a run of zeros is followed by a run of ones). This reduced reverse bias may not be sufficient for the detector to work at speed and may cause signal distortions. In addition, the large 1-V swing may overload the input stage of the subsequent equalizer. More generally, if the maximum permissible input voltage swing is $v_{I.ovl}^{pp}$, then the input overload current is given by $i_{ovl}^{pp} = v_{I.ovl}^{pp}/R$, where R is the front-end resistor.

Is there a way to get a high transimpedance, a large bandwidth, a high input overload current, and low noise all at the same time? Yes, the shunt-feedback transimpedance amplifier, which we treat next, is the answer to this question.

Before leaving this section, it is worth mentioning that the low-impedance front-end has an interesting application in high-speed receivers for which active electronic components such as TIAs and MAs are not yet available. In this case, the received signal is first amplified optically (e.g., with an erbium-doped fiber amplifier [EDFA]) to a fairly high power level and then is detected with a passive low-impedance front-end. For example, given an amplified optical power level of +10 dBm, $\mathcal{R} = 0.8$ A/W, and $R_T = 25$ Ω, the output signal from the low-impedance front-end is about 400 mV$_{pp}$, enough to drive a clock and data recovery circuit (CDR) directly.

5.2.2 Shunt Feedback TIA

Simple Analysis. The circuit of the basic *shunt-feedback TIA* is shown in Fig 5.6. The photodetector is connected to the input of an inverting voltage amplifier, which has a feedback resistor R_F leading from its output to the input. The current from the photodetector flows into R_F, and the amplifier output responds in such a way that the input remains at virtual ground. As a result, the output voltage is $v_O \approx -R_F \cdot i_I$, and thus the transimpedance is approximately R_F. Note that because of the virtual ground presented by the amplifier input, the photodetector reverse bias voltage is fairly independent of the detector current, a prerequisite for good overload behavior.

Now let's analyze the frequency response. For now we assume that the feedback amplifier has an infinite bandwidth; later we drop this assumption. The voltage

114   TRANSIMPEDANCE AMPLIFIERS

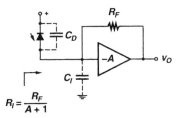

Fig. 5.6 Basic shunt-feedback transimpedance amplifier.

gain of the inverting amplifier is $-A$ and its output resistance is zero. The input capacitance of the amplifier is C_I and the input resistance is taken to be infinite, a good assumption for an FET amplifier. Because the photodetector capacitance, C_D, and the input capacitance, C_I, appear in parallel (from an AC point of view), we can combine them into a single capacitance $C_T = C_D + C_I$. Given these assumptions, we can calculate the frequency-dependent transimpedance as

$$Z_T(s) = -R_T \cdot \frac{1}{1 + s/\omega_p}, \tag{5.12}$$

where

$$R_T = \frac{A}{A+1} \cdot R_F, \tag{5.13}$$

$$\omega_p = \frac{A+1}{R_F C_T}. \tag{5.14}$$

Equation (5.13) confirms our earlier observation that the transimpedance R_T is approximately equal to the feedback resistor R_F, given a gain A much larger than unity. The 3-dB bandwidth of the TIA follows from Eq. (5.14) as

$$BW_{3dB} = \frac{\omega_p}{2\pi} = \frac{1}{2\pi} \cdot \frac{A+1}{R_F C_T}. \tag{5.15}$$

This means that the bandwidth is $A + 1$ times *larger* than that of a high-impedance front-end with resistor R_F and total capacitance C_T. This bandwidth improvement can be understood in terms of the circuit's closed-loop input resistance, R_I (cf. Fig. 5.6). Because of the feedback action, the input resistance is $A + 1$ times smaller than R_F:

$$R_I = \frac{R_F}{A+1}. \tag{5.16}$$

In turn, the pole ω_p is sped up by the factor $A + 1$ over the situation without feedback amplifier.

The low input resistance of the shunt-feedback TIA also improves its input overload current. For a given input signal current, the input voltage swing is $A+1$ times smaller than that of a high-impedance front-end. Conversely, given the maximum permissible input voltage swing $v_{I,ovl}^{pp}$, the input overload current is $A+1$ times larger:

$$i_{ovl}^{pp} = (A+1) \cdot \frac{v_{I,ovl}^{pp}}{R_F}. \tag{5.17}$$

Alternatively, we can express the input overload current in terms of the maximum permissible output voltage swing $v_{O,ovl}^{pp}$, which yields

$$i_{ovl}^{pp} = \frac{A+1}{A} \cdot \frac{v_{O,ovl}^{pp}}{R_F} = \frac{v_{O,ovl}^{pp}}{R_T}. \tag{5.18}$$

The voltage swings v_I^{pp} and v_O^{pp} are limited by a number of mechanisms. For example, an insufficient reverse voltage across the photodetector can cause a slow response. Also, an insufficient voltage drop across a current-source transistor in the feedback amplifier can lead to a reduced bias current and a slow response. Finally, in an implementation with bipolar junction transistors (BJTs), a large voltage swing can cause some base-collector diodes to become forward biased, that is, to enter saturation, resulting in signal distortions. The TIA's overload current is given by either Eq. (5.17) or (5.18), whichever expression is smaller.

It is interesting to note that the shunt-feedback topology also reduces the output resistance of the TIA. So far, we assumed that the feedback amplifier has a zero output impedance, but in a practical implementation, it is nonzero. When the feedback loop is closed, this output impedance is reduced by a factor $A+1$ for frequencies less than $1/(2\pi \cdot R_F C_T)$. [→ Problem 5.4]

The improved bandwidth and input overload current are the main reasons for the popularity of the shunt-feedback TIA. In summary, this circuit features a high transimpedance ($R_T \approx R_F$) and a low input-referred noise current density ($I_{n,TIA}^2 = 4kT/R_F$ plus the noise contributions from the feedback amplifier, which we discuss in Section 5.2.3), similar to those of the high-impedance front-end, but a much better bandwidth and input overload current, comparable with those of the low-impedance front-end. But note that to realize all these advantages, we need a feedback amplifier with the necessary bandwidth, gain, low noise, and so forth.

Effects of Finite Amplifier Bandwidth. In the following, we consider a more realistic feedback amplifier model. Let's replace our infinite-bandwidth amplifier with one that has a single dominant pole with the time constant T_A, a good approximation for a single-stage amplifier. The 3-dB bandwidth of this amplifier, not to be confused with the 3-dB bandwidth of the TIA, is given by $f_A = 1/(2\pi \cdot T_A)$. Now, the open-loop frequency response, $|A_{open}(\omega)|$, has two poles, as shown in Fig. 5.7. The high-frequency pole at $1/T_A$ is due to the feedback amplifier and the low-frequency pole at $1/(R_F C_T)$ is due to the low pass formed by R_F and C_T. In a second-order system like this, we have to watch out for undesired peaking in the closed-loop frequency response. Peaking in the frequency domain corresponds to ringing (or overshoot) in the time domain, which degrades the eye quality. Given the above feedback amplifier model, we can calculate the closed-loop transimpedance as

$$Z_T(s) = -R_T \cdot \frac{1}{1 + s/(\omega_0 Q) + s^2/\omega_0^2}, \quad (5.19)$$

where

$$R_T = \frac{A}{A+1} \cdot R_F, \quad (5.20)$$

$$\omega_0 = \sqrt{\frac{A+1}{R_F C_T \cdot T_A}}, \quad (5.21)$$

$$Q = \frac{\sqrt{(A+1) \cdot R_F C_T \cdot T_A}}{R_F C_T + T_A}. \quad (5.22)$$

In these equations, R_T is the transimpedance at DC and did not change from the case with infinite bandwidth (Eq. (5.13)), ω_0 is the pole (angular) frequency, and Q is the pole quality factor,[3] which controls the peaking and ringing (not to be confused with the Personick Q introduced in Section 4.2). Two values for Q are of particular interest. For $Q = 1/\sqrt{3} = 0.577$, we obtain the so-called *Bessel response*, which has maximally flat group-delay characteristics, $\tau(f)$, and thus the smallest group-delay variation. This response has no peaking in the frequency response (of the amplitude) and produces only a negligible amount of overshoot in the time domain ($\approx 0.43\%$). For $Q = 1/\sqrt{2} = 0.707$, we obtain the so-called *Butterworth response*, which has a maximally flat frequency response, $|Z_T(f)|$, with no peaking. However, this response has a slight amount of peaking in the group-delay characteristics (≈ 0.06 UI at $BW_{3dB} = 0.7B$) and produces a little bit more overshoot in the time domain than the Bessel response ($\approx 4.3\%$). For larger values of Q, peaking, overshoot, and ringing become progressively worse.

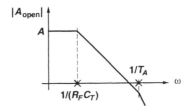

Fig. 5.7 Open-loop frequency response of a TIA with a single-pole feedback amplifier. Note that the frequency axis is in units of $\omega = 2\pi f$.

In the following, we require for our TIA that $Q \leq 1/\sqrt{2}$, which guarantees a flat frequency response and limits overshoot and ringing to less than 4.3%. By setting Eq. (5.22) to less than $1/\sqrt{2}$ and using the approximation $\sqrt{A^2 - 1} \approx A - 1/(2A)$,

---

[3] $Q = 1/(2\zeta)$, where ζ is the damping factor.

which is valid for $A^2 \gg 1$, we find that we have to make the bandwidth of the feedback amplifier larger than

$$f_A \geq \frac{1}{2\pi} \cdot \frac{2A}{R_F C_T}. \tag{5.23}$$

The interpretation of this equation is that the two open-loop poles shown in Fig. 5.7 must be spaced apart by at least $2A$. Equivalently, the feedback amplifier bandwidth (f_A) must be at least a factor two larger than the unity-gain frequency of the open-loop response ($A/[2\pi \cdot R_F C_T]$). If we desire the more conservative Bessel response, the poles must be spaced apart by $3A + 1$. [$\rightarrow$ Problem 5.5]

For the case of a Butterworth response, the 3-dB bandwidth of the TIA is given directly by $BW_{3dB} = \omega_0/(2\pi)$ and Eq. (5.21). When inserting the amplifier time constant necessary for Butterworth response, $T_A = R_F C_T/(2A)$ (from Eq. (5.23)), we arrive at the following TIA bandwidth:

$$BW_{3dB} = \frac{1}{2\pi} \cdot \frac{\sqrt{2A(A+1)}}{R_F C_T}. \tag{5.24}$$

By comparing this equation with Eq. (5.15), we find that the bandwidth *increased* by about $\sqrt{2}$ as a result of making the bandwidth of the feedback amplifier finite. It is remarkable that a system gets faster by making one of its components slower!

Transimpedance Limit. It can be shown easily that the 3-dB bandwidth of the second-order system in Eq. (5.19) is bounded by $BW_{3dB} \leq \omega_0/(2\pi)$, if we require that $Q \leq 1/\sqrt{2}$. When applying this fact to Eq. (5.21) and expressing R_F in terms of R_T with Eq. (5.20), we arrive at the following inequality, which is known as the *transimpedance limit* [95]:

$$R_T \leq \frac{A \cdot f_A}{2\pi \cdot C_T \cdot BW_{3dB}^2}. \tag{5.25}$$

Note that the expression $A \cdot f_A$ in the above equation represents the gain-bandwidth product of the feedback amplifier, which is roughly proportional to the technology parameter f_T (see Section 6.3.2 for more about f_T). Thus, the transimpedance limit has the following interpretation: if we want to double the bit rate (double BW_{3dB}) without using a faster technology (same $A \cdot f_A$ and C_T), then the transimpedance will degrade by a factor of *four*. This is the reason why higher bit-rate TIAs generally have a lower transimpedance. Alternatively, if we want to maintain the same transimpedance at twice the bit rate, one solution is to half C_T and double $A \cdot f_A$, which means that we need a technology that is twice as fast (twice the f_T). [$\rightarrow$ Problem 5.6]

It is instructive to compare Eq. (5.25) with the corresponding equation for the low- and high-impedance front-ends. For both of these front-ends, the transimpedance is simply given by $R_T = 1/(2\pi \cdot C_T \cdot BW_{3dB})$. Thus, as long as the gain-bandwidth product of the feedback amplifier is larger than the required front-end bandwidth, $A \cdot f_A > BW_{3dB}$, the shunt-feedback TIA has an advantage.

We can conclude further from Eq. (5.25) that given a certain technology (for the feedback amplifier and photodetector), the product $R_T \cdot BW_{3dB}^2$ is approximately constant. Note that in contrast to a single-stage voltage amplifier, which has an approximately constant gain-bandwidth product, the TIA has an approximately constant transimpedance-bandwidth-*squared* product. For this reason, the simple transimpedance-bandwidth product ($R_T \cdot BW_{3dB}$) measured in ΩHz, which has been proposed as a figure of merit for TIAs, is inversely proportional to the bandwidth and can assume very large values for low-speed TIAs.

Finally, it should be pointed out that the transimpedance limit has been derived based on the basic shunt-feedback topology of Fig. 5.6 and is not a fundamental limit. Other topologies, in particular the TIA with post amplifier (see Section 5.2.5), the TIA with common-base input stage (see Section 5.2.6), and the TIA with inductive input coupling (see Section 5.2.9) may achieve a higher transimpedance.

Multiple Amplifier Poles. The feedback amplifier model with a single dominant pole often is still too simplistic. In practice, additional amplifier poles are caused by cascode transistors, buffers, level shifters, or, in the case of a multistage amplifier, by the additional stages. The situation is complicated further by the packaging parasitics of the photodetector (cf. Fig. 3.3(b)); in particular, the bond-wire inductance plays an important role at 10 Gb/s and above. Under these circumstances, detailed transistor-level simulations, including the parasitics, are necessary to determine the TIA bandwidth and to make sure peaking in the frequency response and group-delay variations are within specifications.

In the case of a feedback amplifier with multiple poles, each pole contributes some undesired phase shift to the open-loop response, making it more challenging to obtain a stable TIA with little overshoot and ringing. What can we do to promote stability in the presence of multiple amplifier poles? One approach is to place the amplifier poles at a higher frequency than given by Eq. (5.23) such that the phase margin of the open-loop response remains sufficient. Another technique is to add a small capacitor C_F in parallel to the feedback resistor R_F. This capacitor introduces a zero at $1/(R_F C_F)$ in the open-loop response, which can be used to undo some of the phase shift and thus improve the closed-loop stability of the TIA. Note that this zero does not appear in the closed-loop response, and hence it is known as a *phantom zero* [111, 137]. [$\rightarrow$ Problem 5.7]

Earlier we mentioned multistage feedback amplifiers as a possible cause of multiple poles. Knowing the difficulties of stabilizing these amplifiers, what is the motivation to use them? In low-voltage FET technologies, the voltage gain achievable with a single stage is limited to relatively small values. For example, given the quadratic FET model and a resistive load, the voltage gain can be shown to be $A = 2V_R/(V_{GS} - V_{TH})$, where V_R is the DC-voltage drop across the load resistor and $V_{GS} - V_{TH}$ is the FET's overdrive voltage. (A higher gain can be achieved with an active load, but at the expense of a higher parasitic load capacitance and more noise.) With a power supply voltage of 1.5 V, headroom considerations limit V_R to about 0.5 V, whereas $V_{GS} - V_{TH}$ should be at least 0.3 V for speed reasons (cf. Section 6.3.2). Thus, this stage has a gain of less than $3.3\times$. The gain is even lower in submicron technologies where the drain current increases less than quadratically with the gate voltage. Now by cascading multiple stages, it is possible to overcome this gain limitation, which in turn permits the use of a larger feedback resistor, resulting in a higher transimpedance and better noise performance. (Note that in the case of a bipolar technology, an appreciable gain can be obtained with a single stage. It can be shown that this gain is $A = V_R/V_T$, where V_T is the thermal voltage, which is about 25 mV; thus with $V_R = 0.5$ V, the gain is about 20×.)

A Numerical Example. To get a better feeling for numerical values, we want to illustrate the foregoing theory with a 10-Gb/s TIA design example. Figure 5.8 shows the familiar shunt-feedback TIA annotated with some example values. The photodetector and amplifier input capacitance are 0.15 pF each, the feedback amplifier has a low-frequency gain of 14 dB (5×), and the feedback resistor is 600 Ω.

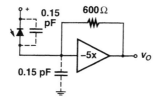

Fig. 5.8 Example values for a 10-Gb/s shunt-feedback TIA.

From these data, we can easily calculate the transimpedance of the TIA at low frequencies:
$$R_T = \frac{5}{5+1} \cdot 600\,\Omega = 500\,\Omega, \quad (5.26)$$
which is equal to 54 dBΩ. This is somewhat lower than R_F, as expected. The TIA bandwidth turns out to be
$$BW_{3dB} = \frac{\sqrt{2 \cdot 5 \cdot (5+1)}}{6.28 \cdot 600\,\Omega \cdot 0.3\,\text{pF}} = 6.85\,\text{GHz}, \quad (5.27)$$
which is suitable for a 10-Gb/s system where the TIA sets the receiver bandwidth. For comparison, the high-impedance front-end with the same transimpedance (500 Ω) and the same total capacitance (0.3 pF), which we discussed in Section 5.2.1, has a bandwidth of only 1.1 GHz. To achieve a flat passband (without peaking) for our TIA, we must make the feedback amplifier bandwidth larger than
$$f_A \geq \frac{2 \cdot 5}{6.28 \cdot 600\,\Omega \cdot 0.3\,\text{pF}} = 8.85\,\text{GHz}. \quad (5.28)$$
Thus, we need a technology in which we can realize an amplifier with the gain-bandwidth product
$$A \cdot f_A = 5 \cdot 8.85\,\text{GHz} = 44\,\text{GHz}. \quad (5.29)$$
The TIA's input impedance at low frequencies is
$$R_I = \frac{600\,\Omega}{5+1} = 100\,\Omega, \quad (5.30)$$

120 TRANSIMPEDANCE AMPLIFIERS

which means that for a maximum permissible input voltage swing of $v_{I.ovl}^{pp} = 0.2\,\text{V}$, the input overload current is $i_{ovl}^{pp} = 2\,\text{mA}$. Under the same condition, the overload current of the 500-Ω high-impedance front-end is 5× lower: $i_{ovl}^{pp} = 0.4\,\text{mA}$. Finally, the input-referred noise current spectrum of our TIA due to R_F is $5.3\,\text{pA}/\sqrt{\text{Hz}}$ (for now, we are neglecting the amplifier noise). This value is comparable with that of the 500-Ω high-impedance front-end.

Table 5.1 summarizes the results from our TIA example and compares them with the low-impedance and high-impedance front-end examples from Section 5.2.1. Note that all overload currents are based on $v_{I.ovl}^{pp} = 0.2\,\text{V}$ and that all noise spectral densities are based on resistor noise only.

Table 5.1 Performance of our shunt-feedback TIA, low-impedance front-end (Low-Z), and high-impedance front-end (High-Z) examples.

Parameter	Symbol	TIA	Low-Z	High-Z
Transimpedance	R_T	500 Ω	50 Ω	500 Ω
3-dB bandwidth	BW_{3dB}	6.9 GHz	11 GHz	1.1 GHz
Overload current	i_{ovl}^{pp}	2 mA	4 mA	0.4 mA
Noise current	$I_{n.TIA}$	5.3 pA$\sqrt{\text{Hz}}$	18 pA$\sqrt{\text{Hz}}$	5.8 pA$\sqrt{\text{Hz}}$

Transimpedance Limit: A Designer's Nightmare. Just as you are getting ready to tape out the 10-Gb/s TIA design you have been working on lately, your boss comes into your cubicle. "We have to implement initiative Leapfrog," she announces. "Can you beef up your 10-gig TIA to 40 gig before tape out on Friday?"

What do you do now? You can't switch to a faster technology, but you could reduce the transimpedance. O.K., you need to boost the bandwidth by a factor 4, from about 7 GHz to 28 GHz, so you start by reducing R_F by a factor 4, from 600 Ω to 150 Ω. Now you've got to check if the feedback amplifier is fast enough to avoid peaking:

$$f_A \geq \frac{2 \cdot 5}{6.28 \cdot 150\,\Omega \cdot 0.3\,\text{pF}} = 35.4\,\text{GHz}. \quad (5.31)$$

Wow, this means a gain-bandwidth product of $5 \cdot 35.4 = 177\,\text{GHz}$, but that's not possible in your 44-GHz technology. But then you realize if you lower the feedback amplifier gain by a factor 2, from 5× to 2.5×, you can do it. Now, $f_A \geq 17.7\,\text{GHz}$ and $A \cdot f_A = 2.5 \cdot 17.7\,\text{GHz} = 44\,\text{GHz}$. That's doable and you get a flat response! Let's check the TIA bandwidth:

$$BW_{3dB} = \frac{\sqrt{2 \cdot 2.5 \cdot (2.5+1)}}{6.28 \cdot 150\,\Omega \cdot 0.3\,\text{pF}} = 14.8\,\text{GHz}. \quad (5.32)$$

This is not enough! Cold sweat is starting to drip from your forehead. So far you gained only a factor 2 or so in speed. You hear your boss' foot steps in the aisle. She is leaving! Good, there will be no more interruptions tonight. In desperation, you lower R_F again, this time from 150 Ω to 50 Ω. Again, peaking is giving you problems and you have to drop the feedback amplifier gain even lower, this time from 2.5× to 1.44×. Soon this amplifier will be a unity-gain buffer! Now you have

$$f_A \geq \frac{2 \cdot 1.44}{6.28 \cdot 50\,\Omega \cdot 0.3\,\text{pF}} = 30.6\,\text{GHz} \tag{5.33}$$

and $A \cdot f_A = 1.44 \cdot 30.6\,\text{GHz} = 44\,\text{GHz}$. Fine, you can design this amplifier. What's the TIA bandwidth now?

$$BW_{3dB} = \frac{\sqrt{2 \cdot 1.44 \cdot (1.44 + 1)}}{6.28 \cdot 50\,\Omega \cdot 0.3\,\text{pF}} = 28.1\,\text{GHz}. \tag{5.34}$$

Hurray, you've got it! Tomorrow, you'll be the "40-Gig Hero." But before leaving work, you quickly check the transimpedance:

$$R_T = \frac{1.44}{1.44 + 1} \cdot 50\,\Omega = 29.5\,\Omega. \tag{5.35}$$

Lousy, 17× less than what you had for your beautiful 10-gig TIA! Could this be the consequence of the transimpedance limit? It would predict a reduction of the transimpedance by 16× for a bandwidth increase of 4×. Anyway, time to go home. You know your boss only cares about 40 gig and that's that.

5.2.3 Noise Optimization

In Section 4.3, we emphasized the importance of the input-referred noise current and its impact on receiver sensitivity. Now, we want to analyze and optimize this important noise quantity for the shunt-feedback TIA. In the process, we also obtain an explanation for the white and f^2-noise components that we have talked about already several times (Sections 4.1, 4.4, and 5.1.4). Additional information about noise in TIAs can be found in [6, 19, 62, 177].

Figures 5.9 and 5.11 show the familiar shunt-feedback TIA with the input stage (or front-end) implemented with an FET and bipolar transistor, respectively. We discuss complete transistor-level circuits in Section 5.3, but for the following (approximate) noise analysis, it is sufficient to consider just the input transistor. The most significant device noise sources are shown. The thermal noise of the feedback resistor, $i_{n.res}$, is present in either implementation. The FET front-end further produces shot noise due to the gate current, $i_{n.G}$, and channel noise, $i_{n.D}$. The bipolar front-end produces shot noise due to the base current, $i_{n.B}$, thermal noise due to the intrinsic base resistance (base-spreading and contact resistance), $i_{n.Rb}$, and shot noise due to the collector current, $i_{n.C}$. As we know, the effect of all these noise sources can be described by a single *equivalent* noise current source, $i_{n.TIA}$, at the input of the TIA (shown with dashed lines).

In the following, we first study the power spectrum $I_{n.TIA}^2(f)$ of this input-referred noise current and then move on to the *total* input-referred noise current, $i_{n.TIA}^{rms}$, which is the relevant quantity for sensitivity calculations. Finally, we discuss how to optimize the noise performance of TIAs.

Input-Referred Noise Current Spectrum. The input-referred noise current spectrum of the TIA can be broken into two major components: the noise from the feedback resistor (or resistors, in a differential implementation) and the noise from the amplifier front-end. Because they usually are uncorrelated, we can write

$$I_{n,TIA}^2(f) = I_{n,res}^2(f) + I_{n,front}^2(f). \quad (5.36)$$

In high-speed receivers, the front-end noise contribution typically is larger than the contribution from the feedback resistor. However, in low-speed receivers, the resistor noise may become dominant. The noise current spectrum of the feedback resistor is white (frequency independent) and given by the well-known thermal-noise equation:

$$I_{n,res}^2(f) = \frac{4kT}{R_F}. \quad (5.37)$$

This noise current contributes directly to the input-referred TIA noise in Eq. (5.36) because $i_{n,res}$ has the same effect on the TIA output as $i_{n,TIA}$. Note that this is the *only* noise source that we considered in Section 5.2.2. We already know from this section that we should choose the highest possible R_F to optimize the TIA's noise performance.

Next, we analyze the noise contribution from the amplifier front-end, $I_{n,front}^2$. The major device noise sources in an FET common-source input stage are shown in Fig. 5.9. The shot noise generated by the gate current, I_G, is given by $I_{n,G}^2 = 2qI_G$ and contributes directly to the input-referred TIA noise. This noise component is negligible for MOSFETs, but can be significant for metal-semiconductor FETs (MESFETs), and heterostructure FETs (HFETs), which have a larger gate-leakage current.

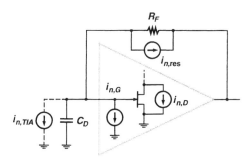

Fig. 5.9 Significant device noise sources in a TIA with FET front-end.

An important noise source in the FET input stage is the channel noise, which is given by $I_{n,D}^2 = 4kT\Gamma g_m$, where g_m is the FET's transconductance and Γ is the channel-noise factor. For MOSFETs, the channel-noise factor is in the range $\Gamma = 0.7$ to 3.0, where the low numbers correspond to long-channel devices. For silicon junction FETs (JFETs), $\Gamma \approx 0.7$, and for GaAs MESFETs, $\Gamma = 1.1$ to 1.75. Now, unlike the other noise sources that we discussed so far, this noise source is not located directly at the input of the TIA and we have to transform it to obtain its contribution to the input-referred TIA noise. A straightforward way to do this transformation is to calculate the transfer function from $i_{n,D}$ to the output of the TIA and divide that by the transfer function from $i_{n,TIA}$ to the output. Equivalently, but easier, we can calculate the implicit transfer function from $i_{n,D}$ to $i_{n,TIA}$ under the condition that the TIA output signal is zero. The implicit transfer function from the input current to the drain current has a low-pass characteristics; therefore, the inverse function, which refers the drain current back to the input, has *high-pass* characteristics. It can be shown that this high-pass transfer function is [57]

$$H(s) = \frac{1 + s R_F C_T}{g_m R_F}, \qquad (5.38)$$

where $C_T = C_D + C_I$ and C_I is the input capacitance of the FET stage at zero output signal, that is, $C_I = C_{gs} + C_{gd}$. Now, using this high-pass to refer the white channel noise, $I_{n,D}^2 = 4kT\Gamma g_m$, back to the input yields

$$\begin{aligned} I_{n,\text{front},D}^2(f) &= \frac{1 + (2\pi f \cdot R_F C_T)^2}{(g_m R_F)^2} \cdot 4kT\Gamma g_m \\ &= 4kT\Gamma \cdot \frac{1}{g_m R_F^2} + 4kT\Gamma \cdot \frac{(2\pi C_T)^2}{g_m} \cdot f^2. \end{aligned} \qquad (5.39)$$

And here, for the first time, we encounter an f^2-noise component. We now understand that it arises from a white-noise source, which became emphasized because of a low-pass transfer function from the input to the source location. Figure 5.10 illustrates the channel-noise component of Eq. (5.39) and the feedback-resistor noise component of Eq. (5.37) graphically. It is interesting to observe that the input-referred channel noise starts to rise at the frequency $1/(2\pi \cdot R_F C_T)$ given by the zero in Eq. (5.38). This frequency is *lower* than the 3-dB bandwidth of the TIA, which is $\sqrt{2A(A+1)}/(2\pi \cdot R_F C_T)$ (cf. Eq. (5.24)). As a result, the output-referred noise spectrum has a "hump," as shown in Fig. 4.2.

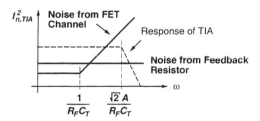

Fig. 5.10 Noise spectrum components of a TIA with FET front-end.

To summarize, we can write the input-referred noise current spectrum of an FET front-end as

$$I_{n,\text{front}}^2(f) = 2q I_G + 4kT\Gamma \cdot \frac{(2\pi C_T)^2}{g_m} \cdot f^2 + \ldots, \qquad (5.40)$$

where we have neglected the first term of Eq. (5.39), which is small compared with the feedback-resistor noise if $g_m R_F \gg \Gamma$. (However, for small values of R_F, this noise can be significant. Another reason to try and make R_F as large as possible!) Besides the noise terms discussed so far, there are several other noise terms that we have neglected. The FET also produces $1/f$ noise, which when referred back to the input turns into f noise at high frequencies and $1/f$ noise at low frequencies. Furthermore, there are additional device noise sources, which also contribute to the input-referred TIA noise such as the FET's load resistor and subsequent gain stages. However, if the gain of the first stage is sufficiently large, these sources can be neglected. [→ Problems 5.8 and 5.9]

The situation for a BJT common-emitter front-end, as shown in Fig. 5.11, is similar to that of the FET front-end. The shot noise generated by the base current, I_B, is given by $I_{n.B}^2 = 2qI_C/\beta$, where I_C is the collector current and β is the current gain of the BJT ($I_B = I_C/\beta$). This white noise current contributes directly to the input-referred TIA noise. Then we have the shot noise generated by the collector current, which is $I_{n.C}^2 = 2qI_C$. This noise current must be transformed to find its contribution to the input-referred TIA noise current. If we neglect R_b, the transfer function for this transformation is the same as in Eq. (5.38), and we find $I_{n.\text{front}.C}^2(f) \approx 2qI_C/(g_m R_F)^2 + 2qI_C \cdot (2\pi C_T)^2/g_m^2 \cdot f^2$. Note how the white shot noise was transformed into a f^2-noise component. Finally, we have the thermal noise generated by the intrinsic base resistance, which is given by $I_{n.Rb}^2 = 4kT/R_b$. This noise current, too, must be transformed to find its contribution to the input-referred TIA noise current. In this case, the high-pass transfer function is $H(s) = R_b/R_F + sR_b C_D$, and thus the noise contribution is $I_{n.\text{front}.Rb}^2(f) = 4kTR_b/R_F^2 + 4kTR_b \cdot (2\pi C_D)^2 \cdot f^2$.

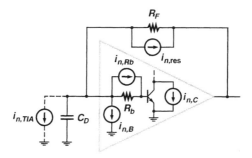

Fig. 5.11 Significant device noise sources in a TIA with bipolar front-end.

To summarize, we can write the input-referred noise current spectrum of a BJT front-end as

$$I_{n.\text{front}}^2(f) = \frac{2qI_C}{\beta} + 2qI_C \cdot \frac{(2\pi C_T)^2}{g_m^2} \cdot f^2 + 4kTR_b \cdot (2\pi C_D)^2 \cdot f^2 + \ldots, \quad (5.41)$$

where we have neglected the first term of $I_{n.\text{front}.C}^2$, which is small compared with the base shot noise if $(g_m R_F)^2 \gg \beta$, and we have also neglected the first term of $I_{n,\text{front},Rb}^2$, which is small compared with the noise from the feedback resistor if $R_F \gg R_b$.

We conclude from Eqs. (5.37), (5.40), and (5.41) that the input-referred noise current spectrum, $I_{n,TIA}^2(f)$, consists mostly of white-noise terms and f^2-noise terms, regardless of whether the TIA is implementation in an FET or bipolar technology. This observation justifies the form of the noise spectrum, $I_{n,TIA}^2(f) = \alpha_0 + \alpha_2 f^2$, which we introduced in Section 4.1.

Throughout this section, we assumed that the TIA is implemented as a single-ended circuit, that is, that there is only one feedback resistor and one input transistor. A differential TIA, as for example the one shown in Fig. 5.31, has more noise sources that must be taken into account. Thus, in general, differential TIAs are noisier than single-ended ones. In particular, if the TIA is balanced (fully symmetrical), the input-referred noise power is twice that given by Eqs. (5.37), (5.40), and (5.41).

Photodetector Impedance. In Section 5.1.4, we pointed out that the input-referred noise current of a TIA depends significantly on the photodetector impedance, which is mostly determined by the photodetector capacitance, C_D. Now, we can see this dependence explicitly in Eqs. (5.40) and (5.41): all the f^2-noise terms depend on either C_D or $C_T = C_D + C_I$.

The textbook approach to model amplifier noise in a source-impedance independent way is to introduce a noise *voltage source* in addition to the noise current source, $i_{n,TIA}$, which we used so far. The noise spectra of these two sources plus their correlation then provides a complete noise model that works for any source impedance. In practice, the calculations associated with this model are quite complex because of the partially correlated noise sources, and we will not pursue this approach here.

To analyze the impact of the photodetector impedance further, we repeat the previous noise calculations for the general photodetector admittance $Y_D(f) = G(f) + jB(f)$, a calculation that is easy to do. Note that if we let $G(f) = 0$ and $B(f) = 2\pi f C_D$, we should get back our old results. If we carry out this generalization for the FET front-end, we find that

$$I_{n,\text{front}}^2(f) = 2qI_G + 4kT\Gamma \cdot \frac{[G(f)R_F + 1]^2}{g_m R_F^2} \qquad (5.42)$$
$$+ 4kT\Gamma \cdot \frac{[2\pi f C_I + B(f)]^2}{g_m} + \dots.$$

Clearly, the second and third noise terms depend on the photodetector admittance. The front-end noise reaches its minimum for the optimum admittance $Y_D(f) = -1/R_F - j2\pi f C_I$ and increases quadratically as we move away from this point. This observation leads us to the idea of *noise matching*. If we can find a matching network, interposed between the photodetector and the TIA, that does not substantially attenuate the signal but transforms the capacitive admittance of the photodetector to a value that is closer to the optimum admittance, then we can improve the noise performance of our TIA. A simple implementation of this idea, which we explore further in Section 5.2.9, is to couple the photodetector to the TIA with a small inductor, as shown in Fig. 5.20(a). At high frequencies, the inductor decreases the susceptance $B(f)$ compared with $2\pi f C_D$, thus improving the noise matching.

Input-Referred RMS Noise Current. Having discussed the input-referred current noise spectrum, we now turn to the *total* input-referred current noise, which is relevant to determine the sensitivity. We can obtain this noise quantity from the spectrum by evaluating the integral in Eq. (5.4). However, more suitable for analytical hand calculations is the use of noise bandwidths or Personick integrals. As we saw in Section 4.4, these methods are equivalent. Let's review the use of noise bandwidths and Personick integrals quickly: if the input-referred noise current spectrum can be written in the form $I_{n,TIA}^2 = \alpha_0 + \alpha_2 f^2$, then the input-referred rms noise current is

$$i_{n,TIA}^{rms} = \sqrt{\alpha_0 \cdot BW_n + \alpha_2/3 \cdot BW_{n2}^3}, \tag{5.43}$$

where BW_n and BW_{n2} are the noise bandwidths. Alternatively, we can write

$$i_{n,TIA}^{rms} = \sqrt{\alpha_0 \cdot I_2 B + \alpha_2 \cdot I_3 B^3}, \tag{5.44}$$

where I_2 and I_3 are the Personick integrals.

A Numerical Example. To illustrate the foregoing theory with an example, let's calculate the noise current for a single-ended 10-Gb/s TIA realized with bipolar transistors. The input-referred noise current spectrum follows from Eqs. (5.37) and (5.41):

$$I_{n,TIA}^2(f) \approx \frac{4kT}{R_F} + \frac{2qI_C}{\beta} + 2qI_C \cdot \frac{(2\pi C_T)^2}{g_m^2} \cdot f^2 + 4kT R_b \cdot (2\pi C_D)^2 \cdot f^2. \tag{5.45}$$

To evaluate this expression numerically, we choose the same values as in our example from Section 5.2.2: $C_D = C_I = 0.15\,\text{pF}$, $C_T = 0.3\,\text{pF}$, and $R_F = 600\,\Omega$. With the typical BJT parameters $\beta = 100$, $I_C = 1\,\text{mA}$, $g_m = 40\,\text{mS}$, $R_b = 80\,\Omega$, and $T = 300\,\text{K}$, we find the spectrum that is plotted in Fig. 5.12. Besides the input-referred noise current spectrum of the TIA shown with a solid line, the contributions from each device noise source are shown with dashed lines. We see that at low frequencies, the noise from the feedback resistor (R_F) dominates, bringing the total spectral density just above $5.3\,\text{pA}/\sqrt{\text{Hz}}$. But at high frequencies, above about 5 GHz, the f^2-noise due to the base resistance (R_b) dominates and makes a significant contribution to the total noise, as we will see in a moment.

Next, to calculate the total input-referred noise current, we use the noise-bandwidth method from Eq. (5.43):

$$\overline{i_{n,TIA}^2} \approx \left(\frac{4kT}{R_F} + \frac{2qI_C}{\beta}\right) BW_n \\ + \frac{1}{3}\left(2qI_C \cdot \frac{(2\pi C_T)^2}{g_m^2} + 4kT R_b \cdot (2\pi C_D)^2\right) BW_{n2}^3. \tag{5.46}$$

With $BW_{3dB} = 6.85\,\text{GHz}$ from our example from Section 5.2.2 and the assumption that the TIA has a second-order Butterworth response, we find with the help of

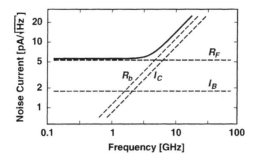

Fig. 5.12 Input-referred noise current spectrum for our bipolar TIA example.

Table 4.6 that $BW_n = 1.11 \cdot 6.85\,\text{GHz} = 7.60\,\text{GHz}$ and $BW_{n2} = 1.49 \cdot 6.85\,\text{GHz} = 10.21\,\text{GHz}$, and we arrive at the following noise value:

$$i_{n.TIA}^{rms} \approx \sqrt{(458\,\text{nA})^2 + (156\,\text{nA})^2 + (502\,\text{nA})^2 + (646\,\text{nA})^2} = 950\,\text{nA}, \quad (5.47)$$

where the terms from left to right are due to R_F, I_B, I_C, and R_b. Note that the two largest contributions to the input-referred rms noise current are from the intrinsic base resistance and the collector shot noise, both having an f^2-noise spectrum.

Finally, for a balanced differential TIA with the same transistor, resistor, and photodetector values, the noise power would be twice as large. As a result, the input-referred rms noise current would be $\sqrt{2}$ times larger, which is $i_{n.TIA}^{rms} \approx 1,344\,\text{nA}$.

Noise Optimization. Now that we have derived analytical expressions for the input-referred rms noise current, we have the necessary tools in hand to optimize the noise performance of a TIA. The noise current of a (single-ended) TIA with an FET front-end follows from Eqs. (5.37), (5.40), and (5.43) as

$$\overline{i_{n.TIA}^2} = \frac{4kT}{R_F} \cdot BW_n + 2q I_G \cdot BW_n + \frac{4kT\Gamma \cdot [2\pi(C_D + C_I)]^2}{3g_m} \cdot BW_{n2}^3 + \ldots, \quad (5.48)$$

where we have expanded $C_T = C_D + C_I$. As we already know, the first term can be minimized by choosing R_F as large as possible. The second term suggests the use of an FET with a low gate-leakage current, I_G. The third term increases with the photodetector capacitance, C_D. As we already know, this term can be minimized by making C_D small or by using noise-matching techniques to reduce the effect of C_D. The third term also increases with the input capacitance, $C_I = C_{gs} + C_{gd}$. However, simply minimizing C_I is not desirable because this capacitance and the transconductance, g_m, which appears in the denominator of the same term, are related as $g_m \approx 2\pi f_T \cdot C_I$. Instead, we should minimize the expression $(C_D + C_I)^2/g_m$, which is proportional to $(C_D + C_I)^2/C_I$ and reaches its minimum at

$$C_I = C_D. \quad (5.49)$$

Therefore, as a rule, we should choose the FET dimensions such that the input capacitance, $C_I = C_{gs} + C_{gd}$, matches the photodetector capacitance, C_D, plus any other stray capacitances in parallel to it. Given the photodetector and stray capacitances, the transistor technology, and the gate length (usually minimum length for maximum speed), the gate width of the FET is determined by this rule.

The noise current of a (single-ended) TIA with a BJT front-end follows from Eqs. (5.37), (5.41), and (5.43) as

$$\overline{i_{n.TIA}^2} = \frac{4kT}{R_F} \cdot BW_n + \frac{2qI_C}{\beta} \cdot BW_n \\ + \frac{2qI_C \cdot [2\pi(C_D+C_I)]^2}{3g_m^2} \cdot BW_{n2}^3 \qquad (5.50) \\ + \frac{4kTR_b \cdot (2\pi C_D)^2}{3} \cdot BW_{n2}^3 + \dots,$$

where we have expanded $C_T = C_D + C_I$. As before, the first term can be minimized by choosing R_F as large as possible. The second term (base shot noise) *increases* with the collector current I_C, whereas the third term (collector shot noise) *decreases* with I_C. Remember that for bipolar transistors, $g_m = I_C/V_T$ where V_T is the thermal voltage, and thus the third term is approximately proportional to $1/I_C$. As a result, there is an optimum collector current for which the total noise expression is minimized. In practice, the bias current optimization is complicated by the fact that $C_I = C_{be} + C_{bc}$ also depends on I_C, modifying the simple $1/I_C$ dependence of the third term. The third and the fourth term both increase with the photodetector capacitance, C_D, and, as we already know, can be minimized by making C_D small or by using noise-matching techniques to reduce the effect of C_D. The fourth term increases with the intrinsic base resistance, R_b, and can be minimized through layout considerations or by choosing a technology with low R_b, such as a heterojunction bipolar transistor (HBT) technology (cf. Appendix D). For transistors with a lightly doped base, such as Si BJTs or SiGe drift transistors, the base resistance decreases with increasing bias current, further complicating the bias current optimization [192]. This decrease in base resistance is the result of a lateral voltage drop in the base layer, which causes the collector current to crowd toward the perimeter of the emitter, that is, closer to the base contact.

Given a choice, should we prefer an FET or bipolar front-end? One study [62] concludes that at low speeds (<100 Mb/s), the FET front-end outperforms the bipolar front-end by a large margin. Whereas at high speeds, both front-ends perform about the same, with the GaAs MESFET front-end being slightly better.

Scaling of Noise and Sensitivity with Bit Rate. How does the input-referred rms noise current of a TIA scale with the bit rate? This is an interesting question because it is closely related to the question of how the sensitivity of a p-i-n receiver scales with the bit rate. What sensitivity can we expect for a receiver operating at 10 Gb/s, 40 Gb/s, or 160 Gb/s?

Let's start with the simple, but inaccurate, assumption that the averaged input-referred noise current density is the same for all TIAs, regardless of speed. In this case the total noise power is proportional to the receiver bandwidth, and thus the bit rate B. Therefore, the input-referred rms noise current is proportional to $\sqrt{B}$. Correspondingly, the sensitivity of a p-i-n receiver should drop by 5 dB for every decade of speed increase, provided the detector responsivity is bit-rate independent. However, by analyzing Table 5.2, which contains noise data of commercially available TIAs, we find that the input-referred rms noise current, $i_{n,TIA}^{rms}$, scales roughly with $B^{0.95}$, corresponding to a sensitivity drop of about 9.5 dB per decade for p-i-n receivers. Finally, the fit to the experimental receiver-sensitivity data presented in [207] (see Fig. 5.13) shows a slope for the p-i-n receiver of about 15.8 dB per decade. Both numbers are significantly larger than 5 dB per decade, which implies that the averaged noise density must increase with bit rate. How can we explain these numbers?

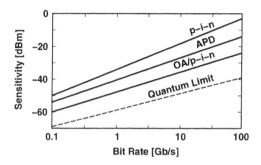

Fig. 5.13 Scaling of receiver sensitivity (at $BER = 10^{-9}$) with bit rate [207].

From Eqs. (5.48) and (5.50), we see that for a given technology and operating point (fixed C_D, C_I, g_m, Γ, R_b, and I_C), many noise terms scale with BW_{n2}^3 and thus B^3. Exceptions are the gate and base shot-noise terms and the feedback-resistor noise term, which scale with BW_n and thus B. However, remember that the feedback resistor, R_F, is *not* bandwidth independent. As we go to higher bit rates, we are forced to reduce R_F. With the transimpedance limit Eq. (5.25) and Eq. (5.20), we can derive that for a given technology (fixed C_D, C_I, and f_T), the feedback resistor, R_F, scales with $1/BW_{3dB}^2$ and thus $1/B^2$. As a result, the feedback-resistor noise term scales with B^3, like many of the other noise terms.[4] Following this analysis and neglecting the base and gate shot-noise terms, we would expect the input-referred rms noise current to be about proportional to $B^{3/2}$. Correspondingly, the sensitivity of a p-i-n receiver should drop by about 15 dB for every decade of speed increase. This number agrees well with the data shown in Fig. 5.13. Note that if we do not require that the technology remains fixed across bit rates, but assume that higher f_T technologies are available at higher bit rates, then the slope of the curve is reduced.

---

[4]The $R_F \sim 1/B^2$ scaling law leads to extremely large feedback-resistor values for low bit-rate receivers (e.g., 1 Mb/s). In practice, dynamic-range and parasitic-capacitance considerations may force the use of smaller resistor values, thus producing more feedback-resistor noise than predicted by the B^3 scaling law at low bit rates [63]. A consequence of this modified scaling law is that low bit rate receivers tend to be limited by the feedback-resistor noise rather than the front-end noise.

For a receiver with an APD or an optically preamplified p-i-n detector, the sensitivity is determined jointly by the TIA noise and the detector noise (cf. Eqs. (4.28) and (4.29)). In the extreme case where the detector noise dominates the TIA noise, we can conclude from Eqs. (4.27), (4.28), and (4.29) that the sensitivity scales proportional to B, corresponding to a slope of 10 dB per decade. The same is true for the quantum limit in Eq. (4.36). In practice, there is some noise from the TIA and the scaling law is somewhere between B and $B^{3/2}$, corresponding to a slope of 10 to 15 dB per decade. The experimental data in Fig. 5.13 confirms this expectation: we find a slope of about 13.5 dB per decade for APD receivers and 12 dB per decade for optically preamplified p-i-n receivers. Note that for a detector-noise limited receiver with a slope of 10 dB per decade, the number of photons per bit (or energy per bit) is independent of the bit rate. However, a receiver with TIA noise, in particular a p-i-n receiver, needs more and more photons per bit (or energy per bit) as we go to higher bit rates.

5.2.4 Adaptive Transimpedance

We now have completed our discussion of the basic shunt-feedback TIA. In the following sections, we explore a variety of modifications and extensions to this basic topology. Although we discuss each technique in a separate section, multiple techniques can often be combined and applied to the same TIA design. We start with a TIA that has an adaptive transimpedance.

Variable Feedback Resistor. The *dynamic range* of a TIA is defined by its overload current, at the upper end, and its sensitivity, at the lower end. For the basic shunt-feedback TIA, both quantities are related to the value of the feedback resistor, and thus the dynamic range can be extended by making this resistor adapt to the input signal strength, as indicated in Fig. 5.14(a) [68, 91, 92, 129].

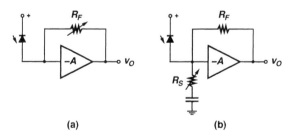

Fig. 5.14 TIA with adaptive transimpedance: (a) variable feedback resistor and (b) variable input shunt resistor.

Let's analyze this approach in more detail. The input overload current, i_{ovl}^{pp}, is given by either Eq. (5.17) or Eq. (5.18), whichever expression is smaller. In either case, the overload current is inversely proportional to the feedback resistor R_F. A similar argument can be made for the maximum input current for linear operation, i_{lin}^{pp}, which also turns out to be proportional to $1/R_F$. The sensitivity, the lower end of the dynamic range, is proportional to the input-referred rms noise current: $i_{\text{sens}}^{pp} \sim i_{n,TIA}^{rms}$. For small values of R_F, when the feedback-resistor noise dominates the front-end noise, the electrical sensitivity, i_{sens}^{pp}, is proportional to $1/\sqrt{R_F}$; for large values of R_F, when the front-end noise dominates, the sensitivity becomes independent of R_F. The optical overload and sensitivity limits following from this analysis are plotted in Fig. 5.15 as a function of R_F on a log-log scale. Now, we make the feedback resistor adaptive: for a large optical signal, R_F is reduced to prevent the high input current from overloading the TIA; for a weak optical signal, R_F is increased to reduce the noise contributed by this resistor. It can be seen clearly from Fig. 5.15 how an adaptive feedback resistor extends the dynamic range over what can be achieved with any fixed value of R_F. As a result of varying R_F the transimpedance

$$R_T = \frac{A}{A+1} \cdot R_F \qquad (5.51)$$

varies too; hence we have a TIA with *adaptive transimpedance*.

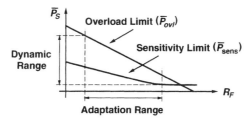

Fig. 5.15 Extension of the dynamic range with an adaptive feedback resistor.

The variable feedback resistor can be implemented with an FET operating in the linear regime, usually connected in parallel to a fixed resistor to improve the linearity and to limit the maximum resistance. The automatic adaptation mechanism can be implemented with a circuit that determines the output signal strength, compares it with a desired value, and controls the gate voltage of the FET such that this value is achieved. Given a DC-balanced NRZ signal with high extinction, the average signal value is proportional to the signal swing, thus permitting an easy way to generate the control voltage. The same control voltage used for offset control, which is derived from the signal's average value (cf. Section 5.2.10), also may be used for transimpedance control [205]. An important consideration for TIAs with an adaptive feedback resistor is their stability. We can see from Eqs. (5.21) and (5.22) that if we vary R_F while keeping A and T_A fixed, both the bandwidth and the quality factor will change. More specifically, if we reduce R_F, the open-loop low-frequency pole at $1/(R_F C_T)$ speeds up, which leads to peaking given a fixed loop gain, A, and a fixed open-loop high-frequency pole, $1/T_A$ (cf. Fig. 5.7 and Eq. (5.23)). In practice, it can be challenging to satisfy the specifications for bandwidth, group-delay variation, and peaking over the full adaptation range. [→ Problem 5.10]

Variable Input Shunt Resistor. An alternative to the TIA with variable feedback resistor is the TIA with variable input shunt resistor, R_S, which is shown in Fig. 5.14(b) [205]. This scheme also extends the dynamic range of the TIA: for a large optical signal, R_S is reduced to divert some of the photodetector current to AC ground, thus preventing the input current from overloading the TIA. (An additional mechanism is required to prevent the DC current from overloading the TIA; cf. Section 5.2.10.) For a weak optical signal, R_S is increased to route more of the photocurrent into the TIA and at the same time reduces the noise contributed by the shunt resistor. As a result of varying R_S, the transimpedance

$$R_T = \frac{A}{A + 1 + R_F/R_S} \cdot R_F \tag{5.52}$$

varies too. As before, the variable shunt resistor can be implemented with an FET operating in the linear regime. Varying the shunt resistor has the advantage over varying the feedback resistor that it is easier to maintain stability and avoid peaking. More specifically, if we reduce R_S, the open-loop low-frequency pole at $1/[(R_S \| R_F)C_T]$ speeds up whereas the loop gain, $AR_S/(R_S + R_F)$, decreases by the same amount, thus maintaining an approximately constant closed-loop response (cf. Fig. 5.7 and Eq. (5.23)).

In general, the bandwidth of TIAs with adaptive transimpedance tends to increase with the magnitude of the input signal, that is, with $1/R_T$. For transmission systems without optical amplifiers, this usually is not a concern. Although the bandwidth increase at high power levels causes the receiver to pick up more noise, the signal is strong and the overall SNR is high. However, in optically amplified transmission systems, the situation is different. There, an increase in received signal power may be accompanied by a similar increase in optical noise because the optical amplifiers near the receiver amplify the signal as well as the noise. The result is an approximately constant (power independent) OSNR at the receiver. Under these conditions, an increase in TIA bandwidth at high power levels is detrimental because it leads to a decrease in electrical SNR and an increase in BER (cf. Eq. (4.32)). A filter, added at the output of the TIA, can stabilize the receiver's bandwidth.

5.2.5 Post Amplifier

High-speed TIAs typically feature outputs with a 50-Ω impedance. Such outputs permit the reflection-free transmission of the output signal over a standard 50-Ω transmission line to the next block such as the main amplifier (cf. Appendix C). The 50-Ω impedance usually is provided by an output buffer that follows the basic shunt-feedback TIA, as shown in Fig. 5.16. If this buffer has a gain larger than one, it acts as a *post amplifier* and boosts the transimpedance of the basic shunt-feedback TIA [113].

It can be shown easily that the overall transimpedance of the circuit in Fig. 5.16 is given by

$$R_T = A_1 \cdot \frac{A}{A+1} \cdot R_F, \tag{5.53}$$

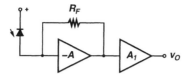

Fig. 5.16 TIA with post amplifier.

where A_1 is the gain of the post amplifier. From this equation, we see that there are two ways to increase the overall transimpedance: (i) increase the feedback resistor, R_F, or (ii) increase the post-amplifier gain, A_1. An important difference between the two, however, is that in the first case, the noise is reduced as the transimpedance is increased, whereas in the second case, the noise remains approximately constant. Thus, we should always try to make R_F as large as possible, or at least large enough such that the feedback-resistor noise becomes small compared with the front-end noise, even if a post amplifier is present. [→ Problem 5.11]

It is interesting to observe that the transimpedance limit, presented in Eq. (5.25), does not apply to a TIA with post amplifier. This can be understood by continuing our numerical example from Section 5.2.2. There, the basic shunt-feedback TIA had a bandwidth of 6.85 GHz combined with a transimpedance of 500 Ω. In the 44-GHz technology, which we assumed for the example, we can build a post amplifier with a gain of two and a bandwidth of 22 GHz. Thus, the TIA with post amplifier has a transimpedance of 1 kΩ, and the bandwidth shrinks very little from the original 6.85 GHz (to about 6.5 GHz).

The post amplifier described here is essentially a "hidden" main amplifier, or at least the first stage of it. Thus, the post amplifier can be implemented with any of the main-amplifier circuit techniques that we cover in Chapter 6.

5.2.6 Common-Base/Gate Input Stage

We know from Eqs. (5.21) and (5.22) that the photodetector capacitance, C_D, influences both the bandwidth and the stability of the basic shunt-feedback TIA. More specifically, if we increase C_T ($= C_D + C_I$), the open-loop low-frequency pole at $1/(R_F C_T)$ slows down, which reduces the TIA bandwidth; alternatively, if we decrease C_T, the open-loop low-frequency pole speeds up, which leads to peaking given a fixed loop gain, A, and a fixed open-loop high-frequency pole, $1/T_A$ (cf. Fig. 5.7 and Eq. (5.23)). To obtain a stable TIA frequency response and a reliable bandwidth for a variety of photodetectors with differing capacitances, we can insert a current buffer in the form of a common-base (or common-gate) stage between the photodetector and the basic shunt-feedback TIA, as shown in Fig. 5.17. The *common-base input stage* (Q_1, R_C, and R_E) isolates the photodetector capacitance C_D from the critical node x [190].

Ideally, the expression for the low-frequency transimpedance, Eq. (5.20), is not affected by this addition because the current gain of the common-base stage is close

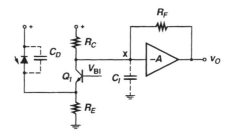

Fig. 5.17 TIA with common-base input stage.

to unity. The new pole introduced by the common-base stage should be placed sufficiently high such that it does not interfere with the frequency response of the shunt-feedback TIA. The low input resistance of the common-base stage, which is about $1/g_m$, helps to satisfy this condition (cf. Section 6.3.2 on cascodes). For example, if Q_1 is biased at a collector current of 2 mA, the resistance into the emitter is approximately 12.5 Ω. Thus, if we limit the total input capacitance (which includes the photodetector capacitance) to less than 1 pF, the pole frequency will be higher than 13 GHz. Besides isolating the photodetector capacitance from node x, the common-base stage also may reduce the capacitive load at node x. The original load of $C_T = C_D + C_I$ is replaced by $C'_T = C_O + C_I$, where C_O is the output capacitance of the common-base stage. If C'_T is smaller than C_T, we can increase R_F to R'_F to move the open-loop low-frequency pole back to its original location: $1/(R'_F C'_T) = 1/(R_F C_T)$. As a result, the transimpedance is increased and the noise contributed by the feedback resistor is decreased. Note that the transimpedance limit, Eq. (5.25), does not apply in its original form but must be modified by replacing C_T with C'_T.

The primary drawback of the common-base stage is that it introduces a number of new noise sources (Q_1, R_C, and R_E) that are located right at the input of the TIA and thus directly impact the input-referred noise current. In practice, these new noise contributions easily may nullify the noise improvement mentioned before. Furthermore, if the current gain of the common-base (or common-gate) stage is less than one, the input-referred noise current of the shunt-feedback TIA is enhanced and its transimpedance is reduced. Finally, the TIA's power consumption is increased when using a common-base input stage.

5.2.7 Current-Mode TIA

Instead of adding a current buffer in front of the shunt-feedback TIA, we may consider replacing the feedback voltage amplifier by a feedback current amplifier, as shown in Fig. 5.18(a). The current amplifier senses the input current, i, with a small resistor, R_S, and outputs the amplified current, Ai, at a high-impedance output. The current amplifier can, for example, be implemented with a current mirror that has an output FET that is A times wider than the input FET, as shown in Fig. 5.18(b). Similar to the current buffer, the current amplifier provides a low input impedance, making the frequency response of the TIA insensitive to the photodetector capacitance, C_D [130, 199]. The use of current buffers and current amplifiers is frequently referred to as *current-mode techniques*.

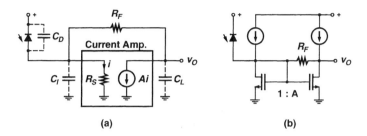

Fig. 5.18 (a) Current-mode TIA and (b) its implementation with a current mirror.

From the TIA circuit and current-amplifier model shown in Fig. 5.18(a), we can calculate the low-frequency transimpedance as

$$R_T = \frac{A}{A+1} \cdot (R_F - R_S/A), \tag{5.54}$$

which is very similar to the result that we obtained for the voltage-mode TIA in Eq. (5.20). The input resistance of the current-mode TIA turns out to be $R_I = R_S/(A+1)$ at low frequencies and $R_I = R_S$ at high frequencies. Because R_S and thus R_I is small, the photodetector capacitance, C_D, as well as the input capacitance, C_I, have little impact on the frequency response of the TIA. In fact, the bandwidth of this current-mode TIA mostly is determined by the output pole, which is given by the feedback resistor, R_F, and the load capacitance, C_L. [→ Problem 5.12]

The primary drawback of the current-mode TIA shown in Fig. 5.18(b) is that it contains more noise sources than the corresponding voltage-mode TIA. In particular, the input FET of the current mirror is located right at the input of the TIA, and thus directly impacts the input-referred noise current.

5.2.8 Active-Feedback TIA

In yet another variation of the shunt-feedback TIA, the voltage amplifier is left in place, but the feedback resistor R_F is replaced by a voltage-controlled current source (a transconductor), as shown in Fig. 5.19(a). This topology is known as an *active-feedback TIA*. The transconductor g_{mF} can, for example, be implemented with an FET, as shown in Fig. 5.19(b). Note that the voltage amplifier must be noninverting to obtain negative feedback through M_F.

From the TIA circuit and transconductor model shown in Fig. 5.19(a), we can calculate the low-frequency transimpedance as

$$R_T = 1/g_{mF}. \tag{5.55}$$

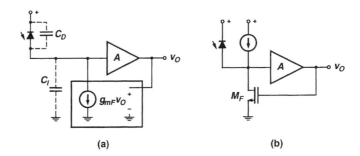

Fig. 5.19 (a) Active-feedback TIA and (b) its implementation with a MOSFET.

The low-frequency input resistance of the active-feedback TIA turns out to be $R_I = 1/(A \cdot g_{mF})$. Note that if we identify $1/g_{mF}$ with R_F, this TIA behaves similar to the shunt-feedback TIA. An advantage of this topology over the shunt-feedback TIA is that the voltage amplifier output is not resistively load by the feedback device (M_F). However, active feedback tends to result in a higher input capacitance (C_I) and more noise than shunt feedback. Furthermore, active feedback with an FET is less linear than shunt feedback with a resistor. The main application of the active-feedback TIA is as a load element in main amplifiers. We discuss this application further in Chapter 6. [→ Problem 5.13]

5.2.9 Inductive Input Coupling

In high-speed receivers, the photodetector and the TIA chip often are located in the same package. This approach is known as *copackaging* and has the purpose of minimizing the interconnect parasitics between the detector and the TIA. The bond wire that typically is used for this interconnect can be modeled by the inductor L_B, as shown in Fig. 5.20(a). Although we may think at first that this inductor should be made as small as possible, it turns out that there is an optimum value for L_B corresponding to an optimum length for the bond wire.

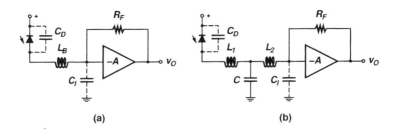

Fig. 5.20 TIA with (a) an inductor and (b) an L-C low-pass network to couple the photodetector to the input.

In Section 5.2.3, we observed that a small series inductor can improve the noise matching and thus reduce the input-referred noise current. Besides this, a small series inductor also can enhance the TIA's bandwidth. We can understand this in a qualitative way as follows: near the resonance frequency of the tank circuit formed by C_D, L_B, and C_I, the current from the photodetector through L_B into the TIA is *enhanced* over the situation without inductor ($L_B = 0$). In other words, the shunting effect of C_D is partly "tuned out" by L_B, causing a more efficient transfer of the photocurrent into the TIA. If we place the resonance near the point where the TIA's frequency response starts to roll off, we can extend its 3-dB bandwidth. The reduction of the input-referred noise current, which we discussed earlier in terms of noise matching, also can be explained by this resonant current gain in the input network. The resonance of the tank circuit occurs approximately at the frequency $1/(2\pi\sqrt{L_B C_D})$, assuming C_I is effectively shorted by the low input resistance, R_I. Thus, the bond-wire inductance that places this resonance near the 3-dB point of the TIA is [108]

$$L_B \approx \frac{1}{(2\pi \cdot BW_{3dB})^2 \cdot C_D}. \tag{5.56}$$

Note that inserting L_B in between the photodetector and the TIA introduces two new poles to the TIA's transfer function. In practice, it can be difficult to coordinate the new and old poles such that the specifications for peaking and group-delay variation are satisfied. [→ Problem 5.14]

The inductor in Fig. 5.20(a) can be replaced by a more general low-pass coupling network, as shown in Fig. 5.20(b). The idea behind this network is to incorporate the parasitic capacitances C_D and C_I into an *L-C* low-pass filter (C_D, L_1, C, L_2, and C_I), which is designed to have a frequency response that enhances the TIA's bandwidth and reduces its input-referred noise current [71]. To demonstrate the potential of this technique, let's make an idealized example in which we assume that the feedback amplifier has an infinite bandwidth and that the detector and input capacitances are equal, $C_D = C_I$. Thus, before inserting a coupling network, the TIA has a first-order frequency response with the bandwidth $1/(2\pi \cdot 2R_I C_D)$, where R_I is the TIA's input resistance (cf. Eqs. (5.15) and (5.16)). Now, let's choose as the coupling network an infinite, lossless, artificial transmission line with all shunt capacitances equal to $C_D = C_I$ and its characteristic impedance equal to R_I. This coupling network has the desirable properties of absorbing the parasitic capacitances C_D and C_I into its end points and preventing signal reflections from the TIA back to the detector. The bandwidth of the TIA is now determined by the cutoff frequency of the artificial transmission line, which is given by $2/(2\pi \cdot R_I C_D)$. Thus, this coupling network improves the bandwidth $4\times$ over the original bandwidth. We explain this transmission-line argument in greater detail in Section 6.3.2, when we discuss inductive interstage networks for broadband amplifiers.

5.2.10 Differential TIA and Offset Control

Differential circuits have a number of important advantages over single-ended circuits. Among the most significant ones are the improved immunity to power-supply and substrate noise as well as the increased voltage swing (cf. Appendix B). For these reasons, *differential TIAs* find application in noisy environments, such as a mixed-signal system on a chip, and in low-voltage systems where the differential output signal provides a larger dynamic range. A differential TIA also facilitates the connection to a differential main amplifier, avoiding the need for a reference voltage. The main drawbacks of differential TIAs are their higher input-referred noise and higher power consumption.

To implement a fully differential optical receiver, we would need not only a differential TIA but also a differential photodetector. Unfortunately, differential detectors are not normally available for the on-off keying (OOK) format and thus most receivers are based on photodetectors that produce a single-ended current signal. As a result, differential TIAs typically have a single-ended input and differential outputs, as shown in Fig. 5.1(b). An important question is about how to interface the single-ended detector with the differential amplifier. There are two major approaches that we call the *balanced TIA* and the *pseudo-differential TIA* in the following.

Balanced TIA. Figure 5.21 shows how the basic shunt-feedback TIA can be turned into a differential topology.[5] For the circuit to be balanced, the unused input, v_{IN}, must be loaded with the same impedance as that presented by the photodetector. One way to do this is to connect a matched dummy photodetector, which is kept in the dark, to the unused input. Alternatively, a small capacitor that matches the photodetector capacitance, $C_X = C_D$, can be connected to this input, as shown in Fig. 5.21. In the case that a common-base/gate input stage is used, it must be placed in front of both inputs to preserve the balance.

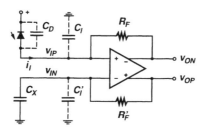

Fig. 5.21 Differential TIA with single-ended photodetector.

The balanced TIA is characterized by excellent noise immunity. Any noise on the power supply or the substrate couples equally strongly to the noninverting as well as the inverting input of the feedback amplifier and thus is suppressed as a common-mode disturbance. The transimpedance, bandwidth, and stability analysis, which we carried out in Section 5.2.2, remains valid for the balanced TIA, if we replace the single-ended

---

[5]In this and the following circuits, we always assume that the differential-output feedback amplifier includes some means to keep the output common-mode voltage at a fixed level. For implementation examples, see Section 5.3.

input voltage, v_I, by the *differential* input voltage, $v_{IP} - v_{IN}$, the single-ended output voltage, v_O, by the *differential* output voltage, $v_{OP} - v_{ON}$, the single-ended feedback-amplifier gain by the *differential* gain, $A = \Delta(v_{OP} - v_{ON})/\Delta(v_{IP} - v_{IN})$, and so forth. In particular, the differential transimpedance is given by $R_T = \Delta(v_{OP} - v_{ON})/\Delta i_I = A/(A+1) \cdot R_F$, as in Eq. (5.20). However, the input-referred rms noise current of the balanced TIA is $\sqrt{2} \times$ larger than that of the corresponding single-ended TIA (cf. Section 5.2.3), which, unfortunately, may reduce the optical receiver sensitivity by up to 1.5 dB.

Note that, because of its single-ended nature, the photodetector does not "see" the differential input resistance $R_I = 2\Delta(v_{IP} - v_{IN})/\Delta(i_{IP} - i_{IN}) = 2R_F/(A+1)$, but the single-ended input resistance $R_I = \Delta v_{IP}/\Delta i_I$, which is about $R_F/2$ (cf. Appendix B.2 for the definition of the differential resistance). As a result, the voltage swing at the photodetector of a balanced TIA typically is larger than that of a single-ended TIA. [$\rightarrow$ Problem 5.15]

Pseudo-Differential TIA. If noise immunity is not a primary concern, we can replace the matched capacitor C_X in Fig. 5.21 by a large capacitor, $C_X \rightarrow \infty$, shorting the unused input to AC ground. This large capacitor disables the AC feedback through R'_F and we end up with essentially a single-ended topology. As a result, the thermal noise contribution of R'_F is eliminated and the input-referred rms noise current is reduced. However, because of the asymmetric input capacitances, power-supply and substrate noise couple differently to the two inputs, causing noise to leak into the differential mode.

The transimpedance, bandwidth, and stability analysis, which we have carried out for the single-ended TIA, remain valid for the pseudo-differential TIA if we replace the single-ended input voltage, v_I, by the single-ended input voltage, v_{IP}, the single-ended output voltage, v_O, by the single-ended output voltage, v_{ON}, the single-ended feedback-amplifier gain, A, by *half* of the differential gain, $1/2 \cdot A = |\Delta v_{ON}/\Delta v_{IP}|$, and so forth. It follows that the single-ended transimpedance is now given by $R_T = |\Delta v_{ON}/\Delta i_I| = A/(A+2) \cdot R_F$. Although only one output is used for internal shunt feedback, both outputs are available to the outside world. Thus, we also can specify the differential transimpedance, which is twice the single-ended one: $R_T = \Delta(v_{OP} - v_{ON})/\Delta i_I = 2A/(A+2) \cdot R_F \approx 2R_F$.

In comparison with the balanced TIA, the pseudo-differential TIA has a somewhat better sensitivity (lower input-referred noise current) but reduced immunity to power-supply and substrate noise. Furthermore, its single-ended input resistance has the lower value $R_I = 2R_F/(A+2)$ compared with $R_I \approx R_F/2$. Note that if a TIA that is designed for the balanced configuration is operated in the pseudo-differential configuration, its pole placement becomes nonoptimal because the feedback amplifier gain effectively is cut in half by AC grounding the unused input. In fact, the resulting pseudo-differential configuration has about twice the differential transimpedance but a lower bandwidth and quality factor.

Offset Control. Besides the asymmetry in input impedance, the single-ended photodetector also causes an asymmetry in the output-signal levels. The noninverting and inverting output signals of the TIA in Fig. 5.21 are vertically offset against each other, as shown in Fig 5.22(a). This can be understood as follows. First, recall the unipolar nature of the photocurrent (cf. Fig. 5.3). Thus, when the photodetector is dark, the input current is close to zero and the two output voltages are about equal (they both assume the output common-mode voltage). When the detector is illuminated, a current starts to flow into R_F, forcing v_{ON} (dashed line) to decrease. Meanwhile, v_{OP} (solid line) has to increase to keep the output common-mode voltage at a fixed level. Note that if the input current were bipolar, swinging symmetrically about zero, no such offset would occur.

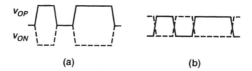

Fig. 5.22 TIA output signals: (a) without and (b) with offset control.

Although this output offset could be suppressed by AC coupling the TIA outputs to the inputs of the next block (usually the main amplifier), it often is preferable to eliminate the offset in the TIA with an *offset control* circuit. By comparing Fig 5.22(a) and 5.22(b), we see that without offset control, only half of the available TIA output swing can be used, whereas with offset control, all of the swing can be used. Thus, offset control improves the dynamic range of the TIA. Figure 5.23 shows a typical offset control circuit. The idea behind this circuit is to remove the average photocurrent from the detector by subtracting the DC current I_{OS}, thus making the current flowing into the TIA swinging symmetrically about the zero level. The control circuit determines the output offset voltage by subtracting the time-averaged (low-pass filtered) values of the two output signals and, in response to this difference, controls the current source I_{OS} such that the output offset becomes zero. Besides the offset control circuit shown in Fig. 5.23, there are several other solutions. For example, the output offset can be determined from the difference of the peak values (instead of the average values) of the two output signals or it can be determined from the average voltage drop across R_F.

To minimize the TIA's input capacitance, we may consider moving the offset-control current source to the unused input and reversing its polarity to $-I_{OS}$. Although this arrangement does eliminate the output offset voltage, it suffers from the drawback that the amplifier's average input common-mode voltage now varies strongly with the received power level, and as a result, the amplifier's common-mode range may be violated at high input power levels.

Although we introduced the offset control mechanism in the context of the differential TIA, it also can be applied to the single-ended TIA. Here again, the offset control mechanism subtracts the average current from the photocurrent, now with the purpose of making the DC component of the output signal independent of the received power level.

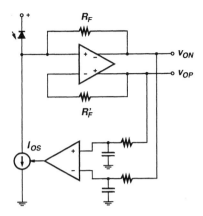

Fig. 5.23 Differential TIA with offset control.

5.2.11 Burst-Mode TIA

How does a burst-mode TIA differ from what we discussed so far? A burst-mode TIA must be able to accept input signals whose amplitude vary significantly from burst to burst (up to 30 dB in PON systems). Furthermore, the bursty input signal is not DC balanced, which precludes offset control mechanisms based on averaging.

Amplitude Control. The large amplitude variations of the input signal point to the use of a TIA with adaptive transimpedance, as discussed in Section 5.2.4. But in contrast to an adaptive continuous-mode TIA, the burst-mode TIA requires a fast adaptation mechanism. Burst-mode systems often provide only a short (e.g., 24 bits) preamble, during which the receiver must adjust its gain and decision threshold before receiving the payload. A fast burst-by-burst adaptation mechanism can be implemented, for example, as follows [204]. Before the burst arrives, the transimpedance is set to its maximum value. Then, when the first one bit of the burst arrives, a peak detector at the output of the TIA detects the amplitude and reduces the transimpedance accordingly. The transimpedance is held constant for the duration of the burst, and at the end it is reset to its maximum value.

An alternative approach, which avoids the need for fast control circuits, is based on an intentionally nonlinear TIA that compresses the dynamic range in a manner similar to a logarithmic amplifier. Figure 5.24 shows an implementation example with a nonlinear feedback network consisting of R_F, R_{F2}, and a diode [17]. For small input signals, the diode is turned off and the transimpedance is determined by R_F. For input signals that produce a voltage drop across R_F that is large enough to forward bias the diode, the feedback resistance reduces to $R_F \| R_{F2}$, thus reducing the transimpedance and preventing the TIA from overloading. The capacitor C_{F2} prevents the open-loop low-frequency pole from speeding up when R_{F2} is switched on and thus avoids peaking.

142 TRANSIMPEDANCE AMPLIFIERS

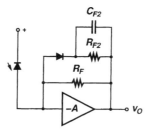

Fig. 5.24 Burst-mode TIA with nonlinear feedback.

Threshold and Offset Control. Another issue in burst-mode receivers is the accurate control of the decision threshold voltage. Because the amplitude of the signal is varying from burst to burst, the decision threshold voltage must be set for every single burst. An incorrectly set threshold level causes pulse-width distortions or the complete loss of data. For single-ended burst-mode TIAs, such as the one shown in Fig. 5.24, threshold control usually is performed by the burst-mode main amplifier (cf. Section 6.3.6). For differential burst-mode TIAs, just as in the case of differential continuous-mode TIAs, we would like to eliminate the output offset voltage to improve their dynamic range. By doing so, we also implicitly define a decision threshold level, namely the crossover voltage $v_{OP} = v_{ON}$, which corresponds to the zero-threshold level of the differential signal. Thus, by performing offset control for a TIA, we also implicitly perform threshold control.

How can we eliminate the output offset voltage of a burst-mode TIA? A simple AC coupling circuit as well as the offset-control circuit based on low-pass filtering in Fig. 5.23 do not work because the received signal lacks DC balance. The offset-control circuit shown in Fig. 5.25, also known as the *adaptive threshold control* (ATC) circuit, eliminates the output offset voltage on a burst-by-burst basis [118, 119]. For now, let's ignore the current source I_{OS}. Before the burst arrives, the peak detector is reset to a voltage equal to the output common-mode voltage of the amplifier. The differential output voltage is now zero. Then, when the first one bit of the burst arrives, v_{OP} increases and v_{ON} decreases. The peak value of v_{OP} is stored in the peak detector and fed back to the inverting input. During the next zero bit, the value of the peak detector appears at the v_{ON} output. Why? Because there is no voltage drop across R'_F (no current), no voltage across the inputs of the feedback amplifier (for a large gain), and no voltage drop across R_F (no photocurrent). Thus, the peak values of both output signals are equal, which means that the output offset has been eliminated. When the entire burst has been received, the peak detector is reset to its initial value.

In terms of transimpedance, bandwidth, and stability, the burst-mode TIA in Fig. 5.25 is similar to the pseudo-differential (continuous-mode) TIA discussed before. In particular, its differential transimpedance is about $R_T \approx 2R_F$. Note that the peak detector output presents an AC ground, once the burst amplitude has been acquired.

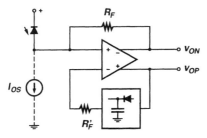

Fig. 5.25 Differential burst-mode TIA with adaptive threshold control.

Chatter Control. Besides amplitude, threshold, and offset control, there is another problem with burst-mode TIAs that occurs during the extended periods of time that may elapse in between bursts. During these dead periods, no optical signal is received, the transimpedance is set to its maximum value, and the decision threshold is set close to zero in anticipation of a burst. Unfortunately, with these settings, the amplified TIA noise crosses the decision threshold randomly, thus generating a random bit sequence called *chatter* at the output of the receiver.

One way to fix this problem is to introduce a small intentional offset voltage at the TIA output. The current source I_{OS} shown in Fig. 5.25 can do just this [118]. Note that the offset voltage must be larger than the peak noise voltage to suppress the chatter, but it must not be too high, either, or the receiver's sensitivity is degraded.

5.2.12 Analog Receiver

Before leaving this section, we briefly look at the world of analog receivers. Such receivers are used, for example, in CATV/HFC applications and in optical links connecting cellular-radio base stations with remote antennas. In contrast to digital receivers, analog receivers must be highly linear to minimize the distortion of the fragile analog signals (e.g., AM-VSB and QAM signals). A simple implementation of an analog receiver is shown in Fig. 5.26(a). It consists of a low-impedance front-end followed by a linear amplifier. Typically, the front-end impedance and the amplifier input impedance are 50 Ω (or 75 Ω in CATV systems) such that standard cables and connectors can be used to assemble the receiver. The linearity of the p-i-n photodetector usually is quite good, but close attention must be payed to the linearity of the amplifier (see Section 8.2.10 for an example of a linear CATV amplifier). The linearity of an analog CATV receiver is specified in terms of the composite second order (CSO) distortion and the composite triple beat (CTB) distortion (cf. Section 4.8). Typical numbers for a good AM-VSB receiver are $CSO < -65$ dBc and $CTB < -80$ dBc at a received optical power of 0 dBm.

Besides linearity, low noise also is an important factor for analog receivers. As we know, the low-impedance front-end shown in Fig. 5.26(a) is rather noisy, but by using a transimpedance amplifier or a matching transformer, the noise performance can be improved. Figure 5.26(b) shows a low-noise receiver front-end with an impedance

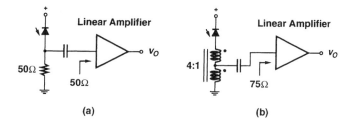

Fig. 5.26 Receivers for analog signals: (a) low-impedance front-end and (b) front-end with matching transformer.

matching transformer [14]. The transformer with a 4:1 turns ratio matches the photodetector impedance of about 1.2 kΩ to the 75-Ω input impedance of the amplifier (16:1 impedance ratio). This technique eliminates the input resistor to ground and the noise associated with it. Furthermore, because this transformer has a current gain of 4×, the noise current from the amplifier is attenuated by the same factor 4× when referred back to the photodetector. In the remainder of this section, we analyze the impact of the front-end noise (including the amplifier noise), shot noise, and laser noise on the receiver's performance. This is an instructive exercise because the results are quite different from what we know from digital receivers.

Noise Analysis. The noise performance of an analog transmission system normally is characterized in terms of the signal-to-noise ratio (SNR), if baseband modulation is used, or carrier-to-noise ratio (CNR), if passband modulation is used (cf. Section 4.2). Assuming a passband system with a sinusoidal carrier, we can calculate the CNR as follows: the average current produced by the p-i-n photodetector is $\mathcal{R}\overline{P_S}$, where $\overline{P_S}$ is the average optical power received and $\mathcal{R}$ is the responsivity of the detector. The amplitude of the sine-wave current produced by the detector is $m\mathcal{R}\overline{P_S}$, where m is the *modulation index*. Thus, the received electrical signal power is $1/2 \cdot (m\mathcal{R}\overline{P_S})^2$. Next, we consider three noise components: (i) the noise power from the front-end and amplifier circuit, which we designate $\overline{i_{n,\mathrm{amp}}^2}$ as usual, (ii) the shot noise power from the p-i-n photodetector, which follows from Eq. (3.5) as $\overline{i_{n,PIN}^2} = 2q\mathcal{R}\overline{P_S} \cdot BW_n$, and (iii) the laser noise known as relative intensity noise (RIN), which is $\overline{i_{n,RIN}^2} = RIN \cdot \mathcal{R}^2\overline{P_S}^2 \cdot BW_n$. We discuss the latter noise in more detail in Section 7.2 (cf. Eq. (7.11)). Now, dividing the signal power by the total noise power reveals:

$$CNR = \frac{1/2 \cdot m^2\mathcal{R}^2\overline{P_S}^2}{\overline{i_{n,\mathrm{amp}}^2} + 2q\mathcal{R}\overline{P_S} \cdot BW_n + RIN \cdot \mathcal{R}^2\overline{P_S}^2 \cdot BW_n}. \tag{5.57}$$

We can discuss the CNR given by this equation in terms of three upper bounds, one for each of the three noise components. If we consider the front-end noise only, $CNR < m^2\mathcal{R}^2\overline{P_S}^2/(2\overline{i_{n,\mathrm{amp}}^2})$; if we consider the shot noise only, $CNR < m^2\mathcal{R}\overline{P_S}/(4q \cdot BW_n)$; and if we consider the RIN noise only, $CNR < m^2/(2RIN \cdot BW_n)$. Figure 5.27 shows the total CNR (solid line) together with these bounds (dashed lines) for values that are typical for an analog CATV application ($m = 5\%$, $\mathcal{R} = 0.9\,\text{A/W}$, $i_{n,\text{amp}}^{rms} = 12\,\text{nA}$, $BW_n = 4\,\text{MHz}$, and $RIN = -150\,\text{dB/Hz}$). Note that each bound depends differently on the received optical power $\overline{P_S}$: the CNR due to the front-end noise increases with $\overline{P_S}^2$, the CNR due to the shot noise increases with $\overline{P_S}$, and the CNR due to the RIN noise is independent of $\overline{P_S}$. We can see that as we increase $\overline{P_S}$, the RIN noise, which increases proportional to the signal, becomes the ultimate limit for the CNR. To achieve the CNR of more than 50 dB required for analog CATV applications, we need (i) a powerful transmitter (e.g., $+10\,\text{dBm}$) such that the received power is in the range -3 to $0\,\text{dBm}$ and (ii) a low-noise laser (e.g., $RIN < -150\,\text{dB/Hz}$). Note that this situation is very different from that of a digital system, where the required SNR is only about 17 dB, or even less if forward error correction (FEC) is used. Thus, a digital receiver can operate in the regime where the front-end noise strongly dominates the RIN and shot noise terms.

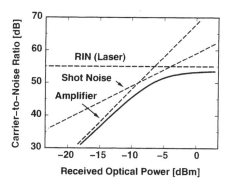

Fig. 5.27 CNR as a function of received optical power for an analog CATV receiver.

5.3 TIA CIRCUIT IMPLEMENTATIONS

In the following, we examine some representative transistor-level TIA circuits, which have been reported in the literature. These circuits illustrate how the design principles discussed in the previous section can be implemented in a broad variety of technologies using different types of transistors such as the metal-semiconductor field-effect transistor (MESFET), the heterostructure field-effect transistor (HFET), the bipolar junction transistor (BJT), the heterojunction bipolar transistor (HBT), and the complementary metal-oxide-semiconductor transistors (CMOS) (cf. Appendix D).

5.3.1 MESFET and HFET Technology

Single-Ended TIA. Figure 5.28 shows a simplified schematic of the GaAs-FET TIA reported in [26, 193]. This single-ended TIA has a bandwidth of 300 MHz and is implemented in a 2-μm GaAs-MESFET technology. The transistors in this circuit are depletion-mode FETs, which means that they conduct current when the gate-source voltage is zero. (In the schematics, we use a thin vertical line from drain to source to distinguish depletion-mode from enhancement-mode devices.)

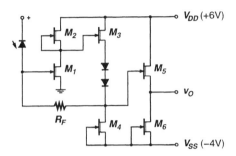

Fig. 5.28 MESFET/HFET implementation of a single-ended TIA based on [193].

The feedback amplifier is implemented with FETs M_1 through M_4. The gain is provided by the common-source stage consisting of M_1 and M_2. Note that M_2 (like M_4 and M_6) has the gate tied to the source and acts as a constant current source. The source-follower (common-drain) stage with M_3 and M_4 buffers the output signal and, with a stack of two Schottky diodes, shifts the DC-voltage to a lower value. The feedback resistor R_F closes the loop around this inverting amplifier. Note that the gate of M_1 is biased through R_F from the output of the amplifier, which explains the need for the level shifter. Another source-follower stage with M_5 and M_6 serves as an output buffer.

In [26, 165] a similar single-ended TIA circuit for 3-Gb/s operation implemented in a 0.5-μm GaAs-MESFET technology has been reported. In contrast to Fig. 5.28, this TIA uses a feedback amplifier with two gain stages. The second stage is a common-gate stage to keep the overall amplifier polarity inverting. This TIA also features an adaptive transimpedance and incorporates an inductive load in the first stage to reduce the noise and to increase the bandwidth.

Differential TIA. Figure 5.29 shows the simplified core of the GaAs-FET TIA reported in [78]. This differential TIA has a bandwidth of 22 GHz and is implemented in a 0.3-μm GaAs-HFET technology.

The differential feedback amplifier is implemented with enhancement-mode HFETs M_1, M_1', M_3, and M_3' and with depletion-mode HFETs M_2, M_2', M_4, and M_4'. The gain is provided by the differential stage consisting of M_1, M_1', the tail current source M_2, M_2', and the load resistors R, R' with series inductors L and L'. The constant tail current together with the linear load resistors guarantee a fixed common-mode output voltage, independent of the differential output voltage. The series inductors reduce the noise and broaden the bandwidth of the stage; we discuss this broadband technique in Section 6.3.2. Each output from the differential stage is buffered by a source follower (M_3, M_3'), which is biased by a current source (M_4,

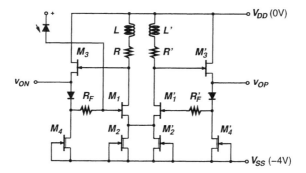

Fig. 5.29 MESFET/HFET implementation of a differential TIA based on [78].

M_4'). Schottky diodes lower the DC output voltage to the two feedback resistors, R_F and R_F', which close the loop around this differential amplifier. In [78], this TIA core is followed by a three-stage post amplifier and a 50-Ω buffer.

5.3.2 BJT, BiCMOS, and HBT Technology

Single-Ended TIA. Figure 5.30 shows the core of the BiCMOS TIA reported in [68]. This single-ended TIA is designed for 1.06-Gb/s NRZ signals and is implemented in a 20-GHz, 1-μm BiCMOS technology (the TIA core shown in Fig. 5.30 uses only BJTs).

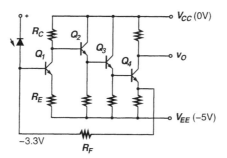

Fig. 5.30 BJT/HBT implementation of a single-ended TIA based on [68].

The feedback amplifier is implemented with transistors Q_1 through Q_4. The gain is provided by the common-emitter stage consisting of Q_1, R_E, and R_C. The ratio of the collector resistor, R_C, to the emitter resistor, R_E, sets the gain. After this gain stage follows a cascade of three emitter followers (Q_2–Q_4), which buffer and level-shift the output signal. For better input-to-output isolation, the last stage has two outputs, one for the feedback signal and one for the output signal. The feedback resistor R_F closes the loop around this inverting amplifier. In [68], this TIA core is followed by a post amplifier, which performs a single-ended to differential conversion, and a 50-Ω output buffer. Furthermore, the transimpedance is made adaptive with three MOSFETs in conjunction with an average detector, a reference voltage generator, and a control amplifier.

In [178], a similar single-ended TIA circuit for 20-Gb/s operation implemented in a 60-GHz SiGe technology has been reported.

Differential TIA with Inductive Input Coupling. Figure 5.31 shows the core of the BiCMOS TIA reported in [71]. This differential TIA is designed for 10-Gb/s NRZ signals and is implemented in a 30-GHz, 0.25-μm BiCMOS technology.

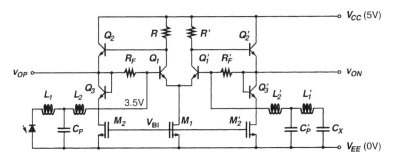

Fig. 5.31 BiCMOS implementation of a differential TIA based on [71].

As it is common practice in BiCMOS technology, the signal path is implemented with BJTs, whereas the bias network is implemented with MOS transistors. This partitioning takes advantage of the BJT's superior high-frequency performance and accurate matching as well as the MOSFET's high output impedance, virtually zero gate current, and low noise. For a noise comparison of MOSFET and BJT current sources, see [139].

The differential feedback amplifier is implemented with the BJTs Q_1 through Q_3, Q'_1 through Q'_3, and the MOSFETs M_1, M_2, and M'_2. The gain is provided by the differential stage consisting of Q_1 and Q'_1, the tail current source M_1, and the polyresistor loads R and R'. The voltage V_{BI} is used to bias all MOSFET current source (M_1, M_2, and M'_2). Each output from the differential stage is buffered by an emitter follower (Q_2, Q'_2). The emitter followers are biased by the MOSFET current sources M_2 and M'_2, which have the diode-connected BJTs Q_3 and Q'_3 in series to keep the drain-source voltage below the breakdown voltage. Two feedback resistors, R_F and R'_F, are closing the loop around this differential amplifier. In [71], this TIA core is followed by another pair of emitter followers and a 50-Ω output buffer.

A point of interest in the circuit of Fig. 5.31 is the L-C coupling network between the photodetector and the TIA input, which improves the TIA's noise and bandwidth (cf. Section 5.2.9). The coupling network consists of the bond-wire inductor L_1, the bond-pad capacitance C_P, and the on-chip spiral inductor L_2. To obtain a balanced TIA configuration, the coupling network and the capacitance of the photodetector, C_X, are replicated at the unused input (cf. Section 5.2.10).

5.3.3 CMOS Technology

Low-Voltage TIA. Figure 5.32 shows the simplified core of the CMOS TIA reported in [186, 187]. This differential TIA is designed for 2.4-Gb/s NRZ signals and is implemented in a 0.15-μm, 2-V CMOS technology. This TIA is part of a single-chip receiver that includes a clock and data recovery circuit (CDR) and a demultiplexer (DMUX). To reject power-supply and substrate noise better, a differential TIA topology was chosen.

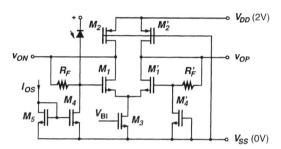

Fig. 5.32 CMOS implementation of a low-voltage TIA based on [186].

The differential feedback amplifier is implemented with the MOSFETs M_1, M'_1, M_2, M'_2, and M_3. The gain is provided by the differential stage consisting of M_1 and M'_1, the tail current source M_3, and the p-MOS load transistors M_2 and M'_2. The latter have grounded gates and operate in the linear regime. The output signals of the differential stage are fed back to the inputs with the feedback resistors R_F and R'_F, which are realized with n-MOS transistors operating in the linear regime. No source-follower buffers are used in this 2-V design because they would cause a low gate-source voltage and thus a low transconductance for the input transistors M_1 and M'_1. This, in turn, would make the TIA more sensitive to substrate noise. M_5 and M_4 form a current mirror, which subtracts the offset control current I_{OS} from the photodetector current. Dummy transistor M'_4 at the unused input balances out the capacitance of M_4. Note that elimination of the output offset voltage to increase the output dynamic range is particularly important in this low-voltage design. In [186], this TIA core is followed by a variable gain amplifier, CDR, and DMUX. The circuits for gain and offset control also are provided on chip.

Common-Gate TIA. Figure 5.33 shows the core of the CMOS TIA reported in [95]. This differential TIA is designed for 2.1-Gb/s NRZ signals and is implemented in a 0.5-μm CMOS technology.

The differential feedback amplifier is implemented with MOSFETs M_1 through M_3 and M'_1 through M'_3. The gain is provided by the differential stage consisting of M_1 and M'_1, the cascode transistors M_2 and M'_2, and the load resistors R and R' with series inductors L and L'. The cascode transistors as well as the series inductors broaden the bandwidth of the stage; we discuss these broadband techniques in Section 6.3.2. Source-followers M_3 and M'_3 buffer the outputs of the differential

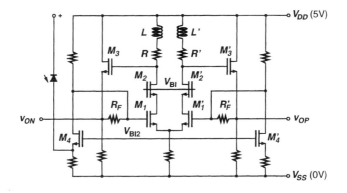

Fig. 5.33 CMOS implementation of a common-gate TIA based on [95].

stage and drive the feedback resistors R_F and R'_F. Transistor M_4 is a common-gate input stage, which decouples the photodetector capacitance from the critical node at the gate of M_1 and thus reduces the dependence of the TIA's bandwidth on the detector capacitance (cf. Section 5.2.6). Transistor M_4 also increases the reverse-bias of the photodetector by V_{DS4}, which improves the speed of the detector. To obtain a balanced TIA configuration, the common-gate input stage is replicated at the unused input. Note that if the value of the photodetector capacitance is known fairly well, this TIA circuit also can be used without the common-gate input stages M_4 and M'_4. Voltages V_{BI} and V_{BI2} bias the common-gate and cascode transistors. In [95], this TIA core is followed by an output buffer to drive off-chip loads.

In [124], a similar differential TIA circuit for 1-Gb/s operation implemented in a 0.25-μm CMOS technology has been reported. In contrast to Fig. 5.33, this TIA uses regulated-cascode input stages, which provide enhanced decoupling of the photodetector capacitance from the TIA summing node [125].

Multistage TIA. Figure 5.34 shows the CMOS TIA reported in [49]. This three-stage TIA is designed for 240-Mb/s NRZ signals and is implemented in a 0.8-μm CMOS technology.

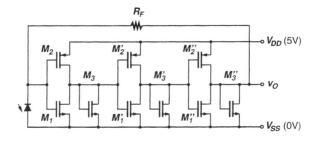

Fig. 5.34 CMOS implementation of a multistage TIA based on [49].

In all the previous examples, the gain of the feedback amplifier was provided by a single stage. In this example, a cascade of three stages is used to achieve a higher gain. Each stage of the feedback amplifier consists of three MOSFETs (M_1–M_3 in the first stage). Transistors M_1 and M_2 act as a push-pull transconductor and M_3 represents the load resistor. Thus, the stage gain is approximately $(g_{m1} + g_{m2})/g_{m3}$, a quantity that is well defined by the device geometries. Because the feedback amplifier has three identical gain stages, the open-loop frequency response of this TIA contains three similar high-frequency poles. To ensure a flat closed-loop frequency response, these poles must be placed at a higher frequency than the single high-frequency pole in Fig. 5.7. Note that this single-ended amplifier must have an odd number of stages to ensure negative feedback through R_F.

5.4 PRODUCT EXAMPLES

Table 5.2 summarizes the main parameters of some commercially available TIA chips. The numbers have been taken from data sheets of the manufacturer that were available at the time of writing. For up-to-date product information, please contact the manufacturer directly. For TIAs with a single-ended output, the transimpedance value, tabulated under R_T, is followed by "(s)." For TIAs with differential outputs, the *differential* transimpedance value is given followed by "(d)." The input overload current, i_{ovl}^{pp}, as well as the maximum input current for linear operation, i_{lin}^{pp}, are specified as peak-to-peak values, as defined in Sections 5.1.2 and 5.1.3.

Comparing parts for different bit rates we find that, in general, higher speed parts exhibit a lower transimpedance, as expected from the transimpedance limit Eq. (5.25). We also find that a few parts, such as the FOA1251B1, have an unusually high transimpedance. These parts have an on-chip post amplifier as discussed in Section 5.2.5. Furthermore, we observe that, in general, higher speed parts have a larger input-referred rms noise current, $i_{n.TIA}^{rms}$, in accordance with our discussion in Section 5.2.3.

The LG1628AXA can sustain a particularly high input overload current owing to its adaptive transimpedance circuit, as the one shown in Fig. 5.14(b), and its offset control circuit, similar to that shown in Fig. 5.23 [205]. Finally, the MAX3866 includes a limiting amplifier besides the TIA, which explains its higher power dissipation compared with the MAX3267.

5.5 RESEARCH DIRECTIONS

The research effort focusing on TIAs can be divided roughly into four areas: higher speed, higher integration, lower cost, and lower noise. In the following, we briefly touch on each of these areas.

Higher Speed. Historically, every new generation of optical telecommunication equipment has been 4× faster than the previous generation. The systems progressed from 155 Mb/s (OC-3) to 622 Mb/s (OC-12) to 2.5 Gb/s (OC-48) to 10 Gb/s

Table 5.2 Examples for high-speed TIA products.

Company & Product	Speed (Gb/s)	R_T (kΩ)	BW_{3dB} (GHz)	$i_{n,TIA}^{rms}$ (nA)	i_{lin}^{pp} (μA)	i_{ovl}^{pp} (mA)	Power (mW)	Technology
Agere LG1628AXA	2.5	11.6 (d)	1.6	300		8.0	728	GaAs HFET
Anadigics ATA30013	2.5	1.8 (s)	2.0	433	600		850	GaAs MESFET
Infineon FOA1251B1	2.5	27.0 (d)	1.8		10	2.2	112	Si BJT
Maxim MAX3267	2.5	1.9 (d)	1.9	500	40	1.0	86	SiGe HBT
Maxim MAX3866	2.5	(d)	1.8	433		2.5	165	
Nortel AB89	2.5	2.5 (d)	2.5	330		2.6	182	Si BJT
Nortel AC89	2.5	4.0 (d)	2.2	270		3.0	325	Si BJT
Agere TTIA110G	10.0	2.0 (d)	10.0	1,500	720	3.0	810	GaAs HFET
AMCC S3090	10.0	1.4 (d)	10.0	1,580		2.2	416	SiGe HBT
Anadigics ATA7601	10.0	1.0 (d)	9.0	1,800	400		340	GaAs HBT
Giga GD19906	10.0	0.5 (s)	10.0	1,200		2.5	750	GaAs HBT
Maxim MAX3970	10.0	0.6 (d)	9.0	1,100	130	2.0	150	SiGe HBT
Nortel D68	10.0	0.5 (s)	10.0			2.5	750	GaAs HBT
Philips CGY2110	10.0	4.0 (d)	9.0			1.8	290	GaAs HFET
AMCC S76800	40.0	0.2 (s)	45.0	4,000		3.0	600	SiGe HBT
Velocium TLA401	43.0	1.5 (d)	40.0	7,000		4.0	1,800	

(OC-192), and so forth. In the world of data communication, the steps taken are even larger: every new generation has been 10× faster than the previous one. Ethernet comes at the speed grades 10 Mb/s, 100 Mb/s, 1 Gb/s, 10 Gb/s, and so forth.

The next speed of great interest is 40 Gb/s (OC-768). Many research groups are aiming at this speed and beyond. To achieve the necessary TIA bandwidth in the range of 30 to 40 GHz, aggressive high-speed technologies are needed. Heterostructure devices, such as HBTs and HFETs, based on compound materials, such as SiGe, GaAs, and InP, are used frequently (cf. Appendix D).

The following papers on high-speed TIAs were published recently:

- In SiGe-HBT technology, a 40-Gb/s, a 35-GHz, and a 45-GHz TIA have been reported in [102], [87], and [86], respectively.

- In GaAs-HFET technology, a 22-GHz TIA has been reported in [78].

- In GaAs-HBT technology, a 40-Gb/s and a 25-GHz TIA have been reported in [181] and [153], respectively.

- In InP-HFET technology, a 49-GHz TIA has been reported in [173]. A 49-GHz photodiode integrated with a 47-GHz distributed amplifier has been reported in [185].

- In InP-HBT technology, 40-Gb/s TIAs have been reported in [163] and [103]. A 56-GHz TIA has been reported in [64].

Higher Integration. Another area of research aims at higher integration by combining the photodetector and the electronic circuits on the same chip. This approach is known as *optoelectronic integrated circuits* (OEIC) [184, 194]. The simplest form of an OEIC is the so-called *p-i-n FET*, which combines a p-i-n photodetector and an FET on the same substrate.

The benefits of the OEIC approach are reduced size, smaller interconnect parasitics, and lower packaging cost. Bringing the photodetector and the TIA on the same chip reduces the critical capacitance C_T at the summing node of the TIA, thus the OEIC approach is particularly attractive for high-speed receivers. On the downside, OEIC receivers often are less sensitive than their hybrid counterparts because compromises must be made when fabricating photodetectors and transistors on the same substrate with the same set of processing steps. However, with the increasing use of EDFAs, this loss in sensitivity may not be a major drawback.

In InP circuit technologies, the base-collector junction of an HBT often can serve as a photodiode, which is sensitive in the 1.3- to 1.6-μm range used for telecommunication. In some MESFET and HFET circuit technologies, a high-quality junction photodiode may not be available. In this case, a *metal-semiconductor-metal photodetector* (MSM) consisting of two back-to-back Schottky diodes may offer a solution. Many silicon circuit technologies offer a medium-speed photodiode, which is sensitive around the 0.85-μm wavelength. For example, in CMOS technologies the junction between the p-drain/source diffusion and the n-well can serve as a photodiode. Such a short-wavelength photodetector combined with a TIA is useful for data-communication applications [152, 203]. Finally, as an alternative approach to the monolithic OEIC, a flip-chip photodetector can be mounted on top of a circuit chip. In this case, the detector and circuit technologies can be chosen (and optimized) independently. For example, a standard silicon circuit technology can be combined with an InP detector, which is sensitive at long wavelengths.

Lower Cost. Another area of research is focusing on the design of high-performance TIAs in low-cost, mainstream technologies. In particular, digital CMOS technology is an attractive choice because it is offered by many foundries and the chips can be fabricated cost effectively on very large wafers.[6] Furthermore, a CMOS TIA design can be combined with other analog blocks and dense digital logic on the a single chip. Such a system-on-a-chip approach reduces the chip count, board space, and power dissipation, and thus further lowers the system cost.

The challenge of designing broadband TIAs in a silicon-based mainstream technology is that the available active and passive devices (e.g., inductors and transmission lines) are of lower quality (slower, noisier, lossier, etc.) compared with those available in a specialized III-V microwave technology. In the case of a system-on-a-chip, noise isolation between the analog and digital sections poses an additional challenge.

A SONET compliant 10-Gb/s TIA implemented in a low-cost 0.25-μm modular BiCMOS technology has been reported in [71]. A *modular* BiCMOS technology is basically a CMOS technology with the addition of a few masks (a module) to provide bipolar transistors. A 2.4-Gb/s, 0.15-μm CMOS TIA integrated with an AGC amplifier, a CDR, and a DMUX on the same chip has been reported in [186]. A 10-Gb/s, 0.18-μm CMOS TIA has been reported in [128].

Lower Noise. Recognizing that low TIA noise translates into high sensitivity, researchers have been studying ways to reduce this noise. As we know, in medium- and low-speed receivers, a large portion of the input-referred TIA noise is due to the feedback resistor(s). Several ways to eliminate these resistors have been investigated. In [140], a noise-free capacitive feedback network is proposed, whereas in [63], a noise-free optical feedback is used. In [56], the shunt-feedback TIA is replaced by an integrate-and-dump arrangement that is free of feedback resistors.

5.6 SUMMARY

The main specifications of the transimpedance amplifier (TIA) are as follows:

- The transimpedance, which we want to be as large as possible to relax the gain and noise requirements for the subsequent main amplifier.

---

[6]However, because the mask cost for deep-submicron CMOS technologies is several hundred thousand U.S. dollars, the cost advantage of these technologies is realized only for very high volumes. If only a few thousand chips are produced, an "exotic" III-V technology may be more cost effective than CMOS.

- The input overload current, which must be large enough to avoid harmful pulse-width distortion and jitter when the maximum optical signal power is received.

- The maximum input current for linear operation, which must be large enough to avoid harmful signal clipping, harmonic, and intermodulation distortions when the maximum optical signal power is received. These distortions are important in applications where linear signal processing (e.g., equalization) is performed or in applications that use a linear modulation scheme (e.g., CATV/HFC).

- The input-referred noise current, which must be as small as possible to obtain high sensitivity. A low input-referred noise current is particularly important for p-i-n receivers. The TIA noise performance can be specified in terms of the input-referred noise current spectrum or the input-referred rms noise current; the latter quantity determines the sensitivity.

- The TIA bandwidth, which is chosen between $0.6B$ and $1.2B$ depending on the bandwidths of the other components in the receiver system.

- The group-delay variation, which must be kept small to minimize jitter and other signal distortions.

The shunt-feedback principle is used in virtually all TIA designs because it simultaneously provides high bandwidth, high transimpedance, high overload current, and low noise. The noise performance of a shunt-feedback TIA can be optimized by choosing a large feedback resistor, minimizing the detector capacitance, matching the TIA input capacitance to the detector capacitance (FET front-end), and carefully choosing the collector current (BJT front-end).

The basic shunt-feedback TIA can be enhanced by making its transimpedance adaptive, which improves the input overload current and the maximum input current for linear operation. A post amplifier can be used to boost the transimpedance. A common-base/gate input stage or a current-mode TIA can be used to make the TIA bandwidth less dependent on the photodetector capacitance. A bond-wire inductance or an L-C filter placed in between the photodetector and the TIA can increase its bandwidth and lower the noise. Differential TIAs are more resilient to power-supply and substrate noise and feature an increased output voltage swing. An offset control circuit can be used to improve the output dynamic range. Burst-mode TIAs require a fast amplitude and threshold (or offset) control mechanism to deal with a bursty input signal of varying amplitude. Analog receivers are optimized for low signal distortions (high linearity) and low noise.

TIAs have been implemented in a wide variety of technologies including metal-semiconductor FET (MESFET), heterostructure FET (HFET), BJT, heterojunction bipolar transistor (HBT), BiCMOS, and CMOS.

Currently, researchers are working on 40-Gb/s TIAs and beyond, TIAs integrated with the photodetector on the same chip, TIAs in low-cost technologies such as CMOS, and ultra-low-noise TIAs.

5.7 PROBLEMS

5.1 Input-Referred RMS Noise Current. A TIA with differential outputs has a differential transimpedance of 1 kΩ and an rms output noise voltage of 1 mV at each output terminal. (a) Assuming that the noise at the two outputs is uncorrelated, how much noise is in the differential mode and how much noise is in the common mode? (b) What is the input-referred rms noise current required to reproduce the single-ended output noise and what is the input-referred rms noise current required to reproduce the differential output noise?

5.2 TIA Dynamic Range. A 10-Gb/s receiver must be able to handle optical input signals in the range from -19 dBm to -10 dBm at $BER < 10^{-12}$. Assume that the photodetector responsivity is 0.8 A/W and that the optical signal has high extinction. (a) What input overload current and input-referred rms noise current should the TIA have? (b) What averaged input-referred noise current density should the TIA have, if its bandwidth is 10 GHz?

5.3 Low-Impedance Front-End. A 50-Ω low-impedance front-end is followed directly by an amplifier with 50-Ω inputs and outputs, gain $A = 40$ dB, noise figure $F = 2$ dB, and noise bandwidth $BW_n = 10$ GHz. (a) How large is the transimpedance of this arrangement? (b) How large is the input-referred rms noise current? (b) How does the optical sensitivity of this front-end compare with a TIA front-end with $i_{n,TIA}^{rms} = 1.4\,\mu A$?

5.4 TIA Output Impedance. Repeat the "simple" (single-pole) analysis of the TIA in Fig. 5.6, but now include a series resistance, R_S, at the output of the otherwise ideal feedback amplifier, A. (a) Calculate the transimpedance, $Z_T(s)$, input impedance, $Z_I(s)$, and output impedance, $Z_O(s)$. (b) What are the bandwidths of these impedances?

5.5 Open-Loop Pole Spacing. For a two-pole TIA, the open-loop pole spacing, $R_F C_T / T_A$, must be $2A$ to obtain a Butterworth response after closing the loop (Eq. (5.23)). (a) Given an arbitrary closed-loop Q value, what is the required open-loop pole spacing? (b) What is the required pole spacing for a Bessel response? (c) What is the required pole spacing for a critically damped response?

5.6 Transimpedance Limit. In the available technology, we can realize voltage amplifiers with a gain-bandwidth product of 44 GHz. Given $C_T = 0.3$ pF and the requirement that the TIA bandwidth must be 70% of the bit rate, what maximum transimpedance values do we expect for the basic shunt-feedback TIA operating at 2.5 Gb/s, 10 Gb/s, and 40 Gb/s?

5.7 Feedback Capacitance. Calculate the transimpedance $Z_T(s)$ of a TIA as shown in Fig. 5.6 but with the addition of a feedback capacitor C_F in parallel to R_F. Assume a single-pole model (with time constant T_A) for the feedback amplifier. (a) What are the expressions for R_T, ω_0, and Q? (b) What value for C_F do we need to obtain a Butterworth response (assume $C_F \ll C_T$)? (c) What is the 3-dB bandwidth of the TIA given a Butterworth response (assume $T_A \ll R_F(C_T + AC_F)$)? (d) Does the transimpedance limit Eq. (5.25) still hold for this case?

5.8 **TIA f^2-Noise Corner.** The "f^2-noise corner" occurs at the frequency where the white noise and f^2 noise are equally strong. (a) Derive an expression for the f^2-noise corner frequency of a TIA with a MOSFET front-end. (b) How is this corner frequency related to the bit rate?

5.9 **TIA with $1/f$ Noise.** Assume that the white channel noise and the $1/f$ noise can be written in the combined form $I_{n,D}^2 = 4kT\Gamma g_m(1 + f_c/f)$, where f_c is the $1/f$-noise corner frequency. Calculate the input-referred noise current spectrum of a TIA with a MOSFET front-end.

5.10 **TIA with Variable Feedback Resistor.** A shunt-feedback TIA has a variable feedback resistor R_F (with negligible parallel capacitance). How does the TIA's frequency response vary for the following cases? (a) The adaptation mechanism controls only R_F, whereas A and T_A remain fixed (assume $T_A \ll R_F C_T$ and $A \gg 1$). (b) The adaptation mechanisms controls A and R_F in a proportional way, whereas T_A remains fixed (same assumptions). (c) The adaptation mechanisms controls A to be proportional to $\sqrt{R_F}$, whereas the gain-bandwidth product of the feedback amplifier remains fixed (same assumptions).

5.11 **TIA Parameters.** You are reviewing a paper that describes a single-ended TIA consisting of a shunt-feedback stage and a post-amplifier stage. The shunt-feedback amplifier and the post amplifier both have a gain of 6 dB. The section on experimental results reports the transimpedance as 62.5 dBΩ and the input-referred noise-current density as varying between 4 and 5 pA/$\sqrt{\text{Hz}}$, depending on frequency. What is your comment to the author?

5.12 **Current-Mode TIA.** Calculate the frequency-dependent transimpedance $Z_T(s)$ of the current-mode TIA shown in Fig. 5.18(a). Assume an idealized current-amplifier model with $R_S = 0$, but take $C_L \neq 0$ into account. How large is the 3-dB bandwidth?

5.13 **Active-Feedback TIA.** Calculate the transimpedance $Z_T(s)$ of the active-feedback TIA shown in Fig. 5.19(a). Assume a single-pole model (with time constant T_A) for the feedback amplifier. (a) What are the expressions for R_T, ω_0, and Q? (b) What time constant T_A is needed to obtain a Butterworth response? (c) What is the 3-dB bandwidth of the TIA given a Butterworth response? (d) Does the transimpedance limit Eq. (5.25) apply to this TIA?

5.14 **TIA with Bond-Wire Inductance.** Calculate the frequency-dependent transimpedance $Z_T(s)$ of the TIA with bond-wire inductance shown in Fig. 5.20(a). Assume that the feedback amplifier has an infinite bandwidth such that the shunt-feedback TIA contributes only a single pole. What conditions must be met to obtain a Butterworth response? (Hint: the third-order Butterworth transfer function has the form $1/[(1 + sT)(1 + sT + s^2T^2)]$.)

5.15 Differential TIA. (a) Calculate the low-frequency relationship between v_{IP} and i_I of the balanced TIA shown in Fig. 5.21. From this result, derive the single-ended input resistance $R_I = \Delta v_{IP}/\Delta i_I$ "seen" by the photodetector. (b) Derive the same single-ended input resistance from the differential input resistance $R_{I.d} = 2\Delta(v_{IP} - v_{IN})/\Delta(i_{IP} - i_{IN}) = 2R_F/(A+1)$ and the common-mode input resistance $R_{I.c} = 1/2 \cdot \Delta(v_{IP} + v_{IN})/\Delta(i_{IP} + i_{IN}) = R_F/2$.

6
Main Amplifiers

In this chapter, we focus on the *main amplifier* (MA). We start by distinguishing two types of MAs, the *limiting amplifier* (LA) and the *automatic gain control amplifier* (AGC amplifier). Then, after introducing the main specifications of the MA and relating them to the system performance, we discuss MA circuit concepts in a general and, as much as possible, technology-independent manner. Subsequently, we illustrate these concepts with practical implementations in a broad range of technologies. We conclude with a brief overview of product examples and current research topics.

6.1 LIMITING VS. AUTOMATIC GAIN CONTROL (AGC)

The purpose of the MA is to amplify the small signal from the transimpedance amplifier (TIA) to a level that is sufficient for the reliable operation of the clock and data recovery (CDR) circuit. The required swing at the output of the MA typically is several 100 mV peak-to-peak. The MA also is known as the *post amplifier* because it follows the TIA in the chain of amplifiers. The MA can be realized as a stand-alone part or it can be integrated with the TIA on the same chip. In the first case, a 50-Ω transmission line typically is used to connect the TIA to the MA, whereas in the second case, this is not necessary. Almost all MAs are fully differential, that is, they feature differential inputs as well as differential outputs (cf. Appendix B for a discussion of differential circuits). Such an MA together with its input and output voltages is shown in Fig. 6.1. The differential input voltage v_I is the difference between the two single-ended input voltages v_{IP} and v_{IN} and, similarly, the differential output voltage v_O is the difference between the two single-ended output voltages v_{OP} and v_{ON}.

160 MAIN AMPLIFIERS

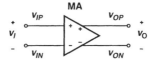

Fig. 6.1 Input and output signals of a fully differential MA.

Depending on the application, it may or may not be acceptable for the MA to introduce nonlinear distortions (cf. Chapter 4). If low distortion is mandatory, an AGC amplifier must be used. If nonlinear distortions can be tolerated, the simpler LA design is preferred.

Limiting Amplifier. For small input signals, most amplifiers display a fairly linear response. For large signals, however, nonlinear effects may distort the output signal. Specifically, a differential stage with a constant tail current has a limited output swing: when all of the tail current is switched through one of the two transistors, severe distortions in the form of clipping set in.

An LA is an amplifier with no special provisions to avoid this naturally occurring clipping or limiting of the output signal. The idealized DC transfer function of an LA is shown in Fig. 6.2(a). For very small input signals, v_I, the amplifier operates in the linear regime (v_O proportional to v_I), but for larger signals, it crosses into the limiting regime ($v_O = $ constant). Although the limiting occurs naturally as a result of a constant tail current or other signal swing constraints, the designer of an LA needs to make sure that the limiting occurs in a *controlled* way. For example, delay variations, which may occur when the LA crosses from the linear into the limiting regime, must be minimized to avoid the generation of jitter.

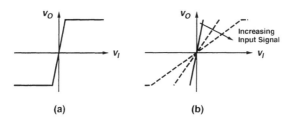

Fig. 6.2 DC transfer function of (a) an LA and (b) an AGC amplifier.

AGC Amplifier. An AGC amplifier consists of a *variable-gain amplifier* (VGA) and an automatic gain-control mechanism that keeps the output swing constant over a wide range of input swings. Whereas the LA starts to distort (limit) for large input signals, the AGC amplifier reduces its gain and thus manages to stay in the linear regime. Figure 6.2(b) shows how the DC transfer function of an AGC amplifier depends on the input signal strength. Nevertheless, for very large signals, the AGC amplifier cannot reduce its gain any further and limiting will occur eventually. The system designer must make sure that the input dynamic range of the AGC amplifier is sufficiently wide to avoid this situation.

Comparison. Note that the distinction between an AGC amplifier and an LA is analogous to the distinction between a TIA with and without adaptive transimpedance, which we discussed in Section 5.2.4. In a way, the LA can be regarded as an AGC amplifier for which the gain is stuck at its maximum value, resulting in a narrow linear regime.

Because an LA does not have a gain-control mechanism, it generally is easier to design than an AGC amplifier. Furthermore, its power dissipation, bandwidth, noise, and so forth often are superior to an AGC amplifier realized in the same technology.

The linear transfer function of the AGC amplifier preserves the signal waveform and permits analog signal processing to be performed on the output signal. Examples for such signal processing tasks are equalization, slice-level steering, and soft-decision decoding, which were discussed in the Sections 4.7, 4.10, and 4.11, respectively. The LA severely distorts the input signal when operating in the limiting regime, causing much of the information in the input signal to be lost. The zero-crossings, however, are preserved exactly as long as the LA is free of offset and memory. For this reason, it is sometimes said that the LA performs the slicing function.

6.2 MA SPECIFICATIONS

In the following, we discuss the main specifications of the MA: the gain, the bandwidth, the group-delay variation, the noise figure, the input dynamic range, the input offset voltage, the low-frequency cutoff, and the AM-to-PM conversion. For a discussion of S parameters, see Appendix C.

Whereas the TIA specifications determine the primary performance of the receiver, such as the sensitivity and the overload limit, the MA specifications have less of an impact. However, insufficient MA specifications may degrade the receiver performance. Thus, in the following, we emphasize the relationship between each specification and the power penalty it causes.[1] As explained in Section 4.5, this relationship permits us to derive expressions for the maximum or minimum permissible value for each specification.

6.2.1 Gain

Definition. The *voltage gain*, A, is defined as the output voltage change, Δv_O, per input voltage change, Δv_I:

$$A = \frac{\Delta v_O}{\Delta v_I}. \tag{6.1}$$

---

[1] All power penalties in this section are derived for an unamplified transmission system with a p-i-n detector (cf. Section 4.5).

Thus, the higher the gain, the more output signal is produced for a given input signal. The gain is specified either on a linear scale or in dBs. In the latter case, the value in dB is calculated as $20 \log A$; for example, a gain of 1,000 corresponds to 60 dB.

When operated with a small AC signal, the MA's transfer function is fully described by the frequency-dependent gain magnitude $|A(f)|$ and the frequency-dependent phase shift $\Phi(f)$ between the input and the output signal. A generalization of the voltage gain defined in Eq. (6.1) to the complex quantity $A = |A| \cdot \exp(j\Phi)$ captures both aspects. Note that this complex gain quantity also can be expressed as $A = V_o/V_i$, where V_o is the output voltage phasor and V_i is the input voltage phasor. The voltage gain A is closely related to the S parameter S_{21}. If the input impedance of the amplifier is equal to the characteristic impedance $R_0 = 50\,\Omega$ ($S_{11} = 0$) and the output of the amplifier is loaded by the characteristic impedance $R_0 = 50\,\Omega$, the two quantities become identical (cf. Appendix C).

The gain usually is specified for a small input signal. In the case of an LA, the signal must be small enough such that it stays in the linear regime of the transfer function, that is, gain compression or limiting do not occur. In the case of an AGC amplifier, the signal must be small enough such that the maximum gain is realized ($A = A_{\max}$), that is, the AGC mechanism doesn't reduce the gain. Besides the maximum gain, the minimum gain and the *automatic gain control range* also are of interest for AGC amplifiers. The AGC range is defined as the ratio of the maximum to the minimum gain. For example, if the minimum gain is 10 dB and the maximum gain is 40 dB, the AGC range is 30 dB.

For amplifiers with differential outputs, the gain can be measured single endedly, with $v_O = v_{OP}$ or $v_O = v_{ON}$, or differentially, with $v_O = v_{OP} - v_{ON}$. It is important to specify which way the gain was measured because the differential gain is 6 dB higher than the single-ended gain. Fortunately, the gain depends little on whether the inputs are driven single endedly or differentially. The input voltage of a differential amplifier is *always* defined as the voltage between the noninverting and inverting input, $v_I = v_{IP} - v_{IN}$, no matter if they are driven single endedly or differentially. The only difference between single-ended and differential drive is that the former excites the common mode, whereas the latter ideally doesn't.

Most high-speed MAs have 50-Ω inputs and outputs. For a correct gain measurement, the signal source impedance must be 50 Ω and the outputs must be properly terminated with 50-Ω resistors.

Power Penalty. Next we derive the power penalty (PP) caused by a finite MA gain and then determine how large we have to make this gain. To discuss the power penalty we refer to the receiver model shown in Fig. 4.1, where the MA is followed by a decision circuit (DEC). If the DEC is perfect, no power penalty is incurred, no matter how large or small the MA gain is. However, real DECs with a finite sensitivity do cause a power penalty in conjunction with the finite MA gain.

Let's define the *decision circuit sensitivity*, also called the *decision threshold ambiguity width* (DTAW), V_{DTA}, as the input voltage swing (measured peak to peak) below which the DEC makes random decisions ($BER = 0.5$). Let's further assume that the DEC operates free of errors if we go just slightly above this voltage (i.e., we assume the DEC is noise free). Figure 6.3 illustrates that we need to increase the DEC input swing, which is the MA output swing v_O^{pp}, by V_{DTA} to obtain the same BER for $V_{DTA} > 0$ as for the ideal case, $V_{DTA} = 0$. The hatched areas represent the receiver noise for the desired bit-error rate (BER), similar to the noise representation in Fig. 4.11. Thus, for a nonzero DTAW, we incur the following power penalty:

$$PP = \frac{v_O^{pp} + V_{DTA}}{v_O^{pp}} = 1 + \frac{V_{DTA}}{v_O^{pp}}. \quad (6.2)$$

For example, with a DEC sensitivity of 10 mV and a voltage swing of 500 mV peak-to-peak into the DEC, we incur a power penalty of 0.086 dB. For the reliable operation of the CDR, we also need enough voltage swing to drive the phase detector. However, the phase detector usually is less demanding than the DEC, and we assume here that the voltage swing required for the CDR is determined by the DEC.

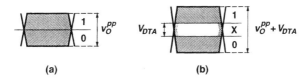

Fig. 6.3 MA output signal with noise for (a) an ideal DEC and (b) a DEC with finite sensitivity.

The power penalty in Eq. (6.2) becomes largest for small values of v_O^{pp}. The TIA supplies the smallest useful signal swing to the MA when it operates at the sensitivity limit. This MA input signal swing can be expressed in terms of the TIA's input-referred rms noise current and its transimpedance using Eqs. (5.1) and (5.5):

$$v_{I,\min}^{pp} = 2Q \cdot R_T \cdot i_{n,TIA}^{rms}. \quad (6.3)$$

The smallest MA output signal swing simply is $v_{O,\min}^{pp} = A \cdot v_{I,\min}^{pp}$, and thus the worst-case power penalty becomes

$$PP = 1 + \frac{V_{DTA}}{2Q \cdot A \cdot R_T \cdot i_{n,TIA}^{rms}}. \quad (6.4)$$

In this derivation, we have assumed that the signals $v_{I,\min}^{pp}$ and $v_{O,\min}^{pp}$ are so small that neither the transimpedance gain nor the MA gain is compressed (or reduced by the AGC). [→ Problem 6.1]

Finally, we can solve Eq. (6.4) for A to get an expression for the minimum permissible MA gain:

$$A \geq \frac{V_{DTA}}{(PP-1) \cdot 2Q \cdot R_T \cdot i_{n,TIA}^{rms}}. \quad (6.5)$$

From this equation, we can conclude that if the DEC has a perfect sensitivity ($V_{DTA} = 0$), no gain is required at all. Furthermore, we can see how a high transimpedance helps to relax the gain requirements for the MA.

Typical Values. Next we calculate some typical numbers based on a 10-mV DEC sensitivity, 0.05-dB power penalty ($PP = 1.0116$), and $BER = 10^{-12}$ ($Q = 7.035$). For a typical 2.5-Gb/s system ($R_T = 3\,\text{k}\Omega$, $i_{n.TIA}^{rms} = 380\,\text{nA}$), we obtain the numerical value

$$A \geq \frac{10\,\text{mV}}{0.0116 \cdot 14.07 \cdot 3\,\text{k}\Omega \cdot 380\,\text{nA}} = 34.6\,\text{dB}. \tag{6.6}$$

And for a typical 10-Gb/s system ($R_T = 1\,\text{k}\Omega$, $i_{n.TIA}^{rms} = 1{,}400\,\text{nA}$) we get

$$A \geq \frac{10\,\text{mV}}{0.0116 \cdot 14.07 \cdot 1\,\text{k}\Omega \cdot 1{,}400\,\text{nA}} = 32.8\,\text{dB}. \tag{6.7}$$

In conclusion, a typical MA has a gain of around 30 to 40 dB.

6.2.2 Bandwidth and Group-Delay Variation

Definition. The MA *bandwidth*, BW_{3dB}, is defined as the (upper) frequency at which the small-signal gain $|A(f)|$ dropped by 3 dB below its midband value. This bandwidth also is called the 3-dB bandwidth to distinguish it from the noise bandwidth.

The bandwidth specification alone does not say anything about the phase of $A(f)$. We know that even if the frequency response $|A(f)|$ is flat up to a sufficiently high frequency, distortions in the form of data-dependent jitter may occur if the phase linearity of $A(f)$ is insufficient. A common measure for phase linearity is the variation of the group delay with frequency. The *group delay*, τ, is related to the phase, Φ, as $\tau(\omega) = -d\Phi/d\omega$.

Note that the definition of bandwidth and group delay is based on the assumption that a sine wave applied to the input of the amplifier produces a sine wave at the output. Although this is the case for a linear amplifier, such as an AGC amplifier, it is not normally the case for an LA, which clips the output signal under normal operation. If the LA is operated in the limiting regime, the concept of bandwidth no longer applies and must be replaced by a large-signal concept such as the switching speed. However, it is always possible to reduce the input signal amplitude to the point where the LA enters the linear regime and then the bandwidth and the group delay are defined as usual. This small-signal bandwidth tends to be a conservative estimate of the LA's large-signal speed.[2]

Typical Values. As discussed in Section 4.6, the MA bandwidth often is made much larger than the desired receiver bandwidth. The receiver bandwidth then is controlled mostly by either the TIA or the filter, whereas the MA has little impact on the bandwidth. Furthermore, the intersymbol interference (ISI) introduced by the MA and its associated power penalty remain small. As a rule of thumb, the MA bandwidth is chosen around $1.0B$ to $1.2B$, nearly twice the recommended receiver

---

[2]This can be explained as follows: although the *bandwidth* of a multistage LA is lower than that of its stages (cf. Section 6.3.1), the *switching speed* of a multistage LA is essentially the same as that of its stages.

bandwidth of $0.6B$ to $0.7B$:

$$2.5\,\text{Gb/s MA}: \quad BW_{3\text{dB}} = 2.5\,\text{GHz} \ldots 3\,\text{GHz}, \quad (6.8)$$

$$10\,\text{Gb/s MA}: \quad BW_{3\text{dB}} = 10\,\text{GHz} \ldots 12\,\text{GHz}. \quad (6.9)$$

Typically, a group-delay variation, $\Delta\tau$, of less than $\pm 10\%$ of the bit period (± 0.1 UI) over the specified bandwidth is required to limit the generation of data-dependent jitter. This corresponds to

$$2.5\,\text{Gb/s MA}: \quad |\Delta\tau| < 40\,\text{ps}, \quad (6.10)$$

$$10\,\text{Gb/s MA}: \quad |\Delta\tau| < 10\,\text{ps}. \quad (6.11)$$

It is interesting to observe that LAs can be designed with a 3-dB bandwidth that is significantly lower than B, because vertical distortions in the eye diagram are clipped away when the amplifier operates in the limiting regime. However, the bandwidth for which $|\Delta\tau| < 0.1$ UI must be around B to avoid large horizontal distortions in the eye diagram (jitter), which are not removed by the LA (cf. Section 4.6).

6.2.3 Noise Figure

The noise generated by the MA adds to the total receiver noise and thus degrades the receiver sensitivity. Because we would like the receiver sensitivity to be determined by the TIA noise alone, we must make the MA noise (when referred to the input of the TIA) small compared with the TIA noise. In the following, we begin by reviewing important noise quantities such as the input-referred noise voltage and the noise figure. Then, we go on to calculate the power penalty caused by the MA noise, which leads up to an expression for the maximum permissible noise figure.

Definition. Figure 6.4 shows a noiseless MA with an *equivalent noise voltage source*, $v_{n.MA}$, connected to one input. This voltage source is chosen such that, together with the noiseless MA, it reproduces the output noise of the actual noisy MA. The voltage provided by the equivalent noise voltage source also is known as the *input-referred noise voltage*. We assume here that the source impedance is known to be $R_S = 50\,\Omega$, and thus a single noise source is sufficient to model the amplifier noise. (In the general case, a noise voltage and a noise current source had to be used.)

The input-referred noise voltage in Fig. 6.4 can be quantified in two different ways:

- **Input-Referred Noise Voltage Spectrum.** We can specify the power spectral density of the input-referred noise voltage, $V_{n.MA}^2(f)$, also known as the *input-referred noise voltage spectrum* for short. Note that this power spectrum is measured in nV^2/Hz and is a function of frequency (the square root of this spectrum, $V_{n.MA}(f)$, is measured in $\text{nV}/\sqrt{\text{Hz}}$).

- **Input-Referred RMS Noise Voltage.** Alternatively, we can specify the *input-referred rms noise voltage*, $v_{n.MA}^{rms}$, which is measured by a single number in μV. Similar to our discussion for the TIA, this rms noise voltage is determined

166   MAIN AMPLIFIERS

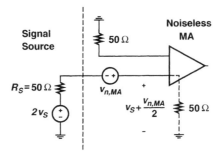

Fig. 6.4   Noise quantities in a differential amplifier with 50-Ω inputs.

by referring the noise spectrum to the output, integrating it at the output, and then referring it back to the input. Thus, we have

$$v_{n.MA}^{rms} = \frac{1}{A_0}\sqrt{\int_0^{>2BW} |A(f)|^2 \cdot V_{n.MA}^2(f)\, df}, \qquad (6.12)$$

where $|A(f)|$ is the frequency-dependent gain of the MA and A_0 is its midband value. For analytical calculations, the integration can be carried out to infinity; for simulations (and measurements), it usually is enough to integrate up to about $2\times$ the MA's bandwidth, after which the contributions to the rms output noise become negligible.

There is an important difference between the noise model of the TIA and the MA. For the TIA, the transimpedance from the equivalent noise current source to the output was the same as the transimpedance from the signal input to the output. This is not so for the MA! Here, the gain from the equivalent noise voltage source to the output is *one half* of the gain from the signal input to the output. This fact can be explained with Fig. 6.4, which shows how the voltage divider formed by the 50-Ω source resistance and the 50-Ω input resistance of the amplifier splits the voltage of the equivalent noise voltage source in half. The noise at the input of the amplifier, sometimes known as the *available input-referred noise voltage*, is

$$v_{n.MA.\mathrm{av}} = \frac{v_{n.MA}}{2}. \qquad (6.13)$$

The important point is that the noise that adds to the MA's input signal, v_S, is the available input-referred noise, $v_{n.MA}/2$, as indicated in Fig. 6.4, and *not* the full input-referred noise. Therefore, the available input-referred noise is the relevant quantity for SNR and sensitivity calculations. In contrast, the full input-referred noise is the relevant quantity for noise figure calculations, as we will see in a moment.

The distinction between input-referred noise and available input-referred noise is familiar to the microwave engineer working with 50-Ω interfaces every day, but it may come as a surprise to the traditional analog designer. In analog circuits with op amps, the amplifier's input impedance is much higher than the source impedance, thus no voltage division takes place and the two noise measures are identical. The same is true for an MA that is integrated with a TIA on the same chip, circumventing an off-chip 50-Ω interface. The input-referred noise specified in MA data sheets usually is the *available* input-referred noise. Sometimes a note clarifies this specification as "output noise divided by midband gain." Circuit simulators usually require that you define the source to which the noise spectrum is referred. Thus, if you define the source like $v_{n.MA}$ in Fig. 6.4, your simulation produces the input-referred noise spectrum. If you want to produce the available input-referred noise spectrum, you can use a simulation setup similar to that shown in Fig. C.1 and refer the noise back to the source v_S in front of the $2\times$ gain block.

We are now ready to define the noise figure. A common definition for the *noise figure*, F, is the ratio of the "total output noise power" to the "fraction of the output noise power due to the thermal noise of the source resistance." Equivalently, the noise figure can be expressed as the ratio of the "input-referred noise power" to the "thermal noise power of the source resistance."[3] If we calculate this ratio in a narrow frequency band, such as 1 Hz, we obtain the so-called *spot noise figure*

$$F(f) = \frac{V_{n.MA}^2(f)}{V_{n.S}^2} = \frac{V_{n.MA}^2(f)}{4kTR_S}, \quad (6.14)$$

where the thermal noise spectrum of the source resistance, $V_{n.S}^2$, was expanded in terms of the source resistance R_S, usually 50 Ω, and the temperature T, which is defined to be 290 K for noise figure calculations. Similarly, if we calculate the ratio defining the noise figure over a wide bandwidth we obtain the *wideband noise figure*

$$F = \frac{\overline{v_{n.MA}^2}}{\overline{v_{n.S}^2}} = \frac{\overline{v_{n.MA}^2}}{4kTR_S \cdot BW_n}, \quad (6.15)$$

where the thermal noise power of the source resistance, $\overline{v_{n.S}^2}$, is calculated by referring the output noise power due to the thermal noise of the source resistance back to the equivalent noise voltage source. The latter quantity can be expressed easily in terms of the MA's noise bandwidth, BW_n, as shown on the right-hand side of Eq. (6.15). Note that the noise figure expressions in Eqs. (6.14) and (6.15) are based on the input-referred noise voltage, not the available input-referred noise voltage. The noise figure is specified either on a linear scale or in dBs. In the latter case, the value in dB is calculated as $10 \log F$. [$\rightarrow$ Problem 6.2]

It is convenient to remember the value for the noise voltage of the source resistance, $v_{n.S}^{rms} = \sqrt{4kTR_S \cdot BW_n}$, given $R_S = 50\,\Omega$ and $T = 290$ K:

$$v_{n.S}^{rms} = 0.8949\,\text{nV}/\sqrt{\text{Hz}} \cdot \sqrt{BW_n}. \quad (6.16)$$

---

[3] Yet another equivalent definition for the noise figure is the ratio of the "input SNR" to the "output SNR," where the input SNR is based on the thermal noise of the source resistance only.

For example, if the output rms noise voltage of an MA with 40-dB gain and 12-GHz noise bandwidth is 20 mV, we conclude that the available input-referred rms noise voltage is $20\,\text{mV}/100 = 200\,\mu\text{V}$, the input-referred rms noise voltage is twice that, $2 \times 200\,\mu\text{V} = 400\,\mu\text{V}$, and the rms noise voltage of the source resistance is $98\,\mu\text{V}$; thus, the noise figure is $10\log(400\,\mu\text{V}/98\,\mu\text{V})^2 = 12.2\,\text{dB}$.

Unfortunately, there is yet another difficulty that is related to the differential inputs of the MA. If we drive the amplifier single endedly as shown in Fig. 6.4, the noise voltage of the source resistance is that of a single 50-Ω resistor, the noise of the termination resistor at the other input counts as amplifier noise. However, if we drive the amplifier differentially, the noise voltage of the source resistance is that of *two* 50-Ω resistors. Thus, the differential noise figure is 3 dB lower than the single-ended noise figure. (The total noise power is the same, but the noise power due to the source resistance has doubled.) Most noise-figure test sets measure only the single-ended noise figure, and thus the noise figure specified in MA data sheets usually is the single-ended one.

What about the outputs of a fully differential MA? Do we also have to distinguish between a single-ended and a differential measurement? Fortunately, there is almost no difference because the multistage topology, which typically is used for MAs, causes the noise voltages at the two outputs to be highly anticorrelated ($v_{n.OP} = -v_{n.ON}$). (Remember that the noise voltages at the two outputs of a differential TIA usually are *not* highly anticorrelated; cf. Section 5.1.4.) If we measure the differential rms output noise voltage of an MA, we find twice the single-ended output noise voltage, but we also have twice the gain; thus, the input-referred rms noise voltage and the noise figure remain the same.

Power Penalty. Now let's calculate the power penalty caused by the MA noise. The total mean-square noise current referred back to the input of the TIA can be written as

$$\overline{i^2_{n,\text{amp}}} = \overline{i^2_{n.TIA}} + \overline{i^2_{n.MA}}, \qquad (6.17)$$

where $\overline{i^2_{n.TIA}}$ is the familiar input-referred noise current of the TIA and $\overline{i^2_{n.MA}}$ is the noise of the MA referred all the way back to the *input of the TIA*. Thus, when adding our noisy MA to the receiver, the input-referred rms noise current goes up from $\sqrt{\overline{i^2_{n.TIA}}}$ to $\sqrt{\overline{i^2_{n,\text{amp}}}}$ and, according to Eq. (4.20), we need to increase the received optical power by the same amount to maintain a constant BER. Therefore, the power penalty due to the MA noise is

$$PP = \sqrt{\frac{\overline{i^2_{n,\text{amp}}}}{\overline{i^2_{n.TIA}}}} = \sqrt{1 + \frac{\overline{i^2_{n.MA}}}{\overline{i^2_{n.TIA}}}}. \qquad (6.18)$$

The noise of the MA referred back to the TIA input can be expressed in terms of the MA's noise figure F and the TIA's transimpedance R_T. We start with the input-referred mean-square noise voltage of the MA, $\overline{v^2_{n.MA}}$. Remember that this noise voltage includes the noise voltage of the source resistance, $\overline{v^2_{n.S}}$, so we have to subtract it out. Then, we divide this result by four to obtain the available input-referred mean-square noise voltage due to the MA only. Again dividing this result by the squared transimpedance, R_T^2, is giving us the desired input-referred mean-square noise current. Finally, we can express the input-referred noise voltage in terms of the noise figure using Eq. (6.15):

$$\overline{i_{n.MA}^2} = \frac{\overline{v_{n.MA}^2} - \overline{v_{n.S}^2}}{4R_T^2} = \frac{(F-1) \cdot \overline{v_{n.S}^2}}{4R_T^2}. \tag{6.19}$$

Combining Eqs. (6.18) and (6.19), we find the power penalty due to the MA noise to be

$$PP = \sqrt{1 + \frac{(F-1) \cdot \overline{v_{n.S}^2}}{4R_T^2 \cdot \overline{i_{n.TIA}^2}}}. \tag{6.20}$$

Solving this equation for F yields the maximum permissible noise figure

$$F \leq 1 + (PP^2 - 1) \cdot \frac{4R_T^2 \cdot \overline{i_{n.TIA}^2}}{\overline{v_{n.S}^2}}. \tag{6.21}$$

From this equation, we can conclude that a high transimpedance helps to relax the noise-figure requirements for the MA. Furthermore, a low-noise TIA also requires a low-noise MA to keep the power penalty low.

Typical Values. Next, we calculate some typical numbers based on a 0.05-dB power penalty ($PP = 1.0116$), and a noise bandwidth $BW_n = 1.2B$. For a typical 2.5-Gb/s system ($R_T = 3\,\text{k}\Omega$, $i_{n.TIA}^{rms} = 380\,\text{nA}$), we obtain the numerical value

$$F \leq 1 + 0.023 \cdot \frac{4 \cdot (3\,\text{k}\Omega)^2 \cdot (380\,\text{nA})^2}{(49.0\,\mu\text{V})^2} = 17.1\,\text{dB}. \tag{6.22}$$

And for a typical 10-Gb/s system ($R_T = 1\,\text{k}\Omega$, $i_{n.TIA}^{rms} = 1{,}400\,\text{nA}$), we get

$$F \leq 1 + 0.023 \cdot \frac{4 \cdot (1\,\text{k}\Omega)^2 \cdot (1{,}400\,\text{nA})^2}{(98.0\,\mu\text{V})^2} = 13.0\,\text{dB}. \tag{6.23}$$

From a wireless designer's perspective, these noise figures are very large compared with what's typical for a *low-noise amplifier* (LNA). However, low noise figures are harder to achieve for a *broadband* amplifier and care must be taken to meet these numbers.

6.2.4 Input Dynamic Range

Definition. The *input dynamic range* of the MA describes the minimum and maximum input signal for which the MA performs a useful function, for example, for which the BER is sufficiently low. The minimum input signal (lower end of the dynamic range) is given by the MA's sensitivity. Similar to the receiver sensitivity definition in Section 4.3, the MA sensitivity is the minimum peak-to-peak signal voltage at the input of the MA necessary to achieve a specified BER ($v_{sens}^{pp} = v_S^{pp}$ @ BER). Because it is the *available* input-referred rms noise voltage, $v_{n.MA.av}^{rms}$, that adds to the input signal voltage, v_S, the sensitivity is given by $v_{sens}^{pp} = 2Q \cdot v_{n.MA.av}^{rms}$. Expressed in terms of the input-referred rms noise voltage (using Eq. (6.13)), the MA sensitivity is

$$v_{sens}^{pp} = Q \cdot v_{n.MA}^{rms}, \quad (6.24)$$

and rewritten in terms of the noise figure (using Eq. (6.15)), it is

$$v_{sens}^{pp} = Q \cdot \sqrt{F \cdot \overline{v_{n.S}^2}}. \quad (6.25)$$

For example, given the input-referred rms noise voltage of 400 μV from the previous example and a required BER of 10^{-12}, the sensitivity of the MA is 2.8 mV.

The definition of the maximum input signal (upper end of the dynamic range) depends on the type of amplifier. An AGC amplifier is supposed to operate linearly, and therefore the maximum input signal swing, v_{lin}^{pp}, is reached when the amplifier starts to limit or distort otherwise. A commonly used criterion is the 1-dB compression point, that is, the signal swing for which the gain drops by 1 dB below its small-signal value. In analog CATV/HFC applications, harmonic and intermodulation distortions are of great significance, and hence v_{lin}^{pp} is defined such that these distortions remain small (cf. Section 4.8). Obviously, these definitions cannot be applied to an LA because it is operated in the nonlinear (limiting) regime on purpose. Thus for the LA, the maximum input signal swing, also called the *input overload voltage*, v_{ovl}^{pp}, is reached when the amplifier produces so much pulse-width distortion and jitter that the specified BER cannot be maintained. For example, in BJT implementations, a large input signal can cause the base-collector diodes to become forward biased, leading to such distortions [38].

Power Penalty. Next, we find the power penalty caused by a finite MA sensitivity and then we determine the MA sensitivity required to keep this power penalty below a given value. We already know the power penalty due to the MA noise from Eq. (6.20), which we can easily convert into the power penalty due to the MA sensitivity by using Eq. (6.25). Assuming that the noise figure is much larger than one ($F \gg 1$), we find that

$$PP = \sqrt{1 + \left(\frac{v_{sens}^{pp}}{2Q \cdot R_T \cdot i_{n.TIA}^{rms}}\right)^2}. \quad (6.26)$$

Solving this equation for v_{sens}^{pp} reveals the required MA sensitivity

$$v_{sens}^{pp} \leq \sqrt{PP^2 - 1} \cdot 2Q \cdot R_T \cdot i_{n.TIA}^{rms} = \sqrt{PP^2 - 1} \cdot v_{I.min}^{pp}. \quad (6.27)$$

This result has an interesting interpretation: the term after the square root is the input voltage swing into the MA when the receiver operates at the sensitivity limit, that is, the quantity that we have called $v^{pp}_{I,\min}$ in Eq. (6.3). Thus, the square-root term tells us how much smaller we have to make the MA sensitivity compared with the minimum input signal into the MA. For example, for a 0.05-dB power penalty ($PP = 1.0116$), the MA sensitivity must be made 6.6× smaller than the minimum input signal into the MA.

Typical Values. Next, we calculate some typical MA sensitivity numbers based on a 0.05-dB power penalty ($PP = 1.0116$), a noise bandwidth $BW_n = 1.2B$, and $BER = 10^{-12}$ ($Q = 7.035$). For a typical 2.5-Gb/s system ($R_T = 3\,\text{k}\Omega$, $i^{rms}_{n.TIA} = 380\,\text{nA}$), we obtain the numerical value

$$v^{pp}_{sens} \leq 7.035 \cdot \sqrt{0.023 \cdot 4 \cdot (3\,\text{k}\Omega)^2 \cdot (380\,\text{nA})^2 + (49.0\,\mu\text{V})^2} = 2.5\,\text{mV}. \quad (6.28)$$

And for a typical 10-Gb/s system ($R_T = 1\,\text{k}\Omega$, $i^{rms}_{n.TIA} = 1{,}400\,\text{nA}$), we get

$$v^{pp}_{sens} \leq 7.035 \cdot \sqrt{0.023 \cdot 4 \cdot (1\,\text{k}\Omega)^2 \cdot (1{,}400\,\text{nA})^2 + (98.0\,\mu\text{V})^2} = 3.1\,\text{mV}. \quad (6.29)$$

The requirements for the upper end of the input dynamic range depend on the maximum output signal from the TIA and can be several volts. A typical input dynamic range for an MA is

$$v^{pp}_{sens} \ldots v^{pp}_{ovl} = 2\,\text{mV} \ldots 2\,\text{V}. \quad (6.30)$$

6.2.5 Input Offset Voltage

Definition. The *input offset voltage*, V_{OS}, is the input voltage for which the output voltage of the MA becomes zero. If the bandwidth of the MA extends all the way down to DC, a DC input voltage can be applied and V_{OS} is the voltage, which forces the output voltage to zero. However, if the gain rolls off at low frequencies (cf. Section 6.2.6), an AC signal must be applied to the input, and V_{OS} is the amplitude of this signal that causes the output signal just to touch the zero level. Alternatively, if the input offset voltage is small, that is, if the offset voltage does not drive the amplifier into the compressive regime, we also can determine the input offset voltage by taking the output offset voltage and dividing it by the midband voltage gain.

A nonzero offset voltage in an LA results in a slice-level error, which in turn causes (i) more bit errors, because it becomes more likely that the noisy signal crosses the off-center slice level, and (ii) pulse-width distortions, because of the finite rise and fall times of the received signal. A small offset voltage in an AGC amplifier is less severe because it can be compensated at the output of the amplifier. In particular, if slice-level steering is used after the AGC amplifier, the offset voltage is eliminated automatically.

Power Penalty. Next, we derive the power penalty caused by an input offset voltage in the LA and then we determine how much offset voltage can be tolerated. In the presence of the input offset voltage V_{OS}, we need to increase the input signal swing, v^{pp}_I, by nearly $2V_{OS}$ to restore the BER to the value without offset. Figure 6.5

172 MAIN AMPLIFIERS

illustrates the steps leading up to this conclusion. The horizontal line in the figure indicates the level at which the signal is sliced. Thus, the power penalty is

$$PP = \frac{v_I^{pp} + 2V_{OS}}{v_I^{pp}} = 1 + \frac{2V_{OS}}{v_I^{pp}}. \qquad (6.31)$$

This, of course, is the same result that we have already found in Eq. (4.52) when we first introduced the concept of power penalty. We see from this equation that the penalty is worst for small input signals. The smallest meaningful input signal to the MA, $v_{I.\min}^{pp}$, has already been stated in Eq. (6.3). Inserting this result into Eq. (6.31) yields the worst-case power penalty:

$$PP = 1 + \frac{V_{OS}}{\mathcal{Q} \cdot R_T \cdot i_{n.TIA}^{rms}}. \qquad (6.32)$$

Solving this equation for V_{OS} reveals the largest permissible input offset voltage

$$V_{OS} \leq (PP - 1) \cdot \mathcal{Q} \cdot R_T \cdot i_{n.TIA}^{rms}. \qquad (6.33)$$

From this equation, we can conclude that a high transimpedance helps to relax the offset requirements for the MA. Once again, a high transimpedance simplifies the MA design!

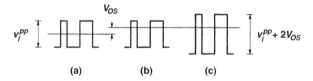

Fig. 6.5 Effect of an input offset voltage in the LA: (a) without offset, (b) with offset, and (c) with offset and increased signal swing to restore the original bit-error rate.

Typical Values. Next, we calculate some typical numbers based on a 0.05-dB power penalty ($PP = 1.0116$) and $BER = 10^{-12}$ ($\mathcal{Q} = 7.035$). For a typical 2.5-Gb/s system ($R_T = 3\,\text{k}\Omega$, $i_{n.TIA}^{rms} = 380\,\text{nA}$), we obtain the numerical value

$$V_{OS} \leq 0.0116 \cdot 7.035 \cdot 3\,\text{k}\Omega \cdot 380\,\text{nA} = 0.093\,\text{mV}. \qquad (6.34)$$

And for a typical 10-Gb/s system ($R_T = 1\,\text{k}\Omega$, $i_{n.TIA}^{rms} = 1{,}400\,\text{nA}$), we get

$$V_{OS} \leq 0.0116 \cdot 7.035 \cdot 1\,\text{k}\Omega \cdot 1{,}400\,\text{nA} = 0.114\,\text{mV}. \qquad (6.35)$$

In conclusion, an LA should have an input offset voltage of less than about 0.1 mV. This is a fairly low offset voltage, even for bipolar implementations. For this reason, MAs typically make use of offset compensation techniques, as we discuss further in Section 6.3.3.

6.2.6 Low-Frequency Cutoff

Definition. The *low-frequency cutoff*, f_{LF}, is defined as the lower frequency at which the small-signal gain $|A(f)|$ dropped by 3 dB below its midband value, as illustrated in Fig. 6.6. A low-frequency cutoff in the receiver response can be caused by a coupling capacitor (AC coupling) between the TIA and the MA or by some types of offset-compensation circuits used in the MA (cf. Section 6.3.3).

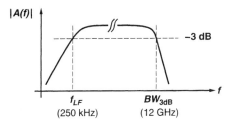

Fig. 6.6 Frequency response of an MA with a low-frequency cutoff.

When receiving a long string of zeros or ones, the output voltage of the amplifier *drifts* as a result of the low-frequency cutoff, as shown in Fig. 6.7. This effect also is known as *baseline wander*. Subsequently, the first few bits after the drift period are sliced with an offset error. Thus, a nonzero low-frequency cutoff has similar effects as a nonzero offset voltage. Specifically, the low-frequency cutoff causes (i) an increase in BER, which varies with the baseline movement and (ii) pulse-width distortions, which also vary with the baseline, that is, data-dependent jitter.

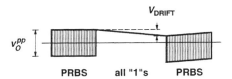

Fig. 6.7 Effect of a low-frequency cutoff: the output signal drifts during a long string of ones and subsequently causes a slice-level error.

Power Penalty. Next, we derive the power penalty caused by the low-frequency cutoff and then determine how low we have to make f_{LF}. What is the longest string of zeros or ones that we will encounter in a data stream? In SONET/SDH systems, which use scrambling as a line code, the run length potentially is unlimited. However, SONET/SDH equipment is tested with a particular bit sequence that puts the system under stress: the so-called "consecutive identical digit immunity measurement" [51]. This sequence consists of a long pseudorandom bit sequence (PRBS) with more than 2,000 bits and 50% mark density followed by 72 consecutive bits of zero; then again, more than 2,000 bits of PRBS followed by 72 bits of one. So it is reasonable to design a SONET/SDH system for the maximum run length $r = 72$. In Gigabit Ethernet systems, which use 8B10B encoding, the runs of zeros or ones are strictly limited to $r = 5$.

Assuming a linear system with a single-pole, high-pass transfer function, we can calculate the drift of the output voltage caused by r consecutive zeros or ones:

$$V_{\text{DRIFT}} = \frac{v_O^{pp}}{2}\left[1 - \exp\left(-2\pi f_{LF} \cdot \frac{r}{B}\right)\right] \approx \frac{v_O^{pp}}{2} \cdot \frac{2\pi f_{LF} r}{B}, \quad (6.36)$$

where v_O^{pp} is the output signals swing and B is the bit rate. The approximation on the right-hand side holds if the time constant $1/(2\pi f_{LF})$ is much larger than the drift time, r/B. Similar to the offset voltage, discussed in Section 6.2.5, the drift voltage causes the power penalty

$$PP = 1 + \frac{2 V_{\text{DRIFT}}}{v_O^{pp}}. \quad (6.37)$$

Inserting the approximation for V_{DRIFT} in Eq. (6.36) into Eq. (6.37) yields

$$PP = 1 + \frac{2\pi f_{LF} r}{B}. \quad (6.38)$$

Solving Eq. (6.38) for f_{LF} reveals the highest permissible low-frequency cutoff:

$$f_{LF} \leq (PP - 1) \cdot \frac{B}{2\pi r}. \quad (6.39)$$

From this equation, we can conclude that the longer the runs are, the lower f_{LF} must be made. [→ Problem 6.3]

Typical Values. Next, we calculate some typical numbers based on a 0.05-dB power penalty ($PP = 1.0116$). For a 2.5-Gb/s SONET system ($r = 72$), we obtain the numerical value

$$f_{LF} \leq 0.0116 \cdot \frac{2.5\,\text{Gb/s}}{6.28 \cdot 72} = 64\,\text{kHz}. \quad (6.40)$$

And for a 10-Gb/s SONET system we get

$$f_{LF} \leq 0.0116 \cdot \frac{10\,\text{Gb/s}}{6.28 \cdot 72} = 257\,\text{kHz}. \quad (6.41)$$

In conclusion, the low-frequency cutoff for a SONET system should be about $40{,}000\times$ lower than the bit rate ($f_{LF} < B/40{,}000$). It the case of a Gigabit Ethernet or Fiber Channel system with $r = 5$, the low-frequency cutoff specification can be relaxed to $f_{LF} < B/2{,}700$, in accordance with [107, p.70].

In practice, the low-frequency cutoff often is set even lower than the numbers derived above (e.g., 2.5 kHz for 2.5 Gb/s and 25 kHz for 10 Gb/s). The cutoff frequency usually is set by an external capacitor, for example, a coupling capacitor (DC block) or a capacitor part of an offset compensation circuit, such as C_1 and C_1' in Fig. 6.28. In this case, there is little cost involved in making this capacitor larger to protect against longer than 72-bit runs and reducing the power penalty to less than 0.05 dB.

6.2.7 AM-to-PM Conversion

There are a variety of effects in the optical fiber that can cause a rapid *amplitude modulation* (AM) of the received signal. For example, the combination of self-phase modulation (SPM) and chromatic dispersion can cause intensity overshoots at the beginning and the end of the optical pulses. Furthermore, stimulated Raman scattering (SRS) in a system that carries multiple wavelength in a single fiber (a WDM or DWDM system) also can cause an amplitude modulation. The SRS effect transfers optical energy from channels with shorter wavelengths to channels with longer wavelengths, as illustrated in Fig. 6.8. Thus, if two one bits are transmitted simultaneously over two different channels, the one in the longer-wavelength channel (λ_1) grows in amplitude as it propagates through the fiber, whereas the one in the shorter-wavelength channel (λ_2) shrinks. However, if a one and a zero bit are transmitted simultaneously over the same two channels, no such effect occurs. The result is a rapid amplitude modulation of the received signal, as shown on the right-hand side of Fig. 6.8. For more information on SPM and SRS, see [5, 136].

Fig. 6.8 Amplitude modulation in a WDM system caused by SRS.

Now, if the MA (or the TIA) exhibits an amplitude-dependent propagation delay, then these amplitude variations are transformed into delay or phase variation. In other words, the amplifier may produce an unwanted *phase modulation* (PM). In particular, an LA may exhibit a significant amount of delay variation when it transitions from the linear regime into the limiting regime. A phase modulation of the output signal is nothing else but jitter. As we know, excessive jitter can cause bit errors and interferes with the clock and data recovery process. Therefore, it is necessary to limit the *AM-to-PM conversion* occurring in the MA.

Definition. AM-to-PM conversion of an MA usually is specified in terms of the maximum delay variation, $\Delta\tau_{AM}$, observed when varying the input-signal swing over the entire dynamic range. Because the actual signal swing varies less than that, the AM-to-PM-induced jitter must be less than $\Delta\tau_{AM}$.

Typical Values. Typically, a delay variation, $\Delta\tau_{AM}$, of less than $\pm 10\%$ of the bit period (± 0.1 UI) is required to limit the generation of jitter. This corresponds to

$$\text{2.5 Gb/s MA:} \quad |\Delta\tau_{AM}| < 40\,\text{ps}, \qquad (6.42)$$
$$\text{10 Gb/s MA:} \quad |\Delta\tau_{AM}| < 10\,\text{ps}. \qquad (6.43)$$

6.3 MA CIRCUIT CONCEPTS

In the following, we discuss MA circuit concepts in a general and, as much as possible, technology-independent manner. This includes the multistage architecture, techniques for broadband stages, offset compensation, and automatic gain control.

6.3.1 Multistage Amplifier

From the discussion in the previous section, we can conclude that the *gain-bandwidth product* (GBW) required for multigigabit MAs is in excess of 100 GHz. For example, a 2.5-Gb/s MA with 30-dB gain and 3-GHz bandwidth requires a GBW of about 100 GHz, similarly, a 10-Gb/s MA with 30-dB gain and 12-GHz bandwidth requires a GBW of about 400 GHz. These GBW numbers are much larger than the f_T of most technologies! Is it possible to build an amplifier with a gain-bandwidth product that is much larger than f_T? Yes, if we use a multistage architecture.

Contrary to op amps, which typically are required to be stable under unity feedback conditions, there is no such stability requirement for MAs. MAs typically are run without feedback across the stages, except for the offset compensation loop, which is so slow that its stability is easy to ensure. For this reason, we don't have to worry about a single dominant pole, and we can cascade multiple stages, as shown in Fig. 6.9. Multiple amplifier stages can boost the gain-bandwidth product of the amplifier (GBW_{tot}) way beyond that of a single stage (GBW_S). How does this work? Let's start with a simple example to develop our intuition. All stages in our sample amplifier are identical and have a brick-wall, low-pass frequency response with a bandwidth of 3 GHz. If we put these stages together, the total gain increases, but the bandwidth remains at 3 GHz. Now, our goal is to build an amplifier with a total gain $A_{tot} = 30$ dB (31.6×) and a bandwidth of 3 GHz, which means that our total gain-bandwidth product must be $GBW_{tot} = 31.6 \times 3$ GHz $= 95$ GHz. Consider these two approaches:

- Single-stage architecture ($n = 1$). In this case, the GBW of the stage is equal to the GBW of the total amplifier: $GBW_S = GBW_{tot} = 95$ GHz.

- Three-stage architecture ($n = 3$). In this case, each stage needs a gain of only 10 dB (3.16×), and thus the gain-bandwidth per stage is $GBW_S = 3.16 \times 3$ GHz $= 9.5$ GHz.

In conclusion, the three-stage design requires 10× less GBW per stage! We also could say that cascading three stages gave us a *gain-bandwidth extension* GBW_{tot}/GBW_S of 10×.

How far can we push this? Could we build our sample amplifier from stages with $GBW_S = 0.95$ GHz? No! As we cascade more and more stages, the GBW requirement per stage is reduced, but even with an infinite number of stages, we still need a stage gain of slightly more than 1.0 to ever reach the 30-dB total gain, and thus a minimum GBW_S of slightly more than 3 GHz is required. We thus conclude that, in our example, the maximum possible gain-bandwidth extension is 95 GHz/3 GHz $= 31.6\times$.

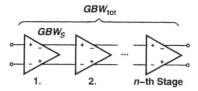

Fig. 6.9 An *n*-stage amplifier with overall gain-bandwidth product GBW_{tot} and stage gain-bandwidth product GBW_S.

We can easily generalize the above example for an arbitrary number of stages, n, and an arbitrary total gain, A_{tot}. In this case, the stage gain is $A_{tot}^{1/n}$, and thus the GBW extension becomes

$$\frac{GBW_{tot}}{GBW_S} = A_{tot}^{1-1/n}. \tag{6.44}$$

This function is plotted in Fig. 6.10 for $A_{tot} = 30\,\text{dB}$ and is labeled "Brick Wall." We can see from Eq. (6.44) with $n \to \infty$ that the maximum GBW extension that can be achieved with a multistage amplifier is given by A_{tot}.

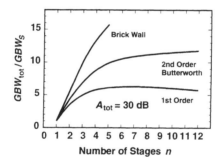

Fig. 6.10 Gain-bandwidth extension as a function of the number of stages.

Real amplifier stages don't have a brick-wall frequency response; they are more likely to have a first- or second-order response. A simple transistor stage with an R-C load has a first-order response, whereas a more complex stage with local feedback or inductive load has a second-order response. The gain-bandwidth extensions for these two cases also are plotted in Fig. 6.10 (for $A_{tot} = 30\,\text{dB}$). For these cases, the GBW extension is less dramatic than for the ideal brick-wall case, but we can still boost the gain-bandwidth product by a respectable 6× or 12×. The reason for this reduced GBW extension is that the total amplifier bandwidth *shrinks* as we cascade more and more stages with a slow (non-brick-wall) frequency rolloff. Furthermore, we can see from the plots that there is an optimum number of stages for which GBW_{tot}/GBW_S reaches the maximum.

178 MAIN AMPLIFIERS

The mathematical expression for the GBW extension as a result of cascading first-order stages turns out to be [28, 55]

$$\frac{GBW_{tot}}{GBW_S} = A_{tot}^{1-1/n} \cdot \sqrt{2^{1/n} - 1}. \tag{6.45}$$

The first term in this expression is identical to Eq. (6.44), the gain-bandwidth extension for brick-wall stages, and the second term describes the bandwidth shrinkage due to the slow frequency rolloff. The optimum number of stages, resulting in the maximum GBW extension, can be derived from Eq. (6.45) as $n_{opt} \approx 2 \ln A_{tot}$, where $A_{tot} \gg \sqrt{2}$ must hold for the approximation to be valid. The optimum stage gain follows as $A_S \approx \sqrt{e} = 4.34\,\text{dB}$. The GBW extension as a result of cascading second-order Butterworth stages ($Q = 1/\sqrt{2}$, no zeros) turns out to be

$$\frac{GBW_{tot}}{GBW_S} = A_{tot}^{1-1/n} \cdot \sqrt[4]{2^{1/n} - 1}. \tag{6.46}$$

The optimum number of stages can be approximated as $n_{opt} \approx 4 \ln A_{tot}$, if $A_{tot} \gg \sqrt[4]{2}$, and the optimum stage gain follows as $A_S \approx \sqrt[4]{e} = 2.17\,\text{dB}$. [$\rightarrow$ Problem 6.4]

For example, if we want to build a 2.5-Gb/s MA with 30-dB gain and 3-GHz bandwidth ($GBW_{tot} \approx 100\,\text{GHz}$) using second-order Butterworth stages, we find that the optimum number of stages is $n_{opt} = 14$ and the corresponding gain-bandwidth extension is 11.7×. However, from a power dissipation and area point of view, this is a rather large number of stages, and we may consider removing some of them. For $n = 4$, the gain-bandwidth extension is still 8.8× and each stage needs a GBW of only around 11 GHz. Such stages can be implemented in a 0.25-μm CMOS technology with $f_T \approx 25\,\text{GHz}$, and thus the whole 2.5-Gb/s MA can be realized in this technology [156].

Bandwidth shrinkage in multistage amplifiers usually is an undesirable effect; however, a shrinkage *reduction* may have an unexpected bandwidth boosting effect. For example, let's look at a four-stage amplifier with a total bandwidth of 1 GHz where each stage consists of a single transistor with an R-C load. Now, we apply shunt peaking to all stages, which, as we discuss in Section 6.3.2, increases the stage bandwidth by a factor 1.7. What happens to the total amplifier bandwidth? It increases to about 2.4 GHz, not 1.7 GHz as we may have thought at first! The reason for this "free lunch" is that shunt peaking turns the first-order stages into second-order stages, and thus the bandwidth shrinkage is reduced by approximately $(2^{1/n} - 1)^{-1/4}$, which is 1.5× for $n = 4$. Actually, it can be shown by simulation that the precise shrinkage reduction is 1.4×, because shunt peaking also introduces a zero. Thus, the four-stage amplifier bandwidth is extended by a total of $1.7 \cdot 1.4 = 2.4\times$.

Finally, it should be pointed out that multistage amplifiers often exhibit a larger group-delay variation than single-stage amplifiers. In particular, if the amplifier is composed of identical stages, the group-delay variations of the individual stages add up in the same direction, effectively multiplying the stage variation by the number of stages. Similarly, a slight peaking in the frequency response of the individual stages may compound to an unacceptable amount of peaking in the multistage amplifier.

Thus, it is important to make sure that the group-delay variation and gain peaking requirements of the multistage amplifier are met.

6.3.2 Techniques for Broadband Stages

Although the multistage architecture greatly relaxes the gain-bandwidth requirements per stage, we still need to make each stage as fast as possible to meet the demanding requirements of multigigabit MAs. In this section, we review the most important techniques for building broadband amplifier stages. Although each technique is described by itself, they can be "mixed and matched" in many ways. For example, the Cherry-Hooper architecture, in widespread use for bipolar MAs and discussed in Section 6.4.2, combines multiple techniques including series feedback and shunt feedback. The bandwidth extensions calculated for the individual techniques, however, may not simply add up when combining multiple techniques. Note that the broadband techniques covered in this section also are applicable to the feedback and post amplifiers of the TIA. Additional information about broadband techniques can be found in [25, 26, 28, 35, 82].

Use Fast Transistors. To obtain fast gain stages, we must optimize the operating point and the geometry of the critical transistors for maximum speed. The speed of a transistor usually is measured by f_T, the frequency where the current gain becomes unity, also known as the *transition frequency*, and $f_{\max}$, the frequency where the power gain becomes unity, also known as the *maximum frequency of oscillation*. In the following, we discuss some important speed optimization techniques, first for bipolar transistors and subsequently for FETs. For more information on speed optimization techniques for bipolar transistors, see [35, 146].

The f_T of a bipolar transistor (BJT or HBT) is given by the following expressions:

$$f_T = \frac{1}{2\pi} \cdot \frac{g_m}{C_{be} + C_{bc}} = \frac{1}{2\pi} \cdot \frac{1}{\tau_F + (C_{je} + C_{jc}) \cdot V_T/I_C}, \quad (6.47)$$

where g_m is the (intrinsic) transconductance, C_{be} and C_{bc} are the base-emitter and base-collector capacitances, τ_F is the carrier transit time, C_{je} and C_{jc} are the parasitic emitter and collector junction capacitances, V_T is the thermal voltage ($V_T = kT/q \approx 25$ mV), and I_C is the collector current. The expression on the left-hand side is written in terms of small-signal parameters and follows directly from the condition that the input admittance for a shorted output, $s(C_{be} + C_{bc})$, has the same magnitude as the transconductance, g_m, when the current gain becomes unity. The expression on the right-hand side restates f_T in terms of physical parameters and the operating point ($g_m = I_C/V_T$, $C_{be} = g_m \tau_F + C_{je}$, $C_{bc} = C_{jc}$).

From the latter expression, we can conclude that a small transit time and low parasitic junction capacitances are necessary to obtain a fast bipolar transistor. In general, n-p-n transistors have a smaller transit time than p-n-p transistors because, among other things, the electron mobility is higher than the hole mobility. Thus, n-p-n transistors are preferred for high-speed applications. Furthermore, the transit time as well as the junction capacitances depend on the emitter area, $A_E = W_E L_E$

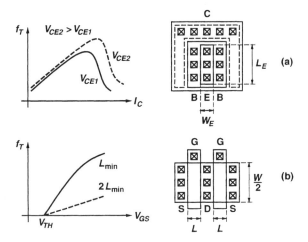

Fig. 6.11 Typical f_T characteristics and layout of (a) a BJT (b) a MOSFET.

(cf. Fig. 6.11(a), right-hand side), which must be selected carefully. A small emitter size leads to a small transistor, and thus to low parasitic junction capacitances, C_{je} and C_{jc}. However, a small emitter size also leads to a high collector current density,[4] I_C/A_E (assuming a fixed collector current). If this density exceeds a critical value, typically in the range of 1 to 6 mA/μm^2, the transit time, τ_F, increases rapidly because of an extension of the base region into the collector region known as *base pushout* or *Kirk effect*. The critical current density, at which the Kirk effect kicks in, increases with the collector-emitter voltage, V_{CE}. For this reason, transistors in high-speed circuits are operated with a comparatively large V_{CE}. Figure 6.11(a) shows how f_T at first increases with the collector current because of the $(C_{je} + C_{jc}) \cdot V_T/I_C$ term in Eq. (6.47). However, when the critical current density is reached, f_T peaks and decreases because of the rapid increase of τ_F. The dashed curve shows how the current where the peaking occurs can be pushed out by increasing V_{CE}. In summary, to operate the transistor near its peak f_T value, we need to choose the optimum emitter area for a given bias current, or for a given emitter area, such as the minimum size permitted by lithography, we need to choose the optimum bias current.

The $f_{\max}$ of a bipolar transistor (BJT or HBT) is given by [183]

$$f_{\max} = \frac{1}{2} \cdot \sqrt{\frac{f_T}{2\pi \cdot R_b C_{bc}}}, \qquad (6.48)$$

---

[4]More precisely, the collector current density is I_C/A_C with the collector area $A_C = (W_E + 2\gamma) \cdot L_E$, where W_E is the emitter width, L_E is the emitter length, and γ is the current spreading from the emitter into the collector. The current spreading effect can be significant in submicron technologies, for example, for $W_E = 0.3\,\mu$m and $\gamma = 0.2\,\mu$m, A_C is 2.3× larger than A_E.

where R_b is the intrinsic base resistance (base-spreading and contact resistance). Thus, to obtain a high f_{max}, we must optimize for a high f_T while minimizing R_b and C_{bc}. The base-spreading resistance can be kept low by using long and narrow emitter stripes (remember, the thin base layer is below the emitter) and contacting the base on both sides, as shown in Fig. 6.11(a). Furthermore, a heavily doped base region, as used in HBTs, also reduces the base resistance (cf. Appendix D).

The f_T of an FET (MESFET, HFET, or MOSFET) is given by the following expressions:

$$f_T = \frac{1}{2\pi} \cdot \frac{g_m}{C_{gs} + C_{gd}} \approx \frac{3}{4\pi} \cdot \frac{\mu_n}{L^2} \cdot (V_{GS} - V_{TH}), \quad (6.49)$$

where g_m is the transconductance, C_{gs} and C_{gd} are the gate-source and gate-drain capacitances, μ_n is the carrier mobility, L is the channel length, V_{GS} is the gate-source voltage, and V_{TH} is the threshold voltage. The expression on the left-hand side is written in terms of small-signal parameters and corresponds directly to Eq. (6.47). The approximation on the right-hand side is written in terms of physical parameters and the operating point and is valid for small overlap capacitances and for low electric fields such that the quadratic FET model applies ($g_m = \mu_n C'_{ox} \cdot W/L \cdot (V_{GS} - V_{TH})$, $C_{gs} = 2/3 \cdot C'_{ox} \cdot WL$, $C_{gd} = 0$).

Equation (6.49) is illustrated graphically with Fig. 6.11(b). For long-channel FETs (low electric fields), f_T increases proportional to the gate overdrive voltage, $V_{GS} - V_{TH}$, as shown with the dashed line ($L = 2L_{min}$). However, for short-channel FETs, carrier velocity saturation causes the curve to flatten out as shown with the solid line ($L = L_{min}$). For MESFETs and HFETs, the curve typically reaches a maximum value after which f_T declines because of the turn on of the Schottky diode (between the gate and channel) and other effects. We conclude that to obtain a fast FET, we must choose (i) an n-channel device because the electron mobility is higher than the hole mobility, (ii) the smallest channel length permitted by lithography, and (iii) the largest possible gate overdrive voltage subject to headroom limitations (for MOSFETs) or the gate overdrive voltage for which f_T reaches its maximum (for MESFETs and HFETs). A typical gate overdrive voltage is around 400 mV.

The f_{max} of an FET (MESFET, HFET, or MOSFET) is given by [142, 183]

$$f_{max} = \frac{1}{2} \cdot \sqrt{\frac{f_T}{2\pi \cdot R_g C_{gd}}}, \quad (6.50)$$

where R_g is the intrinsic gate resistance, and we have neglected the effect of the FET's output conductance. Thus, to obtain a high f_{max}, we must optimize for a high f_T while minimizing R_g and C_{gd}. The gate of MESFETs and HFETs is made from metal and its resistance is generally very low; however, the gate of MOSFETs is made from polysilicon, and its much larger resistance per square can lead to a low f_{max}. In the latter case, the gate resistance can be reduced by breaking wide transistors into several smaller, parallel transistors, resulting in a finger-structure layout. Figure 6.11(b) shows an example where a MOSFET is broken into two half-sized transistors (each one with width $W/2$), which share a common drain region; the two gates and the two source regions must be connected at the metal level. Contacting the gate fingers on both sides further reduces the gate resistance. With these techniques, the gate resistance usually can be made small enough such that f_{max} becomes larger than f_T. (For 90-nm MOSFETs and less, the finger width needed to keep R_g sufficiently low becomes so small that fringing capacitances adversely affect f_T and f_{max} [191].)

Deep submicron CMOS technologies offer f_T's that are comparable with those of fast bipolar technologies. Does this mean that circuits in either technology can operate at about the same speed? No, in general, bipolar technologies have a speed advantage over CMOS technologies even for the same values of f_T and f_{max}. The main reason for this is that the f_T of an FET degrades more easily than the f_T of a bipolar transistor if we take parasitic capacitances, such as wiring and junction capacitances, into account. More specifically, if we add a parasitic capacitance, C_P, from the gate (or the base) to ground and calculate the f_T' of the joint system (transistor plus parasitic capacitance), we find from Eqs. (6.47) and (6.49) for either the FET or the bipolar transistor

$$f_T' = \frac{f_T}{1 + 2\pi f_T \cdot C_P/g_m}. \tag{6.51}$$

From this equation, we see that the transistor with the larger g_m is more resilient to f_T degradations caused by parasitic capacitances. For bipolar transistors, the transconductance is given by $g_m = I_C/V_T$, whereas for FETs, we have $g_m = 2I_D/(V_{GS} - V_{TH})$ or less, if we take carrier velocity saturation into account. Thus, given the same bias currents ($I_C = I_D$), $V_T \approx 25$ mV, and $(V_{GS} - V_{TH})/2 \approx 200$ mV, the bipolar transistor's g_m is around $8\times$ larger than the FET's g_m, and hence the bipolar transistor is significantly more resilient to f_T degradations than the FET. Another way to describe the same situation is to say the f_T of a bipolar transistor is given by the ratio of a "large transconductance" to a "large input capacitance," whereas the f_T of an FET is given by the ratio of a "small transconductance" to a "small input capacitance." Thus, although the ratios may be the same, the latter ratio is more prone to degradation as a result of parasitic capacitances.

Furthermore, the speed of many practical circuits is determined by a combination of the transistor's input capacitance *and* output capacitance, yet f_T and f_{max} measure only the effect of the input capacitance. Thus, depending on the value of the collector-substrate capacitance (for bipolar transistors) or the drain-substrate capacitance (for FETs), the circuit speed may vary even for the same values of f_T and f_{max}.

Boost f_T. The f_T parameter can be related to the time it takes the carriers to travel through the base (or the channel), know as the transit time. For the bipolar transistor, we find the transit time from Eq. (6.47) as $\tau_F = 1/(2\pi f_T)$, if we neglect the parasitic capacitances C_{je} and C_{jc}. For the FET, we can calculate the transit time from the carrier speed, $\mu_n \cdot (V_{GS} - V_{TH})/L$, and the travel distance, L, which yields an expression that is inversely proportional to the f_T expression in Eq. (6.49) (cf. Appendix D). These relationships may suggest that f_T is a fundamental property that cannot be improved on unless we use smaller device geometries or different materials. However, this is not the case, and it is indeed possible to increase f_T by means of circuit techniques.

Remember that f_T is determined by the ratio of the transconductance to the input capacitance under shorted-output conditions. Thus, if we can come up with a circuit that lowers the input capacitance while maintaining the same transconductance, we can increase f_T. The so-called f_T-*doubler* circuit shown in Fig. 6.12(a) does just this [28]. In this circuit, the input voltage between nodes B and E is divided by Q_1 and Q_3 into two equal voltages with node x as the midpoint. The resulting half-voltages are the base-emitter voltages of Q_1 and Q_2, and each is amplified by g_m. Finally, the collector currents of Q_1 and Q_2 are combined at node C. Thus, the transconductance of the overall "super transistor" is the same as that of Q_1 or Q_2 alone, but the base-emitter capacitance, C_{be}, is cut in half. The capacitance reduction can be best understood in terms of the Miller effect[5]: the capacitance C_{be} of Q_1 is not grounded, but rather is connected to node x, which follows the input voltage with the gain $A_X = 1/2$; thus, the input capacitance seen at node B is $(1 - A_X) \cdot C_{be} = C_{be}/2$. In conclusion, the f_T of the "super transistor" is twice that of its constituent transistors. [→ Problem 6.5]

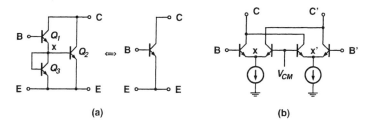

Fig. 6.12 The f_T doubler: (a) principle of operation and (b) practical differential implementation.

Figure 6.12(b) shows a more practical differential implementation of an f_T doubler [195]. Here, the differential input voltage between nodes B and B' is split in two half-sized voltages and is amplified by two separate differential pairs. The bias voltage V_{CM} is set to the input common-mode voltage. The differential output currents from both pairs are combined at the output nodes C and C'. Thus, the overall transconductance is the same as that of a simple differential pair. But unlike in a simple differential pair, the nodes x and x' are not virtual grounds; instead, they follow the nodes B and B' with gain $A_X = 1/2$, respectively. Again, the Miller effect cuts the input capacitance in half, and as a result f_T is doubled. [→ Problem 6.6]

In practice, the f_T-doubler circuits do not exactly double the f_T. One reason is that a parasitic capacitance at node x reduces the gain A_X below $1/2$ and introduces a phase lag. Both effects degrade the Miller reduction of C_{be}. Another reason is that C_{bc}, which also is part of the input capacitance, is not reduced by the Miller effect.

---

[5]The *Miller effect* can be described as follows: a feedback capacitance C across an amplifier with gain A contributes a capacitance equal to $(1 - A) \cdot C$ to the input capacitance. The best known case is that of an inverting amplifier, $A < 0$, where the feedback capacitance is *multiplied* by $|A| + 1$. However, the above equation also holds for $1 > A > 0$, where the capacitance is *reduced* and $A > 1$, where the capacitance is *inverted*.

An important drawback of the f_T doubler is that it has twice the output capacitance (collector-substrate capacitance) compared with a simple transistor. This additional load capacitance may negate some of the f_T doubler's speed advantage. Also note that the f_T-doubled circuit consumes twice the current of the original circuit.

The f_T doubler also can be built with FETs. However, the f_T multiplication obtained may be small when considering parasitic wiring and junction capacitances (cf. Eq. (6.51) for the f_T degradation). Thus, FET-based f_T doublers are most useful for driver stages with very large FETs that present a large input capacitance [31]. Moreover, the unavoidable doubling of the FET output capacitance (drain-substrate capacitance) usually is more detrimental than in the bipolar case, because the ratio of output to input capacitance typically is larger for FETs than for bipolar transistors.

Suppress Capacitances with Series Feedback. A major speed bottleneck in BJTs is the input pole formed by the intrinsic base resistance R_b and the base-emitter capacitance C_{be} (see Fig. 6.13(a)). For now, we neglect the base-collector capacitance, which also contributes to this pole and assume $C_{bc} = 0$. In BJTs, the base is lightly doped, which leads to a relatively high value of R_b and thus a low frequency of the input pole. In HBTs, the base is more heavily doped (cf. Appendix D), and hence the input pole is less of a problem. The base-emitter capacitance can be rather large and is given by $C_{be} = I_C/V_T \cdot \tau_F + C_{je}$. Note that this capacitance is bias dependent and increases with collector current. For example, let's pick a silicon BJT transistor with $R_b = 120\,\Omega$, $C_{be} = 170\,\text{fF}$, $g_m = 40\,\text{mS}$, and $f_T \approx 30\,\text{GHz}$ when operated at $I_C = 1\,\text{mA}$. In this case, the low-pass filter formed by R_b and C_{be} has a bandwidth of only 7.8 GHz, which is much lower than the transistor's f_T.

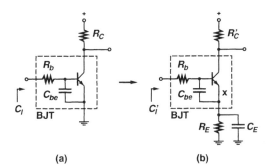

Fig. 6.13 BJT gain-stage (a) without and (b) with series feedback.

A well-known technique to speed up the input pole is the use of *series feedback* [22]. Figure 6.13(b) shows its implementation with an emitter degeneration resistor R_E. Because the emitter voltage at node x is now approximately following the base voltage, the current into C_{be} is reduced, which makes the input capacitance appear smaller. Just like in the case of the f_T doubler, the Miller effect reduces the input capacitance, but unlike in the f_T doubler, g_m also is reduced, and therefore f_T is not boosted. It can be shown that the low-frequency gain from the base to node x is

$A_X = R_E/(R_E + 1/g_m)$, neglecting the effects of the base-emitter resistance and the output conductance. Therefore, the Miller effect reduces the input capacitance from $C_I = C_{be}$ to $C_I' = (1 - A_X) \cdot C_{be}$, which is

$$C_I' = \frac{C_{be}}{g_m R_E + 1}. \tag{6.52}$$

Similarly, the input pole due to R_b and C_{be} is sped up by the same factor $g_m R_E + 1$. Note that the reduced input capacitance also reduces the loading of the previous stage and thus further improves the amplifier bandwidth.

A side effect of adding the emitter resistor, R_E, is that the low-frequency gain of the stage drops by about $g_m R_E + 1$. However, the gain can be restored to its original value simply by increasing the collector resistor, R_C, by the same amount. Unfortunately, this measure also slows down the output pole due to R_C, the load capacitance, and C_{bc}. Increasing I_C and A_E while reducing R_C and R_E by the same factor may result in a better trade off. Note that this re-scaling speeds up the output pole (assuming a constant load capacitance), whereas the gain and the input pole due to R_b and C_{be} remain approximately constant.

Another side effect of the emitter resistor is that it introduces a new high-frequency pole into the transfer function, known as the *emitter pole*. If left uncompensated, this pole reduces the stage bandwidth somewhat. However, adding the emitter capacitor, C_E, as shown in Fig. 6.13(b), introduces a compensating zero. For the capacitance $C_E \approx 1/(2\pi f_T \cdot R_E)$, the zero cancels the emitter pole and the bandwidth is restored.

Continuing our example from above, now let's add the emitter resistor $R_E = 100\,\Omega$. As a result, the input capacitance is reduced by a factor $40\,\text{mS} \cdot 100\,\Omega + 1 = 5$ to $C_I' = 34\,\text{fF}$ and the bandwidth of the R_b-C_{be} low-pass filter goes up to 39 GHz, which is now above the transistor's f_T. To compensate for the lost gain, we can increase R_C by about $5\times$, and to compensate the emitter pole, we can add an emitter capacitor of about 50 fF. [$\rightarrow$ Problem 6.7]

In a variation of the pole-zero cancellation technique explained above, the zero introduced by C_E can be used to boost the stage bandwidth further. Rather than trying to cancel the emitter pole exactly, the zero can be moved to a lower frequency to "push up" the frequency response. This effect can be achieved by increasing the emitter capacitor, C_E, above the value discussed earlier; hence, this technique is known as *emitter peaking*. Note, however, that C_E should not be made too large because the resulting gain peaking and the increased group-delay variations can cause undesirable signal distortions such as jitter.

The inclusion of an emitter degeneration resistor has a number of additional advantages, besides reducing the input capacitance and speeding up the input pole. Among them are: (i) precise control of the stage gain by means of the ratio of two resistors to $|A| = R_C/R_E$, if $R_E \gg 1/g_m$, (ii) improvement of the input resistance, (iii) linearization of the stage's large signal response, and (iv) increase of the input dynamic range in a differential stage. Without emitter degeneration, the input dynamic range is limited to just a few temperature voltages, for example, $3 V_T \approx 75\,\text{mV}$; beyond this range, the tail current is switched entirely to one or the other output.

186 MAIN AMPLIFIERS

In FET circuits, the R_g-C_{gs} input pole usually presents no speed limitation, because the gate resistance can be made sufficiently small with the appropriate layout techniques (short finger structure). Thus, source degeneration is not needed for this purpose. However, there are other uses for source degeneration in FET stages such as controlling the stage gain, linearizing the stage response, reducing the input capacitance, and boosting the bandwidth with the help of source peaking.

Suppress Capacitances with Cascode. In the previous section, we neglected the base-collector capacitance, C_{bc}, which also contributes to the input capacitance and thus to the BJT input pole. Now, we study this contribution and see how we can minimize it. The total input capacitance, C_I, of the BJT stage shown in Fig. 6.14(a) is comprised of two components: (i) the base-emitter capacitance of Q_1, C_{be1}, and (ii) the *Miller capacitance*, which is the base-collector capacitance of Q_1, C_{bc1}, multiplied by the absolute stage gain plus one:

$$C_I = C_{be1} + (|A| + 1) \cdot C_{bc1}. \qquad (6.53)$$

Note that C_{bc1} is located between the input and the *inverting* output of the stage; hence, $A < 0$ and the familiar Miller expression $(1 - A) \cdot C_{bc1}$ also can be written as $(|A| + 1) \cdot C_{bc1}$. The latter expression has the advantage that it doesn't matter whether the stage gain A is defined as positive or negative. Because the Miller capacitance depends on the magnitude of the gain, it can become quite large and may even dominate C_{be}. Of course, the situation for FET stages is analogous and, just as in Eq. (6.53), the total input capacitance can be written as $C_I = C_{gs1} + (|A| + 1) \cdot C_{gd1}$.

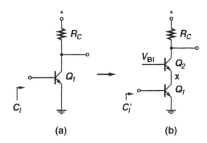

Fig. 6.14 Stage (a) without and (b) with cascode transistor.

The potentially large Miller capacitance can be suppressed by connecting the collector of Q_1 to a low-impedance node, which effectively reduces the gain that multiplies C_{bc1}. This can be done, for example, by stacking the *cascode* transistor Q_2 on top of the main transistor Q_1 as shown in Fig. 6.14(b), where V_{BI} is a bias voltage. Alternatively, the collector of Q_1 can be connected to the input of a transimpedance stage; we discuss this approach in a moment. In the case of the cascode transistor, the input impedance into the emitter of Q_2 is about $1/g_{m2}$ (assuming $R_C \ll 1/g_{o2}$); thus, the voltage gain from the input to node x reduces from $|A| \approx g_{m1} R_C$ to $|A_X| \approx$

g_{m1}/g_{m2}, which is about unity. As a result, the input capacitance is reduced to

$$C'_I = C_{be1} + 2C_{bc1}. \quad (6.54)$$

Note that the DC voltage gain for both stages in Fig. 6.14(a) and (b) is about the same. The cascode transistor acts as a current buffer and passes most of the emitter current directly to the collector. Thus, in both cases the stage gain is given by about $|A| \approx g_{m1} R_C$.

Clearly, the cascode technique is most effective for stages with a high voltage gain where the input capacitance is dominated by the Miller capacitance. This technique also can be applied to FET stages with similar benefits. Besides suppressing the Miller capacitance, the cascode technique has a number of additional advantages: (i) it increases the output resistance at the collector of Q_2 and hence the stage gain as long as the load also is cascoded, (ii) it improves the isolation from the stage output back to the input (unilateralization: $S_{12} \ll S_{21}$), which helps the design of stable amplifiers, and (iii) it can prevent the avalanche breakdown of Q_1 by dividing its collector-emitter voltage between Q_1 and Q_2. Unfortunately, the cascode technique also has a few disadvantages: (i) the cascode transistor, Q_2, introduces an additional high-frequency pole, which may offset some of the bandwidth gained with this technique, and (ii) the cascode transistor reduces the voltage headroom, which is a concern in low-voltage designs.

The simple cascode circuit shown in Fig. 6.14(b) can be extended to the so-called *regulated cascode* circuit [21, 159]. To do this, the base (or gate) of the cascode transistor is connected to an inverting feedback amplifier with gain $|A_F|$, which receives its input signal from node x. The amplifier can be a simple common-emitter (or common-source) stage or a more complex structure. As a result of the feedback action, the impedance at node x is reduced by the factor $|A_F| + 1$ to about $1/[(|A_F| + 1) \cdot g_{m2}]$, which in turn lowers the Miller capacitance. For a large value of $|A_F|$, the input capacitance goes down to $C'_I = C_{be1} + C_{bc1}$. Furthermore, a regulated cascode realized with FETs features an output impedance that is boosted by the factor $|A_F| + 1$ and thus permits the design of very high-gain stages. [→ Problem 6.8]

Suppress Capacitances with TIA Load. A method that has been popularized by Cherry and Hooper [22] not only reduces the Miller capacitance, but also increases the stage bandwidth and produces a second-order stage response, which is advantageous when cascading multiple stages (cf. Section 6.3.1). Figure 6.15 shows the principle: the passive load resistor R_C is replaced by an active load in the form of a transimpedance amplifier (TIA). The TIA consists of the inverting feedback amplifier A_F and the feedback resistor R_F. In the following, we refer to this arrangement as a *gain stage with TIA load*. Note that if the transresistance, R_T, of the TIA is made equal to the load resistor, R_C, both stages in Figs. 6.15(a) and (b) have the same low-frequency gain. Although shown for the example of a BJT stage, the same technique also can be applied to FET stages.

We already discussed the shunt-feedback TIA circuit shown in Fig. 6.15(b) at length in Chapter 5. The role of the photodetector is now played by the input transistor Q_1. Following Eq. (5.13), the transresistance can be written as $R_T =$

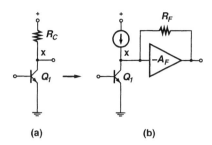

Fig. 6.15 Stage (a) with load resistor and (b) TIA load.

$|A_F|/(|A_F|+1) \cdot R_F$, where $|A_F|$ is the gain of the feedback amplifier and R_F is the feedback resistor. From Eq. (5.16), we know that the TIA circuit presents the low input resistance $R_I = R_F/(|A_F|+1)$ to the node x. With $R_T = R_C$ to make the gain of both stages equal, we can rewrite the input resistance as $R_I = R_C/|A_F|$ and thus, compared with the original stage with load resistor R_C, the pole at node x is sped up by the factor $|A_F|$. Similarly, the gain of Q_1 to node x is reduced from $|A|$ to $|A_X| = |A|/|A_F|$. Thus, the Miller capacitance is reduced, too, and the total input capacitance of the stage with TIA load becomes $C'_I = C_{be1} + (|A|/|A_F|+1) \cdot C_{bc1}$.

For a more detailed analysis of this stage, we write its voltage gain as $A(s) = -g_{m1} \cdot Z_T(s)$, where g_{m1} is the transconductance of Q_1 and $Z_T(s)$ is the transimpedance of the TIA load. Fortunately, in Eqs. (5.19) through (5.22), we already worked out the expression for $Z_T(s)$ for the case of a single-pole feedback amplifier. From these equations, we conclude that the stage gain, $A(s)$, has a conjugate-complex pole pair with the possibility of undesired peaking but also with the advantage of a fast second-order rolloff, which helps to reduce the bandwidth shrinkage in multistage amplifiers. If we choose the Butterworth pole configuration, which results in a maximally flat frequency response, the bandwidth of the stage with TIA load turns out to be

$$BW' \approx \sqrt{|A|} \cdot BW, \qquad (6.55)$$

where BW is the bandwidth of the simple stage with the R_C load and $|A|$ is the overall stage gain. This result is valid if both stages in Fig. 6.15(a) and (b) have the same g_{m1}, the same load capacitance at node x, and the same overall gain. For example, a simple first-order gain stage with the bandwidth $BW = 1\,\text{GHz}$ and the gain $|A| = 4$ (12 dB) can be turned into a second-order Butterworth stage with the bandwidth $BW' = 2\,\text{GHz}$ and the same gain $|A| = 4$ by replacing the resistive load with a TIA load. It is clear from Eq. (6.55) that the bandwidth extension obtained with this technique is most impressive for stages with a high gain. [→ Problem 6.9]

A disadvantage of this topology is its higher power dissipation due to the extra TIA in every stage. In terms of power dissipation and the order of the frequency response, a stage with TIA load is really a "double stage," that is, two consecutive stages. We know from Section 6.3.1 that increasing the number of stages in a multistage amplifier, while keeping its total gain constant, often boosts its bandwidth. Thus, the question arises whether splitting the simple stage in Fig. 6.15(a) into two simple stages, each with gain $\sqrt{|A|}$, would yield the same bandwidth extension as in Eq. (6.55)? It turns out that the bandwidth extension as a result of "stage splitting" is

$$BW' \approx \sqrt{(\sqrt{2}-1)\cdot |A|}\cdot BW, \qquad (6.56)$$

which is similar to Eq. (6.55) in the sense that it also is proportional to $\sqrt{|A|}$, but the total bandwidth of the split stage is about 1.55× smaller than that of the stage with TIA load. Note that the factor 1.55 represents the bandwidth shrinkage when cascading two first-order stages. Thus, we can conclude that the bandwidth extension as a result of the TIA load is like that as a result of stage splitting but without the bandwidth-shrinkage penalty of cascading two stages. [→ Problem 6.10]

So far, we considered the TIA load to be a shunt-feedback TIA, as shown in Fig. 6.15(b), but other TIA topologies can be used as well. In particular, the active-feedback TIA, which we briefly discussed in Section 5.2.8, in an interesting candidate. In the active-feedback TIA, the feedback resistor R_F is replaced by a transconductor g_{mF}, which usually is realized with an FET. This TIA load has the transimpedance $1/g_{mF}$, and thus the stage gain is defined by the transconductance ratio $|A| = g_{m1}/g_{mF}$. If the input and feedback transistors are of the same type, this gain can be made insensitive to process, temperature, and supply voltage variations. It can be shown that the bandwidth extension given in Eq. (6.55) also holds for a gain stage with active-feedback TIA load. [→ Problem 6.11]

Compensate Capacitances with Negative Capacitances. Lowering the input capacitance of a stage reduces the loading of the previous stage and thus helps to improve the amplifier bandwidth. So far, we discussed ways to lower the input capacitance by suppressing the effects of C_{be} and C_{bc} (or C_{gs} and C_{gd} in an FET amplifier). Now we want to look at another approach in which we reduce the existing input capacitance by putting a *negative* capacitance in parallel. But what is a negative capacitance? A negative capacitance has the unusual property that its voltage *drops* when we try to charge it *up*! Although this may sound weird, there are active circuits that can produce such a capacitance.

One way to produce a negative capacitance is by exploiting the Miller effect. To do this, we connect a regular capacitor C_F across a noninverting amplifier with a gain larger than one ($A > 1$). In this case, the Miller capacitance $(1-A)\cdot C_F$ is negative. Figure 6.16 shows how this idea can be applied to reduce the input capacitance of a differential stage. By adding the feedback capacitors C_F and C'_F, the original input capacitance C_I at each input is reduced to

$$C'_I = C_I + (-|A|+1)\cdot C_F, \qquad (6.57)$$

where $|A|$ is the stage gain. Note that the capacitors provide positive feedback. From Eq. (6.57), we see that for $|A| = 1$, the feedback capacitor C_F has no effect. This makes sense because for $|A| = 1$, the output voltage exactly follows the input voltage (as in a voltage buffer), and thus there is no voltage drop across C_F and no current flowing through it. If the gain is increased to $|A| > 1$, then the negative Miller capacitance appears and the input capacitance is reduced.

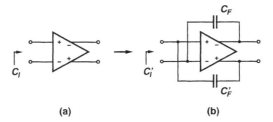

Fig. 6.16 Differential stage (a) without and (b) with negative Miller capacitance.

The generic differential stage in Fig. 6.16(a) can be implemented, for example, with a BJT differential pair. In this case, the parasitic base-collector capacitances, C_{bc}, of the BJTs are multiplied by $|A|+1$ and appear as part of the input capacitance C_I. If we now add cross-coupled capacitors C_F and C'_F, which are equal to C_{bc}, they produce the negative Miller capacitance $(-|A|+1) \cdot C_F$, which mostly neutralizes the effects of C_{bc}; thus, this technique is known as *neutralization*. For improved tracking between the feedback capacitors and C_{bc}, matched dummy transistors can be used to implement C_F and C'_F [35].

A limitation of this technique is that for frequencies approaching the stage bandwidth, the amplitude of the output signal drops and its phase is no longer exactly aligned with the input signal. As a result, the negative Miller capacitance is reduced at these frequencies. Another detrimental effect is that the capacitors C_F and C'_F present an additional load to the output of the stage, which reduces its bandwidth. Finally, if the feedback capacitors are made too large, the overall capacitance at the stage input may become negative, making the amplifier unstable.

Another way to produce a negative capacitance is with a so-called *negative impedance converter* (NIC) connected to a regular capacitor. The NIC inverts the capacitance of the regular capacitor by swapping the terminal voltages while keeping the currents going to the original terminals. A pair of cross-coupled transistors can approximate this function (cf. Fig. 6.55(a)). A negative capacitance produced in this manner can be connected in parallel to the stage inputs to reduce the input capacitance [31]. [→ Problem 6.12]

Reduce the Capacitive Load with Buffering and Scaling. Consider two consecutive stages as shown in Fig. 6.17(a). The total load capacitance seen by the first stage consists of three components: the stage self loading, C_O, the interconnect capacitance, and the next-stage loading, C_I. For simplicity, we ignore the interconnect capacitance in the following analysis. This load capacitance determines the speed of the first stage's output pole(s). For example, if the first stage has a simple resistive load as in Fig. 6.15(a), the frequency of the output pole is about $1/[2\pi \cdot R_C(C_I+C_O)]$, where R_C is the load resistor. Thus, reducing the input capacitance from C_I to C'_I, by any of the techniques discussed so far, increases the bandwidth due to the output

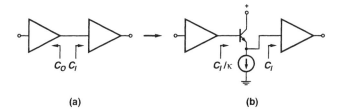

Fig. 6.17 Two consecutive stages (a) without and (b) with interstage buffer.

pole(s) from BW to

$$BW' = \frac{C_I + C_O}{C'_I + C_O} \cdot BW. \tag{6.58}$$

The next-stage loading can be reduced further by inserting buffers in between the stages. Figure 6.17(b) shows how two stages can be decoupled with an emitter-follower buffer. The buffer has an input capacitance, C'_I, which is κ times smaller than the load capacitance, C_I, it is designed to drive

$$C'_I = C_I/\kappa. \tag{6.59}$$

Because the buffer's input capacitance is a fraction of the load capacitance, whereas (ideally) the input and output voltages are the same, the buffer can be regarded as a *capacitance transformer*. But note that a change in the load capacitance requires the resizing of the buffer to realize the corresponding change in input capacitance (and to maintain the same buffer bandwidth). From Eq. (6.58), we conclude that inserting the buffer increases the bandwidth due to the output pole(s) to

$$BW' = \frac{1 + C_I/C_O}{1 + (C_I/C_O)/\kappa} \cdot BW. \tag{6.60}$$

For example, if $C_I/C_O = 1.5$ and we insert a buffer that reduces the load capacitance by $\kappa = 7$, then the output pole is sped up by approximately $2.1\times$.

In practice, the bandwidth gained by this technique is partially offset by the bandwidth lost because of the finite buffer bandwidth and signal attenuation in the buffer. In particular, there is a trade-off between the capacitance transformation ratio, κ, and the buffer bandwidth, BW_B: a large emitter follower with a large g_m has a high bandwidth but also a fairly large input capacitance and thus a low κ, whereas a small emitter follower, which is easily loaded down by the output capacitance, has a low bandwidth but also a small input capacitance and thus a high κ. In the case of a MOSFET buffer stage, this trade-off can be described by $\kappa \approx \kappa_1 \cdot (f_T/BW_B - \kappa_0)$, where κ_0 and κ_1 are constants that depend on the buffer topology (source follower or common source) and the technology [156]. Typically, κ reduces to unity for a bandwidth around $0.4 f_T$ to $0.7 f_T$. [$\rightarrow$ Problems 6.13 and 6.14]

The use of interstage buffers is very popular in bipolar designs. Often two or three consecutive emitter followers are used to boost the capacitance transformation ratio while maintaining a high buffer bandwidth. When using emitter followers, especially when using cascades of emitter followers, it is important to watch out for peaking in the frequency response (or ringing in the time domain). Emitter followers (as well as source followers) tend to have an *inductive* output impedance, which can produce an undesirable resonance with the load capacitance. The output impedance becomes inductive when the sum of the source and base resistance exceeds $1/g_m$, similar to what is happening in an active inductor, which we will discuss shortly. A careful choice of the quiescent current and the use of damping resistors can control the peaking [146]. Interstage buffers also can be used in MOSFET designs, usually in the form of source followers. However, the considerable attenuation and level shifting inherent to an n-MOS source follower in an n-well technology (as a result of the body effect) limit the usefulness of this approach. The headroom problem associated with the large level shifting can be overcome by AC coupling the input or output of the source follower; however, the resulting low-frequency cutoff may not be acceptable and the bottom-plate parasitic capacitance of the AC-coupling capacitor reduces the bandwidth [141].

Another approach to reduce the next-stage loading C_I is to scale down the transistor sizes and currents in the next stage [156]. Figure 6.18(b) shows two stages where the driven stage is made κ times smaller than the driving stage, thus reducing the next-stage loading to $C_I' = C_I/\kappa$, just like in the case of the buffer. Note that when reducing all transistor widths in a MOSFET stage by κ and increasing all resistors values by κ, the capacitances and currents are reduced by κ, whereas the node voltages, gain, and bandwidth ideally remain unaffected. In reality, the bandwidth of the scaled-down stage does shrink somewhat because the wiring capacitance, which we neglected in our analysis, usually does not reduce with κ.

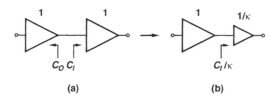

Fig. 6.18 Two consecutive stages (a) without and (b) with inverse scaling

The inverse scaling technique avoids some of the buffer's shortcomings: it does not attenuate the signal, it introduces no new poles limiting the bandwidth, and it causes no level shifting. Because attenuation and level shifting are especially problematic in source-follower buffers, this technique is particularly useful for MOSFET amplifiers. However, when inverse scaling is applied to a multistage amplifier, it leads to either a very large input stage or a very small output stage. For example, if every stage in a four-stage amplifier is made half the size of the driving stage ($\kappa = 2$), the output stage becomes 16× smaller than the input stage. For this reason, the inverse scaling technique mostly is useful for amplifiers that receive the input signal from an off-chip source, which can drive the large input stage, but need to drive only a small on-chip load. Note that in the case of a 50-Ω input, the input capacitance, and thus the size of the input stage, is limited by the required S_{11} specification (cf. Appendix C.2). [→ Problem 6.15]

Tune out Capacitances with Inductors: Shunt Peaking. Until now, we focused on broadband techniques that do not require inductors. However, if inductors are available, they open up a number of new and interesting possibilities that we explore in the remainder of this section.

Figure 6.19(a) shows a simple common-source MOSFET stage with the load resistor R. In Fig. 6.19(b), the inductor L was inserted in series with the load resistor R in an attempt to "tune out" part of the load capacitance $C_L = C_I + C_O$. This method is known as *shunt peaking*. If we insert an inductor with the value

$$L = 0.4 \cdot R^2 \cdot C_L, \qquad (6.61)$$

the bandwidth of the stage increases from BW to about

$$BW' = 1.7 \cdot BW, \qquad (6.62)$$

equivalent to a 70% bandwidth extension. Despite the name shunt *peaking*, no peaking occurs in the frequency response if L is chosen according to Eq. (6.61). However, for larger values of L, undesired peaking sets in, as illustrated in Fig. 6.20. Besides boosting the bandwidth, shunt peaking improves the rolloff characteristics, which helps to reduce the bandwidth shrinkage in multistage amplifiers.

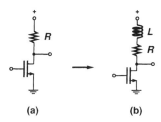

Fig. 6.19 Stage with (a) resistive load and (b) inductive load.

Intuitively, we can describe the bandwidth boosting effect of the inductor as follows: just at the frequency where the gain would ordinarily roll off because of the output pole ($\omega = 1/(RC_L)$), the load impedance starts to go up because of its inductive component ($Z(j\omega) = R + j\omega L$). The increased load impedance boosts the gain and thus compensates for the gain rolloff. (For a corresponding description in the time domain, see Section 8.2.3.) A more scientific explanation is given in Fig. 6.21, which shows the pole/zero movement as a function of the inductor value. (The transfer function underlying this picture is given in the solution to Problem 6.16.) Without the inductor, $L = 0$, there is only a single real pole at $s = -1/(RC_L)$ indicated by an ×. For a small inductor, $L = 0.2R^2C_L$, a real pole/zero pair, indicated by an × and an ○, appears at high frequencies. When increasing the value of L, the original

194 MAIN AMPLIFIERS

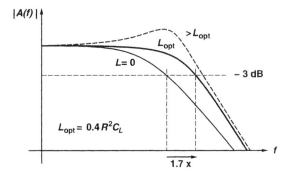

Fig. 6.20 Bode plot showing the effects of shunt peaking.

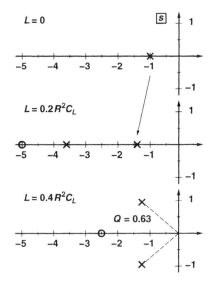

Fig. 6.21 Root-locus plot showing the effects of shunt peaking (s is normalized to $1/(RC_L)$).

pole moves to the left (bandwidth extension!) and the new pole moves to the right, until they eventually "collide" and move into conjugate-complex positions located on a circle. For the optimum value $L = 0.4R^2C_L$, the pole pair has the quality factor $Q = 0.63$ and the frequency response is maximally flat. The conjugate-complex pole pair together with the zero provide the bandwidth extension of 70%. (If the objective is to obtain a maximally flat *group-delay* characteristics, the optimum inductance is $L = 0.32R^2C_L$ and the bandwidth extension is about 60%.) Note that the rolloff initially is second order because of the conjugate-complex poles, but then reverts to first order because of the zero. [→ Problem 6.16]

The inductor can be realized as an on-chip *spiral inductor*, a *bond-wire inductor*, or an *active inductor* (a.k.a. *synthetic inductor*). A simple and compact model for the on-chip spiral inductor is shown in Fig. 6.22(a); a more complex model can be found in [85]. Besides the desired inductance L, there also is the series resistance R_S and the parasitics C_P and R_P, which describe the coupling of the spiral inductor to the substrate. Note that the quality factor of the inductor, which is controlled by R_S and R_P, is uncritical in this application, because the intended R-L series connection has a low quality factor anyway. However, the self-resonance frequency, which is given by $f_{SR} \approx 1/(2\pi\sqrt{LC_P})$ for an inductor with one terminal grounded, limits the useful frequency range of the spiral inductor. As a rule of thumb, the self-resonance frequency should be kept a factor two above the stage bandwidth. For example, consider a metal-4 spiral inductor with 8 turns, an outer diameter of 120 μm, a metal width of 2 μm, and a metal spacing of 1 μm realized in a 0.25-μm CMOS technology. Fitting the S parameters of this spiral inductor to the simple inductor model in Fig. 6.22(a) resulted in $L = 10$ nH, $R_S = 60\,\Omega$, $C_P = 50$ fF, and $R_P = 10\,\Omega$. Thus, the self-resonance frequency of this inductor is at 7.1 GHz, making it useful for shunt-peaking applications up to frequencies of about 3.5 GHz. Unfortunately, on-chip spiral inductors much larger than 10 to 15 nH occupy a very large area and tend to have a low self-resonance frequency. An interesting alternative to implement large inductors is to boost the inductance of a smaller spiral inductor with an active device [112, 165].

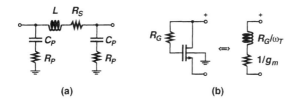

Fig. 6.22 Model of (a) spiral inductor and (b) active inductor.

Figure 6.22(b) shows the implementation of an active inductor and its equivalent circuit. When used for shunt peaking, this circuit can implement both the R and L in Fig. 6.19(b). At low frequencies, the active inductor behaves like a gate-drain connected transistor with an impedance of about $1/g_m$. At high frequencies, the low-pass filter formed by R_G and C_{gs} cuts the gate-drain connection, causing the

196 MAIN AMPLIFIERS

impedance of the circuit to increase similar to that of an inductor. At very high frequencies, around f_T, the impedance tops out and eventually drops again. It turns out that for $R_G > 1/g_m$, the impedance becomes inductive (in a certain frequency range) and the inductance in series with the $1/g_m$ resistance is

$$L \approx \frac{R_G(C_{gs} + C_{gd})}{g_m} = \frac{1}{2\pi} \cdot \frac{R_G}{f_T}. \tag{6.63}$$

Note that this inductance can be conveniently controlled with the gate resistor, R_G, while keeping the series resistance, $1/g_m$, constant. The active inductor can be used for frequencies up to about $f_T/2$. Compared with the spiral inductor, the active inductor is much smaller and more amenable to monolithic integration, but it also is much more noisy and produces a large DC voltage drop, which is a concern in low-voltage designs (cf. Section 6.4.3). [→ Problem 6.17]

In practice, the bandwidth extension achieved with shunt peaking usually is less than the theoretical 70% per stage, either limited by the (spiral) inductor's self resonance or the parasitics and the maximum operating frequency of the active inductor. However, the bandwidth extension for the entire amplifier may exceed 100% (cf. Section 6.3.1).

Absorb Capacitances into Inductive Load Impedance. At this point, you may wonder if there is another load impedance that results in an even higher bandwidth than the R-L load discussed above. In fact there is, and it can be shown that the maximum bandwidth improvement over a stage with resistive load is 2×. Thus, shunt peaking, which is giving us an improvement of 1.7×, is pretty close to the optimum. Although not of great practical importance, it is a useful warm-up exercise for our later discussions to analyze this load impedance and its bandwidth.

A network theorem that was discovered and published by Bode in 1945 says, in a simplified form, that the following relationship holds for any physically realizable passive impedance $Z(f)$:

$$BW \leq \frac{1}{2\pi} \cdot \frac{2}{|Z_0| \cdot C_\infty}, \tag{6.64}$$

where C_∞ is the high-frequency asymptotic capacitance ($C_\infty = \lim_{\omega \to \infty} 1/[j\omega \cdot Z(\omega)]$), $|Z_0|$ is the magnitude of the low-frequency impedance, and BW is the bandwidth over which $|Z(f)| = |Z_0|$ is maintained. (See [16, 201] for a precise formulation of this theorem and its derivation.) Now, if we identify Z with the total load impedance of the stage, including the load capacitance $C_L = C_I + C_O$, we conclude that C_∞ is at least equal to C_L. Furthermore, to maintain the desired low-frequency gain, $|Z_0|$ must be equal to R. Thus, it follows from Eq. (6.64) that the stage bandwidth is limited to $BW \leq 2/(2\pi \cdot RC_L)$, which is twice the bandwidth achieved with a resistive load. [→ Problem 6.18]

How can we realize the load impedance promised to us by Bode? Figure 6.23(b) shows its implementation with an infinite, lossless, artificial transmission line. Such a transmission line has a real characteristic impedance given by

$$Z_{TL} \approx \sqrt{\frac{L}{C}} \quad \text{for} \quad f < f_{\text{cutoff}} = \frac{1}{2\pi} \cdot \frac{2}{\sqrt{LC}}. \tag{6.65}$$

Above the *cutoff frequency*, $f > f_{\text{cutoff}}$, the impedance becomes strongly reactive and signals injected into the transmission line are reflected rather than transmitted.[6] The first shunt C of the transmission line (dashed in Fig. 6.23(b)) is formed by the load capacitance C_L, that is, we can say that the load capacitance of the stage is absorbed into one end of the transmission line. Because all capacitors in the transmission line are equal, we have $C = C_L$. To avoid reflections on the transmission line, we must match its characteristic impedance to the load resistor R, and thus we have $Z_{TL} = R$. With these conditions and Eq. (6.65), we find $L = R^2 C_L$ and the cutoff frequency, which determines the stage bandwidth, follows as $f_{\text{cutoff}} = 2/(2\pi \cdot RC_L)$. Thus, compared with a stage with resistive load, the bandwidth increased from BW to

$$BW' = 2 \cdot BW, \tag{6.66}$$

in agreement with what we concluded earlier from Bode's theorem.

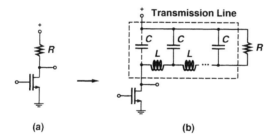

Fig. 6.23 Stage with (a) resistive load and (b) artificial transmission-line load.

Absorb Capacitances into Inductive Interstage Network. In the above discussion, we assumed that the output of the driving stage is directly connected to the input of the driven stage, and we optimized the load impedance attached to this connection. Such a load also is known as a *one-port interstage network*. Now, we generalize this concept to a *two-port interstage network*, where one port connects to the output of the

---

[6]More precisely, the impedance of an artificial transmission line depends on the point where it is cut: full series, mid series, full shunt, or mid shunt. For low frequencies, all four impedance values are close to $\sqrt{L/C}$, as given in Eq. (6.65), but at frequencies approaching f_{cutoff}, they diverge: the mid-series and mid-shunt impedances remain real up to f_{cutoff} but their magnitudes diverge from $\sqrt{L/C}$; the full-series and full-shunt impedances become complex, but their magnitude remains exactly $\sqrt{L/C}$ up to f_{cutoff}. For example, the full-shunt impedance, relevant to Fig. 6.23(b), is given by $Z_{TL} = \sqrt{L/C} \cdot \left(\sqrt{1 - f^2/f_{\text{cutoff}}^2} + j \cdot f/f_{\text{cutoff}}\right)^{-1}$ with the magnitude $|Z_{TL}| = \sqrt{L/C}$ up to f_{cutoff} [201].

driving stage and the other port to the input of the driven stage. An example of such an interstage network is given in Fig. 6.24 in the form of an L-C ladder network [8]. Note that the self-loading capacitance of the driving stage, C_O, and the input capacitance of the driven stage, C_I are absorbed into this network at the input and output ports, respectively. Now, the bandwidth of the stage in Fig. 6.24(b) is determined by the low-pass interstage network, which may be higher than the original bandwidth of the stage in Fig. 6.24(a), $BW = 1/[2\pi \cdot (R_I \parallel R_O)(C_I + C_O)]$. By choosing the appropriate values for L_1, L_2, C, R_O, and R_I, the transfer function of the interstage network can be controlled, for example, to have Bessel or Butterworth characteristics. Of course, the network could be made more complex than the fifth-order ladder structure shown in Fig. 6.24(b). In practice, however, losses in the interstage network limit its order to small values such as 2 to 5. Note that in this technique the signal is propagating *through* the inductors and thus their loss is critical, whereas in the shunt-peaking technique, the inductor is part of the load and loss is of little concern.

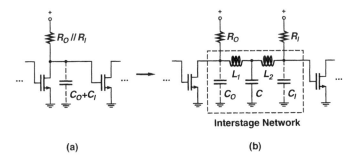

Fig. 6.24 Stages (a) without and (b) with an inductive interstage network.

How much bandwidth extension can we obtain with this technique? We can again use the transmission-line argument introduced earlier to discuss its capability. Let's assume that the input and output capacitances are equal, $C_I = C_O = C_L/2$, and that the load resistor R is connected to the output port of the interstage network, $R_O \to \infty$ and $R_I = R$. Then, we extend the interstage network in Fig. 6.24(b) into an infinite, lossless, artificial transmission line, with all shunt capacitors being equal to $C = C_L/2$.[7] To match the transmission line to the load resistor R, we choose all inductors in the line to be $L = R^2 C_L/2$. The cutoff frequency of the transmission line, which determines the stage bandwidth, follows from Eq. (6.65) as $f_{\text{cutoff}} = 4/(2\pi \cdot RC_L)$. Thus, compared with a stage with resistive load, the bandwidth increased from BW to

$$BW' = 4 \cdot BW. \qquad (6.67)$$

---

[7]If you are concerned about the infinite delay introduced by this interstage network, there is an easy fix: rather than taking the signal from the end of the transmission line, take it from the second shunt C, right after the input, replacing this C with C_I. The load resistor, R, however, stays at the end of the transmission line.

In contrast to the one-port interstage network, we now absorb half of the load capacitance into each end of the transmission line rather than the full capacitance into one end only, hence the additional factor $2\times$. It fact, a more systematic analysis shows that the maximum bandwidth improvement achievable with a two-port interstage networks is even greater than 4, namely $\pi^2/2 = 4.93$ [16]. Also, note that in addition to the bandwidth extension, the higher-order rolloff characteristics of the interstage network helps to reduce bandwidth shrinkage in multistage amplifiers.

As an illustration, let's look at two simple examples of a ladder interstage network. The simplest interstage networks contain just a single inductor. If we let $C_O = C = L_2 = 0$ and $R_I \to \infty$ in Fig. 6.24(b), we end up with just $R_O = R$, $C_I = C_L$, and $L_1 = L$. This case is known as *series peaking*. With the inductor value $L = R^2 C_L/2$, we obtain a second-order Butterworth low-pass response and a bandwidth extension of $\sqrt{2}\times$. Intuitively, we can explain the bandwidth boosting effect of the inductor as follows: just at the frequency where the gain would ordinarily roll off because of the output pole, the L_1-C_I series connection starts to resonate, forcing most of the drain current to flow into C_I rather than R_O. Unfortunately, the 40% bandwidth improvement resulting from series peaking is far from the promised $4\times$ improvement and can be beaten easily with shunt peaking, which provides a 70% improvement (the reason is the additional zero introduced by shunt peaking). In a second example, we let $C = L_2 = 0$, $C_I = C_O = C_L/2$, and $R_I = R_O = 2R$, producing a symmetrical network with the single inductor $L_1 = L$. With the value $L = R^2 C_L$, we obtain a third-order Chebyshev low-pass response and the appreciable bandwidth extension of about $2.2\times$; however, the 1-dB passband ripple of this network may not be acceptable in many applications.

Another type of interstage network is shown in Fig. 6.25. This so-called *T-coil network* consists of two mutually coupled inductors with the same value L and the coupling factor k plus a bridge capacitor C_B. Note that this network looks somewhat like a combination of series and shunt peaking. For the case where $C_I = C_L$ and $C_O = 0$ and the values $k = 1/3$, $L = 3/8 \cdot R^2 C_L$, and $C_B = C_L/8$, we obtain a second-order Butterworth low-pass response and the amazing bandwidth extension of $2\sqrt{2}\times$, which is about $2.8\times$ [82]. Even for the Bessel case ($k = 1/2$, $L = 1/3 \cdot R^2 C_L$, $C_B = C_L/12$), which has minimal group-delay variations, the bandwidth extension is still about $2.7\times$. The T-coil network is particularly useful when driving large capacitive loads ($C_I \gg C_O$). It can be implemented with on-chip coupled spiral inductors [31, 32].

The discussed interstage networks can be used not only to couple amplifier stages, but also to provide broadband coupling between a sensor and the input of an amplifier or the output of an amplifier and a load. An example of this application is given in Section 5.2.9, where an L-C ladder network was used to couple a photodetector to the input of a TIA.

Absorb Capacitances into Transmission Lines: Distributed Amplifier. We have seen how breaking a large capacitance ($C_O + C_I$) into two smaller ones (C_O and C_I) permits us to boost the bandwidth. We can push the envelope further by subdividing each transistor into several smaller ones, thus breaking its capacitance

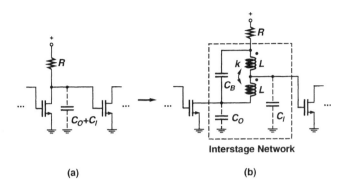

Fig. 6.25 Stages (a) without and (b) with a T-coil interstage network.

into even smaller pieces. This brings us to the most aggressive broadband technique known as the *distributed amplifier* or the *traveling-wave amplifier*, illustrated in Fig. 6.26. Figure 6.26(a) shows an ordinary lumped MOSFET amplifier stage with the load resistor R_D and the gate resistor R_G connected to the bias voltage V_{BI}. In Fig. 6.26(b), the MOSFET of the lumped stage has been split into three smaller ones, each one with the same length but $1/3$ of the original width. All gates and drains are connected together with the inductors L_I and L_O, respectively. Obviously, both stages behave the same way at low frequencies where L_I and L_O act as shorts. But at high frequencies, there is an important difference: all gate and drain capacitances (C_I and C_O) are absorbed into two artificial transmission lines, one at the input and one at the output of the stage. Now, the amplifier bandwidth is determined by the cutoff frequency of these transmission lines. For such a distributed amplifier to work properly, the input and output transmission lines must be phase matched, that is, their delays must be equal. Only then do the amplified signal components at the output add up in phase, producing a flat frequency response. Note that at the left end of the output transmission line, the signal components do not add up in phase, and thus this output is not usable. This end must be terminated with the resistor R_D to prevent reflections to the output. Similarly, the input transmission line must be terminated at the right end with the resistor R_G to prevent reflections.

Clearly, this technique easily can be generalized from three sections, as in the example above, to n sections.[8] Now, let's calculate the bandwidth of a stage with n sections. The characteristic impedance of the output transmission line, Z_{TLO}, is determined by the inductors L_O and the split self-loading capacitances C_O/n (remember, each transistor is $1/n$th the width of the original transistor). Below the cutoff frequency, f_{cutoff}, it can be approximated by the expression for an infinite,

---

[8] We are using the word *sections* rather than *stages* because the gains of the sections *add* to form the total gain. In contrast, the gains of the stages in a multistage amplifier are multiplicative.

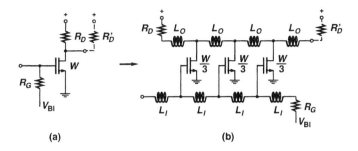

Fig. 6.26 (a) Lumped amplifier and (b) distributed amplifier with artificial transmission lines.

lossless, artificial transmission line (cf. Eq. (6.65)):

$$Z_{TLO} \approx \sqrt{\frac{L_O}{C_O/n}} \quad \text{for} \quad f < f_{\text{cutoff}} = \frac{1}{2\pi} \cdot \frac{2}{\sqrt{L_O \cdot C_O/n}}. \quad (6.68)$$

Requiring that the transmission line impedance, Z_{TLO}, matches the termination resistor, R_D, to avoid reflections from the left end, we find that the inductors have to be $L_O = R_D^2 C_O/n$. The cutoff frequency of the output transmission line thus becomes $f_{\text{cutoff}} = 2n/(2\pi \cdot R_D C_O)$. If we assume that the input transmission line has the same cutoff frequency as the output line (e.g., $C_I = C_O$, $L_I = L_O$, and $R_G = R_D$), this cutoff frequency determines the bandwidth of the distributed stage.[9]

How does this bandwidth compare with that of the lumped amplifier stage in Fig. 6.26(a)? For a fair comparison, both stages must have the same gain and the same total load capacitance, $C_L = C_I + C_O = 2C_O$. The load of the distributed stage consists of two parallel termination resistors: R_D at the backend of the transmission line and R'_D at the output of the transmission line representing an external load or the next stage loading (cf. Fig. 6.26(b)), thus the gain is $A = g_m R_D/2$, where g_m is the total transconductance of all transistors combined. Assuming the same gain, transconductance, and load, the lumped amplifier stage has the bandwidth $BW = 1/(2\pi \cdot R_D C_O)$. Therefore, the bandwidth of the distributed amplifier stage can be written as[10]:

$$BW' = 2n \cdot BW. \quad (6.69)$$

In theory, there is no limit to the bandwidth that can be achieved by increasing the number of sections ($n \to \infty$). In practice, the number of sections is limited to 4

---

[9]Note the following subtle difference between the input and output transmission lines: because the signal amplitude on the output transmission line grows from left to right, the Miller capacitance of the split transistors also increases from left to right. Thus, the shunt capacitors of the input transmission line are not exactly C_I/n, but increase from left to right.
[10]The reader may wonder why the bandwidth of the distributed amplifier does not equal that of the lumped amplifier when $n = 1$. In our model, the distributed amplifier with $n = 1$ still has an inductive input and output network (two L_O and two L_I), whereas the lumped amplifier does not. In our approximate calculations, these networks give a bandwidth extension of $2\times$.

to 7, mostly because of losses in the artificial transmission lines. Losses are caused by the series resistance of the inductors as well as the input resistance and output conductance of the transistors. It can be shown that the gain-bandwidth product of a distributed GaAs FET amplifier is limited to about $f_{\max}$ as a result of the losses associated with R_g and g_o [201]. The use of artificial transmission lines with simple terminations, as shown in Fig. 6.26(b), leads to a fair amount of passband ripple. The gain flatness can be improved by using *half sections* ($L_I/2$ and $L_O/2$) or so-called *m-derived half sections* at the ends of the transmission lines [134, 201]. Also, distributed amplifiers tend to exhibit gain peaking and group-delay variations near the cutoff frequency. Note that although the bandwidth of distributed amplifiers increases with n, the group delay remains at $1/(2\pi \cdot BW)$, independent of n. Optimization procedures can improve the gain and delay uniformity.

In another approach to implement a distributed amplifier, the input and output transmission lines are assembled from short uniform transmission-line segments, as shown in Fig. 6.27. The resulting, periodically loaded transmission lines have a lower characteristic impedance than the unloaded ones. Specifically, the impedance of the capacitively loaded output transmission line is [72]

$$Z_{TLO} \approx Z_{TSO} \cdot \sqrt{\frac{C_{TSO}}{C_{TSO} + C_O/n}}, \qquad (6.70)$$

where Z_{TSO} is the characteristic impedance of the unloaded transmission line, C_O/n is the periodic load capacitance, C_{TSO} is the capacitance of the transmission-line segment, which can be calculated as $C_{TSO} = t_{\text{delay}}/Z_{TSO}$ for the lossless case. In practice, the loaded impedance often must be equal to 50 Ω, which means that the unloaded impedance must be substantially above 50 Ω. Unfortunately, such high-impedance transmission lines that also exhibit a low loss are difficult to realize in standard silicon technologies. Compared with the artificial transmission line in Fig. 6.27(a), the periodically loaded transmission line in Fig. 6.27(b) has a different cutoff behavior, but the bandwidth that can be achieved with these two techniques is similar.

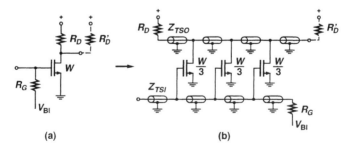

Fig. 6.27 (a) Lumped amplifier and (b) distributed amplifier with periodically loaded transmission lines.

Distributed amplifier stages can be cascaded like any other amplifier stage. When cascading two distributed stages, the output transmission line of the first stage feeds the input transmission line of the second stage. Both transmission lines need to have the same characteristic impedance to avoid reflections. When cascading a lumped and a distributed stage, the lumped stage must provide the necessary termination for the distributed stage.

Another advantage of distributed amplifiers over lumped amplifiers is their superior high-frequency impedance matching. At high frequencies, the lumped input and output capacitances of an amplifier limit the achievable input and output matching (S_{11} and S_{22}), respectively (cf. Appendix C.2). By breaking these capacitances into smaller ones and absorbing them into transmission lines, the above limit can be circumvented. Note that the improvements in high-frequency matching and high-frequency gain are related: with better matching, more signal goes into the amplifier (and comes out of the amplifier), resulting in a better gain. For more information on distributed amplifiers see [201].

6.3.3 Offset Compensation

In Section 6.2.5, we saw that the input offset voltage of an LA should be limited to about 0.1 mV. For larger values, the data signal is sliced so much off center that the receiver sensitivity is substantially degraded. A typical BJT amplifier has a 3σ random offset voltage of a few milli-Volts, whereas a high-speed MOSFET amplifier has an offset voltage of around 10 mV. In either case, this is too much and we need an *offset compensation* scheme to reduce the offset voltage to the required value.

Figure 6.28 shows a popular MA topology that combines the offset-compensation circuit with the input termination [98]. The input signal is AC coupled to the MA with the capacitors C and C'. Two 50-Ω resistors serve as the input termination and also supply a differential DC voltage from the error amplifier A_1, which compensates for the MA's offset voltage. To do this, the error amplifier senses the DC component of the MA's differential output signal, that is, the output offset voltage, by means of the two low-pass filters R_1-C_1 and R'_1-C'_1, and adjusts its output voltage until the output offset voltage becomes zero. Note that in Fig. 6.28, the output impedance of the error amplifier is assumed to be zero; if it is larger than zero, the value of the termination resistors has to be reduced accordingly. For example, in [98], A_1 is implemented with a pair of emitter followers and the termination resistors are 40 Ω.

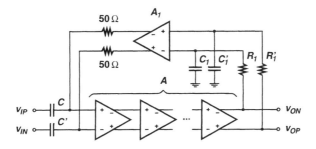

Fig. 6.28 MA with offset compensation circuit.

204 MAIN AMPLIFIERS

Another approach to cancel the offset voltage is shown in Fig. 6.29. Here, the first stage of the MA has two differential inputs: one input is used for the data signal and the second input is used for the offset-compensation voltage. Such a stage can be implemented, for example, with two differential pairs joined at the outputs and is known as a *differential difference amplifier* (DDA).[11] As before, the offset-compensation voltage is supplied by the error amplifier A_1, but now this amplifier only needs to drive the high-impedance DDA inputs rather than the 50-Ω resistors. The input termination for the MA can be implemented with two 50-Ω resistors to ground.

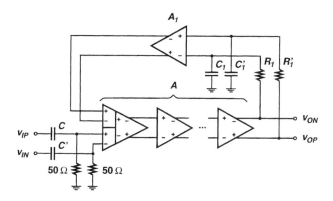

Fig. 6.29 MA with DDA-type offset compensation circuit.

There are two reasons why the circuits in Figs. 6.28 and 6.29 do not completely eliminate the offset voltage: (i) the finite gain of the error amplifier A_1 and (ii) the offset voltage V_{OS1} of the error amplifier. Calculating the output offset voltage of the circuit in Fig. 6.28 and dividing it by the signal gain, A, reveals that the input offset voltage is reduced from V_{OS} to

$$V'_{OS} = \frac{\sqrt{V_{OS}^2 + A_1^2 V_{OS1}^2}}{AA_1 + 1} \approx \sqrt{\left(\frac{V_{OS}}{AA_1}\right)^2 + \left(\frac{V_{OS1}}{A}\right)^2}, \qquad (6.71)$$

where all offset voltages are expressed as 3σ values (assuming zero-mean Gaussian distributions) and V_{OS} and V_{OS1} are assumed to be statistically independent. The approximation on the right-hand side holds if $AA_1 \gg 1$. The result in Eq. (6.71) also applies to the DDA circuit in Fig. 6.29, if the two port gains of the DDA are equal. (If the gain from the offset-compensation port is smaller, A_1 has to be reduced accordingly.) From Eq. (6.71), we conclude that although the original offset voltage, V_{OS}, is suppressed by the low-frequency loop gain, the offset voltage of the error amplifier, V_{OS1}, is suppressed only by A. Thus, unless the offset of the error amplifier

---

[11] The two input ports of a typical CMOS DDA are matched and thus have the same gain [155, 158]. However, in offset-cancellation applications, the port where the offset-compensation voltage is applied often is made to have a lower gain.

is made very small, $V_{OS1} \ll V_{OS}/A_1$, this component now may dominate the total offset. However, because the error amplifier doesn't have to be fast, large transistors with good matching properties can be used to minimize V_{OS1}.

Depending on the amount of offset that must be compensated, A_1 can be an amplifier ($A_1 > 1$), a buffer ($A_1 = 1$), or be left out entirely. For a BJT amplifier, a passive R-C-R structure [143] or a buffer implemented as an emitter follower [98] often is sufficient, whereas MOSFET amplifiers, which have a larger offset, usually require additional loop gain to meet the offset specification. In the case of an AGC amplifier, we have to make sure that the offset voltage is sufficiently suppressed at the lowest gain setting.

Low-Frequency Cutoff. The offset-compensation circuits in Figs. 6.28 and 6.29 not only suppress the unwanted offset voltage, but also some important low-frequency components of the input signal. In other words, these offset-compensation circuits introduce a low-frequency cutoff in the overall MA frequency response (cf. Fig. 6.6). As discussed in Section 6.2.6, an LF cutoff causes baseline wander and data-dependent jitter if the cutoff frequency is too high.

Two structures in the circuit of Fig. 6.28 cause an LF cutoff: (i) the AC coupling capacitors C and C' with the 50-Ω termination resistors and (ii) the offset compensation loop with R_1, R_1', C_1, and C_1'. If we assume that C_1 and C_1' are very large, the LF-cutoff frequency due to the AC coupling is $f_{LF} = 1/(2\pi \cdot 2R_0 C)$, where $R_0 = 50\,\Omega$. If we assume that C and C' are very large, the LF-cutoff frequency due to the offset-compensation loop is

$$f_{LF} = \frac{1}{2\pi} \cdot \frac{AA_1/2 + 1}{R_1 C_1}. \tag{6.72}$$

Note that the expression $AA_1/2$ is the AC loop gain. The factor $1/2$ is caused by the voltage divider formed by the 50-Ω termination resistors and the 50-Ω source resistance. For the DDA-type offset compensation in Fig. 6.29, the AC loop gain is AA_1, and correspondingly, the LF-cutoff frequency is $f_{LF} = (AA_1 + 1)/(2\pi \cdot R_1 C_1)$, if we interpret A as the gain from the offset-compensation port to the amplifier output. [$\rightarrow$ Problem 6.19]

From these results, we can conclude that to obtain a particular cutoff frequency, we need to make the loop bandwidth, $1/(2\pi \cdot R_1 C_1)$, much smaller than this frequency. For example, if the AC loop gain is 100 and we need a cutoff frequency of less than 250 kHz, the loop bandwidth must be made 2.5 kHz or less. The small loop bandwidth can be realized, for example, with feedback capacitors C_F around the error amplifier A_1, thus using the Miller effect to create the large effective input capacitances $C_1 = (A_1 + 1) \cdot C_F$.

As with any feedback system, we have to make sure that the offset compensation loop is dynamically stable (in its differential and common mode). Fortunately, this usually is not a problem because the dominant open-loop pole, which is required to be at a very low frequency to meet the LF cutoff requirements, is far removed from the high-frequency poles, thus permitting a high loop gain without violating phase or gain margins.

Slice-Level Adjustment. As we discussed in Section 4.10, a receiver with an avalanche photodetector (APD) or an optical preamplifier produces significantly more noise on the ones than the zeros, resulting in an eye diagram and noise statistics as illustrated in Fig. 6.30. Under these circumstances, the optimum slice level is somewhat below the midpoint between zeros and ones (the DC component of the received signal). To make the LA slice at this optimum level, we introduce a *controlled* amount of offset voltage, $V_{OS,SA}$.

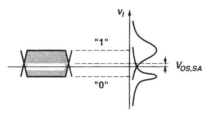

Fig. 6.30 Slice-level adjustment is required if the noise distributions for zeros and ones are unequal.

A simple modification to the offset-compensation circuits in Figs. 6.28 and 6.29 can introduce such an offset voltage. Figure 6.31 shows the offset compensation circuit from Fig. 6.28, but with the error amplifier replaced by a DDA. In a feedback configuration, the ideal DDA forces the two differential input-port voltages to become equal. Thus under steady-state conditions, the DC component of the MA's output voltage, that is, the output offset voltage, becomes approximately equal to V_{SA}. More precisely, we can calculate the systematic input offset voltage of the MA in Fig. 6.31 as

$$V_{OS,SA} = \frac{A_1 V_{SA}}{AA_1 + 1} \approx \frac{V_{SA}}{A}, \qquad (6.73)$$

where amplifiers A and A_1 were assumed to have no systematic offset. Thus, the desired input offset voltage, $V_{OS,SA}$, can be controlled conveniently with the slice-level adjustment voltage V_{SA} applied to the DDA.

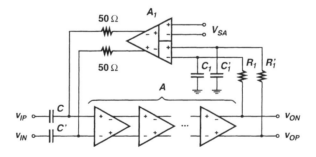

Fig. 6.31 MA with offset compensation and slice-level adjustment.

Alternatively, slice-level adjustment can be implemented in the TIA or, if a linear MA is used, in the decision circuit.

6.3.4 Automatic Gain Control

An AGC amplifier consist of a *variable-gain amplifier* (VGA) and an *automatic gain control* (AGC) mechanism, as shown in Fig. 6.32. Similar to the LA, the VGA is implemented as a multistage amplifier to maximize its gain-bandwidth product. But in contrast to the LA, the gain of the stages is now controlled by the DC voltage V_{AGC}. The amplitude of the VGA's output signal is measured by an *amplitude detector* and is compared with the reference voltage V_{REF} to produce the gain-control voltage V_{AGC}. The speed and stability of the gain-control loop is controlled by the low-pass filter consisting of R and C. For simplicity, Fig. 6.32 shows a single gain-control voltage, V_{AGC}, controlling all the stages. However, in a practical AGC amplifier, multiple gain-control voltages may be used and not all stages may be gain controlled. For example, to optimize the noise performance, it makes sense to stagger the gain control, first reducing the gain of the output stages before reducing the gain of the input stages.

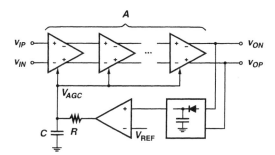

Fig. 6.32 Block diagram of an AGC amplifier.

Let's have a closer look at the two main components of the AGC amplifier: the variable-gain stages and the amplitude detector.

Variable-Gain Stages. The gain of an amplifier stage can be controlled in a number of different ways. When choosing the gain-control method it is important to keep in mind that changing the gain inadvertently changes some other parameters, such as the bandwidth, the input dynamic range, the noise, or the output common-mode voltage, as well. Usually, we would like the bandwidth and the output common-mode voltage to stay constant, whereas the input dynamic range should be largest for the low-gain settings and the noise should be lowest for the high-gain settings.

In the following, we summarize some popular gain-control methods:

- Vary the transconductance g_m of the gain transistor. Because the gain of a simple amplifier stage (without feedback) is given by $g_m \cdot R_L$, varying g_m changes the gain. One way to control the value of g_m is by means of the transistor's bias current. Figure 6.33(a) shows a differential pair with a variable tail current. Reducing the tail current to $I_0 - I_{AGC}$ lowers the g_m of the differential pair and thus the gain of the stage. But unfortunately, reducing the tail current also lowers the voltage drop across the load resistors R and R', and thus the output common-mode voltage shifts upward. To keep this voltage constant, we can add two more variable current sources at the outputs of the stage, as shown in Fig. 6.33(a). Now, the current that is taken away from the tail current, I_{AGC}, is rerouted to the two outputs, keeping the quiescent current through the load resistors at the fixed value $I_0/2$. Alternatively, if we want to increase the tail current to $I_0 + I_{AGC}$, we have to inject the currents $I_{AGC}/2$ into the output nodes to prevent the additional tail current from flowing into the load resistors. A drawback of this stage is that the maximum input voltage for linear operation is reduced as the gain is reduced. This is unfortunate because we expect the largest input signals for the lowest gain settings. The input-referred noise voltage increases somewhat at low gains, but from a signal-to-noise point of view, this should not be a problem. The bandwidth stays approximately constant.

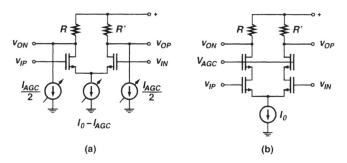

Fig. 6.33 VGA stages based on a variable transconductance.

Another way to lower the value of g_m is by reducing the drain-source voltage, thus pushing the FET into the linear regime. Besides reducing g_m, this method also increases g_o, thus reducing the gain further. A way to implement this scheme is to add cascode transistors as shown in Fig. 6.33(b). The gate voltage of the cascode transistors, V_{AGC}, controls the gain. The gain and the cascode transistors can be combined into a so-called *dual-gate FET*. Because the tail current now remains constant, the output common-mode voltage also remains constant and the input dynamic range depends less on the gain setting. However, this stage can be strongly nonlinear when driven with a large input signal. In particular, at low gain settings, the DC transfer function can be expanding rather than compressive.

- Vary the load resistor R_L. Because the gain of a simple amplifier stage (without feedback) is given by $g_m \cdot R_L$, varying R_L changes the gain. Figure 6.34(a) shows a differential pair with the variable load resistor R_{AGC} ($R_L = R_{AGC} \| 2R$).

Although we could vary the single-ended load resistors R and R' directly, this would alter the output common-mode voltage; hence, it is preferable to introduce the variable differential load as shown. The variable load can be implemented, for example, with an FET operating in the linear regime. This stage has a constant input dynamic range and a good noise performance. Its main drawback is that the bandwidth expands as the gain is reduced.

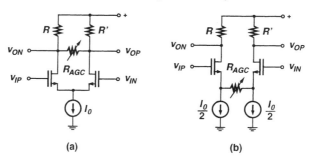

Fig. 6.34 VGA stages based on (a) a variable load and (b) a variable series feedback.

- Vary the amount of feedback. Figure 6.34(b) shows a differential pair with series feedback provided by the variable source degeneration resistor R_{AGC}. In this example, we split the tail current I_0 in two halves; alternatively, a single tail-current source could be connected to the midpoint of R_{AGC}. The advantage of the split current source is that it produces no DC voltage drop across R_{AGC}, thus avoiding potential headroom problems. The advantage of the single current source is that it contributes only common-mode noise to the outputs, thus giving better noise performance. The variable degeneration resistor can be implemented, for example, with an FET operating in the linear regime. This stage has an approximately constant bandwidth, a constant output common-mode voltage, and its dynamic range increases as the gain is reduced, as desired. Furthermore, the degeneration resistor improves the linearity. For these reasons, FET VGA often are based on this approach.

- Build the amplifier stage as an analog multiplier. One input of the multiplier is used for the signal and the other input is used for the gain-control voltage, V_{AGC}. This approach often is used for bipolar VGA stages where voltage-controlled resistors (FETs) are not available. Two- or four-quadrant multipliers based on the Gilbert cell can be used for this purpose. When designed properly, these stages can achieve a gain-independent bandwidth and a large dynamic range. We discuss some implementation examples in Section 6.4.2.

- Switch between two (or more) fixed gains. If we have two stages with different gains, we can digitally select one of the two gains by selectively enabling the corresponding stage. Note that this arrangement corresponds to a multiplier with one analog and one digital input. We discuss an implementation example based on this approach in Section 6.4.2.

Amplitude Detectors. The amplitude detector in Fig. 6.32 can be implemented as a *peak detector* or as a *rectifier*. The principle of a peak detector is shown in Fig. 6.35(a). Whenever the input signal v_I exceeds the output voltage v_O, the diode (assumed to be ideal) turns on and charges the capacitor C to the value of v_I. Note, however, that such a peak detector can respond only to *increasing* amplitudes. To track decreasing amplitudes as well, a decay or reset mechanism must be added. In Fig. 6.35(a), the necessary decay is provided by the current source I, which discharges the capacitor slowly. Unfortunately, this current source also causes the voltage to droop during long strings of zeros. The values for C and I must be chosen carefully to meet the response time and droop specifications of the amplitude detector.

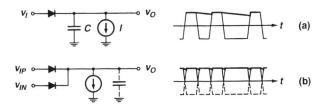

Fig. 6.35 Principle of (a) a peak detector and (b) a full-wave rectifier.

The principle of a full-wave rectifier is shown in Fig. 6.35(b). Rather than taking the peak value, we are now taking the *absolute* value of the input signal. Thus, for an NRZ (or RZ) signal that swings symmetrically about zero, the result is a DC voltage with a ripple caused by the finite rise and fall times. If the input signal is available in differential form, v_{IP} and v_{IN}, as it usually is the case in MAs, the rectifier can be built with two diodes. Assuming ideal diodes, the output voltage v_O is equal to the larger one of the two input voltages, $v_O = \max(v_{IP}, v_{IN})$, which is equal to $|v_{IP}|$ for $v_{IN} = -v_{IP}$. A small capacitor can be added at the output to filter out the ripple. Note that in contrast to the peak detector, this capacitor does not need to hold the output voltage for multiple bit periods but only for the short rise and fall times of the signal. In summary, the full-wave rectifier approach avoids the problem of droop, but it requires differential signals with fast rise/fall times and a low offset.

In the remainder of this section, we briefly look at practical implementations of the peak detector and the rectifier. Figure 6.36(a) shows the BJT peak detector used in [119]. Transistor Q_3 is turned on by the differential stage, Q_1 and Q_2, if the input voltage v_I exceeds v_O, thus charging up the capacitor C. The emitter follower Q_4 buffers the output voltage from the capacitor and also provides a small discharge current, $I = I_E/\beta$, to it. Note that embedding the peak-detector core, Q_3 and C, into the feedback loop compensates for the base-emitter voltage drops of Q_3 and Q_4 and makes v_O track the peaks of v_I with minimal offset. A CMOS peak detector similar to the one shown in Fig. 6.36(a) has been reported in [105]. Additional CMOS peak-detector circuits can be found in [92, 187].

Figure 6.36(b) shows the simplified circuit of the BJT full-wave rectifier used in [144]. The emitter-coupled BJT pair, Q_1 and Q_2, operates as the "max" circuit and also provides high-impedance inputs. Note that the output signal v_O is shifted down

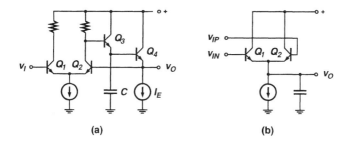

Fig. 6.36 Realization of (a) a peak detector and (b) a full-wave rectifier.

by a base-emitter voltage drop relative to the input signals. One way to compensate for this shift is to also shift V_{REF} in Fig. 6.32 by the same amount. Also note that the differential input signals, v_{IP} and v_{IN}, must have a low offset voltage for the accurate detection of small signals. A CMOS full-wave rectifier similar to the one shown in Fig. 6.36(b) has been reported in [123]. There, the rectifier was combined with the op amp that performs the comparison to V_{REF}; this op amp also compensates for the voltage drop in the rectifier. A CMOS full-wave rectifier using the current-mode approach has been reported in [180].

6.3.5 Loss of Signal Detection

In the event of an accidental fiber cut or a failure of the laser or its power supply, the receiver loses the signal. The receiver must detect this *loss of signal* (LOS) condition to give an alarm and possibly to restore the connectivity automatically. Some MAs include a loss-of-signal detector on the chip to support this function.

The block diagram of an LOS circuit is shown in Fig. 6.37. An amplitude detector monitors the signal strength at the MA output. The implementation of this block was discussed in Section 6.3.4. The measured amplitude is compared with a threshold voltage V_{TH} with a comparator circuit. The threshold voltage is set equal to the amplitude that just meets the BER requirement. The comparator usually exhibits a small amount of hysteresis to avoid oscillations in the LOS output signal. Finally, a timer circuit suppresses short LOS events. For instance, the SONET standard requires that only LOS events that persist for longer than 2.3 μs are signaled [188].

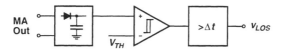

Fig. 6.37 Block diagram of a loss-of-signal detector.

6.3.6 Burst-Mode Amplifier

How does a burst-mode MA differ from a continuous-mode MA? Figure 6.38(a) shows a single-ended continuous-mode LA with an AC-coupled input signal. As a result of the AC coupling, the data signal is sliced at its average value. For a DC-balanced continuous-mode signal, the average value corresponds to the vertical center of the eye and all is well. (We are disregarding the case of unequal noise distributions, which was discussed in Section 6.3.3.) If we use the same AC-coupled LA to process a burst-mode signal, as shown in Fig. 6.38(b), the signal again is sliced at the average value. However, in this case, the average value is nowhere near the vertical center of the eye, and pulse-width distortions, jitter, and a high BER are the undesirable consequences.

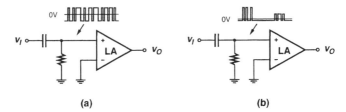

Fig. 6.38 An AC-coupled LA operated with (a) continuous-mode and (b) burst-mode signals.

To avoid this problem, burst-mode receivers often use DC coupling between all of its components (TIA, MA, and CDR). However, a downside of this approach is that all the offset voltages produced by these components must be tightly controlled. The offset voltage in the TIA can be controlled as explained in Section 5.2.11. Then, if the offset introduced by the MA is small, additional offset control may not be necessary. Alternatively, if the TIA has no offset control or is single ended, the MA must perform the offset control function. Figure 6.39 shows an example of an LA with a single-ended input that performs offset control [50, 104]. A peak and bottom detector determine the maximum and minimum values of the input signal, respectively. The average between these two values is obtained with two matched resistors, R and R', which then is used as the decision threshold voltage V_{DTH}. This circuit slices the input data signal at the vertical center of the eye regardless of the signal's average value or amplitude. Note that the peak and bottom detectors must be reset in between bursts to acquire the correct amplitude for each individual burst. Remember that large amplitude variations from burst to burst are common in passive optical networks such BPON and EPON.

Another approach to designing a burst-mode receiver is to AC couple the TIA to the MA, as shown in Fig. 6.38(b), but to use a very small coupling capacitor such that the time constant is much smaller than the bit period ($RC \ll 1/B$). In this case, AC coupling acts as a differentiator, producing short positive and negative pulses at the rising and falling data edges, respectively. Subsequently, the main amplifier is designed as a comparator with hysteresis (positive feedback), which acts as an integrator restoring the original NRZ signal. The main drawback of this approach is

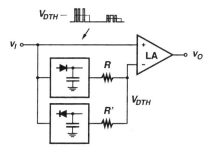

Fig. 6.39 A DC-coupled burst-mode limiting amplifier.

that the high-pass coupling network enhances the received noise, which results in a reduced receiver sensitivity [100].

6.4 MA CIRCUIT IMPLEMENTATIONS

In the following, we examine some representative transistor-level MA circuits, which have been reported in the literature. These circuits illustrate how the design principles discussed in the previous section can be implemented in a broad variety of technologies using different types of transistors such as the metal-semiconductor field-effect transistor (MESFET), the heterostructure field-effect transistor (HFET), the bipolar junction transistor (BJT), the heterojunction bipolar transistor (HBT), and the complementary metal-oxide-semiconductor transistors (CMOS) (cf. Appendix D).

6.4.1 MESFET and HFET Technology

Single-Ended Gain Stage. Figure 6.40 shows a simplified schematic of the GaAs-FET gain stage reported in [26, 27]. This single-ended stage is part of a three-stage amplifier, which has a bandwidth of 3.3 GHz and is implemented in a 1-μm GaAs-MESFET technology. The transistors in this circuit are depletion-mode FETs, which means that they conduct current when the gate-source voltage is zero.

The gain transistor M_1 is driving the active-feedback load composed of FETs M_3 through M_5. FET M_2 provides the bias current for M_1 and M_5. The active-feedback load consists of the source-follower, M_3 and M_4, a stack of four Schottky diodes for level shifting, and the feedback FET M_5. Neglecting the output conductances of M_1, M_2, and M_5 and assuming unity gain for the source follower, this load presents the equivalent resistance $R_L = 1/g_{m5}$ (cf. Section 6.3.2, TIA Load). Therefore, the gain of the stage is about $A = g_{m1}/g_{m5}$, a quantity that can be set accurately by the ratio of the widths of M_1 and M_5. Overall, the active-feedback load provides good gain and operating-point stability. Another source-follower stage with M_6 and M_7, which can be sized independently of the source-follower M_3 and M_4, serves as a buffer to drive the subsequent stage.

214 MAIN AMPLIFIERS

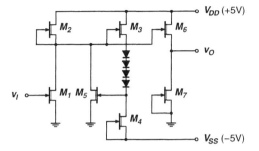

Fig. 6.40 MESFET/HFET implementation of a single-ended gain stage based on [27].

The three-stage amplifier in [27] uses this topology for the second and third stage. The first stage is realized as a common-gate stage, which provides good input matching and low noise. All stages in this single-ended design are AC coupled.

Differential Gain Stage. Figure 6.41 shows a simplified schematic of the GaAs-FET VGA stage reported in [78]. This differential stage is part of a four-stage AGC amplifier, which has a bandwidth of 18 GHz and is implemented in a 0.3-μm GaAs-HFET technology with enhancement- and depletion-mode devices.

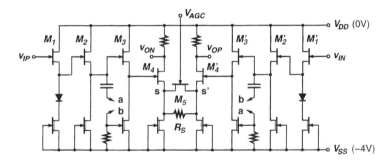

Fig. 6.41 MESFET/HFET implementation of a differential VGA stage based on [78].

The differential input signals v_{IP} and v_{IN} are fed into a cascade of three source followers, M_1, M'_1, M_2, M'_2, M_3, and M'_3, which provides impedance transformation and level shifting. The source followers are implemented with enhancement-mode FETs, whereas the bias current sources are implemented with depletion-mode FETs. The depletion-mode current sources are easier to bias, have lower crosstalk, and reduced supply voltage dependence. The source followers M_3 and M'_3 are enhanced with an R-C coupling network to improve their speed (active source follower). The operation of this network can be understood as follows: when the output of source follower M_3 is rising, the bias current of M_3 is momentarily reduced by means of the control voltage at node b, thus accelerating the transition. This bias current reduction is caused by the falling edge at the output of the source follower M_2', which is coupled to node b through the R-C high-pass network. Similarly, M_2 helps to accelerate the transitions of M_3'. The output signals from the source follower cascade drive the differential pair, M_4 and M_4'. FET M_5 acts as a voltage-controlled resistor, which varies the amount of series feedback and thus the gain of this stage. The source degeneration resistor R_S improves the linearity.

The AGC amplifier in [78] consists of an input buffer, four DC-coupled stages of the type shown in Fig. 6.41, and an output buffer. A passive on-chip R-C-R feedback network reduces the offset voltage (cf. Section 6.3.3). The voltage V_{AGC} is produced by an on-chip AGC circuit.

In [77], a similar differential stage has been reported, which is part of a three-stage LA with a bandwidth of 29.3 GHz implemented in a 0.2-μm GaAs-HFET technology. In contrast to Fig. 6.41, this stage has no gain control, and therefore the degeneration devices R_S and M_5 are removed and nodes s and s' are shorted together. With this modification, the stage operates at its maximum gain. Furthermore, this LA stage uses inductive loads for shunt peaking.

6.4.2 BJT and HBT Technology

Cherry-Hooper Stage. Figure 6.42 shows a simplified schematic of the bipolar gain stage reported in [86]. This stage is part of a three-stage LA, which has a bandwidth of 45 GHz and is implemented in a 105-GHz SiGe-HBT technology. This type of stage is known as a *Cherry-Hooper stage* and has been used successfully since 1963, when the original Cherry-Hooper paper [22] was published (e.g., see [38, 96, 133, 143]).[12]

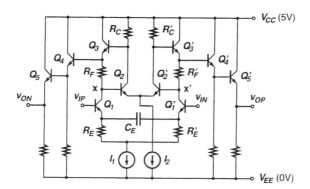

Fig. 6.42 BJT/HBT implementation of a gain stage based on the Cherry-Hooper architecture (e.g., [86]).

---

[12] The paper [22] describes the design of single-ended wideband amplifiers using an alternation of series- and shunt-feedback stages. Later, this concept was extended to a differential topology and to include emitter followers (see Fig. 6.42).

The input signals v_{IP} and v_{IN} drive the differential pair Q_1 and Q'_1 with the emitter degeneration resistors R_E and R'_E. We already discussed in Section 6.3.2 how emitter degeneration (series feedback) reduces the input capacitance and thus speeds up the BJT input pole. The emitter capacitor C_E can be used to introduce a zero, further improving the speed. The load of the input differential pair is presented by the TIA consisting of Q_2, Q'_2, Q_3, and Q'_3 and the shunt-feedback resistors R_F and R'_F. Note the similarity between this TIA and the one shown in Fig. 5.31. We know from Section 6.3.2 how the low input impedance of the TIA speeds up the pole associated with the nodes x and x' and also reduces the Miller capacitance at the input of this stage. Finally, the differential output signals are buffered with the emitter followers Q_4, Q'_4, Q_5, and Q'_5. These buffers reduce the loading, can provide some peaking, and shift the DC voltage to a lower level such that the Q_1 and Q'_1 of the next stage can operate at a higher V_{CE}, which improves their speed.

The degenerated input differential pair has a transconductance of about $1/R_E$ (more precisely, $1/[R_E + 1/g_{m1}]$), whereas the TIA has a transresistance of about R_F (assuming the gain of Q_2 and Q'_2 is much larger than one). The product of these two quantities, $A = R_F/R_E$ (or $A = R_F/[R_E + 1/g_{m1}]$, to be more precise), is the voltage gain of the Cherry-Hooper stage. Because this gain is given primarily by a resistor ratio, it is insensitive to process, temperature, and supply voltage variations. Note that the Cherry-Hooper stage can be viewed as a double stage combining a *transadmittance stage* (TAS) with a *transimpedance stage* (TIS). Thus in a multistage application, we have an alternation between TAS and TIS. It also has been observed that the large impedance mismatch between the TAS, TIS, and emitter followers contributes to the wide bandwidth of this architecture [22, 145].

In [38], a similar Cherry-Hooper stage has been reported, which is part of a three-stage LA with a bandwidth of 15 GHz implemented in a 47-GHz SiGe HBT technology. In contrast to Fig. 6.42, this stage does not have the emitter degeneration resistors R_E and R'_E in the TAS, resulting in the increased stage gain $A = g_{m1}R_F$. Furthermore, the collector resistors R_C and R'_C are each split into a series connection of two resistors R_{C1}-R_{C2} and R'_{C1}-R'_{C2} with the bases of Q_3 and Q'_3 connecting to the midpoints. This modification provides a gain enhancement of $1 + R_{C2}/R_{C1}$ with little impact on the bandwidth.

In [133], another form of the Cherry-Hooper stage has been reported, which is part of a three-stage LA with a bandwidth of 9 GHz implemented in a 0.4-μm BJT technology. In contrast to Fig. 6.42, this stage uses a simplified TIA load without the emitter followers Q_3 and Q'_3. In this case, the noninverted and inverted currents from the TAS and TIS, respectively, flow into the same collector resistors R_C and R'_C, possibly giving rise to a nonmonotonic DC transfer function. To ensure monotonic behavior and proper limiting, it is recommended to make the TIS tail current, I_2, twice as large as the TAS tail current, I_1.

Four-Quadrant Multiplier VGA Stages. How can we control the gain of a Cherry-Hooper stage for use in an AGC amplifier? Unfortunately, bipolar technologies don't offer voltage-controlled resistors that we could use to control R_E or R_F (cf. Section 6.4.1). The solution that we explore next is to turn the Cherry-Hooper stage into an analog multiplier.

Figures 6.43 and 6.44 show the schematics of the bipolar VGA stages reported in [98]. These stages are part of a three-stage AGC amplifier, which has a bandwidth of 10 GHz and is implemented in a 22-GHz BJT technology.

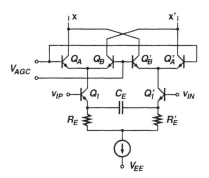

Fig. 6.43 BJT/HBT implementation of a four-quadrant multiplier VGA stage based on [98]. Nodes x and x′ connect to a TIA load as in Fig. 6.42.

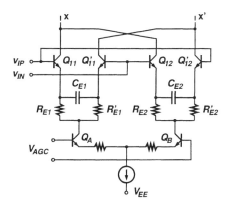

Fig. 6.44 BJT/HBT implementation of a modified four-quadrant multiplier VGA stage based on [98]. Nodes x and x′ connect to a TIA load as in Fig. 6.42.

Figure 6.43 shows how the input differential pair can be turned into a four-quadrant multiplier that has the form of a *Gilbert cell*. The TIA load is the same as that shown in Fig. 6.42. If the gain-control voltage V_{AGC} is large, the output currents from the differential pair Q_1 and Q'_1 are directly routed through Q_A and Q'_A to the nodes x and x′ producing a gain of approximately $A = R_F/R_E$, as in the original Cherry-Hooper stage. However, if V_{AGC} is reduced, the currents to the nodes x and x′ become a combination of the direct contributions through Q_A and Q'_A and the crossed-over contributions through Q_B and Q'_B. These two contributions partly cancel each other, which results in a reduced gain. In particular, for $V_{AGC} = 0$, the cancellation is complete and the gain becomes zero (assuming the transistors Q_A, Q'_A, Q_B, and Q'_B are sized identically). Note that the gain control voltage for this stage always must be kept positive. Should it drop below zero, the gain starts to increase again (and the data signal is inverted), which means that the negative-feedback AGC loop turns into a positive-feedback loop, making the AGC unstable. Also, note that the input dynamic range of this stage, just like that of the original Cherry-Hooper stage, is determined by the voltage drop across the emitter degeneration resistors R_E and R'_E, and thus is independent of the gain-control voltage V_{AGC}. This may be an issue at low gains where the input signals are largest.

Figure 6.44 shows another approach to design a VGA stage with BJTs. Compared with Fig. 6.43, the input ports of this modified four-quadrant multiplier are used differently: the lower inputs are now used for the gain-control voltage, whereas the upper inputs are used for the input signal. The maximum gain of approximately $A_{\max} = R_F/R_{E1}$ is obtained when all of the tail current is directed through Q_A into the differential pair Q_{11} and Q'_{11}. If we reduce V_{AGC} and divert some of the tail current through Q_B into the other pair, Q_{12} and Q'_{12}, the transconductance of the first pair drops, whereas the second pair contributes an inverted signal current to the output. As a result of these two mechanisms, the gain is reduced.

In [98], the first stage uses the topology shown in Fig. 6.43, the second stage uses the topology shown in Fig. 6.44, and the third stage has a fixed gain. The reason for using both types of VGA stages is that they can be made to have complementary characteristics. Whereas the gain of the stage in Fig. 6.43 drops with increasing input voltage, that of the stage in Fig. 6.44 increases. Whereas the bandwidth of the stage in Fig. 6.43 shrinks with decreasing gain, that of the stage in Fig. 6.44 can be made to expand with the appropriate choice of R_{E1}, R'_{E1}, R_{E2}, R'_{E2}, C_{E1}, and C_{E2}. Thus, combining the two stages results in superior linearity (large $v_{\lin}^{pp}$) and a gain-independent bandwidth. In addition to the three gain stages, the AGC-amplifier chip in [98] also includes an input buffer, two output buffers (one to drive the decision circuit and one to drive the clock-recovery circuit), an offset compensation circuit, and a gain-control circuit.

In [114], similar VGA stages have been reported, which are part of a three-stage AGC amplifier with a bandwidth of 32.7 GHz implemented in a 92-GHz SiGe HBT technology.

Two-Quadrant Multiplier VGA Stage. Figure 6.45 shows a simplified schematic of the bipolar VGA stage reported in [178]. This stage is part of a three-stage VGA for a 20-Gb/s application and is implemented in a 60-GHz SiGe graded-base technology.

In contrast to Fig. 6.43, this stage is based on a two-quadrant multiplier. The two quadrants corresponding to $V_{AGC} < 0$ are not needed; in fact, they must be avoided to keep the control loop stable. The stage shown in Fig. 6.45 controls the gain by dumping a variable amount of signal current through Q_B and Q'_B into V_{CC}. Thus, when the stage is completed with a TIA load as in Fig. 6.42, the gain can be varied in the range $A = 0$ to R_F/R_E. A drawback of this topology is that the DC currents

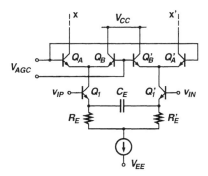

Fig. 6.45 BJT/HBT implementation of a two-quadrant multiplier VGA stage based on [178]. Nodes x and x' connect to a TIA load as in Fig. 6.42.

into the nodes x and x' are varying with the gain-control voltage, producing a variable output common-mode voltage. A bias stabilization circuit can be added to take care of this problem.

The VGA in [178] uses the topology shown in Fig. 6.45, enhanced with a bias stabilization circuit and with $R_E = R'_E = C_E = 0$ for the first stage. The second and third stages have a fixed gain. This VGA is used to drive a decision circuit located on the same chip.

Selectable Gain Stage. Figure 6.46 shows a simplified schematic of the bipolar VGA stage reported in [37, 39]. This stage is part of a three-stage AGC amplifier with a bandwidth of 9 GHz and is implemented in a 50-GHz SiGe technology.

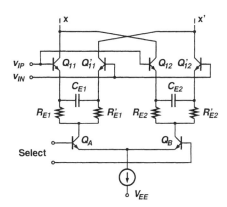

Fig. 6.46 BJT/HBT implementation of a selectable gain stage based on [37, 39]. Nodes x and x' connect to the gain-control circuit (emitters of Q_A, Q_B and Q'_A, Q'_B) in Fig. 6.43 or to the TIA load in Fig. 6.42.

In this approach, the input signals are applied simultaneously to two differential pairs, one with emitter degeneration resistors R_{E1}, R'_{E1} and the other with R_{E2}, R'_{E2}. The differential pair, Q_A and Q_B, acts as a switch that steers the tail current either into the left or the right differential pair, depending on the binary value of the select signal. The active pair together with the TIA load in Fig. 6.42 determine the gain to be either $A = R_F/R_{E1}$ or R_F/R_{E2}. This topology has the advantage that it has a wide input dynamic range when the lower gain is selected (large degeneration resistor), whereas it has a good noise figure when the higher gain is selected (small degeneration resistor).

The first stage of the AGC amplifier in [37, 39] combines the selectable gain circuit in Fig. 6.46 with the the gain-control circuit (Q_A, Q'_A, Q_B, Q'_B) in Fig. 6.43. The select input is used to choose the gain range, whereas the V_{AGC} input is used to control the gain continuously within each range. In the lower range (-7 dB to 7 dB), the maximum input voltage is 1.7 V and the noise figure is 16 dB, whereas in the upper range (7 dB to 20 dB), the maximum input voltage is 0.2 V and the noise figure improves to 12 dB. The second stage features a continuous gain control input only, and the third stage has a fixed gain. The whole AGC amplifier is integrated with a CDR and DMUX on a single chip.

Distributed Amplifier. Figure 6.47 shows a simplified schematic of the bipolar distributed amplifier reported in [10]. This single-ended amplifier has a bandwidth of 74 GHz and is implemented in a 160-GHz InP-HBT technology.

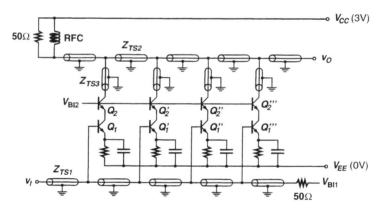

Fig. 6.47 BJT/HBT implementation of a distributed amplifier based on [10].

Each of the four sections of the distributed amplifier is structured as a cascode circuit with Q_1 and Q_2. The cascode technique lowers the input capacitance of each section by suppressing the (section dependent) Miller capacitance of Q_1. Furthermore, it lowers the output conductance of each section, thus reducing the losses in the output transmission line. The emitter-degeneration resistor of Q_1 increases the input resistance of each section, thus reducing the losses in the input transmission line. Furthermore, the resistor improves the linearity and lowers the input capacitance. The input transmission line is formed by the transmission-line segments Z_{TS1}, and the output transmission line is formed by the transmission-line segments Z_{TS2}, which are all realized as coplanar waveguides. The transmission-line segments Z_{TS3} provide a small amount of peaking. An RF choke (RFC) can be used to shunt the back-termination resistor of the output transmission line, thus eliminating the DC voltage drop across this resistor.

The circuit in Fig 6.47 also can be used as a TIA. In this case, the photodetector is connected directly to the input v_I. As in a low-impedance front-end, the detector current is converted first to a voltage by means of the 50-Ω input impedance, then this voltage is amplified by the distributed amplifier.

6.4.3 CMOS Technology

Low-Voltage VGA Stage. Figure 6.48 shows the schematic of the CMOS VGA stage reported in [187]. This stage is part of a four-stage AGC amplifier, which has a bandwidth of 2 GHz and is implemented in a 0.15-μm, 2-V CMOS technology.

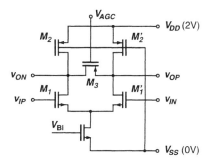

Fig. 6.48 CMOS implementation of a VGA stage based on [187].

The stage consists of the n-MOS differential pair M_1 and M'_1 and the p-MOS load transistors M_2 and M'_2. Note that the p-MOS loads operate in the linear regime. The gain is controlled with the p-MOS transistor M_3, which presents a variable differential load resistance.

In [187], the first three stages of the AGC amplifier have a fixed gain and are implemented as shown in Fig. 6.48, but without the FET M_3. The last stage has a variable gain and is implemented as shown in Fig. 6.48. This AGC amplifier is part of a single-chip receiver, which includes a TIA, a CDR, and a DMUX.

VGA Stage with Replica Biasing. Figure 6.49 shows a simplified schematic of the CMOS VGA stage reported in [47]. This stage is used in a six-stage VGA amplifier for a 480-Mb/s application and is implemented in a 1.2-μm, 5-V CMOS technology.

The gain of this stage is controlled by the variable p-MOS load transistors M_2 and M'_2, which both operate in the linear regime. The replica biasing circuit, on the right-hand side, generates the gate voltage such that the (large-signal) drain-

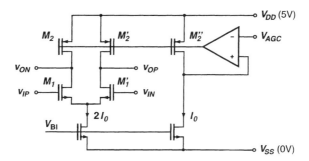

Fig. 6.49 CMOS implementation of a VGA stage based on [47].

source resistance of the replica M_2'' and the loads M_2 and M_2' all become equal to $(V_{DD} - V_{AGC})/I_0$. Thus, a well-controlled (temperature- and process-independent) variable load resistance is created. A potential drawback of varying the single-ended load resistance (as opposed to the differential load resistance, as in Fig. 6.48) is that the common-mode output voltage varies with the gain; in fact, in Fig. 6.49, we have $v_{OCM} = V_{AGC}$.

In [47], eight such VGA amplifiers are used to implement a parallel receiver. The receiver chip also performs the CDR and 1:8 DMUX functions.

Gain Stage with TIA Load. Figure 6.50 shows a simplified schematic of the MOS gain stage reported in [31]. This circuit is used for all stages in a five-stage LA, which has a bandwidth of 9.4 GHz and is implemented in a 0.18-μm, 1.8-V CMOS technology.

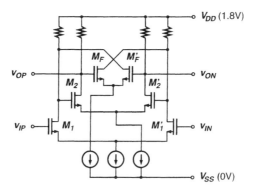

Fig. 6.50 CMOS implementation of a gain stage with active-feedback TIA loads based on [31].

The differential pair M_1 and M_1' is loaded by the TIA consisting of M_2, M_2', M_3, and M_3'. The TIA load is a differential implementation of the active-feedback TIA shown in Fig. 5.19. The differential pair M_2 and M_2' provides the voltage gain and FETs M_F and M_F' constitute the active feedback. Neglecting the effects of the load resistors and the FET's output conductances, the transimpedance of this TIA is $1/g_{mF}$. The active-feedback TIA load and the bandwidth extension that can be achieved with it was discussed in Section 6.3.2.

In [31], the bandwidth of this stage is enhanced further by applying shunt peaking and negative Miller capacitances. The LA chip also contains an input matching network with a T-coil and an offset cancellation circuit.

Gain Stage with Active Inductors. Figure 6.51 shows the schematic of the MOS gain stage reported in [156]. This circuit is used for the last three stages of a four-stage LA, which has a bandwidth of 3 GHz and is implemented in a 0.25-μm, 2.5-V CMOS technology.

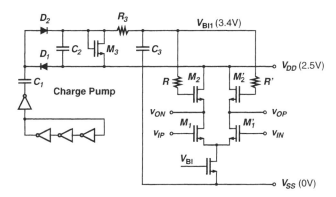

Fig. 6.51 CMOS implementation of a gain stage with active-inductor loads based on [156].

The differential pair M_1 and M_1' is loaded by the active inductors consisting of transistors M_2 and M_2' and resistors R and R'. The active inductor and the bandwidth extension that can be achieved with inductive loads was discussed in Section 6.3.2. To alleviate headroom problems in this low-voltage design, the bias voltage V_{BI1} is set to one n-MOS threshold voltage above V_{DD}. As shown in Fig. 6.51, this bias voltage easily can be generated on chip with a charge pump: a ring oscillator drives a capacitor-diode charge pump (C_1, C_2, D_1, and D_2), producing a voltage above V_{DD}. This voltage is clamped to the desired value by M_3. Oscillator ripples are filtered out with the low-pass network R_3, C_3. A welcome side effect of using active-inductor loads is that the low-frequency gain of the stage is set primarily by the transistor geometry $A = \sqrt{W_1/W_2}$, which is insensitive to process, temperature, and supply voltage variations.

In [156], the active-inductor stages are combined with inverse scaling and buffering to boost the bandwidth of the LA further.

224 MAIN AMPLIFIERS

Common-Gate Stage with Offset Control. Figure 6.52 shows a simplified schematic of the common-gate CMOS stage reported in [156]. This circuit, enhanced with active-inductor loads, is used as the input stage of the four-stage LA mentioned above.

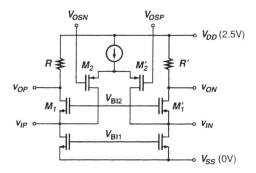

Fig. 6.52 CMOS implementation of a common-gate input stage based on [156].

The common-gate n-MOS transistors M_1 and M_1' provide a low-impedance input, which can be made equal to 50 Ω with the appropriate choice of transistor dimensions and bias currents (constant-g_m biasing). Apart from providing the input termination, the common-gate input transistors also provide an effective electrostatic discharge (ESD) protection. Standard ESD networks are not suitable for high-speed inputs because they limit the signal bandwidth too much. It therefore is advantageous to build the input stage such that the parasitic MOS diodes act as the ESD protection: in Fig. 6.52, there are three parasitic diodes at each input, two to ground and one to V_{DD}. Finally, the input stage also features the p-MOS differential pair M_2 and M_2', which permits to control the offset voltage with $V_{OSP} - V_{OSN}$. A feedback circuit, as shown in Fig. 6.29, can be used to cancel the offset voltage automatically.

6.5 PRODUCT EXAMPLES

Tables 6.1 and 6.2 summarize the main parameters of some commercially available LA and AGC amplifiers. The numbers have been taken from data sheets of the manufacturer that were available at the time of writing. For up-to-date product information, please contact the manufacturer directly. For single-ended amplifiers, the gain value, tabulated under A, is followed by "(s)." For differential amplifiers, the *differential* gain value is given followed by "(d)." The input overload voltage, v_{ovl}^{pp}, as well as the maximum input voltage for linear operation, v_{lin}^{pp}, are specified as peak-to-peak values. The noise-figure values, tabulated under F, are measured single endedly, that is, the source noise is that of a single 50-Ω resistor (cf. Section 6.2.3).

PRODUCT EXAMPLES 225

Table 6.1 Examples for high-speed LA products.

Company & Product	Speed (Gb/s)	A (dB)	BW_{3dB} (GHz)	f_{LF} (kHz)	F (dB)	v_{ovl}^{pp} (V)	Power (mW)	Technology
Agere LG1605DXB	2.5	34 (d)	3.0	2.5	15	1.6	400	GaAs HFET
AMCC S3051	2.5	32 (d)	2.0	2.0	18	1.0	375	
Maxim MAX3265	2.5	49 (d)		2.0		1.2	165	
Philips OQ2538HP	2.5	46 (d)	3.0		14	1.2	270	Si BJT
Agere TLMA0110G	10.0	39 (d)	9.0	25.0	17	2.0	700	GaAs HFET
AMCC S3096	10.0	50 (d)	10.0					SiGe HBT
Maxim MAX3971	10.0	(d)	10.0	40.0		0.8	155	SiGe HBT
OKI GHAD4103	10.0	18 (d)					1,200	GaAs HFET
Bookham P35-5142	40.0	12 (s)	40.0				900	GaAs HFET

Table 6.2 Examples for high-speed AGC amplifier products.

Company & Product	Speed (Gb/s)	A_{max} (dB)	BW_{3dB} (GHz)	f_{LF} (kHz)	F (dB)	v_{lin}^{pp} (V)	Power (mW)	Technology
Nortel AC03	2.5	36 (d)	2.0		9	2.0	500	Si BJT
Nortel AC10	2.5	30 (d)	1.9	25.0	10	0.3	198	Si BJT
Giga GD19902	10.0	26 (d)	10.0				2,800	GaAs HFET
OKI GHAD4102	10.0	16 (d)	10.0				1,300	GaAs HFET
Centellax OAA4MSC	40.0	12 (s)	40.0				1,500	

We note that all listed 2.5- and 10-Gb/s parts have a differential topology. The two single-ended 40-Gb/s parts are implemented as distributed amplifiers. Comparing parts for different bit rates, we find that, in general, higher-speed parts consume quite a bit more power. The reader may wonder about the difference between the two similar Nortel parts: the AC03 requires a 5.2-V power supply, whereas the AC10 runs from the lower 3.3-V supply, which also explains the difference in power dissipation.

6.6 RESEARCH DIRECTIONS

The research effort focusing on MAs can be divided roughly into two categories: higher speed and lower cost. In the following, we briefly touch on these activities.

Higher Speed. It was pointed out in Section 5.5 that many research groups are now aiming at the 40-Gb/s speed and beyond. To this end, fast MAs, with a bandwidth of 40 GHz and more, must be designed. Usually, heterostructure devices, such as HBTs and HFETs, based on compound materials, such as SiGe, GaAs, and InP, are used to reach this goal (cf. Appendix D).

The following papers on high-speed MAs were published recently:

- In SiGe-HBT technology, a 31-GHz and a 32.7-GHz AGC amplifier as well as a 40-Gb/s and a 49-GHz LA have been reported in [87], [114], [150], and [86], respectively.

- In GaAs-HFET technology, a 18-GHz AGC amplifier as well as a 27.7-GHz and a 29.3-GHz LA have been reported in [78], [78], and [77], respectively.

- In GaAs-HBT technology, a 26-GHz VGA has been reported in [153].

- In InP-HBT technology, a 30-GHz LA and a 74-GHz distributed amplifier have been reported in [96] and [10], respectively.

It can be seen that, in terms of speed, the LAs have an advantage over the more complex AGC amplifiers.

Lower Cost. Another area of research is focusing on the design of high-performance MAs in low-cost, mainstream technologies. For the reasons already given in Section 5.5, digital CMOS is of particular interest.

For example, a SONET-compliant 10-Gb/s LA has been implemented in a low-cost "modular BiCMOS" technology [70]. A 2.4-Gb/s, 0.15-μm CMOS AGC amplifier has been reported in [187], and a SONET-compliant 2.5-Gb/s, 0.25-μm CMOS LA has been demonstrated in [156]. 10-Gb/s, 0.18-μm CMOS LAs have been reported in [128] and [31]. The promise of a CMOS main amplifier is that it can be integrated with the CDR, DMUX, and the digital frame processing on single CMOS chip to provide a cost-effective and compact low-power receiver solution.

6.7 SUMMARY

Two types of main amplifiers (MA) can be distinguished:

- The limiting amplifier (LA), which generally is faster, dissipates less power, and is easier to design. However, its strong nonlinearity for large input signals restricts its field of application.

- The automatic gain control (AGC) amplifier, which is linear over a wide range of input amplitudes and thus is suitable for receivers with an equalizer, decision-point steering, a soft-decision decoder, and so forth. However, its more complex design generally results in a lower bandwidth and higher power dissipation.

The main specifications of the MA are as follows:

- The voltage gain, which must be large enough to provide enough voltage swing for the subsequent clock and data recovery (CDR) circuit.

- The bandwidth, which must be large enough to prevent the MA from introducing a noticeable amount of intersymbol interference (ISI).

- The group-delay variation, which must be kept small to minimize jitter and other signal distortions.

- The noise figure, which must be low enough to avoid a noticeable degradation of the receiver sensitivity.

- The MA sensitivity, which must be much smaller than the minimum input signal into the MA to avoid a noticeable degradation of the receiver sensitivity.

- The input overload voltage, which must be large enough to avoid harmful pulse-width distortion and jitter for the maximum input signal into the MA.

- The input offset voltage, which must be much smaller than the minimum input signal into the MA to avoid a harmful slice-level error (especially in LAs).

- The low-frequency cutoff, which must be low enough to avoid a noticeable baseline wander. This is particularly important if signals with long runs of zeros and ones (i.e., scrambled data signals) are used.

- The AM-to-PM conversion, which must be low enough to limit the generation of jitter in the presence of spurious amplitude modulation.

Virtually all MAs are structured as multistage amplifiers because this topology permits the realization of very high gain-bandwidth products ($\gg f_T$). Furthermore, broadband techniques such as series feedback, emitter peaking, cascoding, transimpedance load (with shunt feedback or active feedback), negative Miller capacitance, buffering, scaling, inductive load (e.g., shunt peaking), inductive interstage network, and distributed amplifier are applied to the gain stages to improve their bandwidth and to shape their rolloff characteristics.

228 MAIN AMPLIFIERS

Most MAs include an offset compensation circuit to reduce the random offset voltage below a critical value. Some MAs also permit the introduction of a *controlled* amount of offset to adjust the slice level. This feature is useful to optimize the bit-error rate (BER) performance in the presence of unequal noise distributions for zeros and ones. AGC amplifiers consist of a variable-gain amplifier (VGA), an amplitude detector, and a feedback loop that controls the gain such that the output amplitude remains constant. Some MAs feature a loss-of-signal detector to detect faults such as a cut fiber. Burst-mode amplifiers have special provisions, such as fast decision threshold control, to deal with a bursty signal of varying amplitude and no DC balance.

MAs have been implemented in a wide variety of technologies including metal-semiconductor FET (MESFET), heterostructure FET (HFET), BJT, heterojunction bipolar transistor (HBT), BiCMOS, and CMOS. The Cherry-Hooper architecture, which combines series-feedback and shunt-feedback stages, often is chosen for BJT and HBT implementations.

Currently, researchers are working on 40-Gb/s MAs and beyond, as well as MAs in low-cost technologies such as CMOS.

6.8 PROBLEMS

6.1 Power Penalty. What is the optical sensitivity of a receiver that consists of a photodetector with the responsivity $\mathcal{R}$, a TIA with the input-referred rms noise current $i_{n.TIA}^{rms}$ and the transimpedance R_T, an MA with the gain A, and a DEC with the sensitivity V_{DTA}?

6.2 Noise Figure. (a) Write the noise-figure expression "total output noise power" divided by "fraction of the output noise power due to the thermal noise of the source resistance" in mathematical form. Show that this expression is identical to "input SNR" divided by "output SNR," where the input SNR is based on the thermal noise of the source resistance only. In both cases, take the noise power in the frequency band from f_1 to f_2. (b) Show that for $f_2 - f_1 \to 0$, the noise figure expression from Problem 6.2(a) becomes the spot noise figure in Eq. (6.14). (c) Show that for $f_1 = 0$, $f_2 \to \infty$, the noise figure expression from Problem 6.2(a) becomes the wideband noise figure in Eq. (6.15).

6.3 AC-Coupling Capacitor. An amplifier with the output resistance $R_O = 50\,\Omega$ is AC coupled with capacitor C to another amplifier with the same input resistance $R_O = 50\,\Omega$. (a) What is the low-frequency cutoff, f_{LF}, caused by the capacitor C? (b) What is the power penalty caused by a finite capacitor C? (c) How large should the capacitor C be made for a 2.5-Gb/s and 10-Gb/s system?

6.4 Bandwidth Shrinkage and GBW Extension. (a) Calculate the 3-dB bandwidth of a cascade of n identical second-order Butterworth stages. Normalize this bandwidth to that of a single stage to obtain the bandwidth shrinkage as a function of n. (b) Given a desired gain, A_{tot}, and bandwidth, BW_{tot}, for the n-stage amplifier above, calculate the GBW necessary for each stage

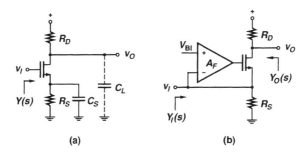

Fig. 6.53 Circuits for Problems 6.7 and 6.8: (a) stage with series feedback and (b) regulated cascode.

(GBW_S). How large is the ratio of the GBW for the total amplifier to the GBW per stage? (c) Derive the number of stages, n_{opt}, that maximizes the gain-bandwidth extension.

6.5 Miller Effect. A capacitor C is connected between an input terminal with voltage v_I and a node with the voltage $A \cdot v_I$. (a) Assuming A is frequency independent, calculate the effective input capacitance, C_I. (b) Assuming A has the frequency dependence $A(s) = A_0/(1+sT)$, calculate the input admittance, $Y(s)$.

6.6 f_T Doubler. It has been suggested that a simple differential pair is an f_T doubler because its differential input capacitance is $C_{be}/2$ (neglecting C_{bc}), whereas its differential transconductance, $g_m = \Delta(i_{C.P} - i_{C.N})/\Delta(v_{BE.P} - v_{BE.N})$, is the same as that of a single transistor. What is wrong with this argument?

6.7 Series Feedback. (a) Calculate the voltage transfer function, $A(s) = V_o/V_i$, for the MOSFET stage with series feedback shown in Fig. 6.53(a). Use a simplified FET model with $C_{gd} = g_{mb} = g_o = R_g = 0$; assume C_{sb} is included in C_S and C_{db} is included in C_L. Derive expressions for the poles and zeros of $A(s)$. (b) What value for C_S results in a single-pole response? (c) Calculate the input admittance, $Y(s)$, for the above circuit assuming the same simplified FET model. Derive expressions for the poles and zeros of $Y(s)$. (d) What are the conditions for a purely capacitive input?

6.8 Regulated Cascode. (a) Calculate the input admittance, $Y_I(s)$, of the MOSFET regulated-cascode circuit shown in Fig. 6.53(b). Use a simplified (DC) FET model with all capacitances set to zero and $g_{mb} = 0$; also assume that the feedback amplifier has the frequency-independent gain A_F. Simplify the resulting expression for $R_D \ll 1/g_o$ and $(|A_F|+1) \cdot g_m/g_o \gg 1$. (b) Calculate the output admittance, $Y_O(s)$, of the regulated-cascode circuit using the same FET and amplifier models as before. Simplify the resulting expression for $(|A_F|+1) \cdot g_m/g_o \gg 1$ and $(|A_F|+1) \cdot g_m R_S \gg 1$. (c) Discuss the advantage of the regulated cascode over the simple cascode. (d) How do these

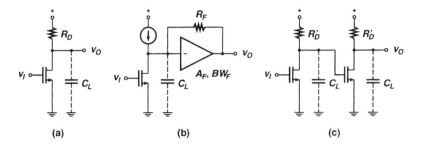

Fig. 6.54 Circuits for Problems 6.9 and 6.10: (a) common-source stage, (b) stage with TIA load, and (c) split stage.

results change if the FET is replaced by a bipolar transistor? Compared with the FET, the bipolar transistor has a nonzero base-emitter conductance equal to g_m/β.

6.9 **Shunt-Feedback TIA Load.** (a) Calculate the bandwidth BW of the simple MOSFET common-source stage shown in Fig. 6.54(a), assuming $R_g = g_o = 0$ and a total capacitance at the output node equal to C_L. (b) Calculate the bandwidth BW' of the MOSFET stage with shunt-feedback TIA load shown in Fig. 6.54(b), assuming the same characteristics for the MOSFET, the same total capacitance C_L at the drain of the MOSFET, the same overall stage gain as before, a single-pole feedback amplifier with gain A_F and bandwidth BW_F, and a Butterworth response for the overall stage. Hint: use the result in Eq. (5.19) for the shunt-feedback TIA load. (c) What is the relationship between BW and BW'? Assume that the gain-bandwidth product of the simple common-source stage and the feedback amplifier in the TIA load are the same.

6.10 **Stage Splitting.** (a) Calculate the bandwidth BW' of the MOSFET amplifier with two identical stages shown in Fig. 6.54(c), assuming the same characteristics for the MOSFET, the same load capacitance C_L, the same overall stage gain as in Problem 6.9(a). (b) What is the bandwidth extension, BW'/BW, as a result of splitting the simple common-source stage of Problem 6.9(a) in two?

6.11 **Active-Feedback TIA Load.** Show that the bandwidth extension given in Eq. (6.55) also holds for a gain stage with active-feedback TIA load. Hint: use the results from Problem 5.13 for the active-feedback TIA load.

6.12 **Negative Capacitance.** (a) Calculate the differential admittance $Y(s) = 1/2 \cdot (I_p - I_n)/(V_p - V_n)$ of the MOSFET NIC with capacitor C shown in Fig. 6.55(a) (cf. Appendix B.2 for the definition of the differential admittance). Use a simplified FET model with $C_{sb} = C_{db} = C_{gd} = g_{mb} = g_o = R_g = 0$ and assume ideal current sources. (b) What is the maximum frequency for which $Y(s)$ represents a negative capacitance and how large is this capacitance?

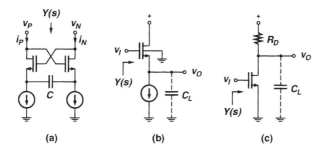

Fig. 6.55 Circuits for Problems 6.12, 6.13, and 6.14: (a) negative capacitance, (b) source follower, and (c) common-source buffer.

6.13 Source Follower and Capacitance-Transformation Ratio. (a) Calculate the voltage transfer function $A(s)$ for the MOSFET source follower shown in Fig. 6.55(b). Use a simplified FET model with $R_g = 0$; assume C_{sb} is included in C_L. The current source is ideal. Derive expressions for the poles and zeros of $A(s)$. What is the bandwidth, BW_B, of this buffer? (b) Calculate the input admittance $Y(s)$ for the above circuit assuming the same FET and current-source model. Derive expressions for the poles and zeros of $Y(s)$. What is the low-frequency input capacitance, C_I, of this buffer? (c) Calculate the capacitance-transformation ratio $\kappa = (C_L - C_{L0})/(C_I + C_{I1})$ of this buffer, where C_{L0} is the self-loading capacitance of the buffer, which includes C_{sb}, and C_{I1} is the wiring capacitance to the buffer input. Express κ as a function of f_T/BW_B, where $f_T = g_m/[2\pi(C_{gs} + C_{gd})]$. Evaluate this function for $A(0) = 0.8$, $C_{gd} = 0.3 C_{gs}$, $C_{L0} = 1.7 C_{gs}$, and $C_{I1} = 0.2 C_{gs}$.

6.14 Common-Source Buffer and Capacitance-Transformation Ratio. (a) Calculate the voltage transfer function $A(s)$ for the MOSFET common-source buffer shown in Fig. 6.55(c). Use a simplified FET model with $R_g = 0$; assume C_{db} is included in C_L. The resistor is ideal. Derive expressions for the poles and zeros of $A(s)$. What is the bandwidth, BW_B, of this buffer? (b) Calculate the input admittance $Y(s)$ for the above circuit assuming the same FET and resistor model. Derive expressions for the poles and zeros of $Y(s)$. What is the low-frequency input capacitance, C_I, of this buffer? (c) Calculate the capacitance-transformation ratio $\kappa = (C_L - C_{L0})/(C_I + C_{I1})$ of this buffer, where C_{L0} is the self-loading capacitance of the buffer, which includes C_{db}, and C_{I1} is the wiring capacitance to the buffer input. Express κ as a function of f_T/BW_B, where $f_T = g_m/[2\pi(C_{gs} + C_{gd})]$. Evaluate this function for $A(0) = -0.8$, $C_{gd} = 0.3 C_{gs}$, $C_{L0} = 0.6 C_{gs}$, and $C_{I1} = 0.2 C_{gs}$.

6.15 Amplifier with Scaled Stages. A scalable first-order stage with scale parameter ξ has the input capacitance $\xi \cdot C_{I0}$, output capacitance $\xi \cdot C_{O0}$, transconductance $\xi \cdot g_{m0}$, and load resistance R_0/ξ. Its power dissipation is $\xi \cdot P_0$, and its input-referred mean-square noise voltage is v_{n0}^2/ξ; note that its gain,

$A_S = g_{m0}R_0$, and unloaded bandwidth, $BW_S = 1/(2\pi R_0 C_{O0})$, are independent of the scale parameter ξ. A multistage amplifier with the input capacitance C_I driving the load capacitance $C_L < C_I$ is assembled from n such stages. (a) Assuming identically scaled stages, $\xi_i = \xi_1$ for $i = 1\ldots n$, what are the amplifier's bandwidth, power dissipation, and input-referred noise? (b) Assuming uniformly down-scaled stages, $\xi_{i+1} = \xi_i/\kappa$ for $i = 1\ldots n-1$, what are the optimum scale factor κ, the amplifier's bandwidth, power dissipation, and input-referred noise? (c) How do these two designs compare given the values $n = 4$, $C_I/C_L = 16$, $C_{I0}/C_{O0} = 1.5$, and $g_{m0}R_0 = 2.5$?

6.16 Shunt Peaking. (a) Calculate the voltage transfer function $A(s)$ for the MOSFET stage with shunt peaking shown in Fig. 6.19(b). Use a simplified FET model with $C_{gd} = g_o = R_g = 0$; assume C_{db} is included in the load capacitance C_L. Derive expressions for the poles and zeros of $A(s)$. (b) A numerical analysis shows that peaking occurs when the poles are conjugate complex with $Q > 0.644$. For what peaking inductor value is the frequency response maximally flat?

6.17 Active Inductor. (a) Calculate the impedance $Z(s)$ for the MOSFET active inductor shown in Fig. 6.22(b). Use a simplified FET model with $C_{sb} = R_g = 0$ (C_{sb} can be accounted for by C_L in a shunt-peaking application). Derive expressions for the dominant pole and the zero of $Z(s)$; assume that the poles are spaced far apart. (b) In which frequency range is $Z(s)$ inductive? (c) What is the condition for the impedance to be inductive at any frequency? (d) What is the inductance value assuming the poles are far away from the zero?

6.18 Bode Theorem. Bode's network theorem about physically realizable passive impedances in the finite frequency range 0 to BW can be written as [16]

$$\int_0^1 \frac{\ln|Z(f)|}{\sqrt{1-(f/BW)^2}} \, d(f/BW) \leq \frac{\pi}{2} \cdot \ln \frac{2}{2\pi \cdot BW \cdot C_\infty}.$$

Simplify this inequality for the case that $|Z(f)| = Z_0$ for frequencies in the range 0 to BW.

6.19 Offset Compensation. (a) Calculate the transfer function $A_{\text{tot}}(s) = V_o/V_s$ of the MA with the DDA-type offset compensation circuit shown in Fig. 6.29. The MA inputs, v_{IP} and v_{IN}, are driven from the voltage source v_S through two series resistors $R_0 = 50\,\Omega$; the MA core and the error amplifier have the frequency independent gains A and A_1, respectively; and the offset compensation input of the DDA has the reduced gain ξA. Derive expressions for the poles and zeros of $A_{\text{tot}}(s)$. (b) What is the LF-cutoff frequency for $C \to \infty$ and $C_1 \to \infty$?

7
Optical Transmitters

Having completed our discussion of the optical receiver, we now turn to the transmitter. We start with a brief discussion of the transmitter specifications. Then, we focus on the devices used for the electrical-to-optical conversion, namely, the laser and the modulator. Their characteristics are important for the driver design as well as the transmission system design.

Types of Modulation. Figure 7.1 illustrates two alternative ways to generate a modulated optical signal. In Fig. 7.1(a), the laser is turned on and off by modulating its current; this method is known as *direct modulation*. In Fig. 7.1(b), the laser is on at all times, a so-called continuous wave (CW) laser, and the light beam is modulated with a kind of optoelectronic shutter, a so-called modulator; this method is known as *external modulation*. Direct modulation has the advantages of simplicity, compactness, and cost effectiveness, whereas external modulation can produce higher-quality optical pulses, permitting extended reach and higher bit rates.

Direct as well as external modulation can be used to produce non-return-to-zero (NRZ) or return-to-zero (RZ) modulated optical signals. However, to produce very high-speed RZ-modulated signals, a cascade of two optical modulators, known as a *tandem modulator*, frequently is used. In this arrangement, the first optical modulator modulates the light from the CW laser with an NRZ signal and the second modulator, which is driven by a clock signal, converts the NRZ signal to an RZ signal in the optical domain.

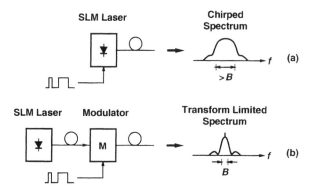

Fig. 7.1 Optical transmitters: (a) direct modulation vs. (b) external modulation.

7.1 TRANSMITTER SPECIFICATIONS

In the following, we look at two important specifications of the optical transmitter: (i) the spectral linewidth and (ii) the extinction ratio. The values that can be achieved for these parameters depends on whether direct or external modulation is used.

Spectral Linewidth. For a perfectly monochromatic light source followed by a perfect intensity modulator, the optical spectrum of the modulated output signal looks like that of an amplitude-modulated (AM) transmitter: a carrier and two sidebands corresponding to the spectrum of the baseband signal. In the case of an NRZ modulation, the spectrum looks as shown in Fig. 7.1(b). The 3-dB bandwidth of one NRZ sideband is about half the bit rate, $B/2$, and thus the full bandwidth, which covers both sidebands, is about equal to the bit rate.[1] If we convert this bandwidth or frequency linewidth to the commonly used wavelength linewidth, we get

$$\Delta\lambda = \frac{\lambda^2}{c}\Delta f \approx \frac{\lambda^2}{c}B, \qquad (7.1)$$

where λ is the wavelength and c is the speed of light in vacuum ($c \approx 3 \cdot 10^8$ m/s). For example, at 10 Gb/s, the frequency linewidth is about 10 GHz, corresponding to a wavelength linewidth of 0.08 nm for $\lambda = 1.55\,\mu$m. In practice, it is difficult to build a transmitter with a linewidth as narrow as this; only some types of external modulators can come close to this ideal. Optical pulses that do have this narrow spectrum are known as *transform limited pulses*.

For most transmitters, the modulation process not only changes the light's amplitude but also its phase or frequency. This unwanted frequency modulation (FM) is called *chirp* and causes the spectral linewidth to broaden as shown in Fig. 7.1(a). The directly modulated laser is a good example for a transmitter with a significant

---

[1] The full bandwidth measured null-to-null is $2B$.

amount of chirp. Mathematically, the effect of chirp on the transmitter linewidth can be approximated by

$$\Delta\lambda \approx \frac{\lambda^2}{c}\sqrt{\alpha^2 + 1} \cdot B, \qquad (7.2)$$

where α is known as the *chirp parameter* or *linewidth enhancement factor* [44, 73]. With the typical value $\alpha \approx 4$ for a directly modulated laser [44], the linewidth of a 10-Gb/s transmitter broadens to about 41 GHz or 0.33 nm. External modulators also exhibit a small amount of chirp, but virtually all types of modulators can provide $|\alpha| < 1$ and some achieve $|\alpha| < 0.1$, thus approaching the transform limited case [44].

So far, we assumed that the *unmodulated* source is perfectly monochromatic (zero linewidth), or at least that the unmodulated linewidth is much smaller than those given in Eqs. (7.1) and (7.2). Single-longitudinal mode (SLM) lasers, which we discuss in Section 7.2, can provide such a source. However, many sources have a much larger linewidth. For example, a Fabry-Perot laser has a typical unmodulated linewidth of about 3 nm, a light-emitting diode (LED) has an even wider linewidth in the range of 50 to 60 nm. For such wide-linewidth sources, the spectrum of the modulation signal and the chirp mostly are irrelevant, and the transmitter linewidth is simply given by

$$\Delta\lambda \approx \Delta\lambda_S, \qquad (7.3)$$

where $\Delta\lambda_S$ is the linewidth of the unmodulated source.

We know from Chapter 2 that optical pulses with a wide linewidth tend to spread out quickly in a dispersive medium such as a single-mode fiber (SMF) operated at a wavelength of 1.55 μm. The power penalty due to this pulse spreading is known as the *dispersion penalty*. Data sheets of lasers and modulators frequently specify this dispersion penalty for a given amount of fiber dispersion. For example, a 2.5-Gb/s, 1.55-μm laser may have a dispersion penalty of 1 dB given a fiber dispersion of 2,000 ps/nm. We know from Section 2.2 that a fiber dispersion of 2,000 ps/nm corresponds to about 120 km of SMF. With Eqs. (2.6) and (2.8), we further can estimate that the linewidth of this laser transmitter must be around 0.1 nm.

We analyze the impact of direct and external modulation on the maximum bit rate and transmission distance in Section 7.4. But as a rough guide, we can say that telecommunication systems at 10 Gb/s and more generally use external modulation, 2.5-Gb/s systems use direct or external modulation depending on the fiber length, and systems less than 2.5 Gb/s generally use direct modulation. Short-reach data communication links operating at the 1.3-μm wavelength, where dispersion in an SMF is small, use direct modulation even at 10 Gb/s.

The narrow linewidth obtained with external modulation not only reduces the dispersion penalty, but also permits a closer channel spacing in a dense wavelength division multiplexing (DWDM) system. To avoid crosstalk, the channels must be spaced further apart than the linewidth of each channel. Current DWDM systems have a channel spacing of 200 or 100 GHz with a trend toward 50 GHz.

Extinction Ratio. Optical transmitters, no matter if directly or externally modulated, do not shut off *completely* when a zero is transmitted. This undesired effect is quantified by the *extinction ratio* (ER), which is defined as follows[2]:

$$ER = \frac{P_1}{P_0}, \qquad (7.4)$$

where P_0 is the optical power emitted for a zero and P_1 the power for a one. Thus, an ideal transmitter would have an infinite ER. The ER usually is expressed in dBs using the conversion rule $10 \log ER$. Typically, ERs for directly modulated lasers range from 9 to 14 dB, whereas ERs for externally modulated lasers can exceed 15 dB [29]. SONET/SDH transmitters typically are required to have an ER in the range of 8.2 to 10 dB, depending on the application.

It doesn't come as a surprise that a finite ER causes a power penalty. Figure 7.2(a) and (b) illustrates how decreasing the ER reduces the optical signal swing, $P_1 - P_0$, even if the average power $\overline{P} = (P_1 + P_0)/2$ is kept constant. To restore the original signal swing, we have to increase the average transmitted power, as shown in Fig. 7.2(c). The power penalty PP due to a finite extinction ratio can easily be derived as [5]

$$PP = \frac{ER+1}{ER-1}. \qquad (7.5)$$

For example, an extinction ratio of 10 dB ($ER = 10$) causes a power penalty of 0.87 dB ($PP = 1.22$).

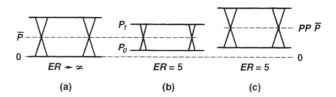

Fig. 7.2 Eye diagram (a) with infinite ER, (b) with $ER = 5$, and (c) with $ER = 5$ and increased average power (1.5×) to restore the original signal swing.

In deriving Eq. (7.5), it was assumed that the ER does not affect the amount of noise at the receiver. This is the case for unamplified p-i-n receivers; however, in systems with an avalanche photodetector (APD) or optical amplifiers, the noise increases when the ER is reduced: (i) the nonzero value of P_0 adds noise on the zeros and (ii) the increase in power to compensate for the finite ER adds noise on the zeros and ones, necessitating an even larger power increase to compensate for this noise. As a result

---

[2] In the literature, ER sometimes is defined as P_1/P_0 and sometimes as P_0/P_1. In this book, we follow [188] and use the former definition, which results in an ER that is larger than one.

of these two mechanisms, the power penalty becomes larger than given in Eq. (7.5). If we take the extreme case where the receiver noise is dominated by the detector (or optical amplifier) noise such that the electrical noise power is proportional to the received signal current, we find [29] that

$$PP = \frac{\sqrt{ER}+1}{\sqrt{ER}-1} \cdot \frac{ER+1}{ER-1}. \tag{7.6}$$

For example, an extinction ratio of 10 dB ($ER = 10$) causes a power penalty of up to 3.72 dB ($PP = 2.35$) in an amplified lightwave systems. [$\rightarrow$ Problem 7.1]

In regulatory standards, the receiver sensitivity usually is specified for the worst-case extinction ratio. Therefore, the corresponding power penalty must be deducted from the sensitivity based on $ER \rightarrow \infty$, as given for example by Eq. (4.21). Typically, 2.2 dB (for $ER = 6$ dB) must be deducted in short-haul applications, and 0.87 dB (for $ER = 10$ dB) must be deducted in long-haul applications [37].

Instead of specifying the *average* transmitter power, the *optical modulation amplitude* (OMA), which is defined as $P_1 - P_0$ and measured in dBm, can be used to measure the transmitter power. The OMA measure is used, for example, in 10-GbE systems. If the transmitter power and the receiver sensitivity are specified in terms of OMA rather than average power, a finite extinction ratio causes much less of a power penalty (no power penalty in the case of constant noise).

7.2 LASERS

In telecommunication systems, the *Fabry-Perot* (FP) laser and the *distributed-feedback* (DFB) laser are the most commonly used lasers. In data communication systems, such as Gigabit Ethernet, the FP laser and the *vertical-cavity surface-emitting laser* (VCSEL) are preferred because of their lower cost. In low-speed data communication (up to about 200 Mb/s) and consumer electronics, *light-emitting diodes* (LEDs) also find application as an optical source. The FP laser, DFB laser, and VCSEL are so-called *semiconductor lasers* or *laser diodes*, whereas the LED is not a laser. Figures 7.3 and 7.4 show photos of a cooled and uncooled DFB laser, respectively.

In the following, we give a brief description of the FP laser, DFB laser, VCSEL, and LED. Then, we summarize the main characteristics of these light sources. More information on lasers and their properties can be found in [5, 73, 183, 184].

Fabry-Perot Laser. The FP laser consists of an optical gain medium located in a cavity formed by two reflecting facets, as shown in Fig. 7.5(a). In a semiconductor laser, the gain medium is formed by a forward biased p-n junction, which injects carriers (electrons and holes) into a thin active region. These carriers "pump" the active region such that an incoming photon can stimulate the recombination of an electron-hole pair to produce a second identical photon. Thus, *stimulated emission* provides optical gain if the bandgap energy, that is, the energy released by the electron-

238 OPTICAL TRANSMITTERS

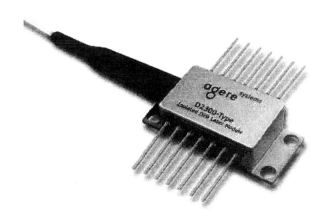

Fig. 7.3 Cooled 2.5-Gb/s DFB laser in a 14-pin butterfly package with single-mode fiber pigtail (2.1 cm × 1.3 cm × 0.9 cm). The pins provide access to the laser diode, monitor photodiode, thermoelectric cooler, and temperature sensor. Reprinted by permission from Agere Systems, Inc.

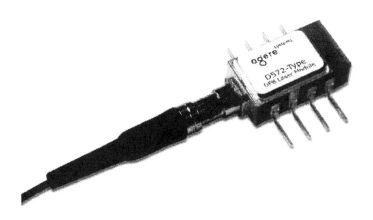

Fig. 7.4 Uncooled 2.5-Gb/s DFB laser in an 8-pin package with single-mode fiber pigtail (1.3 cm × 0.7 cm × 0.5 cm). Reprinted by permission from Agere Systems, Inc.

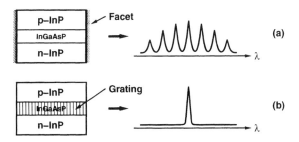

Fig. 7.5 Edge-emitting lasers (schematically): (a) Fabry-Perot laser vs. (b) distributed-feedback laser. The light propagates in the direction of the arrow.

hole pairs, matches the energy of the photons to be amplified.[3] In Fig. 7.5, the active region is a layer of InGaAsP, which can be lattice matched to the surrounding InP layers and the substrate. Interestingly, the bandgap of the $In_xGa_{1-x}As_yP_{1-y}$ compound can be controlled by the mixing ratios x and y to provide optical gain anywhere in the 1.0- to 1.6-μm range. Thus, 1.3-μm as well as 1.5-μm lasers can be based on an InGaAsP active layer. The surrounding p- and n-doped regions are made from InP, which has a wider bandgap than the active InGaAsP material, helping to confine the carriers to the active region. Short-wavelength lasers, operating at the 0.85-μm wavelength, typically use GaAs for the active layer and the wider bandgap AlGaAs material for the p- and n-doped regions (AlGaAs is lattice matched to the GaAs layer and substrate). In practical lasers, the active region usually is structured as a *multiple quantum well* (MQW), resulting in better performance than the simple structure shown in Fig. 7.5. [→ Problem 7.2]

The separation of the two facets, the cavity length, determines the wavelengths at which the laser can operate. If the cavity contains a whole number of wavelengths and the net optical gain is larger than one, lasing occurs. Because the facets in an FP laser are many wavelengths apart (about 300 μm), there are multiple modes satisfying these conditions. FP lasers therefore belong to the class of *multiple-longitudinal mode* (MLM) lasers. As a result, the spectrum of the emitted laser light has multiple peaks, as shown on the right-hand side of Fig. 7.5(a). The corresponding spectral linewidth is quite large, typically around $\Delta\lambda_S = 3$ nm. FP lasers therefore are used primarily at the 1.3-μm wavelength where dispersion in an SMF is low.

Most FP lasers are operated as *uncooled lasers*, which means that their temperature is not controlled and can go up to around 85°C. This mode of operation simplifies the transmitter design and keeps its cost low, but has the drawbacks of varying laser characteristics and reduced laser reliability.

---

[3]Laser is an acronym for "Light Amplification by Stimulated Emission of Radiation." The gain medium also can be used without the facets and then is known as a *semiconductor optical amplifier* (SOA), rather than a laser.

Distributed-Feedback Laser. The DFB laser consists of a gain medium, similar to that in an FP laser, with a built-in grating, which acts as a reflector, as shown in Fig. 7.5(b). The grating can, for example, be implemented by etching a corrugation near the active layer. In contrast to the facets of the FP laser, the grating provides distributed feedback and selects only *one* wavelength for amplification as shown on the right-hand side of Fig. 7.5(b). For this reason, DFB lasers belong to the class of *single-longitudinal mode* (SLM) lasers. The emitted spectrum of an unmodulated DFB laser has a very narrow linewidth, typically $\Delta \lambda_S < 0.001$ nm (<100 MHz). When the laser is directly modulated, the linewidth broadens because of the AM sidebands and chirp, as discussed in Section 7.1.

The *distributed Bragg reflector* (DBR) laser is similar to the DFB laser in the sense that it also operates in a single longitudinal mode, producing a narrow linewidth. In terms of structure, it looks more like an FP laser, however, with the facets replaced by wavelength-selective Bragg mirrors (gratings).

DFB/DBR lasers are suitable for direct modulation as well as for CW sources followed by an external modulator. Because of their narrow linewidth, they are ideal for WDM and DWDM systems. Unfortunately, the wavelength emitted by a semiconductor laser is slightly temperature dependent, for example, a variation of 0.1 nm/$°$C is typical for a DFB laser. Given a DWDM system with a 0.8-nm wavelength spacing (100-GHz grid), the laser temperature must be controlled precisely. Therefore, many DFB/DBR lasers are operated as *cooled lasers*. Such lasers are mounted on top of a *thermoelectric cooler* (TEC), which is controlled by a feedback loop and a thermistor to stabilize the temperature.

Vertical-Cavity Surface-Emitting Laser. The VCSEL emits the light perpendicular to the wafer surface rather than at the edges of the chip (parallel to the wafer surface), as the FP or DFB/DBR lasers do. The VCSEL consists of a gain medium located in a very short vertical cavity (about 1 μm) with Bragg mirrors at the bottom and the top as shown in Fig. 7.6(a). The Bragg mirrors are formed by many layers of alternating high and low refractive-index material. Because of the short cavity length, the longitudinal modes are spaced far apart and just one of them has a net optical gain larger than one, thus the VCSEL also belongs to the class of SLM laser. However, depending on the horizontal size, VCSELs have multiple transverse modes, causing a wider spectral linewidth than the DFB/DBR lasers, typically $\Delta \lambda_S \approx 1$ nm. Also, VCSELs typically are less powerful than DFB/DBR lasers.

The advantage of VCSELs over edge-emitting lasers is that they can be fabricated, tested, and packaged more easily and at a lower cost. However, the very short length of the gain medium requires mirrors with a very high reflectivity to make the net gain larger than one. Currently, VCSELs are commercially available at short wavelengths (0.85-μm band) where fiber loss is appreciably high. Long-wavelength VCSELs are under development. The application of short-wavelength VCSELs mostly is in data communication systems using multimode fiber (MMF).

Light-Emitting Diode. The LED operates on the principle of *spontaneous emission* rather than stimulated emission, and therefore is not a laser. The LED consists of a

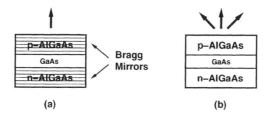

Fig. 7.6 Surface-emitting sources (schematically): (a) vertical-cavity surface-emitting laser vs. (b) light-emitting diode. The light propagates in the direction of the arrow(s).

forward biased p-n junction without any mirrors or gratings, as shown in Fig. 7.6(b). The electron-hole pairs injected into the active region recombine spontaneously, emitting photons with an energy that corresponds to the bandgap energy. As a result, the light is emitted in all directions and it is difficult to couple much of it into a fiber. For example, an LED may couple $10\,\mu\text{W}$ optical power into a fiber, whereas a laser easily can produce 1 mW. Furthermore, because there is no mechanism to select a single wavelength, the spectral linewidth is very wide, typically $\Delta\lambda_S = 50$ to 60 nm. The modulation speed of an LED is limited by the carrier lifetime to a few hundred Mb/s, whereas a fast laser can be modulated in excess of 10 Gb/s.

On the plus side, LEDs are very low in cost, they are more reliable than lasers, and they are easier to drive because they lack the temperature-dependent threshold current typical for lasers. Their application mostly is in short-reach data communication systems using MMF.

I/V Characteristics. From an electrical point of view, the semiconductor laser is just a forward-biased diode. The relationship between the laser current, I_L, and the forward-voltage drop, V_L, is described by the so-called *I/V curve*. Figure 7.7(a) shows an I/V curve that is typical for an edge-emitting InGaAsP laser. For such a laser, the small-signal resistance is normally in the range of 3 to 8 Ω, whereas the forward-voltage drop varies from 0.7 to 2.2 V, depending on the current, temperature, age, and bandgap of the semiconductor materials used. The I/V curve of a VCSEL exhibits a larger small-signal resistance, often in the neighborhood of 50 Ω.

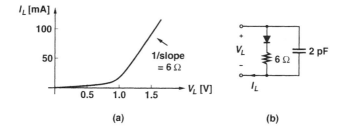

Fig. 7.7 (a) Typical I/V curve of an edge-emitting InGaAsP laser and (b) the corresponding large-signal AC model for a 10-Gb/s part.

A simple large-signal AC model for the semiconductor laser is shown in Fig. 7.7(b); the values shown are typical for a 10-Gb/s, edge-emitting InGaAsP laser. At low currents, the diode determines the forward-voltage drop and the small-signal resistance (V_T/I_L). At high currents, the contact resistance (6 Ω in Fig. 7.7(b)) adds a significant amount of voltage drop and dominates the overall small-signal resistance. It thus is the contact resistance that determines the slope of the I/V curve at normal operating currents. The capacitor (2 pF in Fig. 7.7(b)) models the junction and diffusion capacitance of the forward-biased p-n junction and typically is fairly large, especially when compared with the capacitance of a corresponding photodiode. Compared with edge-emitting lasers, the much smaller VCSELs tend to have a larger contact resistance and a smaller junction capacitance. To model a packaged laser, additional elements, such as the bond-wire inductance, must be added to the equivalent circuit in Fig. 7.7(b).

Some packaged lasers contain a series resistor (e.g., 19 Ω or 44 Ω) to match the 6-Ω laser resistance to that of the transmission line (e.g., 25 Ω or 50 Ω) that connects the laser to its driver. If such a series resistor is used, an RF choke (RFC) usually is included as well to bypass the matching resistor, that is, to apply the bias current directly to the laser. Packaged lasers usually designate one electrode as ground and the other one as RF input. Typically, edge-emitting lasers designate the anode as ground, whereas VCSELs designate the cathode as ground.

L/I Characteristics. The static relationship between the laser current, I_L, and the light output, P_{out}, is described by the so-called *L/I curve*. Figure 7.8 illustrates such an L/I curve schematically together with its dependence on temperature and age. Up to the so-called *threshold current*, I_{TH}, the laser outputs only a small amount of incoherent light. In this regime, the net optical gain isn't large enough to sustain lasing, and thus only spontaneous emission is produced, as in an LED. Above the threshold current, the output power grows approximately linearly with the laser current, and the quantity $\partial P_{out}/\partial I_L$ is known as the *slope efficiency*. Typical values for an edge-emitting InGaAsP MQW laser are $I_{TH} = 10$ mA and a slope efficiency of 0.07 mW/mA. Thus for $I_L = 25$ mA ($= 10$ mA $+ 15$ mA), we obtain about 1 mW of optical output power. VCSELs are characterized by a lower threshold current (<5 mA) and a better slope efficiency, and LEDs have zero threshold current and a much lower slope efficiency.

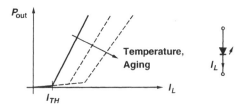

Fig. 7.8 L/I curves for a semiconductor laser.

The slope efficiency is determined by the *differential quantum efficiency* (DQE), which specifies how efficiently electrons are converted into photons. If every electron produces a photon and they all are coupled into the fiber, the slope efficiency is $hc/\lambda q$, in analogy to the responsivity of a photodetector with 100% quantum efficiency (cf. Eq. (3.2)). At the wavelength $\lambda = 1.55\,\mu\mathrm{m}$, this ideal slope efficiency is about 0.8 mW/mA. In practice, however, not all electrons produce photons, some photons are radiated out of the back facet, and not all photons from the front facet couple into the fiber, resulting in an about ten-fold lower slope efficiency. Note that as in the case of the photodetector, the *optical power* is linearly related to the *electrical current* through the device, which means that we carefully have to distinguish between optical and electrical dBs.

As indicated in Fig. 7.8, the laser characteristics, in particular I_{TH}, are strongly temperature and age dependent. For example, a laser that nominally requires 25 mA to output 1 mW of optical power may require in excess of 50 mA at 85°C and near the end of life to output the same power. The temperature dependence of the threshold current is exponential and can be described by [5]

$$I_{TH}(T) = I_{TH0} \cdot \exp\left(\frac{T}{T_0}\right), \qquad (7.7)$$

where I_{TH0} is the (extrapolated) threshold current at 0 K and T_0 is a constant in the range of 50 to 70 K for InGaAsP lasers. The temperature dependence of the laser's slope efficiency is less dramatic, a 30% to 40% reduction when heating the laser from 25°C to 85°C is typical.

Because of the strong temperature and age dependence of the laser's L/I curve, communication lasers usually have a built-in monitor photodiode to measure the optical output power at the back facet of the laser. The current from this photodiode closely tracks the optical power coupled into the fiber with little dependence on temperature and age; a tracking error of $\pm 10\%$ is typical. A feedback mechanism that compares the current from the monitor photodiode with a reference current and controls the laser current accordingly can be used to implement a so-called *automatic power control* (APC). More on this subject in Section 8.2.6.

For analog applications, such as CATV/HFC systems, the *linearity* of the L/I curve is another important laser parameter, besides the slope efficiency and the threshold current. A deviation from the linear characteristics causes harmonic and intermodulation distortions in the analog signal. In the case of a TV signal, these distortions degrade the picture quality. Typically, the laser linearity is specified by the composite second order (CSO) and composite triple beat (CTB) distortion parameters (cf. Section 4.8) measured at a certain laser current and modulation index.

Dynamic Behavior. The dynamic relationship between the laser current and the light output is quite complex. It is fully described by the so-called *laser rate equations*, two coupled, nonlinear, differential equations relating the carrier density and photon density in the laser cavity [5, 83]. It is beyond the scope of this book to discuss these equations. Instead, we just want to give a verbal description of what happens in response to a laser-current pulse ramping up from zero and, after one or more bit periods, returning back to zero, as shown in Fig. 7.9(a).

244 OPTICAL TRANSMITTERS

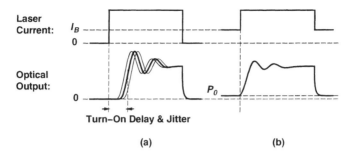

Fig. 7.9 Turn-on delay, jitter, and relaxation oscillations in a directly modulated laser: (a) without bias and (b) with bias.

At first, the carrier density in the gain medium is zero. After the current turn-on, the carrier density builds up until the critical point is reached where the net optical gain becomes unity. Only after this so-called *turn-on delay* (TOD) does the laser begin to produce coherent light. Actually, in addition to the unity gain, some random spontaneous emission aligned with the cavity is needed as well to "ignite" a sustained stimulated emission, and therefore a random *turn-on delay jitter* appears on the rising edge of the optical pulse. As soon as lasing sets in, stimulated emission causes many carriers to recombine, thus producing more photons and optical power. However, the high rate of recombination leads to a reduction in the carrier density and thus a lower optical gain. Because of this back-and-forth between elevated carrier and photon densities, a damped oscillation, called *relaxation oscillation*, can be observed. Finally, when the laser is turned off, the photon density decays rapidly, following an exponential law, because the remaining carriers recombine quickly as a result of stimulated emission.

It is possible to transform the laser rate equations into an equivalent electrical circuit in which the carrier and photon densities both are represented by voltages. Such a circuit then can be combined with the large-signal AC circuit in Fig. 7.7(b), permitting the simulation of the *optical* laser output signal with a circuit simulator such as SPICE [90, 189].

The dynamic behavior of an LED is much simpler than that of a laser because the carrier density only affects the photon density but *not* vice versa. Remember, the photons in an LED are produce by spontaneous recombination of carriers rather than stimulated, photon-induced recombination. As a result, the electrical-to-optical transfer function of an LED can be well approximated by a first-order low-pass function (no relaxation oscillations) and the LED's modulation response is much slower than that of a laser.

Turn-On Delay and Extinction Ratio. From this brief description of the laser dynamics, we can now understand that adverse effects such as turn-on delay, turn-on delay jitter, and overshoot can be ameliorated by always keeping the laser current above zero, even when the laser is turned off. Figure 7.9(b) shows a laser-current pulse where this minimum current, known as bias current, is set to I_B. If we bias the laser close to the threshold current, the carrier density is already built up and the net optical gain is close to unity, yet the photon density is still low. With just a little bit more current, the laser turns on quickly. A simple equation describes the TOD as follows [183]:

$$t_{TOD} = \tau_c \cdot \ln \frac{I_{L,\text{on}} - I_B}{I_{L,\text{on}} - I_{TH}}, \qquad (7.8)$$

where τ_c is the carrier lifetime, which typically is around 3 ns, I_B is the off-current or bias current, and $I_{L,\text{on}}$ is the on-current. For a zero bias current, $I_B = 0$, and the typical numbers $I_{L,\text{on}} = 25$ mA and $I_{TH} = 10$ mA, we obtain a TOD of about 1.5 ns. However, if the laser is biased at $I_B = I_{TH} = 10$ mA, the TOD goes to zero. Note that lasers with a low threshold current have a small turn-on delay, even if they are operated without bias current. In practical transmitters, the laser is almost always biased near I_{TH}, except in some low-speed applications (155 Mb/s and below) where it is possible to predistort the electrical signal to compensate for the TOD (cf. Section 8.3.3).

Biasing not only helps to improve the TOD but also reduces the turn-on delay jitter, the amplitude of the relaxation oscillations, and the optical chirp. However, there also is an important downside to biasing the laser near the threshold current: it lowers the extinction ratio. Even if the laser is not lasing at $I_B < I_{TH}$, it still produces a small amount of incoherent light, P_0, as indicated in Fig. 7.9(b). In high-speed transmitters, the laser often is biased a little bit above I_{TH} to improve its speed, however, at the expense of the ER. Typical ER values for high-speed semiconductor lasers are in the range of 9 to 14 dB [29].

Modulation Bandwidth. The laser's *modulation bandwidth* determines the maximum bit rate for which the laser can be used. This bandwidth is closely related to the frequency of the relaxation oscillations and thus depends on the interplay between the carrier and photon densities as described by the rate equations. The small-signal modulation bandwidth for $I_B > I_{TH}$ can be derived from the rate equations and is proportional to [5]

$$BW \sim \sqrt{I_B - I_{TH}}. \qquad (7.9)$$

Thus, the more the laser is biased above the threshold current, the faster it becomes. For this reason, the laser in high-speed transmitters is biased as much above the threshold current as possible while still meeting the ER specification.

In contrast, the LED's modulation bandwidth mostly is determined by the carrier lifetime, τ_c, and can be expressed as $BW = \sqrt{3}/(2\pi \tau_c)$ [5]. Thus, the LED's speed does not depend strongly on the bias current.

Chirp. In Section 7.1, we introduced the chirp parameter α, which describes the transmitter's spurious frequency modulation (FM). More precisely, the α parameter relates the optical frequency excursion, Δf, to the change in optical output power, P_{out}, as follows [73]:

$$\Delta f(t) \approx \frac{\alpha}{4\pi} \cdot \frac{d}{dt} \ln P_{\text{out}}(t). \qquad (7.10)$$

This relationship holds for direct as well as external modulation. With a positive value for α, as is the case for directly modulated lasers, the frequency shifts slightly toward the blue when the output power is rising (leading edge) and toward the red when the power is falling (trailing edge). Directly modulated lasers produce chirp ($\alpha \neq 0$) because the injected current not only modulates the optical gain and thus the output power, but also the refractive index of the gain medium and thus the optical frequency. [$\rightarrow$ Problem 7.3]

From Eq. (7.10) we can see that slowing the rise and fall times of the laser-current pulses, and thus lowering the rate of change of P_{out}, helps to reduce the amount of chirp [169]. We further can conclude that large relaxation oscillations, causing a large rate of change of P_{out}, exacerbate the chirping problem.

Noise. The stimulated emission of photons in the laser produces a coherent electromagnetic field. However, occasional spontaneous emissions add amplitude and phase noise to this coherent field. The results are a broadening of the (unmodulated) spectral linewidth and fluctuations in the intensity. The latter effect is known as *relative intensity noise* (RIN). A p-i-n photodetector receiving laser light that contains intensity noise produces a corresponding electrical RIN noise, $i_{n.RIN}$, in addition to the fundamental (quantum) shot noise.[4] For a given laser, the electrical RIN-noise power, $\overline{i_{n.RIN}^2}$, is approximately proportional to the received signal power, I_{PIN}^2, [83]:

$$\overline{i_{n.RIN}^2} = RIN \cdot I_{PIN}^2 \cdot BW_n, \tag{7.11}$$

where *RIN* is a parameter characterizing the laser RIN noise measured in dB/Hz. This means that the CW signal-to-noise ratio (SNR) at the receiver due to RIN noise is fixed at $I_{PIN}^2/\overline{i_{n.RIN}^2} = 1/(RIN \cdot BW_n)$ and cannot be improved by increasing the laser power. Note that this situation is very different from that of the detector noise, $\overline{i_{n.PD}^2}$, which was proportional to I_{PD} as opposed to I_{PD}^2. For example, given a laser with $RIN = -135$ dB/Hz, the SNR due to RIN noise is 35 dB in a 10-GHz bandwidth, regardless of power. For the reception of a digital NRZ signal, this is more than enough and RIN noise often can be neglected in the analysis of digital transmission systems. However, in *analog* transmission systems, RIN noise is critical. For example, a CATV/HFC system using AM-VSB modulation typically is limited by RIN noise (cf. Section 5.2.12).

The effect of RIN noise on a digital transmission system can be quantified, as usual, with a power penalty. The RIN noise adds to the noise at the receiver, which means that we need to transmit more power to achieve the same bit-error rate (BER) as without RIN noise. The power penalty due to RIN noise is [5]

$$PP = \frac{1}{1 - Q^2 \cdot RIN \cdot BW_n}. \tag{7.12}$$

---

[4]Some authors include the shot noise as part of the RIN noise (e.g. [6]), but here we keep them separate.

With the example values $1/(RIN \cdot BW_n) = 35\,\text{dB}$ and $BER = 10^{-12}$ ($Q = 7.035$), the power penalty is only $0.068\,\text{dB}$ ($PP = 1.016$). However, if the SNR due to RIN noise approaches $Q^2 = 49.5$ ($16.9\,\text{dB}$), the power penalty becomes infinite. This can be explained as follows: we know from Eq. (4.13) that to receive an NRZ signal with much more noise on the ones than the zeros (as is the case for RIN noise), we need an SNR of at least $1/2 \cdot Q^2$. The SNR due to RIN noise of the *unmodulated* (CW) optical signal is $1/(RIN \cdot BW_n)$, as given by Eq. (7.11). NRZ modulation reduces the signal power by a factor four and the noise power by a factor two, so the SNR due to RIN noise of the *modulated* signal becomes $1/(2 \cdot RIN \cdot BW_n)$. Because the transmitted SNR must be higher than the required SNR at the receiver, we need $1/(RIN \cdot BW_n) > Q^2$, in accordance with Eq. (7.12). [→ Problem 7.4]

In MLM lasers, such as FP lasers, as well as SLM lasers with an insufficient *mode-suppression ratio* (<20 dB), another type of noise called *mode-partition noise* (MPN) also is of concern. This noise is caused by power fluctuations among the various modes (mode competition) and is harmless by itself because the *total* intensity remains constant. However, chromatic dispersion in the fiber can desynchronize the mode fluctuations and turn them into additional RIN noise. Furthermore, a weak optical reflection of the emitted light back into the laser can create additional modes, further enhancing the RIN noise. These RIN-noise enhancing effects may be strong enough to reduce the received SNR below the critical value of $1/2 \cdot Q^2$. This situation manifests itself in a bit-error rate floor, that is, a minimum BER that cannot be reduced, regardless of the transmitted power [5].

Reliability. The *mean-time to failure* (MTTF) of a typical semiconductor laser is between 1 and 10 years, if continuously operated at 70°C. When cooled down to room temperature, its lifetime extends by about a factor $10\times$, to between 10 and 100 years [174]. In comparison, LEDs are much more reliable and have an MTTF of 100 to 1,000 years when operated at 70°C.

These laser reliability numbers may not look so great, but they are much better than those of the early lasers. The first "continuous-wave" semiconductor laser that could operate at room temperature was demonstrated by researchers at the *Ioffe Physical Institute* in Leningrad in 1970. However, the lifetime of these GaAs lasers was measured in seconds; not very much of a continuous wave indeed! It took about seven years of tedious work until *AT&T Bell Laboratories* could announce a semiconductor laser with a 100-year (million hour) lifetime [43].

7.3 MODULATORS

Two types of optical *modulators* commonly are used in communication systems: the *electroabsorption modulator* (EAM) and the *Mach-Zehnder modulator* (MZM). The EAM is small and can be integrated with the laser on the same substrate, whereas the MZM is much larger but features superior chirp and ER characteristics. An EAM combined with a CW laser source is known as an *electroabsorption modulated laser*

(EML). See Figs. 7.10 and 7.11 for photos of a packaged EML module and an MZ modulator, respectively.

In the following, we give a brief description of the two modulators and summarize their main characteristics. More information on modulators and their properties can be found in [1, 3, 4, 44, 73].

Electroabsorption Modulator. An EML consist of a CW DFB laser followed by an EAM, as shown in Fig. 7.12. Both devices can be integrated monolithically on the same InP substrate, leading to a compact design and low coupling losses between the two devices. The EAM consists of an active semiconductor region sandwiched in between a p- and n-doped layer, forming a p-n junction. The EAM works on the principle known as *Franz-Keldysh effect*, according to which the effective bandgap of a semiconductor decreases with increasing electric field. Without bias voltage across the p-n junction, the bandgap of the active region is just wide enough to be transparent at the wavelength of the laser light. However, when a sufficiently large reverse bias is applied across the p-n junction, the effective bandgap is reduced to the point where the active region begins to absorb the laser light and thus becomes opaque. In practical EAMs, the active region usually is structured as an MQW, providing a stronger field-dependent absorption effect (known as the *quantum-confined Stark effect*) than the simple structure shown in Fig. 7.12.

The relationship between the optical output power, P_{out}, and the applied reverse voltage, V_M, of an EAM is described by the so-called *switching curve*. Figure 7.13(a) illustrates such a curve together with the achievable ER for a given switching voltage, V_{SW}. The voltage for switching the modulator from the on state to the off state, the *switching voltage* V_{SW}, typically is in the range of 1.5 to 4 V, and the dynamic ER usually is in the range of 11 to 13 dB [29, 73]. Because the electric field in the active region not only modulates the absorption characteristics, but also the refractive index, the EAM produces some chirp. However, this chirp usually is much less than that of a directly modulated laser with the chirp parameter, $|\alpha|$, typically being smaller than one. A small on-state (bias) voltage around 0 to 1 V often is applied to minimize the modulator chirp [3, 73].

From an electrical point of view, the EAM is just a reverse-biased diode. Thus, when the CW laser is off, the EAM impedance mostly is capacitive. For a 10-Gb/s modulator, the equivalent capacitor is about 0.1 to 0.15 pF, as shown in Fig. 7.13(b). When the CW laser is on, however, the photons absorbed in the EAM generate a photocurrent, pretty much like in a p-i-n photodetector. This current is a function of how much light is absorbed in the EAM, which in turn is a function of the modulation voltage, V_M. Thus, the photocurrent can be described by an equivalent voltage-controlled current source, $I(V_M)$, as shown in Fig. 7.13(b). As a result of this current source, the capacitive load appears shunted by a *nonlinear* resistance, which has a high value when the EAM is completely turned on or off, but assumes a low value during the transition [76, 137].

Packaged EAMs often contain a 50-Ω parallel resistor to match the modulator impedance to that of the transmission line that connects the EAM to its driver (dashed resistor in Fig. 7.13(b)). Nevertheless, it is difficult to achieve good matching over

MODULATORS   249

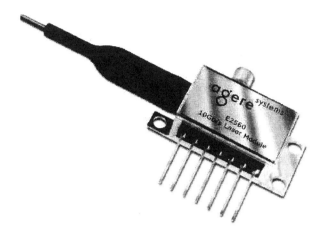

Fig. 7.10 A 10-Gb/s EML module containing a DFB laser and an electroabsorption modulator, with a GPO connector for the RF signal, 7-pin connector for power supply and control, and a single-mode pigtail (2.6 cm × 1.4 cm × 0.9 cm). Reprinted by permission from Agere Systems, Inc.

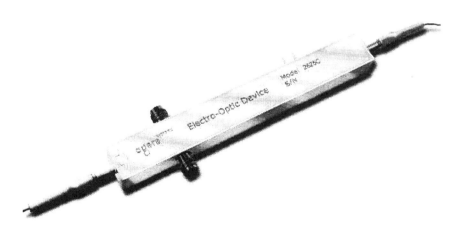

Fig. 7.11 A 40-Gb/s Lithium-Niobate Mach-Zehnder modulator with dual-drive inputs (V-type connectors), DC bias electrodes (pins), and two polarization-maintaining fiber (PMF) pigtails (12 cm × 1.5 cm × 1.0 cm). Reprinted by permission from Agere Systems, Inc.

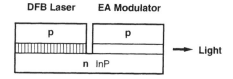

Fig. 7.12 Integrated laser and electroabsorption modulator (schematically).

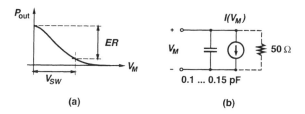

Fig. 7.13 (a) Switching curve and (b) electrical equivalent circuit of an electroabsorption modulator.

all voltage and frequency conditions. At low frequencies, the nonlinear resistance degrades the matching, whereas at high frequencies, the shunt capacitance does the same thing. The accuracy of the matching usually is specified by the S_{11} parameter (cf. Appendix C).

Mach-Zehnder Modulator. Figure 7.14 (left) shows the top view of an MZ modulator. The incoming optical signal is split equally and is sent down two different optical paths. After a few centimeters, the two paths recombine, causing the optical waves to interfere with each other. Such an arrangement is known as an *interferometer*. If the phase shift between the two waves is 0°, then the interference is constructive and the light intensity at the output is high (on state); if the phase shift is 180°, then the interference is destructive and the light intensity is zero (off state). The phase shift, and thus the output intensity, is controlled by changing the delay through one or both of the optical paths by means of the *electrooptic effect*. This effect occurs in some materials such as lithium niobate ($LiNbO_3$), some semiconductors, as well as some polymers and causes the refractive index to change in the presence of an electric field. Figure 7.14 (right) shows a cut view through an MZ modulator based on lithium niobate. Two RF waveguides in the form of coplanar transmission lines produce electrical fields, which penetrate into the two optical waveguides below. The latter are made from titanium (Ti) diffused into the lithium niobate substrate. Such a modulator also is known as a *lithium-niobate modulator*. [→ Problem 7.5]

The modulator shown in Fig. 7.14 is a so-called *dual-drive MZM*, because both light paths are controlled by two *separate* RF waveguides. Dual-drive MZMs, which have two input ports (also known as *arms*), can be driven in a push-pull fashion, which results in essentially zero optical chirp (more on this later). Another type of MZM controls both light paths with a *single* RF waveguide and thus is known as a

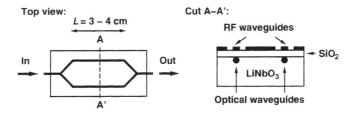

Fig. 7.14 Dual-drive Mach-Zehnder modulator based on LiNbO₃ (schematically).

single-drive MZM. The latter modulator requires only a single input signal, but in general is not chirp free.

Figure 7.15(a) shows the switching curve of an MZM, that is, the optical output power, P_out, as a function of the input voltage, V_M, applied to the RF waveguide. (In the case of a dual-drive MZM, V_M represents the differential voltage between the two input ports.) Note that in contrast to the switching curve of the EAM, this curve is periodic rather than monotonic. The switching curve of an ideal MZM can be described by

$$P_\text{out} = P_\text{in}\left[1 + \cos\left(\pi \cdot \frac{V_M}{V_\pi}\right)\right], \qquad (7.13)$$

where V_π is the switching voltage. For $V_M = 0, 2V_\pi, 4V_\pi$, and so forth, the output power is maximum (constructive interference) and for $V_M = V_\pi, 3V_\pi, 5V_\pi$, and so forth, the output power is zero (destructive interference). Unfortunately, in real MZMs, delay mismatch between the two optical paths causes the switching curve to be shifted horizontally from its ideal position given by Eq. (7.13). Even worse, the horizontal shift is temperature and age dependent, causing the switching curve to *drift*. For this reason, MZMs typically require a bias controller that produces a bias voltage to compensate for this drift. The bias voltage can be added to the RF signal with a bias T; alternatively, some MZMs provide separate electrodes for the bias voltage besides the RF signal input [1].

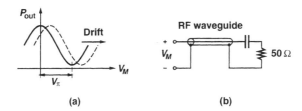

Fig. 7.15 (a) Switching curve and (b) electrical equivalent circuit of a Mach-Zehnder modulator.

The switching voltage, V_π, typically is in the range of 4 to 6 V. A dual-drive MZM can be switched by applying $+V_\pi/2$ and $-V_\pi/2$ to its two input ports, and thus requires a lower voltage swing per port than a single-drive MZM. The switching voltage of an MZM is inversely proportional to the length of the optical paths, L, that is, the product $V_\pi L$ is a constant. For example, given the typical value of 14 Vcm for this product [44], a modulator that is 14 cm long could be switched with a voltage of just 1 V. But such a long modulator would be very slow because the speed of an MZM depends inversely on its length L and the speed mismatch between the electrical and optical waves. For this reason, high-speed modulators are short and require a high switching voltage, making the driver design a formidable challenge.

From an electrical point of view, the MZM is just a transmission line (or, in the case of the dual-drive MZM, two transmission lines). The transmission line impedance typically is around 50 Ω, permitting the use of standard connectors and cables. The back-end of the transmission line usually is AC terminated within the package, as shown in Fig. 7.15(b).

The dynamic ER of a MZM is in the range of 15 to 17 dB and the chirp parameter can be made as low as $|\alpha| < 0.1$ [44]. In fact, with a dual-drive MZM, the chirp parameter can be controlled with the driving voltages. Assuming that the modulator is biased at the midpoint of the switching curve and that the two inputs ports are driven by synchronized signals having the same waveform shape, the chirp parameter turns out to be [4, 44]

$$\alpha = \frac{v^{pp}_{M1} + v^{pp}_{M2}}{v^{pp}_{M1} - v^{pp}_{M2}}, \qquad (7.14)$$

where v^{pp}_{M1} and v^{pp}_{M2} are the (signed) voltage swings at the first and second input port, respectively. We can see that if the two input ports are driven in a push-pull fashion, $v^{pp}_{M1} = -v^{pp}_{M2}$, the chirp theoretically becomes zero. Furthermore, by choosing $v^{pp}_{M1} < -v^{pp}_{M2}$, the chirp parameter of a dual-drive MZM can be made negative. A small negative chirp, such as $\alpha = -0.5$, causes optical pulses in a dispersive medium with $D > 0$ initially to *compress*, rather than expand,[5] leading to an extended reach. Finally, if both input ports are driven by the same signal, $v^{pp}_{M1} = v^{pp}_{M2}$, the optical output signal is purely phase modulated with the intensity remaining constant ($\alpha \to \infty$).

The capability of a dual-drive MZM to modulate the intensity as well as the phase makes it a very versatile device. One interesting use is the generation of a so-called *optical duobinary* signal. This signal has three levels: (i) light on with no phase shift, (ii) light off, and (iii) light on with 180° phase shift. This duobinary signal has the welcome property that its spectral linewidth is half that of a two-level NRZ signal, thus lowering the dispersion penalty. Yet, with proper precoding at the transmitter side, it can be received with a standard two-level intensity detector [206]. Other advanced modulation formats such as *chirped return-to-zero* (CRZ), *carrier-suppressed return-to-zero* (CS-RZ), and *return-to-zero differential phase-shift keying* (RZ-DPSK) also can be generated with MZ modulators [64]. [$\to$ Problem 7.6]

The drawbacks of MZMs, besides the drift problem, the high switching voltage, and the bulky dimensions (see Fig. 7.11), are their rather high optical insertion loss (4–7 dB) and their sensitivity to the polarization of the optical signal.

---

[5]This effect is not described by our approximate Eqs. (2.6) and (7.2) and a more sophisticated theory is needed (e.g., see [5]).

7.4 LIMITS IN OPTICAL COMMUNICATION SYSTEMS

The following material is not required for the subsequent chapter on laser and modulator driver design, but it is helpful for the understanding of optical transmission systems. By combining the transmitter linewidth equations of this chapter with the fiber properties discussed in Chapter 2 and the receiver sensitivity data of Chapter 5, we can now quantify all major limits in an optical communication system: (i) the limits due to chromatic dispersion, (ii) the limits due to polarization-mode dispersion, and (iii) the limits due to fiber attenuation.

Chromatic Dispersion Limits. For a transmitter that produces clean transform-limited pulses, for example, a DFB laser followed by a dual-drive MZM, we can approximate the maximum, dispersion-limited transmission distance by combining the spectral linewidth of Eq. (7.1) with the chromatic pulse-spreading expression of Eq. (2.6) and the spreading limit of Eq. (2.8) [46]:

$$L \leq \frac{c}{2|D| \cdot \lambda^2 \cdot B^2}. \tag{7.15}$$

However, because Eq. (2.6) is not precisely valid for transmitters with a narrow-linewidth source, such as a DFB laser, the above expression is only an approximation. A more precise analysis reveals that the attainable distances are longer than those given by Eq. (7.15), but the dependences on D and B remain the same. A useful engineering rule states that to keep the dispersion penalty below 1 dB, we have to limit the distance to [73]

$$L \leq \frac{17 \, \text{ps}/(\text{nm} \cdot \text{km})}{|D|} \cdot \frac{6{,}000 \, (\text{Gb/s})^2}{B^2} \, \text{km}. \tag{7.16}$$

For example, given an externally modulated 1.55-μm source without chirp transmitting over an SMF, we find from this rule that a 2.5-Gb/s system is limited to a span length of about 960 km and a 10-Gb/s system to about 60 km.[6] Note that the maximum transmission distance diminishes rapidly with increasing bit rates, following a $1/B^2$ law. This is so because the linewidth of a transform-limited transmitter increases with B (Eq. (7.1)), producing proportionally more pulse spreading (Eq. (2.6)), while at the same time the permitted amount of spreading decreases with B (Eq. (2.8)). So we are faced with two problems when going to higher bit rates and hence the $1/B^2$ dependence.

For a transmitter with a narrow-linewidth source that produces a significant amount of chirp, such as a directly modulated DFB laser, the dispersion-limited transmission distance can be approximated by combining Eq. (7.2) with Eqs. (2.6) and (2.8).

---

[6] 2.5-Gb/s systems over more than 1,000 km and 10-Gb/s systems over more than 100 km can be realized if a dispersion penalty larger than 1 dB is tolerated, if a transmitter with prechirp optimization is used, or both.

Compared with Eq. (7.15), the maximum distance is reduced by $\sqrt{\alpha^2+1}$:

$$L \leq \frac{c}{\sqrt{\alpha^2+1} \cdot 2|D| \cdot \lambda^2 \cdot B^2}. \qquad (7.17)$$

Again, a more precise analysis finds a somewhat longer distance, but the dependences on α, D, and B remain the same. The following engineering rule is used often [73]:

$$L \leq \frac{1}{\sqrt{\alpha^2+1}} \cdot \frac{17\,\text{ps/(nm} \cdot \text{km})}{|D|} \cdot \frac{6{,}000\,(\text{Gb/s})^2}{B^2}\,\text{km}. \qquad (7.18)$$

For example, given a directly modulated 1.55-μm DFB laser with $\alpha = 4$ transmitting over an SMF, we find that a 2.5-Gb/s system is limited to a span length of about 230 km, and a 10-Gb/s system is limited to a span length of about 15 km.

For a transmitter with a wide-linewidth source, such as an FP laser or an LED, the dispersion-limited transmission distance can be found by combining Eq. (7.3) with Eqs. (2.6) and (2.8) [46]:

$$L \leq \frac{1}{2|D| \cdot \Delta\lambda_S \cdot B}. \qquad (7.19)$$

For example, given a 1.55-μm FP laser with a 3-nm linewidth transmitting over an SMF, we find that a 2.5-Gb/s system is limited to a span length of about 4 km and a 10-Gb/s system to about 1 km. Note that in Eq. (7.19), the transmission distance diminishes proportional to $1/B$ rather than to $1/B^2$. This is so because in this case, the transmitter linewidth is not affected by the bit rate.

The numerical results of the examples are summarized in Table 7.1. The bit-rate dependence for all three examples (FP = Fabry-Perot laser with $\Delta\lambda = 3$ nm, DFB = directly modulated DFB laser with $\alpha = 4$, and EM = chirp-free external modulator) is plotted in Fig. 7.16. Note that all examples are based on a dispersion parameter of $D = 17\,\text{ps/(nm} \cdot \text{km})$ typical for a standard SMF operated at 1.55 μm. However, it is possible to reduce this value, and thus lengthen the transmission distances, by one of the following methods: (i) concatenate the SMF with a dispersion compensating fiber (DCF) to lower the overall dispersion, (ii) use a dispersion-shifted fiber (DSF) or nonzero dispersion-shifted fiber (NZ-DSF) instead of the SMF, or (iii) operate the system at the 1.3-μm wavelength where dispersion in an SMF is lowest (but loss is higher).

Table 7.1 Maximum (unrepeatered) transmission distances over an SMF at 1.55 μm for various transmitter types based on Eqs. (7.19), (7.18), and (7.16) with $D = 17\,\text{ps/(nm} \cdot \text{km})$.

Transmitter Type	2.5 Gb/s	10 Gb/s
Fabry-Perot laser ($\Delta\lambda = 3$ nm)	4 km	1 km
Distributed feedback laser ($\alpha = 4$)	230 km	15 km
External modulator ($\alpha = 0$)	960 km	60 km

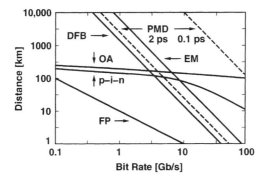

Fig. 7.16 Limits due to chromatic dispersion (FP, DFB, EM), polarization-mode dispersion (PMD, dashed), and attenuation (p-i-n, OA) in an SMF link operated at the 1.55-μm wavelength.

PMD Limit. By combining the equation for pulse spreading due to polarization-mode dispersion (PMD), Eq. (2.4), with the spreading limit $\overline{\Delta T} \leq 0.1/B$ (cf. Section 2.2), we find the PMD-limited transmission distance as

$$L \leq \frac{(0.1)^2}{D_{PMD}^2 \cdot B^2}. \tag{7.20}$$

For example, given the parameter $D_{PMD} = 0.1\,\text{ps}/\sqrt{\text{km}}$ typical for new PMD optimized fiber, we find that a 2.5-Gb/s system is limited to a span length of about 160,000 km and a 10-Gb/s system to about 10,000 km. These are huge spans capable of connecting two continents! However, older, already-deployed fiber with $D_{PMD} = 2\,\text{ps}/\sqrt{\text{km}}$ poses more serious limits: a 2.5-Gb/s system is limited to a span length of about 400 km and a 10-Gb/s system to about 25 km. Note that these numbers are smaller than the dispersion-limited transmission distances for transform-limited pulses.

The bit-rate dependence of the PMD limit, for both values of the PMD parameter, is plotted in Fig. 7.16 with dashed lines.

Attenuation Limit. Given a transmitter launching the power $\overline{P}_{\text{out}}$ into the fiber and a receiver with the sensitivity $\overline{P}_{\text{sens}}$, we easily can derive that fiber attenuation is limiting the transmission distance to [46]

$$L \leq \frac{10}{a} \cdot \log\left(\frac{\overline{P}_{\text{out}}}{\overline{P}_{\text{sens}}}\right), \tag{7.21}$$

where a is the fiber attenuation measured in dB/km. By expanding the log expression, we can rewrite this equation in the more practical and intuitive form

$$L \leq \frac{\overline{P}_{\text{out}}[\text{dBm}] - \overline{P}_{\text{sens}}[\text{dBm}]}{a}. \tag{7.22}$$

For example, given a fiber attenuation of 0.25 dB/km, typical for an SMF operated at 1.55 μm, a launch power of 1 mW (0 dBm), and a 10-Gb/s p-i-n detector with a sensitivity of −18.5 dBm, we find that the system is limited to a span length of 74 km. If we replace the p-i-n photodetector by an APD with the sensitivity of −27.0 dBm, this length increases to 108 km. Finally, for a receiver with an optically preamplified p-i-n detector that has a sensitivity of −36.0 dBm, the span length increases to 144 km.

The bit-rate dependence of the attenuation limit for a p-i-n receiver (p-i-n) as well as an optically preamplified p-i-n receiver (OA) is plotted in Fig. 7.16 based on the sensitivity data from Fig. 5.13. Note that this limit varies much slower with bit rate than the other limits, which can be explained by the log function in Eq. (7.21). The curves in Fig. 7.16 are bases on a fiber attenuation of $a = 0.25$ dB/km; however, with the use of optical in-line amplifiers, the effective value of a can be reduced substantially, thus permitting much longer distances.

7.5 SUMMARY

Two types of modulation are used in optical transmitters:

- Direct modulation, where the laser current is directly modulated by the signal.

- External modulation, where the laser is always on and a subsequent optical modulator is used to modulate the laser light with the electrical signal.

In general, external modulation produces higher quality optical signals with a narrower spectral linewidth and a higher extinction ratio, but also is more costly and bulky than direct modulation.

The following light sources commonly are used in optical transmitters:

- The Fabry-Perot (FP) laser, which is low in cost, but has a wide linewidth that severely limits the transmission distance in a dispersive fiber.

- The distributed-feedback (DFB) laser, which has a very narrow linewidth and thus is an excellent continuous-wave (CW) source for external modulators. Even when modulated directly, the DFB laser has a fairly narrow linewidth (compared with CW operation, it is broadened because of chirp and the spectrum of the modulation signal) and is suitable for long-reach systems.

- The vertical-cavity surface-emitting laser (VCSEL), which is low in cost, mostly is used in data communication systems operating at short wavelengths.

- The light-emitting diode (LED), which is very low in cost but has a very wide linewidth, low output power, and small modulation bandwidth, mostly is used for low-speed, short-reach data communication.

Electrically, all four sources are forward biased p-n junctions.

The following modulators commonly are used in optical transmitters:

- The electroabsorption modulator (EAM), which is small and can be driven with a reasonably small voltage swing. Electrically, it is a reverse-biased p-n junction.

- The Mach-Zehnder modulator (MZM), which generates the highest-quality optical pulses with a controlled amount of chirp and a high extinction ratio. Electrically, it is a (terminated) transmission line.

The maximum transmission distance that can be achieved in an optical communication system is determined by a combination of the chromatic dispersion limit, the polarization-mode dispersion (PMD) limit, and the attenuation limit.

7.6 PROBLEMS

7.1 Power Penalty due to Finite Extinction Ratio. Derive the power penalty caused by a transmitter with a finite extinction ratio. (a) Assume the noise at the receiver is signal independent. (b) Assume the rms noise current at the receiver is proportional to the square root of the signal current.

7.2 Laser Materials. Photodetectors for both the 1.3- and 1.55-μm wavelengths can use the same InGaAs material for the absorption layer. Why do lasers for the 1.3- and 1.55-μm wavelengths require two different InGaAsP compounds for the active layer?

7.3 Chirp Parameter. The chirp parameter α sometimes is defined as

$$\alpha(t) = 2P(t) \cdot \frac{\partial \Phi / \partial t}{\partial P / \partial t}, \tag{7.23}$$

where Φ is the phase and P is the intensity of the electromagnetic field. Show how this definition relates to Eq. (7.10).

7.4 Power Penalty due to RIN Noise. Derive the power penalty caused by the laser RIN noise in a transmission system with a p-i-n receiver. Assume a DC-balanced NRZ signal with high extinction and a signal-independent receiver noise, $\overline{i_{n.\text{amp}}^2}$, in addition to the RIN noise.

7.5 Power Conservation in a Mach-Zehnder Modulator. The MZM uses constructive and destructive interference to modulate the laser light. In the case of destructive interference, light power goes into the modulator but no light power comes out of it. Where does the power go?

7.6 Duobinary Modulation. An NRZ signal with bit rate B is processed (filtered) by adding a delayed copy of the signal to itself. The delay is one bit period or $1/B$. (a) How many levels does the resulting signal have? (b) What is the transfer function $H(f)$ of the "delay and add" function? (c) What is the spectrum of the resulting duobinary signal?

8
Laser and Modulator Drivers

In this chapter, we focus on the *laser driver* and *modulator driver*. We start by introducing the main specifications of these drivers. Then, we discuss laser- and modulator-driver circuit concepts in a general and, as much as possible, technology-independent manner. Subsequently, we illustrate these concepts with practical implementations in a broad range of technologies. We conclude with a brief overview of product examples and current research topics.

8.1 DRIVER SPECIFICATIONS

In the following, we discuss the main specifications of the digital (on-off keying) laser and modulator driver: the modulation and bias current range (for the laser driver), the output voltage range (for the laser driver), the modulation and bias voltage range (for the modulator driver), the power dissipation, the rise and fall times, the pulse-width distortion, the jitter generation, and the eye-diagram mask test. For a discussion of S parameters, see Appendix C.

8.1.1 Modulation and Bias Current Range (Laser Drivers)

The basic input and output signals of a laser driver are shown in Fig. 8.1(a). The differential data inputs (D) are driven with the digital signal to be transmitted. Typically, these inputs are 50-Ω terminated and accept a small differential voltage swing in the range 0.5 to 1.5 V. In the case of a clocked laser driver, clock inputs (CK) also are provided. The clock signal is used by the driver to retime the data signal, resulting in a cleaner laser-current waveform. At the output, the laser driver supplies the modulated current $i_L(t)$ to the laser. In addition to the inputs and output shown in Fig. 8.1(a), inputs for controlling the laser current levels, inputs for compensating pulse-width distortions, and a complementary output usually also are provided.

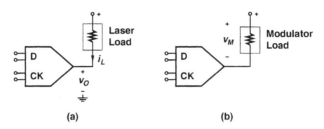

Fig. 8.1 Basic input and output signals of (a) a laser driver and (b) a single-ended modulator driver.

Definition. The *bias current*, I_B, is the current supplied by the laser driver when transmitting a zero (laser off). The *modulation current*, I_M, is the current added to the bias current when transmitting a one (laser on). Therefore, the laser current, i_L, swings between I_B and $I_B + I_M$, as illustrated in Fig. 8.2. These definitions apply to the commonly used DC-coupled laser driver. However, if the driver is AC coupled to the laser, the bias current is defined as the *average* current into the laser and the laser current swings between $I_B - I_M/2$ and $I_B + I_M/2$. Note that in either case, the laser current swing, i_L^{pp}, equals I_M.

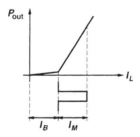

Fig. 8.2 Modulation and bias currents of a DC-coupled laser driver.

The bias and modulation currents of a laser driver are controlled either directly with (analog or digital) trim pots, or by means of a feedback loop using the signal from the laser's monitor photodiode. Typically, the bias current is controlled by an automatic power control (APC) circuit using feedback from the monitor photodiode, whereas the modulation current is set directly with a trim pot.

Typical Values. The modulation current range must be large enough to reach the maximum desired optical output power with a low-efficiency laser under high-temperature and end-of-life conditions. A typical range seen in commercial 2.5- and 10-Gb/s drivers for uncooled lasers is

$$I_M = 10\ldots100\,\text{mA}. \tag{8.1}$$

Similarly, the bias current range must be large enough to cover the threshold current, I_{TH}, of a high-threshold laser under high-temperature and end-of-life conditions. A typical range seen in commercial 2.5- and 10-Gb/s drivers for uncooled lasers is

$$I_B = 0\ldots100\,\text{mA}. \tag{8.2}$$

8.1.2 Output Voltage Range (Laser Drivers)

Although the laser driver's primary function is to generate the laser current, i_L, its output voltage, v_O, also must be considered (see Fig. 8.1(a)). The proper operation of the driver is guaranteed only if the output voltage stays in the permitted *output voltage range*, also called the *compliance voltage*. If the output voltage becomes too small, $v_O < v_{O.\text{min}}$, the laser driver typically produces large pulse-width distortions and jitter because its output transistors are pushed into saturation (BJT) or the linear regime (FET). Furthermore, the modulation and bias currents may drop below their programmed values. If the output voltage becomes too large, $v_O > v_{O.\text{max}}$, the output devices may break down.

The output voltage range constrains the laser loads that can be driven. If the load causes a large voltage drop, the lower limit may be violated. Conversely, some AC coupling schemes use a pull-up inductor and require that the output voltage can swing above the supply voltage, which may conflict with the upper limit of the voltage range. We discuss the impact of DC- and AC-coupled lasers on the output voltage in Section 8.2.1. A laser driver that can operate at a small output voltage also has the advantage that the supply voltage for the load, and with it the power dissipation, can be kept small.

Typical Values. The minimum permissible output voltage of a laser driver, relative to the negative supply voltage, typically is in the range

$$v_{O.\text{min}} = 1.4\ldots2.0\,\text{V}. \tag{8.3}$$

The maximum permissible output voltage usually is in the vicinity of the positive supply voltage. Some drivers allow for a higher voltage to permit a pull-up inductor to use in conjunction with an AC-coupled laser.

8.1.3 Modulation and Bias Voltage Range (Modulator Drivers)

The basic input and output signals of a single-ended modulator driver are shown in Fig. 8.1(b). Similar to the laser driver, we have the differential data inputs (D) and the optional clock inputs (CK) for retiming. At the output, the modulator driver generates the voltage $v_M(t)$ across the modulator. In addition to the inputs and output shown in Fig. 8.1(b), inputs for controlling the modulator voltage levels, inputs for compensating pulse-width distortions, and a complementary output usually also are provided.

Definition. The *modulation voltage*, V_S, is the difference between the on- and off-state voltage supplied by the modulator driver. Note that this voltage equals the *voltage swing* across the modulator, $V_S = v_M^{pp}$. In the case of an electroabsorption modulator (EAM) driver, the *bias voltage* or *DC offset voltage*, V_B, is the voltage supplied by the driver during the on state. In the case of a Mach-Zehnder modulator (MZM) driver, the bias voltage, V_B, is the average voltage (DC component) supplied by the driver. See Fig. 8.3 for an illustration of V_S and V_B in relationship to the switching curves of an EAM and an MZM.

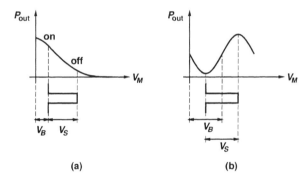

Fig. 8.3 Modulation and bias voltages of (a) an EAM and (b) an MZM driver.

For an EAM, the voltage swing must be equal to or larger than the modulator's switching voltage, V_{SW}, to obtain a sufficient extinction ratio (ER). The bias voltage usually is set to a small value around 0 to 1 V to optimize the chirp parameter α.

For an MZM, the voltage swing must closely match the switching voltage, V_π (or $V_\pi/2$ at each input port for a dual-drive MZM operated in push-pull mode). Because of the sinusoidal switching curve (see Fig. 8.3(b)), the extinction ratio degrades if the voltage swing is smaller or larger than V_π; these conditions are known as *under-* or *overmodulation*, respectively. However, a small amount of overmodulation sometimes is used to improve the rise and fall times of the optical signal. In applications where the MZM modulates the optical phase (0° or 180°) in addition to the intensity (on or off), such as for optical duobinary modulation or return-to-zero differential phase-shift keying (RZ-DPSK), the desired voltage swing is *twice* the switching voltage, $2V_\pi$.

As shown in Fig. 8.3(b), the optimum bias voltage for the MZM (assuming on-off keying) is at the midpoint of the switching curve, also known as the *quadrature point*. Because of path mismatch and drift, this voltage is not known a priori, and the bias voltage range has to span at least one full period, $2V_\pi$. Note that if drift causes the quadrature point to go outside of this range, the bias voltage simply can be reduced by a multiple of $2V_\pi$ because of the periodicity of the switching curve. Usually, an automatic bias controller (ABC) is used to generate V_B such that the modulator is biased at the quadrature point regardless of drift.

Typical Values. Typical single-ended modulation voltage ranges seen in commercial 2.5- and 10-Gb/s modulator drivers are

$$\text{EAM driver:} \quad V_S = 0.2 \ldots 3\,\text{V}, \tag{8.4}$$

$$\text{MZM driver:} \quad V_S = 0.5 \ldots 5\,\text{V}. \tag{8.5}$$

Typical bias voltage ranges are

$$\text{EAM driver:} \quad V_B = 0 \ldots 1\,\text{V}, \tag{8.6}$$

$$\text{MZM driver:} \quad V_B = 0 \ldots 10\,\text{V}. \tag{8.7}$$

The required voltage swing often dictates the driver technology. For example, whereas a 3-V swing usually can be attained with a SiGe technology, a 5-V swing may necessitate a GaAs technology, which has a higher breakdown voltage.

8.1.4 Power Dissipation

The power dissipation of a laser or modulator driver is quite large when compared with other transceiver blocks, such as the transimpedance amplifier (TIA) or main amplifier (MA). As we have seen, the driver must deliver large current or voltage swings into a load resistance that typically is around 25 to 50 Ω. Furthermore, the high switching speed necessary for Gb/s drivers also requires substantial currents in the predriver and the retiming flip-flop.

A low power dissipation is desirable because it reduces the heat generation in the driver IC and the system. Excessive heating in the IC may require an expensive package, and excessive heating in the system may degrade the laser performance or require a large power-consuming thermoelectric cooler to remove the heat. Furthermore, a low power dissipation also reduces the cost of the power supply and the back-up battery, if required.

Definition. Because laser drivers usually have a programmable modulation and bias current and modulator drivers have a programmable modulation and bias voltage, it is important to specify the programmed values when quoting the power dissipation. Manufacturers usually quote the power dissipation for zero modulation and bias currents (or voltages). In this case, the actual power dissipation when driving the laser (or modulator) is significantly larger than the quoted one because the presence of these currents causes additional power dissipation.

It also is important to distinguish between the total power dissipation and the power dissipation in the driver IC alone. Usually, a significant fraction of the total power is dissipated in the laser (or modulator) and the associated matching resistor(s).

Typical Values. The power dissipation of commercial 2.5- and 10-Gb/s laser (or modulator) drivers programmed for zero modulation and bias currents (or zero modulation and bias voltages) typically is in the range

$$P = 0.2 \ldots 1.4\,\text{W}. \tag{8.8}$$

Note that for zero programmed currents, the total power dissipation is equal to that of the driver IC alone, that is, there is no power dissipation in the load. When programmed for typical modulation and bias currents (or voltages), the total power dissipation increases roughly by 0.1 to 1 W.

8.1.5 Rise and Fall Times

Definition. The *rise time* and *fall time* of a laser (or modulator) driver's output signal can be measured in the electrical or optical domain. An oscilloscope can be used to display the electrical signal waveform at the output of the laser or modulator driver. To display the optical signal waveform at the output of the laser or modulator, an optical-to-electrical (O/E) converter must be connected to the input of the oscilloscope. Usually, the rise time, t_R, is measured from the point where the signal has reached 20% of its full value to the point where it has reached 80%. The fall time, t_F, is measured similarly from the 80% point to the 20% point. However, a few manufacturers use 10% and 90% as measurement conditions, and one has to be careful when comparing specifications of different products. In case the signal exhibits over- or undershoot, the 0% and 100% values correspond to the steady-state values, *not* the peak values. See Fig. 8.4 for an illustration of the rise and fall times in the eye diagram. For a discussion of eye diagrams, please refer to Appendix A.

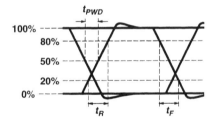

Fig. 8.4 Eye diagram and AC parameters of a laser or modulator driver.

The rise and fall times must both be shorter than one unit interval (1 UI = one bit period). If longer, the driver cannot produce the full swing for a "01010101..." pattern, resulting in intersymbol interference (ISI) and vertical eye closure. It is recommended for a non-return-to-zero (NRZ) system that the total system rise time is kept below 0.7 UI [5]. The driver rise time, t_R, is just one component of the system *rise-time budget*, which also includes the fiber rise time and the receiver rise time (all rise-time components must be added in the square sense). Therefore, the driver rise time must be made significantly shorter than 0.7 UI. However, in laser drivers, the rise time should not be made unnecessarily short to avoid the generation of excessive optical chirp in the laser (cf. Eq. (7.10)) [169].

Typical Values. Typical values for the electrical rise and fall times seen in commercial 2.5- and 10-Gb/s laser or modulator drivers are

$$2.5\,\text{Gb/s:} \quad t_R, t_F < 100\,\text{ps}\ (<0.25\,\text{UI}), \tag{8.9}$$

$$10\,\text{Gb/s:} \quad t_R, t_F < 40\,\text{ps}\ (<0.40\,\text{UI}). \tag{8.10}$$

8.1.6 Pulse-Width Distortion

Definition. An offset or threshold error in the driver circuit may lengthen or shorten the electrical output pulses relative to their ideal width of one unit interval. Furthermore, turn-on delay in the laser may shorten the optical pulses relative to the electrical pulses. The deviation of the pulses from their ideal width is known as *pulse-width distortion* (PWD) and can be measured in the electrical as well as the optical domain. The amount of PWD, t_{PWD}, is defined as the difference between the wider pulse and the narrower pulse divided by two. Figure 8.4 shows how t_{PWD} can be determined from the eye diagram. If the crossing point of the eye is vertically centered, t_{PWD} is zero. Under this condition, the horizontal eye opening is maximized as well.

Many laser and modulator drivers contain a so-called *pulse-width control* (PWC) circuit to compensate for the PWD. An external trim pot connecting to the PWC circuit permits the adjustment of the PWD. In practice, the driver must be trimmed with the desired laser or modulator in place until the crossing point of the optical eye is centered.

A low PWD is desirable because it improves the horizontal eye opening. Furthermore, some clock-recovery circuits in the receiver use the rising and falling edge for phase detection, which requires that both edges are precisely aligned with the bit intervals, that is, the PWD should be small.

Typical Values. Typical values for the electrical PWD seen in commercial 2.5- and 10-Gb/s laser or modulator drivers are below 0.05 UI:

$$2.5\,\text{Gb/s:} \quad t_{PWD} < 20\,\text{ps}, \tag{8.11}$$

$$10\,\text{Gb/s:} \quad t_{PWD} < 5\,\text{ps}. \tag{8.12}$$

The above numbers are for drivers without a pulse-width control circuit or with that feature disabled. If a PWC circuit is present, the adjustment range for t_{PWD} must be specified as well. A typical PWD adjustment range is $\pm 0.20\,\text{UI}\ (\pm 20\%)$.

8.1.7 Jitter Generation

Definition. As we discussed in Section 4.9, data signals in a receiver not only suffer from PWD, but also from timing jitter. Some of this jitter is produced in the transmitter and is known as the *jitter generation* of the transmitter. Jitter generation of the transmitter is determined with a jitter-free transmitter clock, that is, only the intrinsic part of the output jitter is counted. (The effect of clock jitter on the output jitter is measured by the *jitter transfer* parameter.) Similarly, jitter generation of

266 LASER AND MODULATOR DRIVERS

a laser or modulator driver is determined with a jitter-free data and clock signal at the input.

As shown in Fig. 8.5, jitter can be measured in the (electrical or optical) eye diagram by computing a histogram of the time points when the signal crosses a reference level. This level is set to the eye-crossing point where the histogram has the tightest distribution.[1] Note that in the absence of PWD, this level is at 50%. Many sampling oscilloscopes have the capability to calculate and display such histograms. The various types of jitter (deterministic jitter, random jitter, total jitter, etc.) and how they can be quantified (histogram, peak-to-peak, rms, wideband, narrowband, etc.) was discussed in Section 4.9 for the receiver. The same definitions apply to the jitter generated in the transmitter or driver.

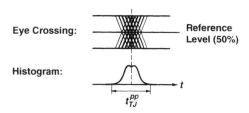

Fig. 8.5 Jitter histogram with deterministic and random jitter.

Jitter in the electrical output signal is caused by noise and ISI from the driver circuit. Reflections on the interconnects also contribute to the jitter. Note that jitter already present in the driver's data or clock input signals also appears at the output and must be subtracted out to obtain the driver's jitter generation. The optical output signal contains additional jitter components produced by the laser or modulator, such as the turn-on delay jitter of a laser.

A low jitter generation is desirable because it improves the horizontal eye opening and makes the clock-recovery process at the receiver more robust. Furthermore, in some types of *regenerators*, the clock signal recovered from the received optical signal is used to retransmit the data. When cascading several such regenerators, the jitter increases because of the jitter generated in each regenerator. Thus, to prevent excessive jitter accumulation along the chain, very tough jitter specifications are imposed on each regenerator.

Typical Values. The jitter generation limits for a SONET transmitter prescribed by the standard [188] (Category II) are 0.01 UI rms and 0.1 UI peak-to-peak:

$$2.5 \, \text{Gb/s:} \quad t_{TJ}^{rms} < 4 \, \text{ps} \quad \text{and} \quad t_{TJ}^{pp} < 40 \, \text{ps}, \tag{8.13}$$

$$10 \, \text{Gb/s:} \quad t_{TJ}^{rms} < 1 \, \text{ps} \quad \text{and} \quad t_{TJ}^{pp} < 10 \, \text{ps}. \tag{8.14}$$

---

[1] Alternatively, the reference level can be set to the switching level of the subsequent device, that is, 50% for a differential device, regardless of the eye crossing. In this case, pulse-width distortion broadens the histogram and appears as a type of jitter, namely duty-cycle distortion jitter.

These values are defined for a jitter bandwidth from 12 kHz to 20 MHz for 2.5 Gb/s and 50 kHz to 80 MHz for 10 Gb/s. This bandwidth is relevant, because high-frequency jitter outside this bandwidth is not passed on to the output of the regenerator (jitter-transfer specification), and therefore does not get accumulated in a chain of regenerators.

The laser or modulator driver's jitter generation must be much lower than the transmitter limits given above, because the driver is only one of several components contributing to the total jitter generation. For example, in a transmitter design half of the total jitter budget may be allocated to the clock from the clock multiplication unit (0.05 UI_{pp}) and the other half to the laser driver, laser, and optics (0.05 UI_{pp}) [45].

8.1.8 Eye-Diagram Mask Test

The so-called *eye-diagram mask test* checks the transmitter signal for many impairments simultaneously such as slow rise and fall times, pulse-width distortion, jitter, ISI, ringing, noise, and so forth. In this test, the (electrical or optical) eye diagram is compared with a mask that specifies regions inside and outside the eye that are off limits to the signal. For example, the SONET OC-48 mask shown in Fig. 8.6 requires that the signal must stay out of the shaded regions to comply with the standard. The rectangle inside the eye diagram defines the required *eye opening*. The regions outside the eye diagram limit the overshoot and undershoot.

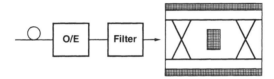

Fig. 8.6 Eye-diagram mask for SONET OC-48.

Often a transmitter not only is required to pass the eye-diagram mask test of the relevant standard, but also to have a sufficient *eye-diagram mask margin*. The eye-diagram mask margin is determined by growing the standard mask until a violation, a so-called *mask hit*, occurs. The margin then is specified as the relative reduction of the permitted regions in percent, for example, a typical eye-diagram mask margin is 20%.

If the signal contains Gaussian random noise, Gaussian random jitter, or both, the mask will always be violated for a long enough measurement. Therefore, it is important to specify the time over which the eye diagram has been measured or the number of samples that have been taken. More precisely, the probability for which a sample falls into the forbidden regions of the mask should be specified.

Electrical eye-diagram mask tests usually are performed directly on the signal, whereas optical eye-diagram mask tests require that the signal is first filtered to suppress effects that would not affect the receiver (e.g., relaxation oscillations of the laser will be attenuated). For example, the SONET standard requires that the optical signal, after O/E conversion, is passed through a fourth-order Bessel-Thomson filter with a 3-dB bandwidth equal to $0.75\,B$ before it is tested against the eye mask (cf. Fig. 8.6). An O/E converter that also performs the filtering required by the standard is known as a *reference receiver*.

8.2 DRIVER CIRCUIT CONCEPTS

In the following, we discuss driver circuit concepts in a general and, as much as possible, technology-independent manner. This includes the current-steering output stage with and without back termination, the predriver with pulse-width control, data retiming, automatic power control, and special techniques for burst-mode and analog drivers.

8.2.1 Current-Steering Output Stage

The output stage of most laser and modulator drivers is based on the *current steering* circuit shown in Fig. 8.7. Although shown with BJTs, the same arrangement also can be used with FETs. Similar to an inverter/buffer from the *current-mode logic* (CML) family, the tail current, I_M, is either switched through the right or left transistor. For this reason, this circuit also is known as a *differential current switch*. To obtain full or near full switching, the differential input-voltage swing, v_I^{pp}, must be sufficiently large. For a BJT current-steering circuit without emitter degeneration, this voltage is around 200 mV. For an FET circuit, it depends on the transistor size and the tail current; to switch a large tail current with a reasonable voltage, wide FETs are required. [→ Problem 8.1]

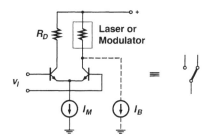

Fig. 8.7 Current-steering circuit for driving a laser or modulator.

The current-steering output stage of Fig. 8.7 is suitable to drive a laser or a modulator. When driving a differential load, such as a dual-drive MZM, both outputs are used. In all other cases, only one output is used and the other output is terminated into a dummy load, R_D. The following important properties of the current-steering circuit make it a good choice for the output stage of drivers (cf. Appendix B):

- The differential design is insensitive to input common-mode noise and power/ground bounce. This is an important prerequisite to achieve low jitter generation in the driver. The differential design further avoids the need for an input reference voltage, and thus prevents pulse-width distortions due to an error in this reference voltage.

- Ideally, the total power-supply current remains constant, that is, it is always I_M $(+I_B)$, no matter if a zero or one is transmitted. The tail current either is routed through the laser/modulator or is dumped into the dummy load, R_D, but it is never switched off. As a result, the generation of power and ground bounce in the presence of parasitic inductances is minimized. On the down side, the power dissipation is twice that necessary to drive the laser/modulator (assuming equal numbers of zeros and ones and $I_B = 0$).

- The voltage across the tail-current source I_M is primarily set by the input common-mode voltage and remains essentially constant for the on and off states. Thus, current overshoots due to the charging of the parasitic capacitance across the tail-current source as well as current variations due to the finite resistance of the tail-current source are relatively small. In a way, the current-steering transistors act as cascode devices for the tail-current transistor.

- The modulation current (or modulation voltage, in case of a modulator driver) can be conveniently controlled by varying the tail current, I_M.

Laser and Modulator Loads. Next, we explore how the current-steering circuit of Fig. 8.7 can drive the various laser and modulator loads. Several typical ways to DC and AC couple these loads to the driver are illustrated in Fig. 8.8 and briefly discussed below:

(a) DC-coupled laser diode. The series resistor, R_S, dampens oscillations due to parasitic inductances and can provide matching to a transmission line, if necessary. The modulation current supplied to the laser is equal to the tail current, I_M. The laser bias current, I_B, can be supplied by an additional current source connected to the output of the driver, as shown with the dashed lines in Fig. 8.7. Alternatively, the bias current can be supplied directly to the cathode of the laser, reducing the voltage drop across R_S. Often an RF choke (RFC) is inserted into the bias line to reduce the capacitive loading of the RF signal.

An important consideration is the voltage drop across the DC-coupled laser load. For example, if the maximum voltage drop across the laser diode is 1.5 V and the maximum current through $R_S = 20\,\Omega$ is 100 mA, then the load drops a total of 3.5 V. This means, that with a minimum permissible laser driver output voltage of 1.5 V, the supply voltage at the laser anode must be at least 5.0 V.

(b) DC-coupled EAM. The parallel resistor, R_P, converts the drive current into a voltage. Thus, the modulation voltage is $V_S = R_P \cdot I_M$. The bias current source, I_B, can be used to generate a bias voltage across the modulator equal to $V_B = R_P \cdot I_B$.

270 LASER AND MODULATOR DRIVERS

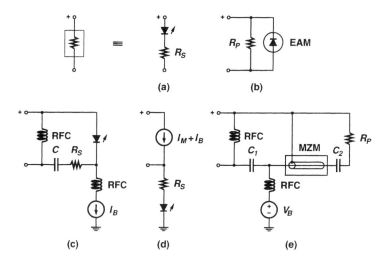

Fig. 8.8 Load configurations for the current-steering output stage: (a) DC-coupled laser, (b) DC-coupled EAM, (c) AC-coupled laser, (d) DC-coupled common-cathode laser, and (e) AC-coupled single drive MZM.

Most EAMs are driven single endedly, like laser diodes. The EAM substrate, which corresponds to the cathode, usually needs to remain at a constant voltage. In particular, if the EAM and the CW laser are integrated on the same substrate, the latter cannot be driven.

(c) AC-coupled laser diode. An RFC pulls the DC output voltage of the current-steering circuit up to the positive supply rail while presenting a high impedance to the RF signal. The RF signal is AC coupled to the laser with capacitor C, which must be large enough to avoid baseline wander in the laser current (cf. Section 6.2.6). The bias current is injected directly into the laser with a second RFC. Note that the tail current, I_M, still sets the laser modulation current, as in the DC-coupled case, but the bias current, I_B, now sets the *average* laser current, as opposed to the off-state laser current.

In this AC-coupled arrangement, the output voltage of the laser driver swings symmetrically around the positive supply voltage. For example, with a maximum modulation current of 100 mA and a total RF load resistance of 25 Ω (e.g., 5-Ω laser diode resistance and 20-Ω series resistor), the swing is 2.5 V. In other words, the maximum voltage drop across the pull-up RFC is only 1.25 V. Thus, with a minimum permissible laser driver output voltage of 1.5 V, the supply voltage must be at least 2.75 V. Whereas a 5-V supply voltage was needed in the DC-coupled case, a 3.3-V supply now becomes possible. Note that in this scheme, the laser driver's maximum permissible output voltage must be at least equal to the supply voltage plus 1.25 V. If this is too high, a resistor in series with the RFC can lower the output voltage.

(d) DC-coupled common-cathode laser diode, such as a vertical-cavity surface-emitting laser (VCSEL). The current source $I_M + I_B$ supplies the on-state laser current when the tail current, I_M, is steered into the dummy load R_D of Fig. 8.7. When the tail current is steered into the laser load, the laser current reduces to its off-state value, I_B.

(e) AC-coupled single-drive MZ modulator. Similar to case (c), a pull-up inductor and AC coupling is used. The modulation voltage is given by $V_S = R_P \cdot I_M$, where R_P is the termination resistor of the MZM. The bias voltage, V_B, is applied to the modulator by means of a bias T (RFC with coupling capacitor C_1). The capacitor C_2 in series with the termination resistor R_P blocks an unnecessary DC current.

A dual-drive MZ modulator also can be driven by the current-steering circuit. In this case, both outputs are used to drive the modulator in a push-pull fashion. Care must be taken that the delays for the two output signals are well matched.

Is the dummy load, R_D, in Fig. 8.7 really necessary or could it be replaced by a short? The purpose of the dummy load is to improve the symmetry of the output stage. An asymmetric load configuration results in an input offset voltage, which can cause PWD. It also results in an undesirable modulation of the voltage across the tail-current source. This voltage modulation together with a parasitic capacitance across the tail-current source causes a momentary error in the drive current, such as an overshoot, degrading the eye. Similarly, a voltage modulation together with a finite resistance of the tail-current source causes undesirable current variations. Both effects also contribute to the power-supply noise in the presence of parasitic inductances. Furthermore, an asymmetric load of the output stage also can affect the predriver symmetry: an asymmetric output stage presents two unequal Miller capacitances to the predriver and produces unequal kick-back noise at its two inputs.

The dummy load can be located on or off chip. If it is located on-chip, it usually is realized with a simple resistor; in the case of laser drivers, a forward-biased diode to match the laser diode sometimes is put in series with the resistor. An off-chip dummy load provides more flexibility in the type of load used (e.g., a resistor with or without pull-up inductor in parallel) and better symmetry, because both outputs experience similar package and routing parasitics.

Switching Speed. When designing high-speed drivers, the switching speed of the current-steering circuit is of critical importance. To achieve fast switching in a BJT current-steering circuit, the emitter area must be chosen carefully: it must be large enough, such that the critical current density is not exceeded (Kirk effect), even for the maximum modulation current, but not too large to keep the input and output capacitances to a minimum (cf. Section 6.3.2). Similarly, in an FET circuit, the channel length must be chosen as small as possible for high speed (cf. Section 6.3.2), but not too small to obtain the necessary drain-source breakdown voltage. Then, the FET width must be chosen large enough to ensure full switching at the given input voltage swing, but not too large, to keep the input and output capacitances to a minimum.

Because the output stage of a laser or modulator driver operates at high current levels, it is important that all interconnect traces and contact areas are made large enough to satisfy the electromigration rules of the chosen technology. For example, to support an average current of 100 mA through aluminum interconnects, it may be necessary to make the traces as wide as 100 μm. However, because these traces have a substantial parasitic capacitance, they should be kept as short as possible and made not wider than necessary. Copper interconnects, if available, offer some mitigation.

From the above observations, we can conclude that the output time constant of the driver increases with the maximum required modulation current (or output swing): larger currents require larger transistors and wider metal traces, both resulting in a larger parasitic output capacitance, yet the output resistance typically remains at either 25 Ω or 50 Ω. Therefore, the MZM driver, which needs to deliver a large output voltage swing at high speed, is one of the most challenging drivers to design.

After the above speed optimizations, further improvements can be achieved with shunt-peaking inductors at the outputs and by distributing the output stage. We discuss shunt peaking of a current-steering circuit in Section 8.2.3 and give an example of a distributed output stage in Section 8.3.2.

Interconnect Inductances. The two output lines of the current-steering circuit conduct rapidly changing currents, hence parasitic inductances, L, in these lines, such as on-chip traces, bond wires, wire frames, and so forth, cause a voltage drop equal to $\Delta v(t) = L \cdot di/dt$. This voltage drop, which can be 1 V or more, eats into the headroom, requiring either a higher supply voltage or a driver with a smaller compliance voltage. To minimize this drop, the parasitic inductances in the output lines must be kept as small as possible. Furthermore, nearby parallel bond wires for the two driver outputs (or the driver output and the power supply line to the on-chip dummy load) are recommended. Such bond wires have mutual magnetic coupling, which reduces the effective inductance seen by the differential output signal. Also, nearby placement of these two complementary outputs reduces the transient noise, which is electrically and magnetically coupled to other nodes. Besides dynamic voltage drops, the parasitic inductances in the output lines also can resonate with the laser and driver capacitances and can produce ringing and jitter. Adding a series resistor, R_S, helps to reduce the quality factor of the parasitic inductor and thus dampens the ringing. An R-C shunt network from the laser cathode to ground can provide additional damping.

A related issue is the parasitic inductance of the interconnect between the driver chip and the laser or modulator. This inductance also can cause a significant voltage drop, can produce ringing, or both. A standard remedy is the use of a transmission line for this interconnect. In a transmission line, the parasitic wire inductance is counterbalanced by a distributed capacitance such that the characteristic impedance becomes real. The direct connection shown in Fig. 8.7 therefore only is practical if either the rate of current change or the parasitic inductances are small. This is the case for low-speed drivers (small di/dt) and drivers that are copackaged with the laser or modulator (small L). The term *copackaging* indicates that the laser or modulator is located in close proximity to the driver and that the two components are interconnected by means of short wire bonds or flip-chip bonds. For high-speed drivers where the load is more than a few millimeters away, a transmission line becomes imperative. The implications of using a transmission line is our next topic. [→ Problem 8.2]

8.2.2 Back Termination

When using a transmission line to connect the driver to the load, undesirable reflections may occur at its ends. To avoid reflections from the load end of the transmission line back into the driver, the laser or modulator must be matched to the characteristic impedance of the transmission line. EAMs can be matched to a 50-Ω transmission line with a 50-Ω parallel resistor (R_P in Fig. 8.8(b)). Laser diodes, which have a typical resistance of about 5 Ω, can be matched to a 25-Ω transmission line with a 20-Ω series resistor (R_S in Fig. 8.8(a)), or to a 50-Ω transmission line with a 45-Ω series resistor.

To keep the power dissipation low, 25-Ω transmission lines generally are preferred for laser drivers: driving 100 mA into a 25-Ω load dissipates 0.25 W in this load (disregarding the laser's nonlinear I/V characteristics), whereas driving the same current into a 50-Ω load dissipates 0.5 W. Furthermore, a lower supply voltage can be used when driving a 25-Ω load as opposed to a 50-Ω load at a given current level, thus power dissipation in the driver IC also is reduced. From a power perspective, it would be best to eliminate the transmission line and the matching resistor entirely and to drive the laser directly. However, as we have discussed before, in high-speed applications this requires copackaging of the driver and the laser to minimize the interconnect inductance.

Open Collector/Drain. Figure 8.9(a) shows the same current-steering output stage as in Fig. 8.7, but now the laser or modulator load, R_L, is driven through a transmission line. If the load impedance matches the characteristic impedance of the transmission line exactly, this arrangement works very well. However, if there is a mismatch, a reflected wave is generated at the load propagating back into the driver. This may occur, for example, in the case of an EAM load, where R_L is bias dependent and cannot match the line impedance under all conditions. Furthermore, at high frequencies the EAM load is capacitive, resulting in a mismatch to the real line impedance. Now, when the reflected wave arrives back at the driver, it sees a high impedance (an open) and consequently is reflected again (nearly unattenuated) forward into the load. These undesirable *double reflections* may degrade the driver's extinction ratio and jitter performance. A driver stage that is unable to absorb reflections coming back from the load is know as an open-collector (or open-drain) stage. In general, such a stage can be used only if a good matching at the load end is guaranteed.

Passive Back Termination. The problem of double reflections can be resolved by incorporating a *back termination* into the driver. A simple way to do this is shown in Fig. 8.9(b), where a termination resistor, R_T, matching the characteristic impedance of the transmission line has been added to the current-steering output stage. Now, any wave that may come back because of a load mismatch is absorbed by this resistor,

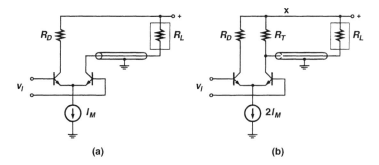

Fig. 8.9 Current-steering output stage for driving a laser or modulator through a transmission line: (a) without back termination (b) with passive back termination.

thus preventing a second reflection into the load. However, the price to pay for this feature is a doubling of the supply current: only about half of the tail current reaches the load and performs a useful function; the other half is "burned up" in the back termination resistor, R_T. For this reason, the tail current in Fig. 8.9(b) is shown as $2I_M$. In fact, if the load is a laser diode, even more than twice the modulation current is needed because of the laser's nonlinear I/V characteristics. [→ Problem 8.3]

The power dissipation of the driver can be reduced to some extent by choosing a back termination that is larger than the characteristic impedance of the transmission line. For example, to drive a 50-Ω transmission line, R_T may be chosen to be 75 or 100 Ω. Although the back matching at low frequencies is degraded by this modification, it may not be significantly impacted (or even improved) at high frequencies where parasitic capacitances and inductances play an important role.

Note that despite the use of current steering, the supply current drawn by the driver from the positive supply (node x in Fig. 8.9(b)) is *not* constant: when the tail current is dumped into the dummy resistor, the current through x is $2I_M$, but when the current is directed into the load R_L, the current through x is only I_M. In the presence of parasitic inductances, these rapid current transients cause large ripples on the supply voltage (node x), which in turn degrade the quality of the output signal. To resolve this issue, the dummy resistor R_D can be split in two parallel resistors, R'_T and R'_D, mimicking the parallel connection of R_T and R_L at the other output. Now, the combined supply current to R_T and R'_T is constant, and little ripple is produced on this supply node. A separate supply voltage terminal is needed for resistor R'_D. Similar to the open-collector case, the current through this supply terminal has large transients and should be placed close to the output terminal.

To improve the rise and fall times of the output stage in Fig. 8.9(b)), a small inductor can be inserted in series with the back termination resistor, R_T. An intuitive explanation of how the inductor improves the edge rate is as follows: during the (high-frequency) signal transitions, the inductor momentarily assumes a high impedance disconnecting the back termination resistor from the output. As a result, twice the regular output current, $2I_M$, becomes available to charge or discharge the output capacitances. To maintain symmetry, a similar inductor should also be inserted into the dummy back termination resistor, R'_T at the other output. In practice, these two inductors to the positive supply rail can be implemented conveniently with bond wires. We discuss this inductive speed-enhancement technique further in the following section on the predriver.

Active Back Termination. A method to protect against double reflections without wasting too much power is given by the so-called *active back termination*. Figure 8.10 shows the schematic implementation of a laser or modulator driver stage that incorporates an active back termination [138]. In contrast to the *passive back termination* discussed before, the back termination resistor, R_T, is not connected to the positive supply voltage, but to an AC voltage generated by a replica stage. The waveform of this AC voltage is such that under normal operation, that is, without reflections, the voltage drop across R_T is zero and no power is wasted in this resistor.

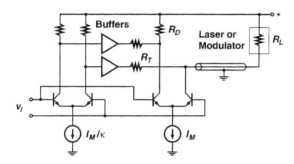

Fig. 8.10 Current-steering output stage with active back termination for driving a laser or modulator.

To save power, the replica current-steering circuit in Fig. 8.10 runs from a κ times smaller current than the output stage. To obtain the same output voltage swing as the output stage, the load resistors are increased by the same factor κ. The outputs of the replica are buffered by unity-gain voltage buffers (zero output impedance is assumed here) before driving the back termination resistors, R_T. Note that the voltage signal at the lower output buffer is the *intended* output signal of the driver. Thus, in the absence of reflections, the voltage drop across R_T is zero; however, in the presence of reflections, they appear only at the driver output (right side of R_T) but not at the replica output (left side of R_T), and thus drop over R_T, where they are absorbed. [→ Problem 8.4]

A slight complication arises when adding a bias current source, I_B, to the driver stage in Fig. 8.10. Without any precautions, some of the bias current will flow into the termination resistor, R_T, thus increasing the power dissipation. However, by adding a correspondingly scaled current source, I_B/κ, to the replica stage, the DC current through R_T again becomes zero.

8.2.3 Predriver

As we have seen, the transistors in the driver's output stage have to switch large currents of around 100 mA and therefore must be made quite large. As a result, their input capacitance also becomes quite large. An on-chip circuit block, such as a retiming flip-flop (cf. Section 8.2.5), may not be able to drive this large capacitance at the required speed. Similarly, if the output stage is driven from off chip, the large capacitance may degrade the input matching parameter S_{11} to an unacceptable value (cf. Appendix C). Another issue is the input voltage swing necessary to switch the output stage. An on-chip circuit block, such as a retiming flip-flop, may not produce a sufficiently large swing. Similarly, if the output stage is driven from off chip, the driver's minimum input swing specification may not be met. Note that for an FET output stage, large devices are required to achieve a small switching voltage, but these devices also have a large input capacitance. Vice versa, a low input capacitance implies small devices, resulting in a high switching voltage. To resolve this dilemma, a so-called *predriver* generally is used to drive the output stage. The predriver must be able to drive a large capacitive load while keeping its input capacitance low. It also must provide sufficient voltage gain to ensure full, or near full, switching of the output stage. In general, the voltage swing provided by the predriver and the transistor sizes in the output stage must be jointly optimized for best rise and fall times of the driver [94].

Figure 8.11 shows an example of a simple bipolar predriver. At the output, the emitter followers Q_3 and Q'_3 provide the necessary low impedance to drive the output stage at high speed. Attention must be payed to the inductive output impedance of the emitter followers, which may cause ringing with the capacitive load of the output stage (cf. Section 6.3.2). The voltage gain of the predriver is provided by the current-steering circuit Q_2 and Q'_2. Compared with the output stage, the transistors of this stage are smaller and operate at a lower tail current. Finally, the emitter followers Q_1 and Q'_1 at the input provide level shifting and further reduce the load presented by the predriver. The input network with the two adjustable current sources are discussed in the following section on pulse-width control. In a practical implementation, multiple cascaded emitter followers may be used at the input and output to provide more impedance transformation and level shifting [147]. Furthermore, multiple gain stages can be used to provide more gain and output voltage swing [94].

Besides emitter- and source-follower buffers, the following two techniques can be used to drive the large capacitive load of the output stage. (i) A negative capacitance can be connected to the outputs of the predriver to compensate for some of the positive capacitance presented by the output stage. The negative capacitance can be synthesized, for example, by inverting a regular capacitance with a negative impedance converter (NIC) [31]. (ii) Inductive techniques can be used to resonate out some of the capacitance of the output stage. For example, inductive loads can be used at the outputs of the predriver [79], or inductive interstage networks, such as T-coil networks, can be used to couple the predriver to the output stage [31]. Finally, the output stage can be distributed to absorb the large input and output capacitances into transmission lines, thus presenting a mostly real impedance to the predriver [202].

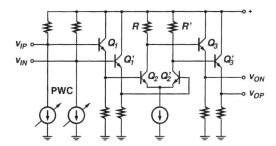

Fig. 8.11 A simple predriver with pulse-width control.

The voltage gain of the predriver can be provided by one or multiple gain stages. Most of the broadband techniques discussed in Section 6.3.2 also apply to the predriver design. (An exception is the inverse scaling technique, which does not apply to the predriver because it must drive a heavy load.) We can summarize these techniques as follows:

- Feedback techniques:
    - Series feedback (emitter/source degeneration, f_T doubler)
    - Shunt feedback (TIA load)
    - Cascode transistor
- Load reduction techniques:
    - Interstage buffer (emitter/source follower(s))
    - Negative capacitance (negative Miller capacitance or capacitor with NIC)
- Inductive techniques:
    - Inductive load (shunt peaking)
    - Inductive interstage network (series peaking, T-coil network)
    - Distributed amplifier (based on inductors or transmission-line segments)

For example, the Cherry-Hooper stage, which combines series and shunt feedback, can be used to provide gain in high-speed predrivers [154]. Shunt peaking is an effective way to enhance the rise and fall times of a current-steering circuit. It can be applied to the predriver circuit in Fig. 8.11 by adding small inductors in series with the load resistors R and R'. These inductors can be realized with on-chip spirals, bond wires, or active inductors, as we discussed in Section 6.3.2.

How does shunt peaking improve the rise and fall times of a current-steering circuit? Figure 8.12 shows a single-ended model of the current-steering circuit with shunt-peaking inductor L and load capacitance C. Let's first set $L = 0$, thus the load consists only of the resistor R and the capacitance C. Assuming that the tail current is switched by a square-wave signal, as shown on the left-hand side of Fig. 8.12, the 20% to 80% rise and fall times of the output pulse can easily be found as

$$t_R = t_F = \ln(0.8/0.2) \cdot RC = 1.39 \cdot RC \quad \text{for} \quad L = 0. \quad (8.15)$$

The output pulse for this case is shown in the upper trace of Fig. 8.12. Next, if we set $L = 0.4R^2C$ to enable shunt peaking, it can be shown that the rise and fall times improve to

$$t_R = t_F = 0.85 \cdot RC \quad \text{for} \quad L = 0.4 \cdot R^2C. \quad (8.16)$$

The output pulse for the latter case is shown in the lower trace of Fig. 8.12. It can be seen that besides the improvement in rise and fall times by about 40%, there also is a slight over- and undershoot (about 2.6% for $L = 0.4R^2C$), which, however, usually is harmless. An intuitive explanation of how shunt peaking improves the edge rate is as follows: when the current source is switched on (falling edge at the output), the inductor at first acts like an open and all of the current is used to discharge C, rather than some of it flowing into R. When the current source is switched off (rising edge at the output), the inductor, which is now "pumped up" with current, charges C more rapidly than R alone.

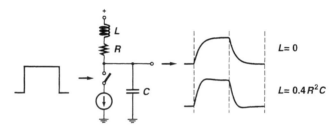

Fig. 8.12 Improvement of the rise and fall times with shunt peaking.

As we have said, the main functions of the predriver are to drive a large capacitive load and to provide enough voltage swing to fully switch the output stage. When designing a predriver, however, several additional points must be considered. (i) The modulation current in the output stage usually is programmable and hence the voltage swing necessary to switch the output stage is not fixed (we assume here an FET output stage or a bipolar output stage with emitter degeneration). Although supplying a large voltage swing to the output stage guarantees switching for all currents, it may lead to distortions in the output signal at small modulation currents. For example, the large voltage transients from the predriver may couple through the base-collector (or gate-drain) capacitance to the output, thus corrupting a small output signal. To avoid such distortions, the predriver's output swing often is made variable such that it increases and decrease with the driver's modulation current. This feature can be implemented, for example, with a variable tail current in the current-steering circuit of the predriver [24, 94, 147]. (ii) The predriver's output common-mode voltage is critical and must be well controlled [24]. A common-mode voltage that is too low pushes the tail-current source, I_M, into saturation (or the linear regime, in the case of an FET source). A common-mode voltage that is too high unnecessarily limits the output voltage range of the driver. (iii) The waveforms produced by the predriver must be such that one of the two transistors in the output stage always is on. If both transistors were to switch off momentarily, the voltage at the common-emitter node would drop and with the capacitance at the common-emitter node cause a current overshoot at the beginning of the next pulse.

8.2.4 Pulse-Width Control

As we mentioned in Section 8.1.6, many laser and modulator drivers include a pulse-width controller to compensate for pulse-width distortions (PWDs) in the optical output signal. Such PWDs can be caused, for example, by an offset voltage in the driver or an asymmetry between the turn-on and turn-off delay of the laser or modulator. Note that although most driver circuits are differential, the laser and EAM are single-ended devices thus introducing an asymmetry. Depending on the bias current, lasers exhibit a turn-on delay that is longer than the turn off delay (cf. Section 7.2). The nonlinear switching curve of EAMs also can produce PWD.

Most pulse-width controllers operate by introducing a variable offset voltage at the input of the predriver. This offset voltage not only can compensate for the random offset voltage of the predriver, but also can predistort the electrical output signal such that after electrical-to-optical (E/O) conversion the optical pulses are free of PWD. Figure 8.13 shows how the pulse width can be predistorted with an offset voltage. At the top, two complementary signals that are free of PWD are shown (solid and dashed traces). By offsetting the inverting and noninverting signals the crossover points can be shifted in time, as shown in the middle. The amount of shift can be controlled with the offset voltage. Finally, at the bottom, the signal is regenerated with a limiter to produce a clean signal with the desired PWD.

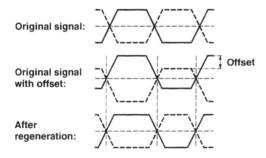

Fig. 8.13 Operation of the pulse-width control circuit.

Figure 8.11 shows a simple way to implement this scheme with two adjustable current sources, labeled PWC, at the input of the predriver. These current sources introduce the variable offset voltage and the subsequent current-steering circuit in the predriver acts as the limiter. Note that the offset voltage cannot be made too large for the current-steering circuit to obtain enough differential signal to switch fully.

Therefore, the amount of PWD that can be compensated also is limited. A drawback of this scheme is that the PWD depends on the edge-rate of the signal at the input of the predriver, which may not be well defined. [→ Problem 8.5]

8.2.5 Data Retiming

In high-speed laser and modulator drivers, the input data signal often is retimed (or resynchronized) with a clean clock signal before being fed to the predriver. A flip-flop located in front of the predriver, as shown in Fig. 8.14, can perform this function. The benefits of *data retiming* are the elimination of pulse-width distortion and jitter from the input data signal. However, jitter in the clock signal does appear undiminished at the output of the driver. It therefore is important that the clock source used for retiming has a very low jitter. For example, SONET compliant laser/modulator drivers typically require a clock source with less than 0.05-UI peak-to-peak jitter.

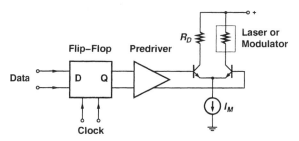

Fig. 8.14 Block diagram of a laser/modulator driver with retiming flip-flop.

Retiming flip-flops usually are implemented with current-mode logic (CML), which is based on nested and cascaded current-steering circuits. The advantages of CML are its high speed, its low sensitivity to common-mode and power-supply noise, and its substantially constant supply current, which minimizes power and ground bounce. It has been found that MOS CML gates are 2 to 3× faster than CMOS gates in the same technology and with the same fanout [36]. If realized with bipolar transistors, CML also is known as *emitter-coupled logic* (ECL); if realized with FETs, it also is known as *source-coupled FET logic* (SCFL).

Figure 8.15 shows a typical bipolar implementation of a CML (or ECL) flip-flop. This master-slave flip-flop consists of a cascade of two identical D latches. Using the D latch on the left-hand side as an example, its operation can be explained as follows: if Q_1 is turned on and Q'_1 is turned off by the clock signal, the D latch acts as an amplifier, passing the input logic state to the output by means of Q_2, Q'_2 and Q_4, Q'_4. Conversely, If Q'_1 is turned on and Q_1 is turned off, the D latch acts as a regenerator storing the previous logic state by means of positive feedback through Q_3, Q'_3, and Q_4, Q'_4. In the latter case, the output state is independent of the input state. The D-latch signal at the load resistors R and R' is level-shifted and buffered with transistors Q_4 and Q'_4 such that the common-mode voltage is suitable to drive the second D latch. This second D latch is clocked from the inverted clock such that when the first latch is in amplification mode, the second one is in regeneration mode and vice versa.

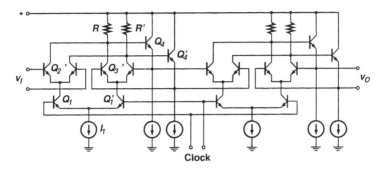

Fig. 8.15 BJT/HBT implementation of the retiming flip-flop.

An important design parameter for CML flip-flops is the output voltage swing (or logic swing). The single-ended output voltage swing is given by the product $R \cdot I_1$, where I_1 is the tail current and R is the value of the load resistors. It turns out that the speed of an FET CML gate improves with increasing voltage swing, thus the voltage swing should be made as large a possible while keeping the FETs in the saturated regime [36]. (For submicron FETs, however, the speed saturates at a certain voltage swing because of the FET's departure from the square law, which holds only for small overdrive voltages.) A typical single-ended swing for an MOS CML gate is 400 to 500 mV. In contrast, the speed of a BJT CML gate largely is independent of the voltage swing and thus can be made smaller; typical values are in the range of 150 to 300 mV. Another important consideration is the input voltage swing necessary to switch the CML gate fully. This swing must be *smaller* than the output swing to permit cascading of multiple gates such as the two D latches in the flip-flop. The same condition also ensures an infinite hold time of the D latches resulting in a static flip-flop. To optimize the flip-flop speed, the f_T of the transistors must be maximized (FETs with minimum channel length; BJTs with optimum collector current density) and the current levels, device sizes, and load resistors of the D latches must be scaled according to their load. [→ Problem 8.6]

Several speed-enhancement techniques can be applied to the basic CML flip-flop shown in Fig. 8.15 or its equivalent implementation with FETs [121, 122]. Similar to the predriver, shunt-peaking inductors can be added in series with the load resistors, R and R', usually in the form of active or spiral inductors, to boost the speed. To lower the capacitive load at the load resistors, the size of the holding transistors, Q_3 and Q'_3, and their bias current can be reduced. However, if these transistors are made too small, or left away entirely, the flip-flop cannot hold its state indefinitely and thus a minimum clock frequency is required for proper operation, that is, it has become a *dynamic flip-flop*.

The flip-flop in Fig. 8.15 requires a full-rate clock, of which only one clock edge (rising or falling) is used for retiming. For very high bit-rate systems, a retiming circuit operating from a half-rate clock and using both clock edges is an interesting alternative. Such a circuit can be built with two parallel flip-flops, which sample the same data signal but are clocked 180° out of phase; a subsequent 2:1 multiplexer combines the two output data streams into one [79].

8.2.6 Automatic Power Control (Lasers)

As we discussed in Section 7.2, the laser's L/I characteristics are strongly temperature and age dependent. Hence, an automatic power control (APC) mechanism usually is required to stabilize the output power of the transmitter. Many laser drivers contain the APC circuit integrated with the driver on the same chip. Figure 8.16 shows a simple APC circuit suitable for a continuous-mode laser driver. A monitor photodiode with good temperature, age, and coupling stability generates a current that is proportional to the transmitted optical power. This current is converted into a voltage and is low-pass filtered with R and C. The resulting voltage at node x, which is proportional to the average optical power, is compared with the reference voltage V_{REF} by an op amp. The op amp output voltage, V_{BC}, then controls the laser's bias current, I_B, such that the desired average optical power, as set by V_{REF}, is obtained. Because only the *average* photodiode current matters, a slow and low-cost monitor photodiode is sufficient.

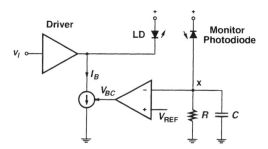

Fig. 8.16 Automatic power control for continuous-mode laser drivers.

Depending on the time constant RC, the power-control mechanism just described may cause an undesirable optical baseline drift when transmitting long strings of zeros or ones. The APC loop not only keeps the average output power constant, but also *suppresses* low-frequency components of the transmitted signal, which manifests itself in the aforementioned baseline drift. The situation is similar to that described in Section 6.3.3, where the offset-compensation loop caused a low-frequency cutoff in the frequency response of the main amplifier (MA). The power penalty associated with this impairment depends on the low-frequency cutoff, f_{LF}, of the transmitter and is identical to that derived for MAs in Section 6.2.6:

$$PP = 1 + \frac{2\pi f_{LF} r}{B}, \qquad (8.17)$$

where r is the maximum run length and B is the bit rate. To keep this power penalty small, the low-frequency cutoff typically is chosen in the kHz range. The APC filter bandwidth, given by $1/(2\pi \cdot RC)$ in Fig. 8.16, must be even smaller than that because the closed-loop low-frequency cutoff is given by the open-loop bandwidth *times* the loop gain. Time constant enhancement techniques can be used to achieve this low bandwidth without the need for large external capacitors [115]. [→ Problem 8.7]

Dual-Loop APC. The circuit shown in Fig. 8.16 also is known as a *single-loop APC*, because only the bias current, and not the modulation current, is controlled in response to the optical output signal by means of a feedback loop. If the threshold current of the laser were the *only* parameter that changes with temperature and age, the single-loop APC would work flawlessly, as illustrated in Fig. 8.17. However, because the laser's slope efficiency changes as well, some undesirable effects come into play. Recall that the single-loop APC tries to keeps the *average* output power, $\overline{P}$, constant. Thus, if the slope efficiency decreases, the power for zeros must increase and the power for ones must decrease, as shown in Fig. 8.18(b). In other words, the extinction ratio degrades. Alternatively, if the slope efficiency increases, the current for zeros drops below the laser's threshold current, as shown in Fig. 8.18(c), resulting in PWD and jitter. [→ Problem 8.8]

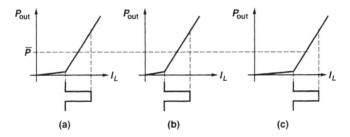

Fig. 8.17 Single-loop APC: (a) nominal threshold current, (b) reduced threshold current, and (c) increased threshold current.

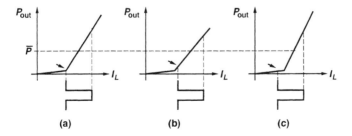

Fig. 8.18 Single-loop APC: (a) nominal slope efficiency, (b) reduced slope efficiency, and (c) increased slope efficiency.

Before turning to the dual-loop APC, let's examine two alternatives for controlling the modulation current. In many lasers, the slope efficiency and the threshold current degrade in a correlated manner. More specifically, the inverse slope efficiency varies approximately linearly with the threshold current when changing the temperature. Thus, the aforementioned problems with the single-loop APC can be alleviated by letting the APC loop control the bias *and* the modulation currents in a fixed proportion [88]. For example, if the APC increases the bias current by ΔI_B, the modulation current is automatically increased by $\Delta I_M = 0.2 \cdot \Delta I_B$. In another approach, the modulation current is controlled in response to the (laser) temperature, a major factor influencing the slope efficiency. This scheme can be implemented with a thermistor sensing the temperature followed by a circuit controlling the modulation current as a linear or nonlinear function of the temperature. In the latter case, a digital look-up table can be used to store the nonlinear function. However, the age dependence of the slope efficiency is not accounted for by this approach, which leads to a lower extinction ratio toward the end of life. [→ Problem 8.9]

A drawback of both approaches discussed so far is that they require knowledge of the laser characteristics: in the first case, the relationship between the slope-efficiency and threshold-current degradations must be known; in the second case, the temperature dependence of the slope efficiency must be known. The so-called *dual-loop APC* avoids this problem by controlling the bias current and the modulation current based on *optical* measurements taken directly from the laser [174]. In one dual-loop approach, the bias current is controlled by the average optical power, just like in the single-loop APC, whereas the modulation current is controlled by the swing of the optical signal. The latter can be determined from the monitor photodiode signal with a peak-detector circuit similar to that shown in Fig. 8.22. A drawback of the peak-detector scheme, in its simplest form, is that it requires a monitor photodiode that is fast enough to follow the bit pattern. However, by waiting for strings of consecutive zeros and ones before measuring the peak value, or by calculating a correction to the measured peak value, or both, the bandwidth limitation of the photodiode can be partially circumvented [100]. In another dual-loop approach, the on-state laser current is modulated with a small-signal low-frequency tone. This pilot tone is easily detected by the (slow) monitor photodiode in the optical signal and its amplitude, which is a measure of the slope efficiency, can be used to control the modulation current.

In practice, a single-loop APC often is sufficient because the slope efficiency is less dependent on temperature and age than the threshold current. This especially is true for cooled lasers, which are operated at a controlled temperature for reasons of wavelength stabilization or reliability.

Mark-Density Compensation. The APC circuit shown in Fig. 8.16 works well for DC-balanced data signals, or more generally, for signals with a constant mark density. (Remember, a DC-balanced signal has an average mark density of one half.) But imagine what happens if we send nothing but zeros to the driver. Will the transmitter turn off? No! An optical power meter connected to the output of the transmitter will always show the same power no matter if we transmit all zeros, all ones, or a pseudorandom bit sequence (PRBS)! The reason is, of course, that the APC circuit keeps the *average* output power constant under all circumstances. So, for an all-zero sequence, the transmitter simply outputs a constant optical power, half-way in between that for a zero and a one. To cope with varying mark densities, a so-called *mark-density compensation* circuit can be added to the APC. Such a circuit varies the reference voltage V_{REF} in proportion to the mark density of the transmitted signal [135, 174]. If the mark density drops to less than 50%, V_{REF} is reduced, if the mark density rises to more than 50%, V_{REF} is increased. Now, if we send nothing but zeros to the driver, the mark density is detected to be zero; thus, V_{REF} is reduced to zero and the transmitter turns off. We discuss a mark-density compensation circuit when we examine an APC for burst-mode transmitters.

Fortunately, all major transmission standards (SONET/SDH, Gigabit Ethernet) do have DC-balanced data signals, and the simple APC circuit in Fig. 8.16 usually is adequate.

8.2.7 End-of-Life Detection (Lasers)

As we discussed in Section 7.2, the mean-time to failure (MTTF) of a continuously operated laser at 70°C is limited to about 1 to 10 years. For this reason, some standards, such as SONET [188], require that the laser's *end-of-life* (EOL) condition automatically is detected and reported. This feature can be implemented easily by using the monitor photodiode, which is present already for the purpose of APC. Figure 8.19 shows how the APC circuit of Fig. 8.16 can be extended to generate a binary EOL signal. A voltage comparator compares the voltage at node x with a reference voltage, which is lower than V_{REF}; for example, in the circuit of Fig. 8.19, this reference voltage is $V_{REF}/2$. The comparator detects when the APC mechanism fails to keep the output power at the desired level and activates the EOL alarm when the power drops below a set value; for example, in the circuit of Fig. 8.19, this set value is half of its nominal value.

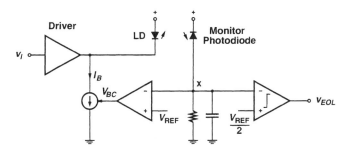

Fig. 8.19 End-of-life detection circuit for continuous-mode laser drivers.

Instead of the binary end-of-life detector, an *aging monitor* can be used, which produces a continuous estimate of the laser's age. Such an estimate can be obtained, for example, by monitoring the laser's threshold current and referring it to a given reference temperature. The bias current, $I_B(T)$, can serve as an estimate for the threshold current at the operating temperature, T, which also is measured. Then, Eq. (7.7) can be applied to refer the threshold current at temperature T to the desired reference temperature.

A related feature available on some laser drivers is the so-called *slow-start* or *smooth-start circuit*. Its purpose is to prevent damaging current spikes to the laser during the power-up phase. Such spikes could occur, for example, when the modulation and bias current sources already are active, but the APC circuit is not yet operational. The slow-start feature can be implemented, for example, by keeping the modulation and bias currents disabled (turned off) for a short period after the power up.

8.2.8 Automatic Bias Control (MZ Modulators)

As we discussed in Section 7.3, the switching curve of Mach-Zehnder modulators suffers from voltage drift with temperature and age. Hence, an *automatic bias control* (ABC) mechanism usually is required to obtain stable operation [44, 135]. A typical ABC circuit is shown in Fig. 8.20 [1]. In this scheme, the driver's output voltage swing is modulated with a small-signal low-frequency tone at, for example, 1 kHz. Modulating the tail current, I_M, of the driver's output stage is an easy way to include this pilot tone. An optical splitter at the output of the MZ modulator directs a small amount of the optical signal, for example, 10%, to a monitor photodiode. The photodiode signal, which contains the pilot tone to a degree that depends on the MZM operating point, is amplified with a TIA and mixed with the original pilot tone. The DC component of the mixer's output voltage is a measure of the detected tone's amplitude and phase. The mixer's output voltage is low-pass filtered (LPF) and amplified to produce the bias voltage, which is then supplied to the MZM by means of a bias T (RFC and coupling capacitor C). Given enough loop gain, the negative feedback provided by this ABC circuit automatically adjusts the bias voltage until the amplitude of the detected tone is minimized. The direction of the bias adjustment is given by the phase relationship between the detected and the original pilot tone. [→ Problem 8.10]

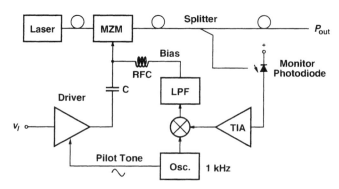

Fig. 8.20 Automatic bias controller for an MZ modulator.

Figure 8.21 illustrates how the amplitude and phase of the pilot tone in the optical signal depends on the MZM's bias voltage. When the bias voltage is correctly adjusted to the quadrature point of the switching curve, the tone does *not* appear in the optical output signal, as shown in Fig. 8.21(a). The reason for this suppression is that the switching curve is flat at both the on and off operating points and thus the optical output power is insensitive to small voltage variations (indicated by three parallel lines in the signals of Fig. 8.21). However, for an incorrect bias voltage, the on and off operating points move into the steep parts of the sinusoidal switching curve and the pilot tone appears in the optical output signal, as shown in Fig. 8.21(b). Note that the phase of the optical pilot tone changes by 180°, depending on whether the operating points are to the right or left of their correct positions.

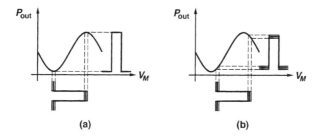

Fig. 8.21 Automatic bias control for an MZ modulator: (a) correct bias and (b) incorrect bias.

Older MZMs exhibited so much voltage drift that, after an extended period of operation, the upper or lower limit of the bias voltage generator was reached. In the ABC circuit of Fig. 8.20, these upper and lower limits are given by the output voltage range of the amplifier that is part of the LPF block. If one of these limits is reached, the bias generator needs to be reset. Note that because of the periodicity of the MZM switching curve, the bias voltage always can be reset by the amount $\pm 2n \cdot V_\pi$, where n is an integer, without affecting the on and off operating points. Fortunately, much progress in reducing the rate of bias drift has been made in recent years and MZMs may now operate for longer than 20 years without requiring a reset [44].

8.2.9 Burst-Mode Laser Driver

The burst-mode laser driver differs from its continuous-mode cousin, which we discussed so far, in the following respects: (i) the burst-mode laser driver typically must provide a very high interburst extinction ratio and (ii) the APC must operate correctly for a bursty, non-DC-balanced, data signal. In the following, we discuss these two differences.

Extinction Ratio. Burst-mode transmitters that are used in point-to-multipoint networks must keep their residual light output in between bursts to a very low level. For example, in a passive optical network (PON) with 32 subscribers sharing a single fiber to the central office, 31 idle subscribers are "polluting" the shared medium with residual light output. This background light reduces the received signal swing at the central office for the one subscriber that is transmitting a burst, especially if this subscriber is located far from the central office. Typically, the interburst ER for a burst-mode laser driver is required to be over 30 dB. The ER requirement within a burst is around 10 dB, similar to that of a continuous-mode transmitter.

To achieve this high ER, the laser bias current must be set to a very low value in between bursts, that is, it must be well below the laser's threshold current, or even zero. During transmission of a burst, the bias current can be increased to reduce the laser's turn-on delay and jitter. If no bias current is used at all, a so-called *turn-on delay compensation* mechanism to reduce pulse-width distortion normally is needed. Even if the bias current is turned on for the duration of the burst, turn-on delay compensation may be needed for the first bit of the burst [100]. To reach the 30-dB extinction ratio quickly at the end of the burst, that is, before another subscriber starts to transmit its burst, a shunt transistor across the laser diode can be used to rapidly drain the carriers from the laser [117]. In Section 8.3.3, we discuss an implementation example of a burst-mode laser driver with turn-on delay compensation.

Automatic Power Control. The average power control circuit in Fig. 8.16, which we discussed in Section 8.2.6, does not work correctly for burst-mode signals. Because the transmitted signal is not DC balanced, keeping the *average* output power constant does not lead to a constant *peak* power (power during the one bits). Figure 8.22 shows one approach to implement a burst-mode APC [50]. First, the current pulses from the monitor photodiode are converted to voltage pulses with a broadband TIA, then a peak detector finds the *peak* value that corresponds to the power transmitted during the one bits. (Depending on the polarity of the TIA front-end, the peak detector is realized as a bottom-level or top-level hold.) The laser's output power is adjusted by means of a negative feedback loop until the peak detector output voltage equals the reference voltage, V_{REF}.

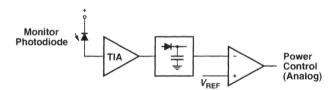

Fig. 8.22 Burst-mode APC based on a peak detector.

Unfortunately, the peak-detector approach of Fig. 8.22 has several shortcomings. (i) During long idle periods, when no bursts are transmitted, the output voltage of the analog peak detector drifts away from the actual peak value, causing an output power error at the beginning of the subsequent burst. (ii) A fast monitor photodiode that can operate at the bit-rate speed is required. (iii) The power dissipation in the TIA and peak detector, which also must operate at bit-rate speed, can be large.

Another approach, based on an integrate-and-dump circuit and digital storage, avoids these shortcomings. Figure 8.23 shows the basic operation of this APC

[160, 162]. The power level is stored in a digital up/down counter, which can hold the precise power level as long as the power supply is up; therefore, output power errors at the beginning of the burst are avoided. The power level is increased or decreased in between bursts with the "Burst Clock" signal. The count direction is given by the up/down signal "u/d̄," which is generated by the integrate-and-dump circuit: before the burst starts, capacitor C is discharged by briefly closing the reset switch S_R, setting the voltage at node x to zero, $v_X = 0$ (the *dump* phase). During the burst, the photodiode current, $i_{PD}(t)$, charges the capacitor C (the *integrate* phase). Simultaneously, the reference current I_{REF}, which represents the desired peak current, is modulated by switch S_D with the transmitted data producing $i_{REF}(t)$. While $i_{PD}(t)$ charges the capacitor, $i_{REF}(t)$ discharges it. At the end of the burst, the capacitor is charged to the following voltage:

$$v_X = \frac{1}{C} \int_{\text{burst}} [i_{PD}(t) - i_{REF}(t)] \, dt = \frac{1}{C} \left(I_{PD} \cdot \frac{n_1}{B} - I_{REF} \cdot \frac{n_1}{B} \right), \quad (8.18)$$

where I_{PD} is the peak value of the photodiode current, n_1 is the number of ones in the burst, and B is the bit rate. In deriving the second expression in Eq. (8.18), we assumed that $i_{PD} = 0$ during the transmission of zeros, as it would be the case for a transmitter without laser bias current. The comparator then compares the voltage at node x with 0 V, the reset voltage. As we can see from Eq. (8.18), the polarity outcome of this comparison is independent of C, B, and n_1; it is only affected by the difference between the photodiode peak current and the reference current, $I_{PD} - I_{REF}$, as desired. If this difference is positive, the power is too high and the counter is stepped down; if it is negative, the power is too low and the counter is stepped up. Note that the integration period in Eq. (8.18) is uncritical and could be made shorter than one burst or extended to include multiple consecutive bursts without changing the result.

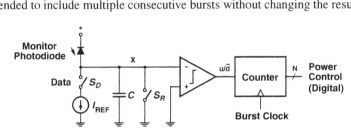

Fig. 8.23 Burst-mode APC based on an integrate-and-dump circuit and digital storage.

The photodiode capacitance, which was limiting the bandwidth in the peak-detector approach of Fig. 8.22, now can serve as the integration capacitor, C, or a part thereof. Thus, there is no need for a fast monitor photodiode. In general, the burst-by-burst operation of the integrate-and-dump approach permits a slower and lower power implementation when compared with the bit-by-bit operation of the peak-detector approach. Also, note that the circuit in Fig. 8.23 performs mark-density compensation, that is, it works accurately even for bursts that contain an unknown and varying amount of zeros and ones. As we discussed in Section 8.2.6, this compensation is achieved by varying the reference level in proportion to the mark density.

In this implementation, the reference level happens to be I_{REF} rather than V_{REF}, and the switch S_D is used to modulate this level with the mark density. If the mark density within a burst is known to be always equal to 50%, for example, because line coding is used, the switch S_D can be omitted and the current source I_{REF} can be replaced by $I_{REF}/2$. [→ Problem 8.11]

End-of-Life Detection. Just as for continuous-mode laser drivers, EOL detection can be implemented with a circuit that is very similar to the APC circuit. But for burst-mode laser drivers, where the transmitted data consists of discrete bursts, it is possible to use a single circuit and time multiplex it to do both the APC and the EOL functions: during a first burst, the reference current in Fig. 8.23 is set to I_{REF} and the circuit performs APC; during a second burst, the reference current is reduced to, say, $I_{REF}/2$ and the same circuit performs EOL detection. (The up/down counter needs to be disabled during EOL detection.) The advantage of this multiplexing scheme is that it performs two functions at the power consumption of one [161, 162].

8.2.10 Analog Laser/Modulator Driver

Analog optical transmitters are used, for example, in CATV/HFC applications as well as for fiber links between cellular-radio base stations and remote antennas. In contrast to digital transmitters, analog laser/modulator drivers must be highly linear, that is, the transmitted optical power must vary linearly with the driver's input voltage. The linearity of a CATV transmitter is specified in terms of composite second order (CSO) and composite triple beat (CTB) distortions (cf. Section 4.8). Typical numbers for a good AM-VSB transmitter are $CSO < -60\,\text{dBc}$ and $CTB < -65\,\text{dBc}$. Furthermore, analog transmitters must be low in noise, for example, the carrier-to-noise ratio (CNR) of an AM-VSB transmitter should be near 50 dB (cf. Section 4.2). Often, the largest noise component is the relative intensity noise (RIN) generated by the laser (cf. Section 7.2), whereas the driver circuit itself is less noise critical.

The implementation of a simple analog laser driver is shown in Fig. 8.24. The current source I_B supplies the bias current through an RFC to the laser. A linear amplifier injects the RF signal current through capacitor C into the laser. The magnitude of the RF current is given by the amplifier's output voltage, v_O, and the combined resistance of the matching resistor (44 Ω) and the laser diode (6 Ω typical), which in our example is equal to 50 Ω (it is assumed that the impedance of C is low at the frequencies of interest). Thus, the total laser current, i_L, is related to the amplifier's output voltage by the linear equation $i_L = I_B + v_O/50\,\Omega$, where v_O is taken to be average free. For an overall linear system, the optical output power of the laser also must be linearly related to the laser current i_L. To achieve this goal, the laser current must stay within the linear part of the laser's L/I curve. The low end of this range is near the threshold current, whereas the high end depends on nonlinear leakage currents, which reduce the slope efficiency at high currents. Apart from the nonlinearity of the L/I curve, dynamic laser nonlinearities must be considered as well.

Figure 8.25 shows the simplified schematic of a linear wideband amplifier for CATV applications with a bandwidth of about 900 MHz [131]. Such amplifiers typ-

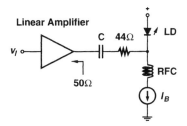

Fig. 8.24 Simple analog laser driver.

ically are implemented as hybrid modules in thin film technology. The necessary linearity of the amplifier is obtained with a differential topology, also known as push-pull topology, and with linear feedback networks. The differential topology reduces even-order distortions by virtue of its symmetry, and the linear feedback networks suppress the remaining odd-order distortions. In the circuit of Fig. 8.25, a transformer converts the single-ended input signal to differential form and also provides impedance matching. The balanced signal from the transformer drives a differential pair, which is biased at a rather high current to minimize distortions (class A amplifier). The resistors R_S, R'_S and R_F, R'_F provide linear series and shunt feedback to improve the linearity. At the output, a second transformer converts the balanced signal back to single-ended form and also provides impedance matching.

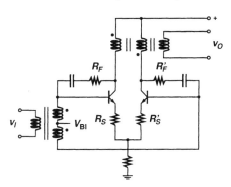

Fig. 8.25 Linear wideband amplifier.

Like digital transmitters, analog optical transmitters can make use of direct modulation or external modulation. As we know, external modulation has the advantage of low optical chirp, which is important when transmitting at the 1.55-μm wavelength. Furthermore, external modulation permits the use of a CW laser that is optimized for low RIN noise. For example, the neodymium doped yttrium-aluminum-garnet laser (Nd:YAG) can generate 50 to 200 mW of optical power at 1.3 μm with $RIN < -165$ dB/Hz [106]. Thus, besides the laser driver discussed above, modulator drivers for EAMs and MZMs also are of interest. Unfortunately, external modulators are inherently nonlinear, which makes the use of a linearization technique necessary. In the following, we discuss two important linearization techniques: *optical feedforward linearization* and *predistortion linearization*. Linearization techniques also can be applied to laser drivers to improve the yield (more tolerance to nonlinear leakage currents), or to permit higher optical output power levels, or both.

Optical Feedforward Linearization. Optical feedforward linearization capitalizes on the fact that it is easier to build a linear receiver than a linear transmitter. Figure 8.26 illustrates the principle of operation. A first transmitter (TX1) converts the electrical input signal v_I into an optical signal by means of external or direct modulation. Unavoidably, this transmitter introduces some undesirable distortions: s + d = signal + distortions. A highly linear receiver (RX) picks up a fraction of the transmitted signal through an optical power splitter. The received signal is compared with the original signal, v_I, to determine the distortions contained within it. The inverted distortions then are converted to an optical signal with a second transmitter (TX2) and added to the main signal with an optical power combiner. At the output of the combiner, the optical signal ideally is free of distortions: s + d − d = s. Note that the optical compensation signal from TX2 must be at a different wavelength than the main signal from TX1 to avoid coherent interference.

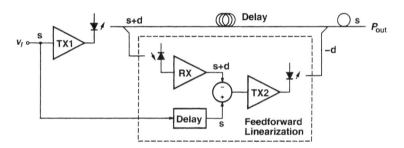

Fig. 8.26 Principle of optical feedforward linearization.

Optical feedforward linearization has the advantage that it can compensate any type of nonlinearity and can automatically track changes in the nonlinearity resulting from temperature variations and aging. In principle, feedforward linearization can even suppress the noise produced by transmitter TX1. A limitation of this approach is the nonlinearity (and noise) introduced by TX2, which is not compensated. For better linearity, direct modulation normally is preferred for TX2. Also, because the optical main signal and the compensation signal must be at different wavelengths, chromatic dispersion in the fiber desynchronizes the signals resulting in larger distortions for longer fibers. In general, the optical feedforward system is quite complex and requires careful tuning of the gains and delays. Nevertheless, an average distortion reduction of 20 dB has been achieved with feedforward linearization applied to a directly modulated transmitter [41]. Similarly, 12- to 20-dB distortion reduction has been achieved for a transmitter with an EAM [54].

Predistortion Linearization. Another method to reduce distortions is predistortion linearization. In this approach, the inverse function of the transmitter's nonlinearity is used to predistort the analog signal before it is fed into the transmitter. Ideally, the predistortion nonlinearity and the transmitter nonlinearity cancel each other, resulting in an optical output signal that is linearly related to the original input signal. A difficulty of this linearization scheme is that it requires that the mathematical form of the nonlinearity is known. Also, in practice, it is difficult to make the two nonlinearities track each other over temperature variations and aging.

The MZ modulator is a good candidate for predistortion linearization because its nonlinearity is well known (a sinusoid), consistent, and stable. Without linearization, the MZ modulator's CTB performance for a 60-channel system with a modulation index of 3.6% is limited to about -39 dBc because of odd-order distortions [106]. But even without linearization, low even-order distortions and good CSO performance can be obtained if the modulator is biased at the inflexion point of its nonlinear switching curve. Note that for the MZM's sinusoidal switching curve, the inflexion point coincides with the quadrature point. The predistortion circuit shown in Fig. 8.27 approximates the predistortion function $v_O = V_O \cdot \arcsin(i_I/I_I)$ required to linearize an MZ modulator [106]. The input currents, $I + i_I$ and $I - i_I$, are provided by a predriver with differential current outputs. The diode loads at the input and the diodes degenerating the emitters of the differential pair perform the predistortion. The differential output voltage, v_O, drives the MZ modulator.

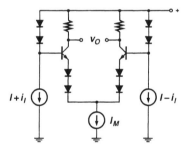

Fig. 8.27 Predistortion circuit for MZ modulators.

A controller, similar to the ABC circuit shown in Fig. 8.20, can be used to automatically adjust the MZM's bias and the predistortion circuit's gain for optimum performance. More specifically, two small-signal pilot tones can be injected into the driver to produce distortion products in the optical output signal at known frequencies. A first negative feedback loop adjusts the bias voltage until the second-order distortion products are minimized, while a second loop adjusts the predistorter's operating point until the third order products are minimized. By combining a predistortion circuit with such a controller, a CTB of better than -65 dBc and a CSO of better than -70 dBc has been achieved [106].

Predistortion linearization also can be applied to laser drivers. In the approach taken in [13], artificial distortions are generated by feeding two slightly mismatched

294 LASER AND MODULATOR DRIVERS

and nonlinear amplifiers with the analog input signal and its inverse. Summing the outputs of the two amplifiers yields a variety of distortion products while suppressing the fundamentals. To align the artificially generated distortions with the laser distortions such that they are equal in magnitude but opposite in phase, they are passed through an attenuator, phase shifter, and frequency filter. Then, they are added to a delayed copy of the original input signal to produce the predistorted signal for driving the laser. With the laser in place, the amplitude, phase, and frequency tilt of the predistorter are adjusted until the CSO/CTB performance is optimum.

8.3 DRIVER CIRCUIT IMPLEMENTATIONS

In the following, we examine some representative transistor-level laser- and modulator-driver circuits that have been reported in the literature. These circuits illustrate how the design principles discussed in the previous section can be implemented in a broad variety of technologies using different types of transistors such as the metal-semiconductor field-effect transistor (MESFET), the heterostructure field-effect transistor (HFET), the bipolar junction transistor (BJT), the heterojunction bipolar transistor (HBT), and the complementary metal-oxide-semiconductortransistors (CMOS) (cf. Appendix D).

8.3.1 MESFET and HFET Technology

Open Drain Modulator Driver. Figure 8.28 shows the schematic of the GaAs-FET output stage reported in [182]. This stage is part of a 10-Gb/s EAM driver and is implemented in a 0.35-μm GaAs-PHFET technology. The transistors in this circuit are depletion-mode FETs, which means that they conduct current when the gate-source voltage is zero.

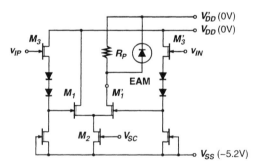

Fig. 8.28 MESFET/HFET implementation of an EAM-driver output stage based on [182].

The output stage consists of an FET current-steering circuit, M_1 and M_1', which dumps the tail current from M_2 either into the positive power supply (no dummy load is used) or into the EAM load. The driver has an open drain output (no back termination) and connects to the EAM load either directly, as shown in the figure, or through a transmission line. The external load consists of a single-ended EAM with a 50-Ω resistor, R_P, in parallel. In Fig. 8.28 and all subsequent driver schematics, we make a distinction between the on-chip supply, V_{DD}, and the off-chip supply, V'_{DD}, to which external components, such as the load, connects. This distinction clarifies the flow of on-chip and off-chip supply currents. The output voltage swing of the modulator driver can be adjusted with the voltage V_{SC}, which controls the tail current from M_2. The output bias voltage of the modulator driver, as shown, is zero, but it can be made nonzero by injecting a DC current into the EAM load.

The FETs M_1 and M'_1 in the output stage have a channel width of 400 μm and present a considerable input capacitance. Thus, a two-stage predriver is used to drive this output stage. Each predriver stage consists of a current-steering circuit followed by a source-follower pair (only the source-follower pair of the second stage, M_3 and M'_3, is shown in Fig. 8.28). The chip further includes an input buffer with an on-chip 50-Ω termination.

In [94], a similar open-drain output stage has been reported, which is part of a 10-Gb/s EAM driver implemented in a 0.2-μm GaAs-PHFET technology. In contrast to Fig. 8.28, this implementation brings both drain outputs off chip. When driving a single-ended EAM load, an external dummy resistor must be connected to the unused drain.

Modulator/Laser Driver with Back Termination. Figure 8.29 shows a simplified schematic of the GaAs-FET output stage reported in [79]. This stage is part of a 40-Gb/s EAM driver and is implemented in a 0.2-μm GaAs-HFET technology with enhancement- and depletion-mode devices.

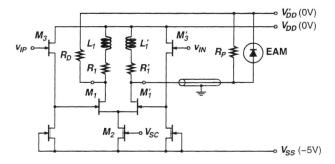

Fig. 8.29 MESFET/HFET implementation of an EAM-driver output stage with back termination based on [79].

The driver is connected to the EAM load through a 50-Ω transmission line. To avoid double reflections on the transmission line, both sides are terminated. The termination on the modulator side is provided by the external 50-Ω resistor R_P and the termination on the driver side, the back termination, is implemented with the on-chip resistor R'_1. The value of R'_1 is chosen to be 100 Ω, rather than 50 Ω, to reduce

296 LASER AND MODULATOR DRIVERS

the power dissipation while still achieving an effective back termination. To balance the output stage and to keep the on-chip supply voltage, V_{DD}, quiet, the dummy load in the left branch of the current-steering circuit exactly mirrors the load in the right branch, that is, it consists of an on-chip 100-Ω resistor, R_1, and an external 50-Ω resistor, R_D. The peaking inductors, L_1 and L'_1, improve the rise and fall times of the output signal.

This output stage is preceded by a predriver, which consists of a cascade of three source-follower pairs, followed by a current-steering circuit, followed by another cascade of three source-follower pairs (only the last pair, M_3 and M'_3, is shown in Fig. 8.29). Both cascades of source followers include R-C high-pass coupling networks to speed up the signal transitions, similar to the MA stage shown in Fig. 6.41. Two on-chip resistors connecting to the output nodes (not shown in Fig. 8.29) can be used to introduce a bias voltage across the modulator. The driver chip in [79] also contains a retiming flip-flop, which operates from a half-rate clock (20 GHz).

The modulator driver circuit in Fig. 8.29 also is capable of driving a single-drive or dual-drive MZM. In the latter case, the external dummy resistor, R_D, is removed and each output is used to drive one port of the MZM. The same circuit also can act as a laser driver, if a bias current source, similar to Q_5 in Fig. 8.32, is added.

Modulator/Laser Driver with Active Back Termination. Figure 8.30 shows a simplified schematic of the GaAs-FET output stage reported in [138]. This stage is part of a 10-Gb/s laser/modulator driver and is implemented in a 0.25-μm GaAs-PHEMT technology.

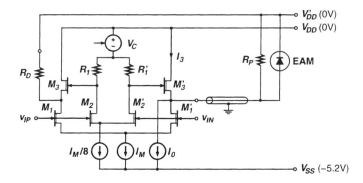

Fig. 8.30 MESFET/HFET implementation of an EAM-driver output stage with active back termination based on [138].

The modulator driver in Fig. 8.30 connects to the EAM load through a 50-Ω transmission line. To avoid double reflections on the transmission line, both sides are terminated. The termination on the modulator side is provided by the external 50-Ω resistor R_P, whereas the back termination is implemented with an *active* circuit. As usual, the driver generates the output signal with a current-steering circuit, M_1 and M'_1. An on-chip dummy resistor, R_D, is included at the drain of M_1. A scaled-down replica (1:8) of the output stage, M_2 and M'_2, generates a copy of the intended output voltage (without reflections). The source followers M_3 and M'_3 buffer the replica signal and are sized such that their output impedance ($\approx 1/g_{m3}$) matches the transmission line impedance. The output impedance of the source follower M'_3 acts as the back termination and absorbs possible reflections. The current I_3 through M'_3 is kept at the constant value I_0 by means of a feedback circuit implemented with an op amp (not shown) that controls the voltage V_C. The power dissipated in this active back termination circuit is $P = (V_{DD} - V_{SS}) \cdot (I_0 + I_M/8)$, where I_0 is the bias current for M'_3 (e.g., 10 mA) and $I_M/8$ is the current in the scaled down replica (e.g., 12.5 mA). For comparison, a passive back termination with a resistor equal to the load resistor dissipates the much larger power $P = (V_{DD} - V_{SS}) \cdot I_M$, where I_M is the modulation current (e.g., 100 mA) and no bias current has been assumed.

The driver IC described in [138] also includes on-chip terminated input buffers for the data and clock signals, a retiming flip-flop, and a pulse-width control circuit. The chip can be configured as a modulator driver driving a 50-Ω load or as a laser driver driving a 25-Ω load. In the latter case, two output stages are connected in parallel. The chip also provides a programmable DC current, which can be injected into the laser or modulator load to add a bias. Note that the feedback circuit that keeps I_3 equal to I_0 also prevents such a bias current from entering the active back termination (M'_3).

8.3.2 BJT and HBT Technology

Open Collector Laser Driver. Figure 8.31 shows a simplified schematic of the bipolar output stage reported in [148]. This stage is part of a 2.3-Gb/s laser driver and is implemented in an 8-GHz Si-BJT technology.

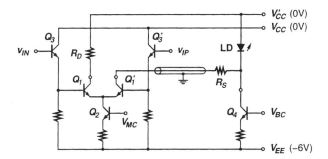

Fig. 8.31 BJT/HBT implementation of a laser-driver output stage based on [148].

The output stage consists of a bipolar current-steering circuit, Q_1 and Q'_1, with both collector outputs coming off chip. One output is terminated with an external dummy resistor, R_D, and the other drives the laser load through a 25-Ω transmission line. The 20-Ω series resistor, R_S, matches the laser impedance of about 5 Ω to the transmission line. Even in applications without a transmission line, a small series resistor R_S is desirable because it dampens the ringing and jitter caused by parasitic inductances (e.g., due to bond wires) in conjunction with laser and driver capacitances. The modulation current, I_M, is supplied by the tail-current source consisting of Q_2 and an emitter degeneration resistor. The magnitude of the modulation current can be controlled with the voltage V_{MC}. Similarly, the laser bias current, I_B, is supplied by the current source consisting of Q_4 and its emitter degeneration resistor; the magnitude of I_B can be controlled with the voltage V_{BC}. Note that the modulation and bias currents are provided through two separate terminals (pins). In this way, the bias current can be injected *after* the series resistor, R_S, avoiding an unnecessary voltage drop across this resistor and averting a potential headroom problem. This output stage is driven by a cascade of two emitter-follower pairs (only the last pair, Q_3 and Q'_3, is shown in Fig. 8.31). The emitter followers present a low output impedance to the output stage and provide level shifting to increase the collector-emitter voltage of Q_1 and Q'_1, thus boosting their maximum f_T (cf. Section 6.3.2).

In [154], a similar open-collector output stage has been reported, which is part of a 10-Gb/s laser driver implemented in a 55-GHz GaAs-HBT technology. In contrast to Fig. 8.31, a cascoded current-steering circuit is used and the laser is connected directly to the driver output, avoiding the transmission line and the series resistor R_S. This direct laser connection permits the driver to run from a lower supply voltage, thus reducing the power dissipation. The chip in [154] also contains an on-chip terminated input buffer and a predriver based on the Cherry-Hooper stage (cf. Section 6.4.2).

Laser/Modulator Driver with Back Termination. Figure 8.32 shows a simplified schematic of the bipolar output stage reported in [147]. This stage is part of a 10-Gb/s laser/modulator driver and is implemented in a 25-GHz Si-BJT technology.

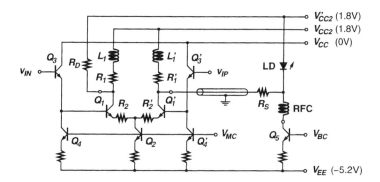

Fig. 8.32 BJT/HBT implementation of a laser-driver output stage with back termination based on [147].

For this driver, the characteristic impedance of the transmission line (and the load) is chosen to be 50 Ω, which permits it to operated as a laser or modulator driver. When used as a laser driver, as shown in Fig. 8.32, a series resistor, $R_S \approx 45\,\Omega$, matches the laser to the transmission line. To avoid double reflections on the transmission line, the driver incorporates the back termination resistors R_1 and R'_1. Ideally, these resistors should match the 50-Ω transmission line; however, as a compromise between matching quality and power dissipation, they were increased to 100 Ω. Note that these resistors add to the power dissipation by absorbing not only a good part of the modulation current from Q_2, but also some of the bias current from Q_5. The emitter degeneration resistors R_2 and R'_2 are distributed among the emitter fingers to ensure an even current distribution within the transistors Q_1 and Q'_1. The peaking inductors, L_1 and L'_1, improve the rise and fall times of the output signal and are realized with overlength bond wires. For versatility, for symmetry of the output stage, and to keep the on-chip supplies V_{CC} and V_{CC2} quiet, both outputs are brought off chip: when operating as a laser or EAM driver, one output is terminated with an external dummy resistor, R_D; when driving a dual-drive MZM, both outputs are used. In Fig. 8.32, the laser bias current from Q_5 is applied to the laser cathode through an RFC to minimize the capacitive loading of this high-speed node.

The input buffer and predriver for the output stage consists of a cascade of three emitter-follower pairs, followed by a current-steering circuit, followed by another three emitter-follower pairs (only the last pair, Q_3 and Q'_3 with current sources Q_4 and Q'_4, is shown in Fig. 8.32). To obtain a clean eye diagram over a wide range of modulation currents, the tail current of the predriver's current-steering circuit as well as the bias currents of several emitter followers are varied with the modulation current. Each bias current has a component that varies proportional to the modulation current (e.g., the currents provided by Q_4 and Q'_4) as well as a component that is constant (not shown in Fig. 8.32). The input buffer and predriver of this chip also include means to control the pulse-width distortions and to equalize the rise and fall times at the laser/modulator load. The output stage runs from a higher supply voltage (V_{CC2} and V'_{CC2}) than the rest of the chip, permitting a higher output voltage swing or a higher current into the 50-Ω load without increasing the power dissipation of the input buffer and predriver. At the maximum swing, the collector-emitter voltage of Q_1 and Q'_1 exceeds the open-base breakdown voltage of 3.7 V; however, the low output impedance of the predriver pushes the breakdown voltage to a higher value (cf. Appendix D.3).

In [66], a similar output stage has been reported, which is part of a 40-Gb/s EAM driver implemented in a 130-GHz InP-HBT technology.

Modulator Driver with Built-In MUX. Figure 8.33 shows a simplified schematic of the bipolar output stage with built-in multiplexer reported in [97]. This circuit is part of a 40- to 50-Gb/s EAM driver and is implemented in a 72-GHz SiGe-BJT technology.

The circuit shown in Fig. 8.33 combines the data-multiplexer and modulator-driver function into a single high-speed stage. Therefore, this circuit also is known as a *power MUX*. The CML-type multiplexer combines two 20-Gb/s data streams into a single 40-Gb/s data stream, which then is used to drive the modulator. A 20-GHz clock signal is needed at the select input of the multiplexer. The power MUX works as follows: the constant current from Q_4 (40–50 mA) is steered by a first CML switch, Q_1 and Q'_1, to either one of two CML data switches, Q_2, Q'_2 or Q_3, Q'_3, depending on the select signal. Then, the selected CML data switch steers the current either

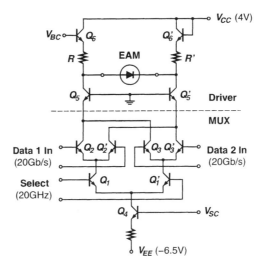

Fig. 8.33 BJT/HBT implementation of a differential 40-Gb/s EAM-driver output stage with built-in MUX based on [97].

into the right or left branch, depending on the input data. The current from the MUX passes through the cascode transistors Q_5 and Q'_5, which prevent breakdown-voltage violations and reduce the capacitance at the driver's output. Finally, the MUX current drops over the load R, Q_6 or R', Q'_6, where the driving voltage for the EAM is produced. The inductive load impedance presented by the emitters of Q_6 and Q'_6 improves the rise and fall times. The driver chip is directly connected to the EAM with low-inductance wire bonds (no transmission lines). The output voltage swing and output bias voltage can be adjusted with V_{SC} and $V_{CC} - V_{BC}$, respectively. The chip in [97] also includes on-chip terminated input buffers and predrivers for the two CML data switches, Q_2, Q'_2 and Q_3, Q'_3. Each buffer/predriver consists of two emitter-follower pairs, followed by a current-steering circuit, followed by another three emitter-follower pairs.

The advantages of the power MUX approach over a conventional full-rate 40-Gb/s modulator driver are as follows. (i) Phase shifts and jitter in the input data signals are suppressed by the MUX. Note that the MUX has a similar effect as a retiming flip flop. (ii) The rise and fall times of the output signal are improved and can, to some extent, be controlled by the clock signal swing. (iii) The half-rate predrivers for the data signals are less critical.

The circuit in Fig. 8.33 drives the EAM differentially and thus requires a symmetrical modulator, that is, it must be possible to drive both electrodes of the EAM independently (no shared electrode with the CW laser) and they must have small and similar capacitances to ground. The advantage of driving the modulator differentially is that only half of the voltage swing is required at each output. For example, a 1-V_{pp}

signal at each output produces a 2-V_{pp} signal across the EAM. However, at the time of writing, such modulators do not seem to be commercially available.

In [166], a similar output stage has been reported, which is part of a 40-Gb/s EAM driver implemented in the same 72-GHz SiGe-BJT technology. In contrast to Fig. 8.33, this driver operates at the full rate of 40 Gb/s and thus has no built-in multiplexer.

Modulator Driver with Distributed Output Stage. Figure 8.34 shows a simplified schematic of the distributed output stage reported in [202]. This stage is part of a 10-Gb/s MZ modulator driver and is implemented in a 50-GHz GaAs-HBT technology.

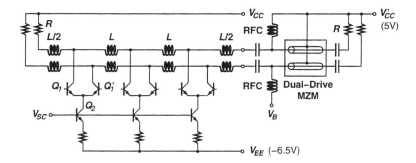

Fig. 8.34 BJT/HBT implementation of a distributed MZM-driver output stage based on [202]. (The input transmission lines and the emitter followers driving Q_1, Q'_1, ... are not shown.)

An important speed limitation of the current-steering circuit, especially when used as part of a high-swing MZM driver, is the time constant formed by the parasitic output capacitance of the large transistors and the load resistor. To circumvent this limitation, the large output capacitance can be distributed into an artificial transmission line. This is done by splitting the current-steering transistor pair into n smaller pairs (first pair: Q_1 and Q'_1) and connecting them with inductors as shown in Fig. 8.34. With the appropriate value for the inductors L, a pair of artificial transmission lines matched to the termination resistors R and the impedance of the MZM are formed. Now, the speed is limited by the cutoff frequency of the artificial transmission lines, which can be made high by choosing a large number of sections, n. A second pair of artificial transmission lines (not shown in Fig. 8.34) provides the differential input signals to the n sections of the distributed output stage. Of course, this technique is the same that we already discussed for distributed amplifiers in Section 6.3.2.

In [202], the number of sections was chosen to be five ($n = 5$). Each section consists of a current-steering circuit (first section: Q_1 and Q'_1) driven by a pair of emitter followers (not shown in Fig. 8.34). The emitter followers lower the input capacitance to the same level as the output capacitance, permitting the use of the same inductor value for the input and output transmission lines. Furthermore, the emitter followers lower the input conductance and thus reduce the loss of the input transmission lines. The distributed output stage is driven by a lumped predriver with two stages. Each predriver stage consists of an emitter-follower pair followed by a current-steering circuit. A further advantage of the distributed output stage is its superior output matching (S_{22}) at high frequencies, which helps to avoid double reflections on the transmission lines to the MZM.

The driver can be AC coupled to a dual-drive MZM as shown on the right-hand side of Fig. 8.34. Because of the push-pull configuration, a voltage swing of only one half of the switching voltage, V_π, is needed per output. The output swing can be adjusted with the voltage V_{SC}, which controls the tail currents (first section: Q_2) of the output stage as well as the currents in the predriver. The bias voltage V_B is supplied to the MZ modulator through a bias T (RFC with coupling capacitor). The low-frequency pilot tone needed for the ABC (cf. Section 8.2.8) can be superimposed on V_{SC}.

8.3.3 CMOS Technology

Predriver. Figure 8.35 shows a simplified schematic of the CMOS predriver reported in [31]. This stage is part of a 10-Gb/s laser/modulator driver and is implemented in a 0.18-μm CMOS technology.

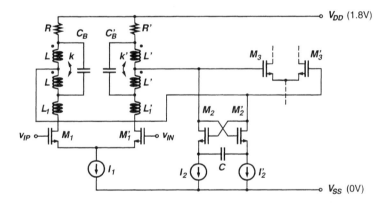

Fig. 8.35 CMOS implementation of a predriver based on [31].

The predriver consists of a current-steering circuit, M_1 and M_1', which is loaded by a combination of the resistors R and R', the T-coil networks L, C_B and L', C_B', and the series inductors L_1 and L_1'. The MOSFETs M_2 and M_2' form a negative impedance converter (NIC), which inverts the capacitance C and thus present the predriver with a negative load capacitance equal to about $-C$. This negative capacitance compensates about 30% of the output stage's large input capacitance (from M_3 and M_3'). The remaining capacitance is driven with the help of the T-coil networks, which provide a significant bandwidth boost (cf. Section 6.3.2). The T-coil networks are realized with on-chip coupled inductors. The inductors L_1 and L'_1 provide series peaking, which further increases the bandwidth.

In [31], this predriver is followed by a current-steering output stage, M_3 and M'_3, with passive 75-Ω back termination. The predriver and the output stage form one of three identical "driver slices" that constitute the reconfigurable laser/modulator driver chip. When configured as a laser driver, all three slices are enabled and operate in parallel; when configured as a modulator driver, only a single slice is enabled.

Burst-Mode Laser Driver. Figure 8.36 shows a simplified schematic of the CMOS low-power laser driver reported in [162]. This circuit is part of a 155-Mb/s burst-mode laser driver for PON applications and is implemented in a 0.5-μm CMOS technology.

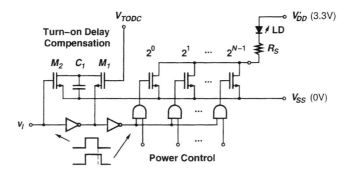

Fig. 8.36 CMOS implementation of a low-power burst-mode laser driver based on [162].

In contrast to the drivers discussed so far, this driver uses *current switching* instead of current steering. This means that during the transmission of a zero, the modulation current is shut off rather than steered into a dummy load. As a result, the average power dissipation is reduced by a factor two. This scheme is particularly suitable for burst-mode laser drivers because it automatically powers down the output stage during idle periods, that is, when no bursts are transmitted. For example, if the average burst activity is 10%, the total power savings are 20×. Furthermore, the driver in Fig. 8.36 operates with *zero* laser bias current. This mode of operation ensures the high interburst extinction ratio required for burst-mode drivers (cf. Section 8.2.9) and saves additional power.

As we know, operating a laser without bias current results in a turn-on delay and turn-on delay jitter (cf. Section 7.2). The turn-on delay can be compensated by predistorting the pulse width of the input data signal. The simple turn-on delay compensation circuit shown in Fig. 8.36 delays the falling data edge by an amount equal to the laser's turn-on delay. On the falling edge of v_I, the output of the first inverter is loaded by C_1 (through M_1), causing a delay of this edge. The amount of delay can be controlled with the gate voltage of M_1, V_{TODC}. On the rising edge of v_I, M_2 discharges C_1 rapidly preventing a similar delay of the latter edge. Unfortunately, the turn-on delay jitter cannot be compensated in a similar manner because of its

304   LASER AND MODULATOR DRIVERS

random nature. But in low-speed applications up to about 155 Mb/s, it is unlikely to cause problems.

The laser driver in [162] also features a p-MOS shunt transistor across the laser diode (not shown in Fig. 8.36) to suppress an optical tail at the end of the burst. A digital APC circuit controls the output power by selectively enabling some of the N parallel current-switching transistors shown in Fig. 8.36. Because these transistors have widths proportional to 2^n with $n = 0 \ldots N - 1$, they form a built-in binary D/A converter. As discussed in Section 8.2.9, the N-bit word is generated with an up/down counter controlled by an integrate-and-dump circuit. The chip further includes an end-of-life detector with selectable threshold.

LED Driver. Figure 8.37 shows a simplified schematic of the CMOS LED-driver output stage with preemphasis reported in [171]. This stage is part of a 125-Mb/s Ethernet transmitter (100Base-FX) and is implemented in a 0.5-μm CMOS technology.

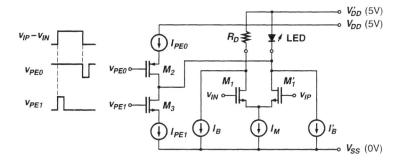

Fig. 8.37   CMOS implementation of a LED-driver output stage with preemphasis based on [171].

The current source I_M supplies the modulation current of typically 70 mA. This current source has a built-in positive temperature coefficient to compensate for the negative temperature dependence of the light-emitting diode's (LED's) slope efficiency. A current-steering circuit, M_1 and M'_1, modulates I_M and drives the LED. An external dummy resistor terminates the unused output of the driver. To enhance the speed of the inherently slow LED, a small bias current, I'_B, and a preemphasis scheme are used. The preemphasis works as follows: during the first 2.5 ns of the 8-ns bit period, an additional current, I_{PE1}, is forced into the LED by means of switch M_3 to improve the optical rise time. Similarly, at the end of the bit period, M_2 turns on to discharge the LED with I_{PE0}, thus improving the optical fall time.

The LED driver in [171] is part of a large mixed-signal chip for Ethernet fiber (100Base-FX) to twisted-pair (100Base-TX) media conversion. It also includes the main amplifier and clock and data recovery circuit for the optical receiver.

8.4 PRODUCT EXAMPLES

Tables 8.1 and 8.2 summarize the main parameters of some commercially available laser and modulator drivers. The numbers have been taken from data sheets of the manufacturer that were available at the time of writing. For up-to-date product information, please contact the manufacturer directly. Rise and fall times, tabulated under $t_{R,F}$, are measured from 20% to 80%. The compliance voltage of laser drivers, $v_{O,\min}$, is measured relative to the negative power-supply terminal. The power dissipation numbers quoted for the laser drivers are under the condition of *zero* modulation and bias currents. Similarly, the power dissipation numbers quoted for the modulator drivers are for *zero* modulation and bias voltages, except for the numbers enclosed in parenthesis, which are quoted for the maximum voltage swing. For modulator drivers with a single-ended output, the maximum voltage swing, tabulated under $V_{S,\max}$, is followed by "(s)." For modulator drivers with differential outputs, the *differential* voltage swing is given followed by "(d)."

From these tables, we see that the power consumption of current 2.5- and 10-Gb/s laser and modulator drivers is rather high. Even with zero output signal, it is in the range of 0.2 to 1.4 W. For a typical output current (or voltage swing), this value increases roughly by 0.1 to 1 W. Note that most modulator drivers listed are implemented in a GaAs technology. GaAs FETs are preferred over other high-speed devices such as SiGe BJTs because of their higher breakdown voltage. The reader may wonder about the difference between the similar 2.5- and 10-Gb/s laser drivers from Agere: the LG1627BXC and TCLD0110G both contain a retiming flip-flop, whereas the LG1625AXF and TLAD0110G do not.

8.5 RESEARCH DIRECTIONS

The research effort focusing on laser and modulator drivers can be divided roughly into four areas: higher speed, higher integration, lower power, and lower cost. In the following, we briefly touch on these areas.

Higher Speed. It was pointed out in Section 5.5 that many research groups are now aiming at the 40-Gb/s speed and beyond. To this end, fast modulator drivers, possibly with integrated MUX, must be designed. Usually, heterostructure devices, such as HBTs and HFETs, based on compound materials, such as SiGe, GaAs, and InP, are used to reach this goal (cf. Appendix D).

The following papers on high-speed laser and modulator drivers were published recently:

- In SiGe-HBT technology, a 23-Gb/s modulator driver with 3.5-V_{pp} single-ended swing, a 40-Gb/s EAM driver with 1.0-V_{pp} single-ended swing, and a 50-Gb/s EAM driver also with 1.0-V_{pp} single-ended swing have been reported in [167], [166], and [97], respectively.

Table 8.1 Examples for high-speed laser driver products.

Company & Product	Speed (Gb/s)	$t_{R,F}$ (ps)	$I_{M,\max}$ (mA)	$I_{B,\max}$ (mA)	$v_{O,\min}$ (V)	Power (mW)	Technology
Agere LG1625AXF	2.5	90	65	40		520	GaAs HFET
Agere LG1627BXC	2.5	100	85	60		730	GaAs HFET
Infineon S896A006	2.5		60	60	1.5	165	Si BJT
Maxim MAX3867	2.5	90	60	100	2.0	205	
Nortel YA08	2.5	150	80	100	1.4	350	Si BJT
Philips OQ2545HP	2.5	120	60	100	3.0	350	Si BJT
Agere TLAD0110G	10.0	30	100	120	1.7	780	GaAs HFET
Agere TCLD0110G	10.0	25	100	100		780	GaAs HFET
Maxim MAX3930	10.0	29	100	100	1.6	540	SiGe HBT

Table 8.2 Examples for high-speed modulator driver products.

Company & Product	Speed (Gb/s)	$t_{R,F}$ (ps)	$V_{S,\max}$ (V)	$V_{B,\max}$ (V)	Power (mW)	Technology
Agere LG1626DXC	2.5	90	3.0 (s)	1.5	730	GaAs HFET
Agere TMOD0110G	10.0	30	5.4 (d)		1,400	GaAs HFET
Giga GD19901	10.0	45	6.0 (d)		(3,200)	GaAs HFET
Giga GD19903	10.0	45	10.0 (d)		(8,600)	GaAs HFET
Maxim MAX3935	10.0	34	3.0 (s)	1.2	550	SiGe HBT
OKI KGL4115F	10.0	40	2.7 (s)	1.0	1,300	GaAs HFET
AMCC S76803	48.0	7	7.0 (d)			SiGe HBT
Velocium AUH232	43.0	8	8.0 (s)		1,250	GaAs HFET

- In GaAs-HFET technology, a 20-Gb/s laser driver has been reported in [196]; 40-Gb/s modulator drivers with a single-ended swing of $2.9\,V_{pp}$ and $6\,V_{pp}$ have been reported in [79] and [172], respectively.

- In InP-HBT technology, a 20-Gb/s modulator driver with $4.0\text{-}V_{pp}$ single-ended swing and a 40-Gb/s EAM driver with $2.2\text{-}V_{pp}$ single-ended swing have been reported in [89] and [66], respectively.

Higher Integration. Another area of research aims at higher integration by combining the laser diode, monitor photodiode, and the driver circuit on the same chip, creating a so-called *optoelectronic integrated circuit* (OEIC). For example, a complete transmitter consisting of a 1.5-μm distributed feedback (DFB) laser and an HFET driver circuit have been integrated on a single InP substrate [84]. However, it is a challenge to combine laser and circuit technologies effectively into a single one because of the significant structural differences between lasers and transistors. For example, lasers require mirrors or gratings for their operation, whereas transistors don't. As a result, transmitter OEICs are not as far advanced as receiver OEICs.

An alternative to the above-mentioned monolithic OEICs is the integration of lasers and drivers by means of flip-chip technology. An important advantage of this flip-chip OEIC approach is that the technologies for the laser chip and the driver chip can be chosen (and optimized) independently, thus avoiding the compromises of monolithic OEICs.

Lower Power. With increasing miniaturization of the transceiver modules, the heat generated by the driver becomes a more serious problem. When using an uncooled laser, the heat from the driver may degrade the laser's performance and lifetime; when using a cooled laser, the thermoelectric cooler must work extra hard to remove the heat from the laser *and* the driver. Therefore, low-power laser drivers are a subject of great interest.

The power dissipation and the associated heating can be reduced by lowering the supply voltage and coupling the driver directly to the laser, that is, avoiding a transmission line and the losses due to matching and termination resistors. Copackaging techniques can be used to keep the package and interconnect parasitics small. To obtain a good eye quality in direct-coupled high-speed drivers, it is important to model the L-C parasitics accurately and to dampen them sufficiently to minimize ringing and jitter.

Lower Cost. Another area of research is focusing on the design of high-performance drivers in low-cost, mainstream technologies. For the reasons already given in Section 5.5, digital CMOS is of particular interest.

For example, 10-Gb/s, 0.18-μm CMOS laser drivers have been reported in [128] and [31]. A laser driver for a fiber-to-the-home system must be very low cost to be competitive with traditional telecom services and low power to minimize the size and cost of the back-up battery. Such a CMOS burst-mode laser driver consuming only 15 mW has been reported in [162].

308 LASER AND MODULATOR DRIVERS

8.6 SUMMARY

The main specifications of digital laser and modulator drivers are as follows:

- The modulation and bias current ranges for laser drivers, which must be large enough to operate the desired laser under worst-case conditions. In particular, uncooled lasers require large current ranges.

- The output voltage range (or compliance voltage) for laser drivers. The low end of this range should be as low as possible to permit DC coupling of the laser while maintaining a low supply voltage.

- The modulation and bias voltage ranges for modulator drivers, which must be large enough to operate the desired modulator under worst-case conditions. In particular, high-speed Mach-Zehnder (MZ) modulators require a large modulation voltage (or voltage swing).

- The power dissipation, which should be as low as possible to save power and limit undesirable heat generation.

- The rise and fall times, which must be short compared with the bit period. However, the rise time of laser drivers should not be too short to limit the generation of optical chirp.

- The pulse-width distortion, which usually is compensated with an adjustable pulse-width control circuit.

- The jitter generation, which must be very low for SONET compliant transmitter.

In addition, some standards, such as SONET, require that the transmitted optical signal complies with a given eye mask.

The output stage of most laser and modulator drivers is based on the current-steering circuit, which has the following advantages: high switching speed, low noise generation, low noise sensitivity, and programmability of the output signal swing. DC or AC coupling can be used to connect the current-steering output stage to the laser or modulator load. AC coupling permits a lower supply voltage, but requires more external components. The driver can be connected to the laser or modulator either directly (e.g., through a short bond wire) or through an impedance-matched transmission line (with or without back termination). The use of a transmission line permits a larger distance between the driver and the laser or modulator.

A predriver, which provides voltage gain and a low-impedance output, normally is used to drive the large output stage. Pulse-width control to compensate for pulse-width distortions usually is implemented by introducing an adjustable offset voltage at the input of the predriver. A flip-flop for data retiming can be used to reduce jitter and pulse-width distortions at the driver output. In laser drivers, an automatic power control (APC) circuit uses negative feedback from the monitor photodiode to keep the optical output power, and optionally the extinction ratio (ER), constant. Similarly, for MZM drivers, an automatic bias control (ABC) circuit is required to stabilize the operating point of the MZ modulator. Some laser drivers feature an end-of-life detector, which issues an alert that the laser must be replaced soon. Burst-mode laser drivers require a very high interburst ER and a special APC circuit that operates correctly for a bursty data signal. Analog laser/modulator drivers must be highly linear to minimize signal distortions and thus often incorporate a linearization scheme.

Laser and modulator drivers have been implemented in a wide variety of technologies including metal-semiconductor FET (MESFET), heterostructure FET (HFET), BJT, heterojunction bipolar transistor (HBT), BiCMOS, and CMOS.

Currently, researchers are working on 40-Gb/s modulator drivers and beyond, drivers integrated with the laser or modulator on the same chip, low-power laser drivers, as well as laser and modulator drivers in low-cost technologies such as CMOS.

8.7 PROBLEMS

8.1 Switching a Current-Steering Circuit. (a) Calculate the differential voltage necessary to completely switch an FET current-steering circuit (without source degeneration). The tail current is I_M and the FETs can be described by the quadratic model $I_D = \mu_n C'_{ox}/2 \cdot W/L \cdot (V_{GS} - V_{TH})^2$. (b) Calculate the differential voltage necessary to switch a BJT current-steering circuit (without emitter degeneration) such that 99% of the tail current flows into one output. The tail current is I_M, and the BJTs can be described by the model $I_C = I_{C0} \cdot \exp(V_{BE}/V_T)$.

8.2 Interconnect Inductance. A 5.2-V, 10-Gb/s laser driver has 30-ps rise and fall times (measured from 20% to 80%) and is programmed for a modulation current of 50 mA. The DC-coupled laser and the series resistor together drop 2.5 V (when the laser is on) and the driver has a compliance voltage of 1.5 V. What is the maximum inductance that can be tolerated in the driver-to-laser interconnection?

8.3 Current Efficiency for Passive Back Termination. A laser driver is implemented with a passive back termination, R_T that matches the characteristic impedance of the transmission line, R_0. The laser's I/V characteristics can be modeled as $V_L = V_{TH} + R_{LD} \cdot I_L$. To provide matching with the transmission line, a resistor $R_S = R_0 - R_{LD}$ is used in series with the laser. The tail current of the output stage is I'_M and the bias current, which is injected into the laser through an RFC, is I'_B. (a) What fractions of I'_M and I'_B end up doing useful work in the laser? (b) How large are these fractions given $R_0 = 25\,\Omega$, $R_{LD} = 6\,\Omega$, and $V_{TH} = 1\,V$?

8.4 Passive vs. Active Back Termination. Calculate the output voltage, v_O, as a function of the input voltage, v_I, and the output current, i_O, for the three idealized circuits shown in Fig. 8.38. How do these circuits relate to a driver with passive and active back termination?

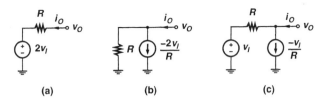

Fig. 8.38 Three implementations of a voltage controlled source with output resistance R.

8.5 Pulse-Width Controller. A pulse-width controller operating according to the principle illustrated in Fig. 8.13 receives an input signal with the differential voltage swing v_I^{pp} and the symmetrical 20% to 80% rise/fall times $t_R = t_F$. (a) Given the offset voltage V_{OS}, what is the pulse-width distortion, t_{PWD}, that it can compensate? Assume that $V_{OS} \ll v_I^{pp}$. (b) Given $t_R = t_F = 0.3$ UI and $V_{OS} = -0.1 v_I^{pp} \ldots 0.1 v_I^{pp}$, what is the range of PWD that can be compensated?

8.6 Speed of MOS CML. A MOS CML inverter, consisting of an n-MOS differential pair, load resistors R and R', and a tail-current source I_1, is loaded by another identical inverter. The differential output voltage swing of the inverter is v_O^{pp} and the differential input switching voltage (peak-to-peak) is $v_{I,\min}^{pp}$. (a) Assuming the quadratic MOS model and considering only the gate-source capacitance, calculate the 20% to 80% rise/fall times in response to a square-wave input signal. (b) Assuming the "large-signal gain" $v_O^{pp}/v_{I,\min}^{pp}$ is set to its minimum value of one, how does the speed depend on the logic swing, v_O^{pp}?

8.7 Power Penalty due to Finite APC Loop Bandwidth. A single-loop APC with a simple R-C loop filter, as shown in Fig. 8.16, is used to control the average output power of a transmitter. Assume that the gain A of the op amp is frequency independent, a change in laser current ΔI_L causes the monitor-photodiode current to change by $\xi \cdot \Delta I_L$, and a change in bias control voltage ΔV_{BC} causes the bias current to change by $g_m \cdot \Delta V_{BC}$. What is the low-frequency cutoff of the transmitter and how large is the associated power penalty?

8.8 Extinction Ratio and Slope Efficiency. A laser driver with a single-loop APC has a constant modulation current I_M, whereas the bias current I_B is controlled such that the average optical power remains at $\overline{P}$. (a) How does the extinction ratio depend on the slope efficiency of the laser? Assume that the optical power from the laser is $P = \xi \cdot (I_L - I_{TH})$ for laser currents above threshold, $I_L > I_{TH}$, and zero for laser currents below threshold. (b) What value does ER assume if the slope efficiency drops by 30% and the original extinction was perfect? (c) What nominal ER value is required such that $ER > 8.2$ dB is guaranteed even if the slope efficiency drops by 10%?

8.9 Modulation Current Control. The slope efficiency and the threshold current of a laser have been measured as a function of the temperature and are given in Table 8.3. A laser driver with a single-loop APC controls the bias current of this laser such that the average optical power is held at -3 dBm. Excluding the dual-loop APC approach, how can the modulation current be controlled such that the ER remains substantially constant?

Table 8.3 Laser characteristics for Problem 8.9.

Temperature (°C)	Slope Efficiency (mW/mA)	Threshold Current (mA)
−40	0.08	2
20	0.07	10
80	0.05	35

8.10 **Automatic Bias Control for MZMs.** In an ABC circuit for MZMs as shown in Fig. 8.20, the pilot tone oscillator generates a sine wave with the frequency $\omega/2\pi$ and an amplitude of 1 V; the mixer has a gain of 20 dB/V. Assuming the MZM transmits a long string of ones, the output signal from the TIA is $v_O = V_O[1 \pm \xi \cdot \sin(\omega t)]$. What is the signal at the mixer output and what is the resulting bias voltage?

8.11 **Mark-Density Compensation.** Show mathematically that if the mark density within each burst always is 50%, the switch S_D in the burst-mode APC circuit of Fig. 8.23 can be omitted and I_{REF} can be replaced by $I_{REF}/2$.

Appendix A
Eye Diagrams

The eye diagram is an intuitive graphical representation of electrical and optical communication signals. The quality of these signals (the amount of intersymbol interference [ISI], noise, and jitter) can be judged from the appearance of the eye. Eye diagrams frequently are used in the literature to document signals in optical receivers and transmitters. In the following, we explain how to produce eye diagrams from measurements and simulations. We also discuss how to determine the eye openings and eye margins of an eye diagram.

Definition. The waveform of a communication signal, such as a non-return-to-zero (NRZ), a return-to-zero (RZ), or a 4-level pulse amplitude modulation (PAM-4) signal, can be turned into an *eye diagram* or *eye pattern* by folding the time axis modulo a whole number of bit (or symbol) intervals. For example, in Fig. A.1, the waveform of an NRZ signal with mild ISI is folded modulo a two-bit interval. To do that, the waveform is first cut into two-bit segments: half a bit on the left, a full bit in the center, and half a bit on the right. Because the ISI in our example is limited to just one bit to the right and left, there are essentially eight distinct segments corresponding to the three-bit binary words: 000, 001, 010, 011, 100, 101, 110, and 111 (see the left-hand side of Fig. A.1). In the case of a signal with stronger ISI, more segments with distinct shapes exist and must be taken into account. Next, all these segments are superimposed, as shown on the right-hand side of Fig. A.1, resulting in the eye diagram.

314 EYE DIAGRAMS

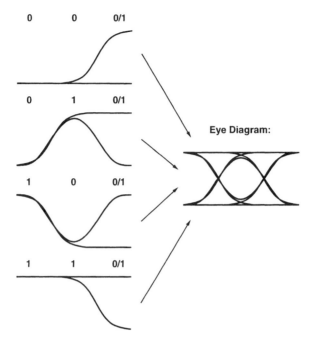

Fig. A.1 Construction of an eye diagram by superimposing the waveforms corresponding to all possible bit sequences.

An important advantage of the eye diagram over the linear signal waveform is that all possible bit transitions can be displayed in a compact representation.

Measurement. The setup shown in Fig. A.2 can be used to display an eye diagram on an oscilloscope. A pulse pattern generator produces an NRZ data signal and a clock signal. The data signal usually encodes a pseudorandom bit sequence (PRBS), which can be produced with a feedback shift register, as shown in Fig. 1.4. The data signal is passed through the device under test (DUT) and the output signal is fed to the vertical input of an oscilloscope. To display the eye diagram on the oscilloscope, it must be triggered from the clock signal, not the data signal. Usually, the bit-clock signal from the pattern generator is used for this purpose, as shown in Fig. A.2. Alternatively, a phase-locked loop (PLL) can be used to recover a periodic trigger signal from the output signal of the DUT. However in this case, the eye diagram will be somewhat different: on one hand, some low-frequency jitter produced in the DUT is suppressed because it is tracked by the PLL, on the other hand, some new jitter produced in the PLL is added to the eye. Some oscilloscopes have the option to display the frequency at which a certain point in the eye is reached with a color code, so-called *color grading*. Figure A.3 shows the eye diagram of an NRZ signal obtained with a sampling oscilloscope.

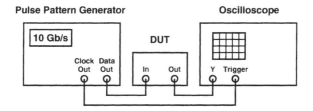

Fig. A.2 Measurement of an eye diagram with an oscilloscope.

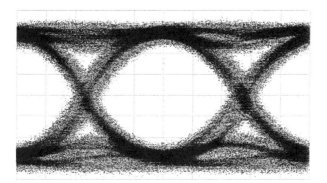

Fig. A.3 Eye diagram of an NRZ signal measured with a sampling oscilloscope. Darker regions are sampled more often.

Simulation. To produce an eye diagram with a circuit simulator such as SPICE—without the need for a specialized post processor—the following method can be used. First, generate a linear ramp voltage with a period of two bit intervals and rapid fall time. Then, plot the data signal against this ramp voltage instead of the time axis as usual. This trick will take care of the folding of the data signal waveform. A drawback of this simple method is that the ramp has a finite fall time, creating spurious trace-back lines across the eye diagram. This problem can be solved by generating a pulse voltage that is always zero except for the trace-back period, where it assumes a large value. When this voltage is added to the data signal, the trace-back lines move outside of the eye diagram and can be "clipped away" by choosing an appropriate plotting window. The following Celerity[1] code illustrates how to produce an eye diagram:

```
* PRBS input signal (10 Gb/s, 30-ps rise/fall time, and 1-V swing)
VI        VI        VGND                TABLE(0  &IV
+                                       15P   &PV0    85P   &PV0
+                                       115P  &PV1    485P  &PV1
+                                       515P  &PV0    585P  &PV0
+                                       615P  &PV1    685P  &PV1
+                                       715P  &PV0    985P  &PV0
```

[1] Celerity is a SPICE-like circuit simulator from Cadence Design Systems, Inc.

```
+                                           1015P  &PV1  1185P  &PV1

                    :

+                                           24115P  &PV1  24385P  &PV1
+                                           24415P  &PV0  24585P  &PV0
+                                           24615P  &PV1  24885P  &PV1
+                                           24915P  &PV0)
.SET       &IV       0
.SET       &PV0     -0.5
.SET       &PV1     +0.5
* Ramp and trace-back signals for the eye diagram
VRMP       VRMP      VGND                   PULSE(0 196P
+                                           100P 196P 4P 0 200P)
RRMP       VRMP      VGND                   1MEG
VTRB       VTRB      VGND                   PULSE(0 1
+                                           296P 0.1P 0.1P 3.8P 200P)
RTRB       VTRB      VGND                   1MEG
* Device under test
X1         VO        VGND      VI     VGND  DUT
* Adding the trace-back voltage to the output signal
ETRB       VO1       VO        VTRB   VGND  10
* Plot VO1 as a function of VRMP to show the eye diagram
.TR        5P        25N
.OUT       VRMP      VRMP      VGND
.OUT       VO1       VO1       VGND
```

In the above code, a 10-Gb/s PRBS signal is generated with a TABLE function. A ramp and trace-back signal with a 200-ps period are generated with a PULSE function each. Note that the ramp signal is set to zero during the first bit period to suppress the plotting of spurious transients at the beginning of the simulation. The PRBS signal is applied to the input of the device under test DUT. At the output, the controlled voltage source ETRB is used to add the trace-back signal, which avoids the trace-back lines across the eye diagram. Finally, the resulting signal VO1 is plotted against the ramp VRMP with a plotting tool such as Advplot[2] to obtain the eye diagram. The eye diagrams in Figs. 4.11 and A.4 have been produced in this manner.

Note that circuit simulators normally do not include random noise when performing a time-domain (transient) simulation. Therefore, eye diagrams produced in this way do not show the effect of noise or random jitter; only ISI and deterministic jitter (including pulse-width distortion) can be seen.

Eye Openings and Eye Margins. The *vertical eye opening* and the *horizontal eye opening* are important characteristics of the eye diagram that aid in quantifying the signal quality. The vertical eye opening is measured at the sampling instant and is expressed as a percentage of the full eye height (not including over- or undershoots). The horizontal eye opening is measured at the slice level and is expressed as a percentage of the bit interval. Sometimes the complementary terms *vertical eye closure*

---

[2] Advplot is a waveform plotter from Cadence Design Systems, Inc.

and *horizontal eye closure* are used instead. Eye closure and eye opening add up to 100%. This sound easy enough, but there are some important details that must be considered.

In the case of an eye diagram *without* noise and random jitter, the openings can be determined in a straightforward way, as illustrated in Fig. A.4(a). The vertical eye closure is caused by ISI, and the horizontal eye closure is caused by deterministic jitter (including pulse-width distortion). It is important to recognize that the eye closure may depend on the sequence length of the PRBS. The sequence length, typically between $2^7 - 1$ and $2^{31} - 1$, is the number of bits put out by the pulse pattern generator before the PRBS repeats itself. In particular, if the device under test has a low-frequency cutoff, the eye closure will become worse with increasing sequence length (cf. Section 6.2.6). Therefore, the sequence length must always be specified when presenting an eye diagram or quoting eye openings.

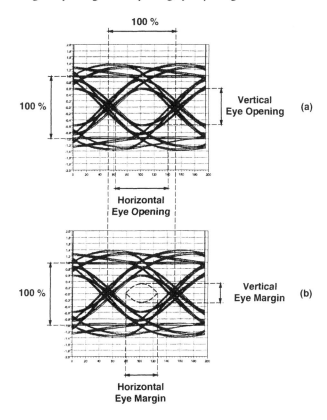

Fig. A.4 (a) Eye openings in the noise-free eye diagram and (b) eye margins in the noisy eye diagram.

In the case of an eye diagram *with* noise and random jitter, we have an additional complication. For Gaussian (unbounded) noise, any opening we may specify eventually will be violated; we just have to wait long enough. Therefore, we have to use a statistical definition for the eye opening. Each point in the eye diagram can be interpreted as a decision point (defined by the sampling instant and the decision threshold) and thus has a bit-error rate (BER) associated with it. As a result, we can plot contours of constant BERs inside the eye. Figure A.4(b) shows one such contour with a dashed line. Clearly, the lower the BER, the smaller becomes the area enclosed by the contour. If we make a decision inside a contour defined by a certain reference BER, the BER will always be less than this reference BER. Now, we can define the eye openings in the noisy eye as the openings of these contours.

To make a distinction between the eye openings in the noisy and noise-free eye, we call the eye openings in the noisy eye *eye margins*. The *vertical eye margin* and *horizontal eye margin* are shown in Fig. A.4(b). If the eye margins for a given reference BER are zero, only a perfect decision circuit could recover the data at the desired reference BER. However, if the eye margins are larger than zero, then the decision circuit is permitted to have some decision-threshold and sampling error while still meeting the desired BER; hence the name *margin* is appropriate. Note that when quoting the eye margins, we need to specify not only the PRBS sequence length but also the reference BER.

BERT Scan. Eye margins are best measured with a so-called *BERT scan*. For this procedure, a *bit-error rate test set* (BERT), consisting of a pulse pattern generator and an error detector, is connected to the device under test (DUT), as shown in Fig. A.5. The error detector slices the data signal at the decision threshold V_{DTH} and samples it at the instant t_S (cf. the eye diagram in Fig. 4.21). Then, the recovered bits are compared with the transmitted bit sequence to determine the BER, which is displayed on the error detector. Both the decision threshold V_{DTH} and the sampling instant t_S are adjustable.

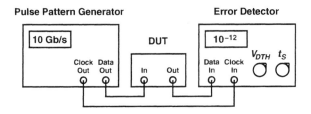

Fig. A.5 Measurement setup for a BERT scan.

A horizontal BERT scan is performed by setting V_{DTH} to the center of the eye and scanning t_S horizontally across the eye. The resulting curve is shown schematically in Fig. A.6(a). The BER is low when sampling at the center of the eye and goes up when approaching the eye crossings to the left and right; hence this curve is known as the *bathtub curve*. The horizontal eye margin is the separation of the two points on the left and right side of the eye where the bathtub curve assumes a specified BER value. For example, in the 10-GbE standard, the horizontal eye margin is specified for a BER of 10^{-12}.

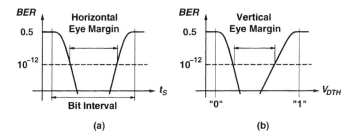

Fig. A.6 Bathtub curves resulting from (a) a horizontal BERT scan and (b) a vertical BERT scan.

Similarly, a vertical BERT scan is performed by setting the sampling instant t_S to the center of the eye and scanning the decision threshold V_{DTH} vertically across the eye. The resulting bathtub curve is shown in Fig. A.6(b). Like before, the vertical eye margin is obtained by comparing this curve with a reference BER level.

Appendix B
Differential Circuits

Most analog circuits in optical transceivers are implemented as *differential* circuits, that is, the critical signals in these circuits are represented by the *difference* of two voltages rather than a single voltage to ground. Differential circuits have several important advantages over single-ended ones:

- *Reduced sensitivity to system noise.* If a differential circuit is balanced (fully symmetrical), power-supply noise, substrate noise, and other system noise sources affect only the common-mode signal, whereas the information-bearing differential signal ideally is left undisturbed (cf. Section B.3).

- *Reduced generation of transient noise.* If a differential circuit is balanced (fully symmetrical), any positive voltage or current transient is accompanied by a corresponding negative transient. With the proper coupling symmetries and current routings, the effects of these transients on other signals ideally cancel each other.

- *Improved amplifier stability.* A consequence of the previous two points is that unwanted coupling between the stages of a multistage amplifier is reduced. Thus, concerns about instabilities caused by spurious feedback through power-supply lines, and so forth are mitigated.

- *Improved voltage swing.* Whereas a single-ended signal is limited to the range of 0 to V_{DD}, a differential signal can cover the range of $-V_{DD}$ to V_{DD}, where V_{DD} is the power-supply voltage. Thus, a differential signal can have twice the voltage swing of a single-ended signal.

- *Reduced second-order nonlinearities.* If a differential circuit is balanced (fully symmetrical), the polarity of the input signal and the polarity of the output signal can be reversed together (by swapping the appropriate terminals) without affecting the input-to-output transfer function. Mathematically, we have $v_O = f(v_I) \Rightarrow -v_O = f(-v_I)$, which means that the transfer function is *odd*. Thus, ideally, no even-order nonlinearities are present and no even-order harmonics and intermodulation products are generated.

- *Improved speed.* Partially differential designs (e.g., with differential input, but single-ended output) often incorporate current mirrors as differential-to-single-ended converters. These mirrors can be avoided in fully differential designs, giving these circuits a speed advantage (higher bandwidth). Furthermore, in fully differential circuits, voltage inversions are free. They can be implemented by simply crossing over the signal wires; thus, no delay is incurred and no circuitry is required.

B.1 DIFFERENTIAL MODE AND COMMON MODE

Definition. Two terminals are required for each input or output of a differential circuit. The voltages at these terminals are called *terminal voltages* and usually are designated v_P and v_N. Because the information-bearing signal is contained in the difference $v_P - v_N$, it is convenient to introduce a new voltage representation called the *mode voltages*. The *differential-mode voltage* v_D and the *common-mode voltage* v_{CM} are defined as

$$v_D = v_P - v_N \quad \text{and} \quad v_{CM} = 1/2 \cdot (v_P + v_N). \tag{B.1}$$

The differential mode contains the information, and the common-mode is the orthogonal component void of information. Figure B.1 illustrates this with an example where the information-bearing signal is a non-return-to-zero (NRZ) signal and the common-mode voltage is some slowly varying interferer. We can see how a clean NRZ signal emerges after subtracting the two corrupted terminal voltages (to extract the differential mode).

The concept of differential mode and common mode is in widespread use throughout the electrical engineering community, but sometimes different names are used. The telephone engineer likes to use the terms *transversal mode* and *longitudinal mode* (don't ask me why) or sometimes *metallic mode* and *longitudinal mode* (the differential voltage appears between the two metallic telephone wires, whereas the common-mode voltage appears between the wires and ground). The microwave engineer frequently refers to the *odd mode* and *even mode*.

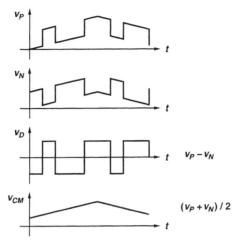

Fig. B.1 Example of terminal voltages (v_P, v_N) and mode voltages (v_D, v_{CM}).

Coordinate Transformation. A useful way to visualize the relationship between terminal voltages and mode voltages is to view them as a *coordinate transformation*. This transformation is illustrated graphically in Fig. B.2. The black dot represents a physical voltage constellation, which has the character of a vector. Given a coordinate system (or base), this vector can be described by coordinates. The two coordinate systems drawn in Fig. B.2 correspond to a measurement of the terminal voltages ($v_P = 2$, $v_N = 4$) and a measurement of the mode voltages ($v_D = -2$, $v_{CM} = 3$), respectively. The coordinate transformation relating the two coordinate systems may at first look like a 45° rotation, but actually is a more general affine transformation.

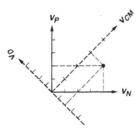

Fig. B.2 Terminal voltages (v_P, v_N) and mode voltages (v_D, v_{CM}) are related by a coordinate transformation.

Common-Mode vs. DC Component. A common mistake is to confuse the common-mode voltage with the DC component of a voltage signal. Let's clarify these terms: the common-mode voltage is the instantaneous average of *two terminal voltages*, whereas the DC component of a signal is the average of a single voltage *over time.* Figure B.3(a) illustrates this difference with an example. The confusion originates from the fact that these two quantities become identical under the following conditions: (i) the common-mode voltage is time independent and (ii) the time average of the differential-mode voltage is zero (no offset). Figure B.3(b) illustrates this degenerate situation with an example.

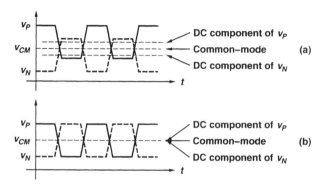

Fig. B.3 Common-mode vs. DC component: (a) where they are different and (b) where they are the same.

B.2 THE MODES OF CURRENTS AND IMPEDANCES

Differential and Common-Mode Currents. So far, we were talking about differential and common-mode voltages, but what about differential and common-mode currents? Figure B.4 shows a circuit block with the terminal voltages v_P and v_N and the *terminal currents* i_P and i_N. At first, we may consider a straightforward generalization of the voltage mode definitions and define the differential current as $i_D = i_P - i_N$ and the common-mode current as $i_{CM} = 1/2 \cdot (i_P + i_N)$. Although this definition does lead to a mathematically consistent description, there are difficulties with the physical interpretation of these currents and the resulting impedances. Instead, the definition that is most commonly used is [15]

$$i_D = 1/2 \cdot (i_P - i_N) \quad \text{and} \quad i_{CM} = i_P + i_N. \quad \text{(B.2)}$$

These currents have a direct physical interpretation (see Fig. B.4). The *common-mode current* simply is the current that flows into the ground node. If this common-mode current is zero, we have $i_P = -i_N$ and thus the *differential-mode current* is the current that flows "through" the differential port ($i_D = i_P = -i_N$).

Differential and Common-Mode Impedances. Now that we have defined the voltage modes and the current modes, it is a simple step to define the modes of impedances (or admittances). For a single-ended port, the impedance Z_{se} is defined by Ohm's law as $V_{se} = Z_{se} \cdot I_{se}$, where V_{se} and I_{se} are the phasors of the port voltage

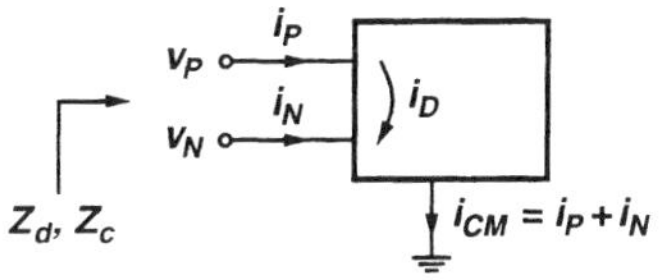

Fig. B.4 Differential and common-mode currents.

and current, respectively. A straightforward generalization to matrix form results in

$$\begin{pmatrix} V_d \\ V_{cm} \end{pmatrix} = \begin{pmatrix} Z_d & Z_{dc} \\ Z_{cd} & Z_c \end{pmatrix} \cdot \begin{pmatrix} I_d \\ I_{cm} \end{pmatrix}, \tag{B.3}$$

where all voltages and currents as now written as phasors. In general, we need four numbers to describe the impedance of a differential port. However, if the circuit is balanced, as it usually is the case, the two mode-conversion components become zero, $Z_{dc} = Z_{cd} = 0$, and we are left with only the differential impedance, Z_d, and the common-mode impedance, Z_c. In this case, we have $Z_d = V_d/I_d$ and $Z_c = V_{cm}/I_{cm}$, or if written in terms of the terminal voltages and currents,

$$Z_d = 2 \cdot \frac{V_p - V_n}{I_p - I_n} \quad \text{and} \quad Z_c = 1/2 \cdot \frac{V_n + V_n}{I_p + I_n}. \tag{B.4}$$

Figure B.5 illustrates this result with a few balanced resistor circuits and the corresponding values for the differential and common-mode resistances. In Fig. B.5(a), we have two independent 50-Ω resistors to ground characterized by a differential resistance of 100 Ω and a common-mode resistance of 25 Ω. It is easy to show that in the case of two independent impedances, Z_{se}, the following relationships hold: $Z_d = 2 \cdot Z_{se}$, $Z_c = 1/2 \cdot Z_{se}$, and $Z_{se} = Z_c + 1/4 \cdot Z_d$. In Fig. B.5(b), we have a floating 100-Ω resistor, which is characterized by a differential resistance of 100 Ω and an infinite common-mode resistance, as we would expect. Note that if we had defined the differential-mode current as $i_P - i_N$, then the differential resistance would be only 50 Ω, offending the intuition of an electrical engineer. Finally, in Fig. B.5(c), we have a single 50-Ω resistor connecting both terminals to ground, which is characterized by a zero differential resistance and a common-mode resistance of 50 Ω, as we would expect. Note that if we had defined the common-mode current as $(i_P + i_N)/2$, then the common-mode resistance would be 100 Ω.[1]

B.3 COMMON-MODE AND POWER-SUPPLY REJECTION

Definition. Figure B.6 shows all the terminals, including the power supply terminals, of two differential amplifiers. The amplifier in Fig. B.6(a) has a single-ended

---

[1] Although most authors define the differential impedance as in Eq. (B.4), there is some variability in the definition of the common-mode impedance. For example, in [34], the differential-mode impedance is defined as in Eq. (B.4), but the common-mode impedance is defined as $Z_c = (V_p + V_n)/(I_p + I_n)$.

326   DIFFERENTIAL CIRCUITS

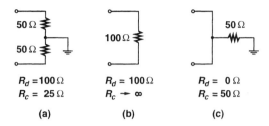

Fig. B.5 Comparison of differential and common-mode resistance for three balanced circuits.

output, and the amplifier in Fig. B.6(b) has a differential output. The complete input-voltage vector is given by (i) the differential voltage, $v_D = v_P - v_N$, which is amplified, (ii) the common-mode voltage, $v_{CM} = (v_P + v_N)/2$, which is suppressed, (iii) the positive power supply v_{DD}, and (iv) the negative power supply v_{SS}, which all may vary with time. Both amplifiers produce an output voltage v_O, in the first case as a single-ended voltage to ground and in the second case as a differential voltage ($v_O = v_{OP} - v_{ON}$). Furthermore, the differential-output amplifier produces a common-mode output voltage $v_{OCM} = (v_{OP} + v_{ON})/2$, not shown in Fig. B.6, which normally is set to a fixed value by means of a common-mode feedback mechanism.

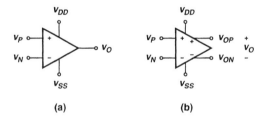

Fig. B.6 Signal and supply voltages of an amplifier with (a) single-ended and (b) differential output.

The DC transfer characteristics of both amplifiers can be described by the function $v_O = f(v_D, v_{CM}, v_{DD}, v_{SS})$. If we linearize this function around the operating point $\{v_O = 0, v_{CM} = V_{CM}, v_{DD} = V_{DD}, v_{SS} = V_{SS}\}$, we arrive at the following expression:

$$v_O = A_d(v_D - V_{OS}) + A_{cm}(v_{CM} - V_{CM}) + A_{dd}(v_{DD} - V_{DD}) + A_{ss}(v_{SS} - V_{SS}), \quad (B.5)$$

where A_d is the differential gain, A_{cm} is the common-mode gain, and A_{dd} and A_{ss} are the positive and negative power supply gains, respectively. V_{OS} is the offset voltage that must be applied to the input to bring the output to zero. Voltages V_{CM}, V_{DD}, and V_{SS} describe the operating point for which all the other parameters are determined.

Rewriting the above equation as an AC small-signal equation yields

$$V_o = A_d(s) \cdot V_d + A_{cm}(s) \cdot V_{cm} + A_{dd}(s) \cdot V_{dd} + A_{ss}(s) \cdot V_{ss}. \quad (B.6)$$

All voltages now are expressed as phasors V_o, V_d, V_{cm}, V_{dd}, and V_{ss}. The four gain parameters A_d, A_{cm}, A_{dd}, and A_{ss} now are frequency dependent (AC equation), and the offset and operating-point voltages disappear (small-signal equation).

The *common-mode rejection ratio* (*CMRR*) and *power-supply rejection ratios* (*PSRR$^+$* and *PSRR$^-$*) can be defined in terms of the gain parameters A_d, A_{cm}, A_{dd}, and A_{ss} as follows:

$$CMRR(s) = \frac{A_d(s)}{A_{cm}(s)}, \quad PSRR^+(s) = \frac{A_d(s)}{A_{dd}(s)}, \quad PSRR^-(s) = \frac{A_d(s)}{A_{ss}(s)}. \quad (B.7)$$

With these definitions, we can rewrite Eq. (B.6) in the intuitively pleasing form

$$V_o = A_d(s)\left(V_d + \frac{V_{cm}}{CMRR(s)} + \frac{V_{dd}}{PSRR^+(s)} + \frac{V_{ss}}{PSRR^-(s)}\right). \quad (B.8)$$

A Limitation of Amplifiers with Single-Ended Output. Oftentimes, Eq. (B.6) is idealized by setting all gains, except the differential gain, to zero

$$V_o = A_d(s) \cdot V_d. \quad (B.9)$$

However, this behavior is *physically impossible* in the case of a single-ended output amplifier! The reason is that the choice of ground potential is no longer arbitrary for Eq. (B.9): a change in ground potential changes the (single-ended) left-hand side of the equation, but leaves the (differential) right-hand side unaffected. If you want to impress your colleagues with you knowledge of physics, tell them that Eq. (B.9) is not invariant under a *gauge transformation* of the electrical potential—a fundamental physical symmetry. A closer inspection of Eq. (B.6) reveals that the gauge invariance is satisfied, if the following relationship holds:

$$A_{cm}(s) + A_{dd}(s) + A_{ss}(s) = 1. \quad (B.10)$$

Therefore, it is impossible to make all the undesirable gains zero as assumed in Eq. (B.9). If we rewrite Eq. (B.10) in terms of CMRR and PSRRs we find that

$$\frac{1}{CMRR(s)} + \frac{1}{PSRR^+(s)} + \frac{1}{PSRR^-(s)} = \frac{1}{A_d(s)}. \quad (B.11)$$

For example, if the differential gain is 60 dB and the common-mode rejection ratio as well as the negative power-supply rejection ratio both are infinite (perfect), then we can conclude that the positive power-supply rejection ratio *must be* 60 dB and cannot be infinite. Note that no such limitation exists for amplifiers with differential outputs. For a more detailed discussion of this result and its consequences, see [157].

Appendix C
S Parameters

Scattering parameters, or *S parameters* for short, commonly are used to characterize high-frequency circuits with 50-Ω ports. For example, the 50-Ω output port of a transimpedance amplifier (TIA) can be characterized by the S_{22} parameter, the 50-Ω input and output ports of a limiting amplifier (LA) can be characterized by the S_{11} and S_{22} parameter, respectively, and the 50-Ω input port of a laser driver can be characterized by the S_{11} parameter. The S_{11} and S_{22} parameters tell us how closely the port impedance matches that of an ideal 50-Ω transmission line. If the matching is perfect, $S_{11} = S_{22} = 0$, or, if expressed in dB, $S_{11} = S_{22} \rightarrow -\infty\,\text{dB}$. In the following, we have a closer look at the S parameters.

C.1 DEFINITION AND SIMULATION

Why S Parameters? High-frequency circuits typically are characterized by S parameters rather than the more familiar Y or Z parameters. The reason for preferring S parameters is that they are easier to measure. (i) Whereas Y and Z parameters must be measured under open- and short-circuit conditions (possibly leading to instability), S parameters are measured under natural 50-Ω terminated conditions. (ii) Whereas Y and Z parameters require the measurement of voltages and currents, S parameters are based on ratios of incident and reflected traveling waves, which can be measured easily and accurately with directional couplers. In practice, a so-called *network analyzer* (NWA) is used to measure the S parameters.

Definition. The key idea behind the S parameters is to decompose each port voltage (e.g., v_I) and port current (e.g., i_I) into an *incident* and *outgoing* component that correspond to the incident and outgoing waves on a transmission line.[1] The decomposition is done such that (i) the superposition of the incident and outgoing voltage components equals the port voltage (e.g., $v_{I,\text{incident}} + v_{I,\text{outgoing}} = v_I$), (ii) the superposition of the incident and outgoing current components equals the port current (e.g, $i_{I,\text{incident}} - i_{I,\text{outgoing}} = i_I$), and (iii) the voltage-to-current ratio of each component is equal to R_0 (e.g., $v_{I,\text{incident}}/i_{I,\text{incident}} = v_{I,\text{outgoing}}/i_{I,\text{outgoing}} = R_0$). R_0 is known as the *characteristic impedance* of the transmission line and typically is set to 50 Ω (purely resistive). We assume $R_0 = 50$ Ω for the rest of this discussion.

We are now ready to define the four S parameters of a two port, $S_{\mu\nu}$ with $\mu = 1, 2$ and $\nu = 1, 2$. The second index, ν, indicates which port is excited with an incident wave and the first index, μ, indicates which port is used to measure the outgoing wave. Depending on whether the outgoing wave appears at the same port as the incident wave or at a different port, it is called the *reflected* or the *transmitted* wave, respectively. By convention, port 1 is the input port (v_I, i_I) and port 2 is the output port (v_O, i_O) (cf. Fig. C.1). The S parameters are defined as the following ratios:

$$S_{11} = \frac{v_{I,\text{reflected}}}{v_{I,\text{incident}}} = \frac{1/2 \cdot (v_I - R_0 \cdot i_I)}{1/2 \cdot (v_I + R_0 \cdot i_I)}, \tag{C.1}$$

$$S_{21} = \frac{v_{O,\text{transmitted}}}{v_{I,\text{incident}}} = \frac{1/2 \cdot (v_O - R_0 \cdot i_O)}{1/2 \cdot (v_I + R_0 \cdot i_I)}, \tag{C.2}$$

$$S_{12} = \frac{v_{I,\text{transmitted}}}{v_{O,\text{incident}}} = \frac{1/2 \cdot (v_I - R_0 \cdot i_I)}{1/2 \cdot (v_O + R_0 \cdot i_O)}, \tag{C.3}$$

$$S_{22} = \frac{v_{O,\text{reflected}}}{v_{O,\text{incident}}} = \frac{1/2 \cdot (v_O - R_0 \cdot i_O)}{1/2 \cdot (v_O + R_0 \cdot i_O)}. \tag{C.4}$$

The first expression in the above equations is the definition in terms of incident and outgoing voltage components, and the second expression is its expansion in terms of port voltages and currents, which follows from the decomposition rules stated earlier. Note that a definition based on incident and outgoing *current* components would yield the identical S parameters.[2]

To make these definitions more concrete, consider the test circuit shown in Fig. C.1(a). The 2× voltage amplifier at the input compensates for the voltage di-

---

[1] In the literature, the wave traveling backwards on the transmission line usually is termed the *reflected wave*. Here we choose the term *outgoing wave* because this wave not only consists of a reflected component, but also of the transmitted component(s) from the other port(s).
[2] In the literature, S parameters usually are defined in terms of the so-called *power waves a* and *b*. The power waves *a* and *b* are proportional to our incident and outgoing voltage components, respectively. Note that the power waves are measured in $\sqrt{W}$ (not W) and the dB value of their ratios, the S parameters, are calculated as $20 \log S_{\mu\nu}$.

vision between the source and load impedances under ideally matched conditions. Although not strictly necessary, the use of this amplifier leads to a simpler, more intuitive explanation and an easy way to simulate S parameters. Following Eqs. (C.1) and (C.2), we can write the complex and frequency dependent S_{11} and S_{21} parameters in terms of the port-voltage phasors, V_i and V_o, and the port-current phasors, I_i and I_o. Now, the test circuit permits us to simplify these expressions as shown on the right-hand side:

$$S_{11}(s) = \frac{V_i - R_0 \cdot I_i}{V_i + R_0 \cdot I_i} = \frac{V_i - V_s}{V_s}, \qquad (C.5)$$

$$S_{21}(s) = \frac{V_o - R_0 \cdot I_o}{V_i + R_0 \cdot I_i} = \frac{V_o}{V_s}, \qquad (C.6)$$

where V_s is the phasor of the source voltage.

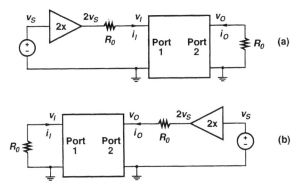

Fig. C.1 Test circuits to determine the two-port S parameters: (a) S_{11} and S_{21} parameters and (b) S_{22} and S_{12} parameters.

The S_{11} parameter is known as the *input reflection coefficient* because it describes what fraction of an incident wave traveling on an ideal 50-Ω transmission line is reflected back from the input port. Equivalently, the S_{11} parameter is a measure of how close the input impedance is to the ideal 50-Ω value. From Fig. C.1(a), we see that if the input impedance is exactly 50 Ω, V_s first is doubled by the test circuit and then is divided by two by the voltage divider formed by R_0 and the input impedance of the two port. Thus we have $V_i = V_s$, and with Eq. (C.5), it follows that $S_{11} = 0$. Similarly, if the input impedance is a short ($V_i = 0$), then $S_{11} = -1$, and if the input impedance is an open ($V_i = 2V_s$), then $S_{11} = 1$. The magnitude of S_{11} is always less than or equal to one, and its dB value (20 log $|S_{11}|$) is always negative (assuming the input impedance is positive). The inverse value of $|S_{11}|$ is known as the *input return loss* and its dB value ($-20 \log |S_{11}|$) is always positive. Sometimes the symbol Γ_{in} is used to indicate the same quantity as S_{11} ($\Gamma_{in} = S_{11}$). If we know the input impedance $Z(s) = V_i/I_i$ of the two port, we can easily calculate $S_{11}(s)$ by applying the transformation

$$S_{11}(s) = \frac{Z(s) - R_0}{Z(s) + R_0}, \qquad (C.7)$$

which follows directly from the first expression in Eq. (C.5). For example, with $Z = 55\,\Omega$, we find that $S_{11} = 0.048$ (-26.4 dB), corresponding to an input return loss of 26.4 dB.

The S_{21} parameter is known as the *forward transmission coefficient* or the *gain* and is closely related to the loaded voltage gain $A(s) = V_o/V_i$. However, unlike the loaded voltage gain, S_{21} also depends on the quality of the input matching. From Eqs. (C.5) and (C.6) we can derive that

$$S_{21}(s) = [1 + S_{11}(s)] \cdot A(s). \tag{C.8}$$

Thus, for a perfect input matching ($S_{11} = 0$), S_{21} becomes identical to the loaded voltage gain. The magnitude of S_{21} can be smaller or larger than one, and its dB value is calculated as $20 \log |S_{21}|$.

The S_{22} and S_{12} parameters play the same role as the S_{11} and S_{21} parameters, if we swap the input and output ports. This symmetry is evident from Eqs. (C.1) through (C.3). With Eqs. (C.3) and (C.4) and the test circuit in Fig. C.1(b), we can determine the S_{22} and S_{12} parameters as

$$S_{22}(s) = \frac{V_o - R_0 \cdot I_o}{V_o + R_0 \cdot I_o} = \frac{V_o - V_s}{V_s}, \tag{C.9}$$

$$S_{12}(s) = \frac{V_i - R_0 \cdot I_i}{V_o + R_0 \cdot I_o} = \frac{V_i}{V_s}. \tag{C.10}$$

The S_{22} parameter (also $\Gamma_{out} = S_{22}$) is known as the *output reflection coefficient*, and its inverse value is known as the *output return loss*. The S_{12} parameter is known as the *reverse transmission coefficient* or the *isolation*.

Simulation. The test circuits in Fig. C.1 can be used directly to determine the S parameters with a circuit simulator such as SPICE. The following Celerity[3] code illustrates how to determine the S_{11} and S_{21} parameters as a function of frequency:

```
* Test setup for the simulation of S11 and S21
VS      VS      VGND                            AC  1
E1      VS2     VGND    VS      VGND            2
R1      VS2     VIC                             50
C1      VIC     VI                              1U
C2      VO      VOC                             1U
R2      VOC     VGND                            50
* Two-port device under test
X1      VO      VGND    VI      VGND            DUT
* Plot (VI-VS) and VO as a function of frequency
.AC     DEC     100     10MEG  100G
.OUT    VS11    VI      VS
.OUT    VS21    VO      VGND
```

The above code closely follows the circuit diagram shown in Fig. C.1(a). Two DC blocks, C1 and C2, were inserted in series with the 50-Ω resistors to prevent the test

---

[3] Celerity is a SPICE-like circuit simulator from Cadence Design Systems, Inc.

circuit from disturbing the operating point of the device under test DUT. The values of these capacitors must be large enough such that, at the frequencies of interest, their impedance is much smaller than 50 Ω. The S_{22} and S_{12} parameters can be determined with a similar piece of code that is based on the circuit diagram shown in Fig. C.1(b).

C.2 MATCHING CONSIDERATIONS

To get a feeling for the S_{11} and S_{22} parameters and to understand better how to meet the associated specifications, we analyze the matching properties of a few simple circuits. In the following, we pick an input port and the associated S_{11} parameter as an example, but our conclusions also are valid for an output port and the associated S_{22} parameter.

R-C Parallel Circuit. Let's assume that the input impedance of our two port can be modeled by the parallel connection of a resistor R and a capacitor C (cf. Fig. C.2, right-hand side). For example, the resistor may represent a polyresistor for the input termination and the capacitor may represent the gate-source capacitance of the input FET together with the pad capacitance. At the chip level, that is, when neglecting packaging parasitics, this simple model often is reasonably accurate.

The S_{11} parameter of this R-C parallel circuit can be calculated analytically by plugging $Z(s) = R/(1 + sRC)$ into Eq. (C.7), which results in

$$S_{11}(s) = S_0 \cdot \frac{1 - s/\omega_z}{1 + s/\omega_p}, \qquad (C.11)$$

where

$$S_0 = \frac{R - R_0}{R + R_0} \approx \frac{\Delta R}{2R_0}, \qquad (C.12)$$

$$\omega_z = \frac{R - R_0}{RR_0C} \approx \frac{2S_0}{R_0C}, \qquad (C.13)$$

$$\omega_p = \frac{R + R_0}{RR_0C} \approx \frac{2}{R_0C}. \qquad (C.14)$$

In these equations, S_0 is the low-frequency value of S_{11}, $\Delta R = R - R_0$ is the deviation of the resistor from its ideal value, ω_z is the (angular) frequency of the zero, and ω_p is the (angular) frequency of the pole. The approximations on the right-hand side are valid under the assumption that $\Delta R \ll R_0$.

The magnitude of the S_{11} parameter in Eq. (C.11) can be plotted as a function of frequency. Figure C.2 shows an asymptotic representation of this function. At low frequencies, S_{11} is equal to S_0 and the quality of the impedance matching is determined by the deviation of the resistor from its ideal value, ΔR. Around frequency $f_z = \omega_z/2\pi$, the S_{11} parameter starts to increase until about $f_p = \omega_p/2\pi$, where it saturates at 0 dB ($S_{11} = -1$) because the capacitor acts like a short at these frequencies. Alternatively, we can plot the S_{11} parameter of Eq. (C.11) in the complex plane with frequency as a parameter, as shown in Fig. C.3. This representation is known as the *Smith chart*. For the example in Fig. C.3, we assumed that R is slightly larger than R_0 such that S_0 is slightly larger than zero. The outer circle of the Smith chart indicates that the magnitude of S_{11} cannot become larger than one. The smaller circle shows the loci of constant conductance ($1/Z = 1/R_0 + jB$ for any susceptance B), and the two arcs mark the loci of constant susceptance ($1/Z = G \pm j \cdot 1/R_0$ for any conductance G).

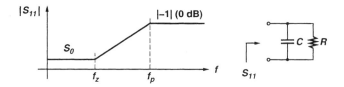

Fig. C.2 Plot of $|S_{11}(f)|$ for an R-C parallel circuit (on a log-log scale).

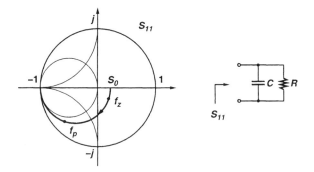

Fig. C.3 Admittance Smith chart of $S_{11}(f)$ for an R-C parallel circuit.

In Fig. C.4, the S_{11} parameter of a parallel circuit with $R = 55\,\Omega$ and $C = 1\,\text{pF}$ has been obtained by simulation using the method described in Section C.1. The solid line shows the magnitude of S_{11} in dB as a function of frequency. As expected, the low-frequency value of S_{11} is $-26.4\,\text{dB}$, a direct consequence of the 55-Ω resistor, and for high frequencies, S_{11} tends to 0 dB because of the shunt capacitor. The -3-dB value for S_{11} is reached at the pole frequency $f_p \approx 1/(\pi R_0 C) \approx 6.4\,\text{GHz}$. Note that this frequency is identical to the 3-dB bandwidth of the input transfer function, that is, the bandwidth of the transfer function from the voltage source v_S (see Fig. C.1) to the voltage across the capacitor C (v_I in this case), which represents the input voltage of the on-chip circuit. The solid line in Fig. C.5 shows the magnitude of this transfer function.

Bounds for R and C. Given this simple R-C input-port model, it is straightforward to derive bounds for R and C that need to be satisfied to meet $|S_{11}|$ specifications. For a given maximum value for $|S_{11}|$ at DC ($= |S_0|$), we find from Eq. (C.12) that R

MATCHING CONSIDERATIONS   **335**

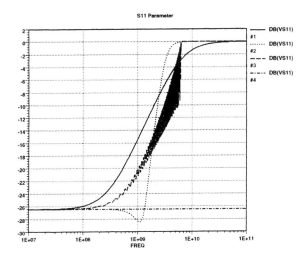

Fig. C.4 Simulated $|S_{11}(f)|$ for various matching networks with termination resistor $R = 55\,\Omega$ and on-chip capacitance $C = 1$ pF. #1 (solid), bare chip; #2 (dotted), simple L-C package model; #3 (dashed), artificial transmission line; #4 (dash-dotted), T-coil network.

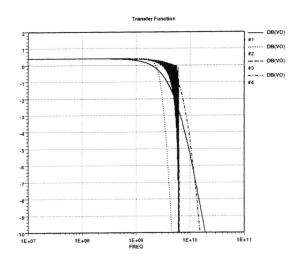

Fig. C.5 Simulated input transfer function for various matching networks with termination resistor $R = 55\,\Omega$ and on-chip capacitance $C = 1$ pF. #1 (solid), bare chip; #2 (dotted), simple L-C package model; #3 (dashed), artificial transmission line; #4 (dash-dotted), T-coil network.

must be in the range

$$R = R_0 \cdot \frac{1 - |S_0|}{1 + |S_0|} \dots R_0 \cdot \frac{1 + |S_0|}{1 - |S_0|}. \tag{C.15}$$

Assuming $|S_0| \ll 1$, we can simplify this expression to

$$R \approx R_0 \cdot (1 - 2|S_0|) \dots R_0 \cdot (1 + 2|S_0|). \tag{C.16}$$

For example, if $|S_{11}|$ is required to be less than -20 dB at DC, then R must be in the range of 41 to 61 Ω, which is approximately 50 $\Omega \pm 20\%$.

For a given maximum value for $|S_{11}|$ at the frequency $f > 0$, we find from Eqs. (C.11) through (C.14) that C must be smaller than

$$C = \sqrt{\frac{|S_{11}(f)|^2 - S_0^2}{1 - |S_{11}(f)|^2}} \cdot \frac{R + R_0}{2\pi f R R_0} \approx \sqrt{\frac{|S_{11}(f)|^2 - S_0^2}{1 - |S_{11}(f)|^2}} \cdot \frac{1}{\pi f R_0}, \tag{C.17}$$

where the approximation on the right-hand side is valid for $\Delta R \ll R_0$. If we pick an $|S_{11}|$ value that is small compared with one ($|S_{11}|^2 \ll 1$) and large compared with its low-frequency value ($|S_{11}|^2 \gg |S_0|^2$), that is, if we pick a value on the rising section of $|S_{11}(f)|$ located between the frequencies f_z and f_p, we can simplify the maximum permissible capacitance further to

$$C \approx \frac{|S_{11}(f)|}{\pi f R_0}. \tag{C.18}$$

For example, if $|S_{11}|$ is required to be less than -10 dB at 10 GHz, then C must be less than 212 fF, assuming $S_0 = 0$.

Packaging Parasitics. In the following, we analyze how bond-wire inductances and other packaging parasitics alter the S_{11} parameter of the simple R-C parallel circuit discussed above. Figure C.6(a) shows a simple package model, consisting of a parasitic capacitance C' and a bond-wire inductance L, followed by our familiar R-C model for the on-chip impedance. If we choose $C' = C$ and $L = 2R_0^2 C$, the packaging parasitics together with the on-chip impedance form a third-order Butterworth low-pass filter with the 3-dB bandwidth $1/(2\pi R_0 C)$ (assuming $R \approx R_0$). The dotted line in Fig. C.4 shows the simulation result for $|S_{11}(f)|$ of this network with $R = 55 \Omega$, $C = C' = 1$ pF, and $L = 5$ nH; the dotted line in Fig. C.5 shows the magnitude of the input transfer function. We see that the packaging parasitics *improve* the input return loss at low frequencies (below approximately 2.4 GHz); however, at high frequencies, the input return loss degrades rapidly. Furthermore, the 3-dB bandwidth is reduced by a factor 2× to approximately 3.2 GHz, when compared with the case without packaging parasitics.

The same general behavior can be observed for different values of C' and L and for more realistic package models: (i) an improvement of S_{11} at low frequencies, (ii) a rapid degradation of S_{11} at high frequencies, and (iii) a reduction in bandwidth. Note that for some package types, such as the thin quad flat pack (TQFP), all pins

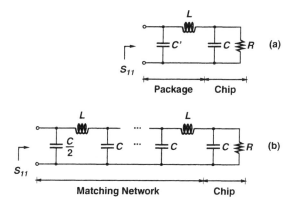

Fig. C.6 An R-C parallel circuit seen through (a) a simple package model and (b) an infinite, lossless, artificial transmission line.

have similar parasitics, whereas for others, such as the ball grid array (BGA), the parasitics vary strongly from pin to pin (or ball to ball). In the latter case, the pins for high-speed input/output (I/O) signal must be selected carefully.

Now that we have seen that a simple L-C network in front of the on-chip R-C impedance can improve the return loss, we may wonder about the theoretical limit of this improvement when using a suitable matching network. Such a matching network is shown in Fig. C.6(b), where the simple L-C section of Fig. C.6(a) was expanded into an infinite, lossless, artificial transmission line. If we choose $L = R_0^2 C$, the transmission line is approximately matched to $R \approx R_0$ and has a cutoff frequency of $1/(\pi R_0 C)$. The wiggly dashed line in Fig. C.4 shows the simulation result for $|S_{11}(f)|$ of such a network with 100 L-C segments, $R = 55\,\Omega$, $C = 1\,\text{pF}$, and $L = 2.5\,\text{nH}$; the wiggly dashed line in Fig. C.5 shows the magnitude of the input transfer function.[4] Compared with the simple R-C parallel circuit, the input return loss improved for frequencies up to the cutoff frequency of about 6.4 GHz, however, above that frequency, the transmission line become totally opaque and the input return loss becomes 0 dB. Note that the 3-dB bandwidth, which also is given by the cutoff frequency, is identical to that of the simple R-C parallel circuit. A more general theoretical analysis reveals that no matching network can improve the S_{11} parameter beyond the integral value [134]

$$\int_0^\infty \ln \frac{1}{|S_{11}(\omega)|}\, d\omega \leq \frac{\pi}{RC}. \qquad (C.19)$$

This inequality is known as the *Bode-Fano limit* for a parallel R-C load.

---

[4]The wiggles are caused by the mismatch between the full-shunt impedance of the transmission line and R near the cutoff frequency; they can be reduced by inserting a half section or an m-derived half section.

T-Coil Network. At this point, it may appear that given the on-chip load capacitance C, a good input return loss can be achieved only for frequencies below $1/(\pi R_0 C)$. However, the T-coil network shown in Fig. C.7 can drive the capacitance C while providing a *frequency independent* return loss at the input port. The T-coil network consists of two mutually coupled inductors with the same value L and the coupling factor k, plus a bridge capacitor C_B. Note that in this network, the termination resistor R and the on-chip capacitance C are not connected in parallel. The frequency-independent return loss of this network can be understood as follows: at low frequencies, the inductors short the input to the resistor R, whereas at high frequencies, the bridge capacitor performs the same function. A more detailed analysis shows that, given the right component values, the input impedance stays at R, regardless of frequency. This is the case for the values $k = 1/3$, $L = 3/8 \cdot R_0^2 C$, and $C_B = C/8$, for which the T-coil network together with the R and C form a second-order Butterworth low-pass filter with the 3-dB bandwidth $\sqrt{2}/(\pi R_0 C)$ (assuming $R \approx R_0$) [82]. The horizontal dash-dotted line in Fig. C.4 shows the simulation result for $|S_{11}(f)|$ of the T-coil network with $R = 55\,\Omega$, $C = 1\,\text{pF}$, $C_B = 0.125\,\text{pF}$, and $L = 0.9375\,\text{nH}$; the dash-dotted line in Fig. C.5 shows the magnitude of the input transfer function from the input port to the voltage across C. Now, the input return loss is frequency independent and its magnitude is simply controlled by the deviation of R from its ideal value. Furthermore, the 3-dB bandwidth is improved by a factor $\sqrt{2}\times$ to approximately 9.0 GHz when compared with the simple R-C parallel circuit.

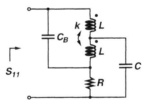

Fig. C.7 A T-coil network for driving the load capacitance C.

In practice, the T-coil network can be implemented, for example, with adjacent parallel bond wires, which provide mutual coupling [18]. In this case, the termination resistor R and the bridge capacitor C_B are located off-chip; furthermore, to obtain the correct coupling polarity between the bond wires, a differential port is required. Alternatively, the T-coil network can be implemented on-chip with interwound spiral inductors [32]. In this case, capacitor C represents the input capacitance of the on-chip circuit including devices for electrostatic discharge (ESD) protection; however, C does not include the bond-pad capacitance, which for this implementation is located at the input of the T-coil.

C.3 DIFFERENTIAL S PARAMETERS

Definition. The four S parameters defined in Section C.1 describe a two-port device such as a single-ended amplifier. A differential amplifier with two inputs and two outputs is a four-port device and is described fully by sixteen S parameters arranged in a 4×4 matrix. Although the concept of two-port S parameters can be extended in a straightforward way to four ports, where each port corresponds to a terminal, it is more convenient and practical to use the so-called *mixed-mode S parameters* [15]. In this representation, the four terminal ports (1, 2, 3, and 4) are replaced by four "virtual ports," namely the differential-mode input port (d1), the common-mode input port (c1), the differential-mode output port (d2), and the common-mode output port (c2). If we write the mixed-mode S parameters in the form $S_{mn\mu\nu}$, where $n\nu$ is the (virtual) port where the incident wave is applied and $m\mu$ is the (virtual) port where the outgoing wave is measured, then the 4×4 S matrix is

$$S = \begin{pmatrix} S_{dd11} & S_{dd12} & S_{dc11} & S_{dc12} \\ S_{dd21} & S_{dd22} & S_{dc21} & S_{dc22} \\ S_{cd11} & S_{cd12} & S_{cc11} & S_{cc12} \\ S_{cd21} & S_{cd22} & S_{cc21} & S_{cc22} \end{pmatrix}. \qquad (C.20)$$

For example, S_{cd21} is the ratio of the transmitted wave at the common-mode output port (c2) to the incident wave at the differential-mode input port (d1).[5] Of the sixteen S parameters, the most significant ones are located in the upper left 2×2 submatrix. This submatrix is known as S_{dd} and describes the important differential-mode behavior of the four-port. The other three 2×2 submatrices describe the usually undesired common-mode and mode-conversion behavior.

In analogy to Fig. C.1(a), we can use the test circuit shown in Fig. C.8(a) to excite the differential-mode input port (d1) and determine the differential S parameters

$$S_{dd11}(s) = \frac{V_i - V_s}{V_s}, \qquad (C.21)$$

$$S_{dd21}(s) = \frac{V_o}{V_s}, \qquad (C.22)$$

where $V_i = V_{ip} - V_{in}$ is the differential input-voltage phasor, $V_o = V_{op} - V_{on}$ is the differential output-voltage phasor, and V_s is the differential source-voltage phasor. The $2\times$ gain block in the test circuit must not produce a common-mode output signal such that $V_{icm} = 1/2 \cdot (V_{ip} + V_{in}) = 0$. In analogy to Fig. C.1(b), the $S_{dd22}(s)$ and $S_{dd12}(s)$ parameters can be obtained by exciting the differential-mode output port.

The test circuit shown in Fig. C.8(b) can be used to excite the common-mode input port (c1) and determine the common-mode S parameters

$$S_{cc11}(s) = \frac{V_{icm} - V_s}{V_s}, \qquad (C.23)$$

---

[5] Note that the codes for the incident and outgoing virtual ports are interleaved in the index of the S parameter. This may be confusing at first, but has become a standard practice in the literature.

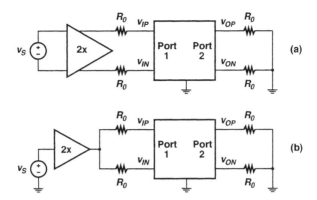

Fig. C.8 Test circuits to determine four-port mixed-mode S parameters: (a) S_{dd11} and S_{dd21} parameters and (b) S_{cc11} and S_{cc21} parameters.

$$S_{cc21}(s) = \frac{V_{ocm}}{V_s}, \tag{C.24}$$

where $V_{icm} = 1/2 \cdot (V_{ip} + V_{in})$ is the common-mode input-voltage phasor, $V_{ocm} = 1/2 \cdot (V_{op} + V_{on})$ is the common-mode output-voltage phasor, and V_s is the (single-ended) source-voltage phasor. Similarly, the $S_{cc22}(s)$ and $S_{cc12}(s)$ parameters can be obtained by exciting the common-mode output port.

Relationship Between Single-Ended and Differential S Parameters. Given a fully differential amplifier, we can measure the S parameters with a single-ended or differential test set-up. Figure C.9 shows the test set-ups for both types of measurement. A common measurement dilemma is that the amplifier has been designed for a differential application, and hence the differential S parameters are of relevance; however, because most network analyzers support only two-port measurements, the single-ended S parameters are measured. In such a situation, it is important to understand the relationship between the single-ended and differential S parameters.

The main difference between the two test set-ups is that the differential set-up (Fig. C.9(b)) excites only the differential-mode input of the amplifier, that is, the common-mode input voltage is zero, whereas the single-ended set-up (Fig. C.9(a)) excites the differential-mode as well as the common-mode input of the amplifier. As a result, the value of the single-ended S_{11} parameter is a mixture of S_{dd11}, S_{dc11}, S_{cd11}, and S_{cc11}. It turns out that if the circuit under test is balanced (fully symmetrical), all mode-conversion S parameters become zero ($S_{dc\mu\nu} = S_{cd\mu\nu} = 0$) and the single-ended S_{11} parameter can be written as

$$S_{11,se} = \frac{S_{dd11} + S_{cc11}}{2}. \tag{C.25}$$

This relationship is illustrated in Fig. C.10, which shows three balanced input circuits together with their differential, common-mode, and single-ended S_{11} parameters. We

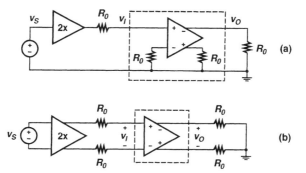

Fig. C.9 Measurement of (a) single-ended and (b) differential S parameters of a fully differential amplifier.

learn from this example that knowing the single-ended S_{11} parameter, which is easy to measure, doesn't let us infer the differential S_{11} parameter, unless we have knowledge of the circuit topology.

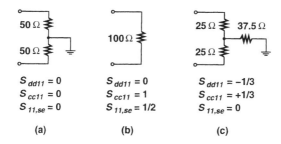

Fig. C.10 Comparison of differential, common-mode, and single-ended S_{11} parameters for three balanced input circuits.

Now what about the differential and single-ended S_{21} parameters? How do they relate? Similar to what we said above, if the circuit under test is balanced, all mode-conversion S parameters vanish and the single-ended S_{21} parameter can be written as

$$S_{21,se} = \frac{S_{dd21} + S_{cc21}}{2}. \tag{C.26}$$

If we assume further that the circuit under test strongly suppresses common-mode signals such that $S_{cc21} \ll S_{dd21}$, then we can simplify the above equation to $S_{21,se} = S_{dd21}/2$. This, of course, is just a statement of the fact that the single-ended gain is half of the differential gain.

Appendix D
Transistors and Technologies

Broadband circuits for optical communication are realized in a wide variety of technologies. For medium- and low-speed applications, standard silicon technologies, which offer MOSFETs, BJTs, or both are preferred because of their cost advantage. However, for high-speed applications, silicon-germanium (SiGe), gallium-arsenide (GaAs), or indium-phosphide (InP) technologies, which offer fast heterostructure transistors in the form of HFETs and HBTs, become necessary.

In this Appendix, we cover the basic operation of the most important transistor types (MOSFET, MESFET, HFET, BJT, and HBT); we also describe the salient features of the technologies most commonly used for optical communication circuits. For more information on this subject, see [101, 109, 126, 183, 184].

D.1 MOSFET AND MESFET

MOSFET Fundamentals. A stylized cross-section through an n-channel *metal-oxide semiconductor field-effect transistor* (MOSFET) is shown in Fig. D.1(a). Its basic operation can be explained as follows. First, the amount of negative charge, Q_n, in the channel, indicated by the "−" symbols in Fig. D.1, is controlled by the

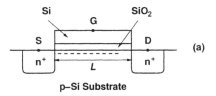

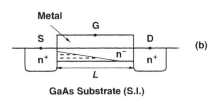

Fig. D.1 (a) Silicon MOSFET and (b) GaAs MESFET (schematically).

voltage at the gate (G). Assuming a small drain-source voltage, we can write

$$Q_n = C_{gs} \cdot (V_{GS} - V_{TH}), \tag{D.1}$$

which is just the charge/voltage relationship of an ideal capacitor with the modification that the charge buildup starts only after the gate-source voltage, V_{GS}, exceeds the threshold voltage, V_{TH}. The gate-source capacitance, C_{gs}, further can be expressed in terms of the device dimensions as $C_{gs} = C'_{ox} \cdot WL$, where C'_{ox} is the gate-oxide capacitance per unit area, W is the channel width, and L is the channel length. Now, the amount of charge in the channel, Q_n, determines the current flow from the source (S) to the drain (D). This can be understood by recognizing that the channel current is equal to the channel charge, Q_n, divided by its transit time, τ. The latter is $\tau = L/v_n$, where L is the channel length and v_n is the electron velocity. Furthermore, at low fields, the electron velocity is given by $v_n = \mu_n E$, where μ_n is the electron mobility and E is the electric field along the channel. This field can be written as $E = V_{DS}/L$, assuming the drain-source voltage, V_{DS}, is small. Putting these expressions together, we find the relationship between the channel current, I_D, and the channel charge, Q_n, for an n-MOS device operated at a small V_{DS} to be

$$I_D = \frac{\mu_n}{L^2} \cdot V_{DS} \cdot Q_n. \tag{D.2}$$

Combining Eqs. (D.1) and (D.2) leads to the well known MOSFET I/V relationship for the *linear regime* (a.k.a. the *ohmic regime* or the *triode regime*)

$$I_D = \mu_n C'_{ox} \cdot \frac{W}{L} \cdot (V_{GS} - V_{TH}) \cdot V_{DS}, \quad \text{if} \quad V_{DS} \ll V_{GS} - V_{TH}. \tag{D.3}$$

In summary, for low drain-source voltages, the FET behaves like a voltage-controlled conductance: $g_o = \partial I_D/\partial V_{DS} \sim (V_{GS} - V_{TH})$.

Next, let's consider a larger drain-source voltage: $V_{DS} > V_{GS} - V_{TH}$. In this case, the voltage drop along the channel causes the channel charge to thin out toward the drain, and just short of the drain, the channel comes to an end. Now, the total channel charge is only about half of the expression given in Eq. (D.1) and the electrical field along the channel is approximately given by $E = (V_{GS} - V_{TH})/L$, that is, it is independent of V_{DS}. As before, the drain current can be obtained by dividing the channel charge, Q_n, by the transit time, $\tau = L/(\mu_n E)$, which results in

$$I_D = \frac{\mu_n}{L^2} \cdot (V_{GS} - V_{TH}) \cdot Q_n. \tag{D.4}$$

Combining half of Eq. (D.1) with Eq. (D.4) leads to the well known MOSFET I/V relationship for the *saturated regime*

$$I_D = \frac{\mu_n C'_{ox}}{2} \cdot \frac{W}{L} \cdot (V_{GS} - V_{TH})^2, \quad \text{if} \quad V_{DS} > V_{GS} - V_{TH}. \tag{D.5}$$

In summary, for sufficiently high drain-source voltages, the FET behaves like a voltage-controlled current source (a transconductance).[1]

Note that the drain current in Eq. (D.5) depends *quadratically* on the gate-source voltage. This is so because an increase in gate-source voltage causes *two* current enhancing effects: (i) the channel charge, Q_n, increases and (ii) the transit time, τ, diminishes at the same time. In submicron FETs, however, the field along the channel easily becomes so strong that the electron velocity saturates at $v_{n.\text{sat}}$. Under this condition, the transit time, $\tau = L/v_{n.\text{sat}}$, becomes independent of the bias voltages and the drain current saturates at $I_D = v_{n.\text{sat}}/L \cdot Q_n$. As a result, submicron devices often enter the saturated regime before V_{DS} reaches $V_{GS} - V_{TH}$, and their I/V relationship then changes from quadratic to *linear*

$$I_D = v_{n.\text{sat}} C'_{ox} \cdot W \cdot (V_{GS} - V_{TH}), \quad \text{if} \quad V_{DS} > v_{n.\text{sat}}/\mu_n \cdot L. \tag{D.6}$$

For example, $v_{n.\text{sat}}/\mu_n$ for silicon is about 1 to 2 V/μm, which means that a 0.1-μm MOSFET saturates at a drain-source voltage of 0.1 to 0.2 V, even if $V_{GS} - V_{TH}$ is larger than that.

The speed of an FET often is quantified by its unity current-gain frequency (a.k.a. transition frequency), f_T. At low electric fields, we find the g_m of a saturated n-MOS device from the quadratic MOSFET model in Eq. (D.5) as $g_m = \partial I_D / \partial V_{GS} = \mu_n C'_{ox} \cdot W/L \cdot (V_{GS} - V_{TH})$. With the idealized assumptions $C_{gs} = 2/3 \cdot C'_{ox} \cdot WL$ and $C_{gd} = 0$, we can derive f_T as (cf. Eq. (6.49))

$$f_T = \frac{1}{2\pi} \cdot \frac{g_m}{C_{gs} + C_{gd}} \approx \frac{3}{4\pi} \cdot \frac{\mu_n}{L^2} \cdot (V_{GS} - V_{TH}). \tag{D.7}$$

We see that f_T is determined by the mobility of the carriers in the channel, the channel length, and the bias conditions. Because the electron mobility is higher than the hole

---
[1] Here we neglected the *channel-length modulation*, which makes the saturated drain current somewhat dependent on V_{DS}, especially for short channel lengths.

mobility, n-MOS devices are faster than p-MOS devices. It is interesting to note that the above expression is inversely proportional to the transit-time expression used to derive Eq. (D.4); thus, $f_T \sim 1/\tau$. At high electric fields, the mobility degrades and the saturated electron velocity, $v_{n.\text{sat}}$, becomes important in determining the speed. From the submicron MOSFET model in Eq. (D.6), we find that $f_T \sim v_{n.\text{sat}}/L$, which means that for large enough bias voltages, f_T saturates. The maximum frequency of oscillation, $f_{\max}$, strongly depends on the intrinsic gate resistance, R_g, (cf. Eq. (6.50)) and thus on the gate material (metal vs. polysilicon) and the layout style. A layout with multiple short gate fingers contacted on both sides can reduce the R_g of polysilicon gates. (For additional information about f_T and $f_{\max}$, see Section 6.3.2.)

In MOSFETs, the gate is isolated from the channel by a thin layer of oxide (SiO_2), hence the name *metal-oxide-semiconductor* (MOS). The dielectric gate oxide ensures that the gate leakage current is virtually zero. As the letter 'M' in MOS indicates, early MOSFETs had metal gates; however, most modern MOSFETs have polysilicon gates, which permit the drain and source regions to be self-aligned to the gate during fabrication.

The source/channel/drain system is isolated from the substrate by a depletion zone formed by the reverse-biased junction between the n-source/channel/drain and the p-doped substrate. A side-effect of this junction is that the channel is capacitively coupled to the substrate, which acts as an unwanted backgate to the transistor.

Silicon-MOSFET Technology. Most silicon-MOSFET technologies offer n-channel as well as p-channel transistors, and thus are known as *complementary metal-oxide-semiconductor* (CMOS) technologies. CMOS technology is ideal for large digital circuits: CMOS logic circuits feature a very low static power dissipation and can be packed very densely on the chip. For these reasons, virtually all digital VLSI chips are implemented in CMOS technology. CMOS technology is widely available, highly developed (small feature size), mature (high yield), low in cost (assuming high volumes to divide the mask costs), and well characterized.

Naturally, it is very desirable to implement analog high-speed circuits for optical communication in the same CMOS technology that is used for VLSI chips. With this approach, the analog front-end circuits and the digital back-end processing can be integrated on a single chip, leading to a compact and low-cost solution. Often only the n-channel devices are used in the analog high-speed circuits; the p-channel devices are significantly slower and typically are used for biasing purposes only.

The speed of modern submicron MOSFETs is formidable and increases as the minimum feature size is reduced in future generations. For example, an n-channel transistor in a typical 0.25-μm technology has $f_T \approx 25$ GHz, and a 0.15-μm transistor has $f_T \approx 50$ GHz (at $V_{GS} - V_{TH} = 400$ mV).[2] For the latest 90-nm technology, an f_T of around 120 GHz is achieved [191]. The maximum frequency of oscillation, $f_{\max}$, usually can be made larger than f_T by choosing the appropriate device layout.

---

[2]Because f_T usually is measured for a high bias voltage where the carrier velocity is saturated, f_T scales approximately like $1/L_{\min}$, where $L_{\min}$ is the minimum feature size of the technology.

A drawback of submicron silicon MOSFETs is their low breakdown voltage; typically, the drain-source voltage is the most critical voltage. As the channel length is reduced, the supply voltage must be reduced as well to avoid breakdown. For example, circuits in a 0.25-μm technology typically operate from a supply voltage of 2.5 V, whereas 0.15-μm circuits operate from 1.5 V.

MESFET Fundamentals. A stylized cross-section through an n-channel *metal-semiconductor field-effect transistor* (MESFET) is shown in Fig. D.1(b). In this device, the channel between the source (S) and the drain (D) is formed by the lightly doped n-type material (n$^-$). The gate (G) metal and the channel material form a Schottky contact, which partly depletes the channel region. The depth of this depletion zone and thus the amount of free charge remaining in the channel are controlled by the gate voltage. Similar to the MOSFET, the device behaves as a voltage-controlled conductance or a voltage-controlled current source, depending on the operating regime. The MOSFET equations Eqs. (D.3), (D.5), and (D.6) also approximately describe the MESFET. (Actually, the MESFET equivalent of C'_{ox} is gate bias dependent, resulting in a different set of equations. However, after a Taylor series approximation around $V_{GS} = V_{TH}$, often used to simplify the MESFET equations, the *form* of the MES- and MOSFET equations become the same.) In contrast to the MOSFET, the gate is in direct contact with the semiconductor material, hence the name *metal-semiconductor* (MES) is used. The reverse-biased Schottky diode provides an isolation between the gate and the channel, but the gate leakage current of a MESFET usually is not negligible. On the plus side, the gate resistance of MESFETs is very low because the gate is made from metal rather than polysilicon.

Similar to the MOSFET, the drain current in the saturated regime grows quadratically for small gate-source voltages. However, for larger voltages, the drain current grows more slowly and eventually saturates at a constant value. This effect is caused by the increased gate leakage as the Schottky contact becomes forward biased (gate turn-on) as well as by velocity saturation of the carriers. A result of this current saturation effect is that the transconductance, $g_m = \partial I_D / \partial V_{GS}$, reaches a maximum value, after which it declines.

Depending on the amount of channel doping, it is possible to build devices that conduct current for $V_{GS} = 0$ V, so-called *depletion-mode MESFETs*, as well as devices that are completely turned off for $V_{GS} = 0$ V, so-called *enhancement-mode MESFETs*. Note that to turn an enhancement-mode MESFET on, the Schottky diode must be slightly forward biased, which makes the gate leakage problem worse.

GaAs-MESFET Technology. Because the low-field electron mobility in GaAs is about six times higher than in silicon, GaAs MESFETs reach higher speeds than silicon FETs for the same channel length and bias voltage. As we know, at high electric fields, the mobility degrades and the saturated electron velocity, $v_{n.\text{sat}}$, becomes important in determining the speed. Here, too, GaAs has a speed advantage over silicon. For example, the 0.2-μm GaAs-MESFET technology used in [80] has a peak f_T of 80 GHz and the circuits can operate from ± 5-V power supplies. GaAs MESFETs have fairly high breakdown voltages, which do not depend strongly on the channel

348 TRANSISTORS AND TECHNOLOGIES

length. In general, materials with a wider bandgap have a higher breakdown voltage because more energy is needed to create an electron/hole pair by means of impact ionization. The GaAs-MESFET technology is mature, widely available, and allows for fairly high integration levels.

Another advantage of GaAs over silicon is the very high resistivity of the undoped substrate material: 10^8 Ω cm for GaAs versus 100 Ω cm for silicon. The GaAs substrate therefore is said to be a *semi-insulating substrate* (S.I.). This type of substrate greatly reduces parasitic capacitances and provides intrinsic isolation between the devices. On the downside, the semi-insulating substrate causes a poorly controlled backgating effect. Like in the MOSFET, the electric field on the backside of the channel has an unwanted influence on the drain current. Whereas in the silicon MOSFET this field is well controlled by the voltage between the source and the semiconducting substrate (the backgate), in the GaAs MESFET this field depends on the voltages of the adjacent devices and the capacitive coupling through the semi-insulating substrate to the channel. For this reason, this effect also is known as *sidegating*. As a result of sidegating, good transistor matching and low offset voltages are hard to achieve in GaAs-MESFET technologies. Another drawback is the about three times lower thermal conductivity of the GaAs substrate over a silicon substrate.

Because of the lack of a good native oxide for GaAs (and other compound semiconductors), GaAs MOSFETs are not feasible and the MESFET structure must be used. This leads to another drawback of the GaAs-MESFET: the gate leakage current caused by the MES Schottky contact is in the range of 1 nA to 1 μA. The leakage can be controlled to some extent by choosing a gate metal, such as a Ti/Pt/Au layered structure, that forms a good (high) Schottky barrier with the GaAs semiconductor. Finally, GaAs-MESFET technologies typically offer only n-channel devices, that is, the technology is not complementary. This restriction limits the circuit topologies that can be used.

D.2 HETEROSTRUCTURE FET (HFET)

HFET Fundamentals. Stylized cross-sections through n-channel HFETs in two different material systems are shown in Fig. D.2. Like MESFETs, HFETs have a metal gate (G) that is in direct contact with the semiconductor material, thus forming a Schottky contact. However, two dissimilar semiconductor materials are used underneath the gate: (i) a doped wide-bandgap material known as the *donor layer* and (ii) an undoped narrow-bandgap material known as the *channel layer*.[3] Because of these dissimilar materials, the device is known as *heterostructure field-effect transistor* (HFET). The channel, which resides in the channel layer, as indicated by the "−" symbols in Fig. D.2, is connected to the source (S) and drain (D) regions by means of two n^+ implants (dashed lines) penetrating the donor layer. Devices are isolated

---

[3] In practice, a thin layer of undoped wide-bandgap material, known as the *buffer layer*, is placed in between the donor and the channel layer to separate the heterojunction from the donor layer.

from each other by interrupting the donor-layer and the channel by a mesa etch or an isolation implant (not shown in Fig. D.2).

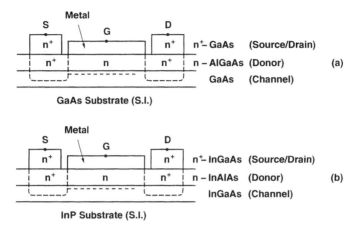

Fig. D.2 HFETs in two different material systems: (a) on GaAs substrate and (b) on InP substrate (schematically).

The heterostructure causes the donor layer to deplete and the resulting free electrons accumulate into a thin layer at the upper side of the undoped channel layer, as shown in Fig. D.2. The band diagram in Fig. D.3 illustrates how the bandgap discontinuity creates a potential well, which fills with the free charge. As in all FET types, the gate voltage controls the amount of charge in the channel, which then controls the channel current. The principle advantage of the HFET is that it achieves a higher speed, compared with the MESFET, because the channel resides in the *undoped* material with lower ionized-impurity scattering and improved electron mobility.

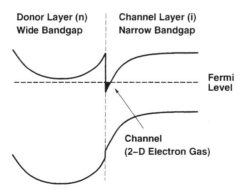

Fig. D.3 Band diagram for the donor and channel layers of an HFET.

Similar to the MESFET, the drain current in the saturated regime grows quadratically for small gate-source voltages. However, for larger voltages, the drain current grows more slowly and eventually saturates at a constant value. This effect is caused by velocity saturation of the carriers and by an impairment of the channel confinement, which causes carriers to spill over into the donor layer, where the velocity is much lower. A result of this current saturation effect is that the transconductance, $g_m = \partial I_D/\partial V_{GS}$, reaches a maximum value, after which it declines.

Another advantage of the HFET over the MESFET is that the wide-bandgap material underneath the gate improves the Schottky barrier height and thus reduces the gate leakage current. Furthermore, HFETs are known to have a better noise figure than MESFETs. Similar to MESFETs, HFETs suffer from poorly controlled backgating and poor device matching (GaAs and InP substrates both are semi-insulating). Typically, HFET technologies offer only n-channel transistors; however, depletion- and enhancement-mode devices are sometimes available.

The HFET also is known by several other names such as *high electron-mobility transistor* (HEMT), *2-dimensional electron-gas field-effect transistor* (2DEG-FET or TEGFET), and *modulation-doped field-effect transistor* (MODFET). In the basic HFET, the layer materials are chosen to be lattice matched with the substrate to avoid strain. However, more advanced HFETs use materials with a small lattice mismatch (<8%), which is acceptable as long as the mismatched layer remains thin. This type of HFET is know as *pseudomorphic heterostructure field-effect transistor* (PHFET) or *pseudomorphic high electron-mobility transistor* (PHEMT). PHFETs have better performance than the basic HFETs because there are more choices for the donor and channel layer materials.

Several naming conventions for the HFET material system are in use. Sometimes the substrate material is used to name the technology (e.g., GaAs-HFET technology). However, there is more than one HFET material system that can be grown on a particular substrate, especially if lattice matching is not strictly required; furthermore, the substrate material determines the transistor characteristics only partially. Thus to be more specific, the materials of the major layers (e.g., donor and channel layers) sometimes are used to name the technology (e.g., AlGaAs/GaAs-HFET, AlGaAs/InGaAs/GaAs-HFET, or even $Al_{0.15}Ga_{0.85}As/In_{0.15}Ga_{0.85}As/GaAs$ technology). In this book, we use the simpler, substrate-based naming convention.

GaAs-HFET Technology. A typical GaAs HFET is shown schematically in Fig. D.2(a). The wide-bandgap donor layer consists of n-doped AlGaAs, which can be lattice matched to the GaAs channel layer and the GaAs substrate underneath (for any Al fraction, $Al_xGa_{1-x}As$). Alternatively, the lattice mismatched InGaAs material (the In fraction has to be less than 25% to prevent excessive strain, e.g., $In_{0.15}Ga_{0.85}As$) can be used for the channel layer resulting in a pseudomorphic GaAs HFET. The GaAs-HFET and GaAs-PHFET technologies have the important advantage that they achieve a very high speed and high breakdown voltages at the same time. For example, the 0.25-μm GaAs-PHEMT technology used in [138] has an f_T of 60 GHz and the circuits can operate from a supply voltage of 5.2 V. The 0.15-μm

GaAs-PHEMT technology reviewed in [191] reaches the peak values $f_T = 110\,\text{GHz}$ and $f_{\max} = 180\,\text{GHz}$ and has a drain-source breakdown voltage of more than 6 V.

InP-HFET Technology. A typical InP HFET is shown schematically in Fig. D.2(b). The wide-bandgap donor layer consists of n-doped InAlAs and the channel layer underneath consists of undoped InGaAs, which both can be lattice matched to the InP substrate (the In fraction in the channel layer must be 53% to lattice match the substrate: $\text{In}_{0.53}\text{Ga}_{0.47}\text{As}$). This technology achieves an even greater speed than the GaAs-HFET technology because of the very high electron mobility and saturated electron velocity in the InGaAs channel. However, breakdown voltages typically are lower. For example, the 0.15-μm InP-HEMT technology used in [173] has an f_T of 167 GHz (at $V_{GS} - V_{TH} = 670\,\text{mV}$), and the circuits can operate from a 3.2-V supply voltage. The 0.1-μm InP-HEMT technology reviewed in [191] reaches the peak values $f_T = 175\,\text{GHz}$ and $f_{\max} = 300\,\text{GHz}$ and has a drain-source breakdown voltage of more than 2 V.

Another advantage of the InP-HFET technology specific to optical communication applications is that it permits the integration of long-wavelength optoelectronic devices. For example, a metal-semiconductor-metal (MSM) photodetector sensitive in the 1.3- to 1.55-μm wavelength range may be integrated on the same chip by reusing the gate metal. A drawback of the InP technology is the present lack of large substrates.

D.3 BIPOLAR JUNCTION TRANSISTOR (BJT)

BJT Fundamentals. A stylized cross-section through a vertical n-p-n BJT is shown in Fig. D.4. Its basic operation can be described as follows. The forward-biased base-emitter (B-E) junction injects electrons from the top into the p-doped base layer. At the bottom of this thin base layer, the electrons, which now are minority carriers, are picked up by the reverse-biased collector-base (C-B) junction, where they produce the collector current. The amount of injected current, and thus the collector current, I_C, can be controlled with the base-emitter voltage, V_{BE}. As a side effect, a certain amount of holes are injected in the reverse direction from the base into the emitter and contribute to an undesirable base current. Because of the critical role played by electrons and holes crossing through p-n junctions, this transistor is known as a *bipolar junction transistor* (BJT). Similar to the FET in the saturated regime, the BJT behaves like a voltage-controlled current source (transconductor). However, unlike the FET, its operation is governed by the laws of thermodynamics (Fermi-Dirac distribution), which results in an *exponential* relationship between the output current and the control voltage. Assuming that the base-emitter junction is forward biased ($V_{BE} > 0$) and the collector-base junction is reverse biased ($V_{BC} \leq 0$), the BJT operates in the *active regime*, and its well-known I/V relationship is

$$I_C = I'_{C0} \cdot A_E \cdot \exp\left(\frac{V_{BE}}{V_T}\right), \tag{D.8}$$

where I'_{C0} is the (extrapolated) collector current density at $V_{BE} = 0$, A_E is the emitter area given by $A_E = W_E L_E$, and V_T is the thermal voltage given by $V_T = kT/q \approx 25\,\text{mV}$.[4]

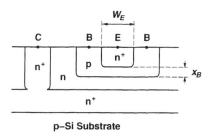

Fig. D.4 Vertical silicon BJT (schematically).

As indicated in Fig. D.4, the emitter (E) is heavily doped (n^+) and can be accessed directly from the top. The base (B) is lightly doped, relative to the emitter, to improve the electron injection efficiency. A high injection efficiency means that the emitter current consists mostly of electrons injected into the base region rather than holes injected into the emitter, thus keeping the base current low. Furthermore, the base is made thin (small x_B) to speed up the transit time and minimize carrier recombination in the base, which also contributes to the base current. However, a thin base also causes a high base-spreading resistance, which degrades the transistor's speed and noise performance. The collector (C) is lightly doped, which reduces the associated junction capacitance, and is contacted by a heavily doped (n^+) buried layer to provide a low-resistance path to the collector terminal. Isolation between the devices can be implemented with p-type material (junction isolation) or silicon oxide (dielectric isolation) separating the collector regions (not shown in Fig. D.4). The transistor shown in Fig. D.4 is known as a *vertical BJT*, because the n-p-n sequence progresses orthogonal to the chip surface. The alternative is a *lateral BJT* with a horizontal n-p-n sequence; however, lateral devices usually suffer from lower speed and lower current gain because the critical base width, x_B, now is controlled by lithography, making the attainable x_B much bigger.

The base current, I_B, of a BJT is approximately proportional to the collector current and can be written as

$$I_B \approx \frac{I_C}{\beta}, \tag{D.9}$$

where β is the current gain of the BJT ($\beta = \partial I_C / \partial I_B$), which typically is in the range of 50 to 150. In practice, β degrades for low and high collector currents, and the approximation in Eq. (D.9) holds only for intermediate currents. Combining Eqs. (D.8) and (D.9), we find the input resistance of a BJT to be $R_I = \partial V_{BE}/\partial I_B =$

---

[4]Here we neglected the *Early effect*, which makes the collector current somewhat dependent on V_{CE}.

$\beta \cdot V_T/I_C$. This resistance is in the kΩ range and thus much lower than the input resistance of an FET.

The speed of a BJT often is quantified by its unity current-gain frequency (a.k.a. transition frequency), f_T, and its maximum frequency of oscillation, $f_{\max}$. From Eq. (D.8), we find that $g_m = \partial I_C/\partial V_{BE} = I_C/V_T$, and with $C_{be} = \tau_F \cdot I_C/V_T + C_{je}$ and $C_{bc} = C_{jc}$, we can derive (cf. Eq. (6.47))

$$f_T = \frac{1}{2\pi} \cdot \frac{g_m}{C_{be} + C_{bc}} = \frac{1}{2\pi} \cdot \frac{1}{\tau_F + (C_{je} + C_{jc}) \cdot V_T/I_C}. \quad (D.10)$$

We see that f_T is determined by the carrier transit time, τ_F, the junction capacitances C_{je} and C_{jc}, and the collector current, I_C. An important difference to the FET is that the carrier transit time through the base is controlled by *diffusion* rather than drift in an electrical field. The transit time, τ_F, is proportional to x_B^2/D_n, where x_B is the base thickness and D_n is the diffusion constant. Thus, the transit time is bias independent to a first approximation.[5] A more detailed analysis reveals that if the current density I_C/A_E exceeds a critical value, the transit time increases rapidly because of an extension of the base region into the collector region known as *base pushout* or *Kirk effect*. (The same effect also causes a reduction of the current gain.) Despite these differences, the speed of FETs as well as BJTs are improved by high-mobility semiconductor materials because the diffusion constant and the mobility are linked by the Einstein relationship $D_n = \mu_n V_T$. The higher electron mobility compared with the hole mobility also is the reason why n-p-n devices are faster than p-n-p devices. The maximum frequency of oscillation, $f_{\max}$, depends strongly on the intrinsic base resistance, R_b, (cf. Eq. (6.48)), and thus can be controlled by the base doping and the layout style: a narrow emitter-stripe layout with a sufficient number of base contacts leads to a low R_b. (For additional information about f_T and $f_{\max}$, see Section 6.3.2.)

In switching and limiting amplifier applications, the BJT may enter a regime where the base-emitter and the collector-base junctions both become forward biased. This operating condition is known as the *saturated regime* of the BJT—not to be confused with the saturated regime of an FET. In this case, both junctions inject electrons into the base region, flooding it with charge. Unfortunately, when the BJT returns from the saturated to the active regime, it takes some time to clear out this excess charge. Therefore, the saturated regime must be avoided in high-speed applications.

At high collector-emitter voltages, avalanche multiplication in the reverse-biased collector-base diode sets in and causes the collector current to increase rapidly. This effect is known as *avalanche breakdown*. Two extreme cases can be identified. (i) If the base is driven from a low-impedance source (and we neglect the intrinsic base resistance), the excess collector current consists only of the avalanche current generated in the collector-base diode. Thus, the emitter-collector breakdown voltage for

---

[5] It is interesting to observe that the BJT's input capacitance (C_{be}) is bias dependent, whereas its transit time is not (ignoring the Kirk effect). In contrast, the FET's the input capacitance (C_{gs}) is constant, whereas its transit time is bias dependent (ignoring velocity saturation).

the shorted-base case, BV_{CES}, is similar to the breakdown voltage of the collector-base diode (with open emitter): $BV_{CES} \approx BV_{CBO}$. The avalanche breakdown of the collector-base diode can be described by the multiplication factor $M(V_{CB})$, which multiplies the collector current; if $M(V_{CB})$ becomes much larger than one, breakdown occurs. (ii) If the base is driven from a high-impedance source or if the base is left open, the situation is more complex. In this case, the avalanche current generated in the collector-base diode pulls up the base voltage, thus producing an *amplified* avalanche current at the collector. More precisely, the base is pulled down by the regular base current I_C/β and pulled up by the avalanche current $(M-1) \cdot I_C$. Breakdown occurs if the pull-up current exceeds the pull-down current, that is, if $M(V_{CB}) > 1 + 1/\beta$, which is just barely more than one. Clearly, the collector-emitter breakdown voltage for the open-base case, BV_{CEO}, is lower than that for the shorted-base case. For a BJT embedded in a practical circuit, the breakdown voltage depends on the exact driving conditions and occurs somewhere in between the extreme values of BV_{CEO} and BV_{CES}. The prediction of the precise breakdown voltage is complicated further by the base-spreading and contact resistances. The lateral voltage drops across these resistances tend to reduce the breakdown voltage. To predict the breakdown voltage accurately, a distributed 3-dimensional model or a multitransistor model must be used [151]. Finally, note that BJTs optimized for high-speed operation tend to have a lower breakdown voltage. More generally, it has been found that there is a limit, known as the *Johnson limit*, to the product $f_T \cdot BV$, which depends mostly on the device material; for silicon, its value is about 100 to 200 GHzV, whereas for InP, it is about 500 to 1,000 GHzV.[6]

BJTs have several important advantages over FETs: (i) their speed is determined by epitaxial growth or diffusion (vertical feature), rather than by lithography (horizontal feature), leading to higher speeds at modest processing requirements; (ii) their exponential I/V characteristics leads to a higher transconductance at a given bias current; (iii) their current-drive capability per chip area is better; (iv) their $1/f$ noise is lower; and (v) their matching properties are superior. However, there are some notable drawbacks as well: (i) the BJT's base current is much larger than the FET's gate current; (ii) the BJT takes a long time to recover after leaving the saturated regime; and (iii) although digital logic circuits can be implemented in BJT technologies using emitter-coupled logic (ECL) or transistor-transistor logic (TTL), they consume a large amount of static power and cannot achieve the high packing density known from CMOS.

Silicon-BJT Technology. Silicon-BJT technologies are mature, widely available, and well characterized. They typically offer n-p-n as well as the complementary p-n-p devices, giving the circuit designer more options. Silicon-BJT technologies provide fairly fast devices, even with modest lithographic resolutions. For example, the 0.8-μm lithography silicon-BJT technology (with 0.4-μm effective emitter

---

[6]More recently, the Johnson limit for silicon has been reevaluated, and it was found to be higher than previously thought, namely around 500 GHzV [110].

width) reviewed in [145] has $f_T = 27\,\text{GHz}$, $f_{\max} = 34\,\text{GHz}$ (at $V_{CE} = 1\,\text{V}$ and $I_C/A_E = 0.75\,\text{mA}/\mu\text{m}^2$), and the open-base collector-emitter breakdown voltage is 3.7 V. Nevertheless, with a careful design, circuits can be made to operate from a 5.2-V power supply.

At the expense of additional masks and a higher process complexity, the BJT and CMOS technologies can be combined into a so-called *BiCMOS technology* offering BJT as well as MOS transistors. This mix gives the circuit designer the best of both worlds; for example, the BJTs can be used for high-speed analog circuits, and the MOSFETs can be used for the digital CMOS logic.

D.4 HETEROJUNCTION BIPOLAR TRANSISTOR (HBT)

HBT Fundamentals. Stylized cross-sections through vertical n-p-n HBTs in three different material systems are shown in Fig. D.5. The basic layer structure (emitter-base-collector) is the same as for the BJT; however, two dissimilar semiconductor materials are used to form the emitter-base junction: the emitter is made from a material with a wider bandgap than the base material. Because at least one junction is composed of two dissimilar materials, this device is known as a *heterojunction bipolar transistor* (HBT).

The principal advantage of the heterojunction is that a good electron injection efficiency can be obtained (high β), even if the base (B) is heavily doped (p^+) and the emitter (E) is lightly doped. The reason for this effect is illustrated with the band diagram for a forward-biased emitter-base heterojunction in Fig. D.6. The potential barrier for electrons going from the emitter to the base is much lower than the barrier for holes going from the base to the emitter; thus, most of the emitter current is carried by electrons. (The undesirable spike in the conduction band of Fig. D.6 can be reduced with a *graded* heterojunction.) Because the base is heavily doped, the base-spreading and contact resistance, R_b, is reduced, which improves $f_{\max}$ (cf. Eq. (6.48)) and the noise performance. Furthermore, because of the lightly doped emitter, the emitter-base junction capacitance, C_{je}, is reduced, which improves f_T (cf. Eq. (6.47)) and $f_{\max}$. The collector (C) remains lightly doped and is contacted by a heavily doped (n^+) subcollector (or buried layer). HBT devices can be isolated from each other by a mesa etch or an isolation implant.

Additional speed can be gained by gradually varying the material composition of the base from the emitter to the collector (graded base). This measure grades the bandgap (wider at the emitter, narrower at the collector) and provides a built-in electric drift field, which reduces the carrier transit time, τ_F, across the base region. Other advantages of HBTs over BJTs are their higher permissible collector-current density, I_C/A_E, before f_T degrades because of the Kirk effect and their higher Early voltage (i.e., higher output resistance). A peculiarity of HBTs with dissimilar E-B and C-B junctions is an offset voltage between the collector and the emitter that must be overcome before a collector current starts to flow.

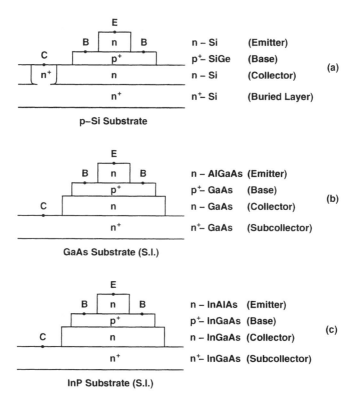

Fig. D.5 HBTs in three different material systems: (a) with SiGe base, (b) on GaAs substrate, and (c) on InP substrate (schematically).

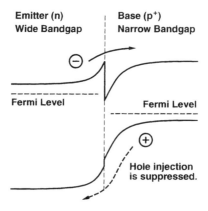

Fig. D.6 Band diagram for the forward-biased emitter-base junction of an HBT.

Similar to the situation with HFETs, the naming of the HBT material system can be based on the substrate material (e.g., GaAs-HBT technology) or the sequence of the major layer materials (e.g., AlGaAs/GaAs-HBT or SiGe-HBT technology).

SiGe-HBT Technology. A typical SiGe HBT is shown schematically in Fig. D.5(a). The base is made from the narrow-bandgap SiGe material (typically the Ge fraction is around 25%: $Si_{0.75}Ge_{0.25}$), whereas the emitter is made from regular silicon. The lattice mismatch between the silicon and SiGe layers creates some strain; however, if the SiGe layer is kept sufficiently thin, this strain is acceptable and may even improve the carrier mobility in the base. SiGe transistors achieve a high speed, but often suffer from a fairly low breakdown voltage. For example, the 0.6-μm lithography SiGe-HBT technology (with 0.3-μm effective emitter width) reviewed in [145] has $f_T = 72\,\text{GHz}$, $f_{\max} = 74\,\text{GHz}$ (at $V_{CE} = 1\,\text{V}$ and $I_C/A_E = 2\,\text{mA}/\mu\text{m}^2$), and an open-base collector-emitter breakdown voltage of only 2.7 V. Nevertheless, with a careful design, circuits can be made to operate from a 5.2-V supply voltage. The 0.18-μm SiGe-HBT technology reviewed in [191] reaches $f_T = 160\,\text{GHz}$ and $f_{\max} = 150\,\text{GHz}$ at $V_{CE} = 1.5\,\text{V}$ and $I_C/A_E = 6.0\,\text{mA}/\mu\text{m}^2$ and has an open-base collector-emitter breakdown voltage of more than 2 V.

An important advantage of the SiGe-HBT technology is its compatibility with the highly developed silicon technologies. Particularly attractive is an integration with the CMOS technology to form a SiGe-BiCMOS technology. For maximum compatibility with existing technologies, the Ge fraction in the base is sometimes lowered and graded from 0% at the emitter to about 10% at the collector. The resulting transistors don't have an emitter-base heterojunction and thus are not "true" SiGe HBTs. Such transistors are referred to as SiGe *drift transistors*.

A drawback of SiGe, as well as other silicon technologies, is the semiconducting substrate, which causes increased wiring parasitics and losses in bonding pads, on-chip transmission lines, spiral inductors, and so forth. (Remember, GaAs and InP technologies offer semi-insulating substrates.)

GaAs-HBT Technology. A typical GaAs HBT is shown schematically in Fig. D.5(b). The emitter is made from the wide-bandgap AlGaAs material, whereas the base is made from regular GaAs. This technology achieves a very high speed and high breakdown voltages at the same time. For example, the GaAs-HBT technology with a 1.4-μm wide emitter reviewed in [153] has $f_T = 60\,\text{GHz}$, $f_{\max} = 111\,\text{GHz}$ (at $V_{CE} = 1.5\,\text{V}$ and $I_C/A_E = 0.5\,\text{mA}/\mu\text{m}^2$), and a shorted-base collector-emitter breakdown voltage of more than 5 V; nevertheless, the circuits can operate from a 7.5-V power supply.

Because of the AlGaAs/GaAs material system, the forward voltage drop of the base-emitter diode is fairly high (1.3–1.4 V), making this technology less power efficient than others.

InP-HBT Technology. A typical InP HBT is shown schematically in Fig. D.5(c). The emitter is made from the wide-bandgap InAlAs (or InP) material, whereas the base is made from the narrow-bandgap InGaAs material, both of which can be lattice matched to the InP substrate ($In_{0.53}Ga_{0.47}As$, $In_{0.52}Al_{0.48}As$). This technology achieves even higher speeds than the GaAs-HBT technology because of the superior carrier transport properties of InGaAs. The simple InP-HBT, as shown in Fig. D.5(c), suffers from a low collector-emitter breakdown voltage (e.g., 2.5 V), but by modifying the collector material to form a second heterojunction at the collector, it can be increased to appreciable values. Such an HBT with two heterojunctions is known as a *double heterojunction bipolar transistor* (DHBT). For example, the InP-HBT technology with a 3-μm wide emitter (2.2-μm effective width) used in [66] reaches $f_T = 130\,\text{GHz}$, $f_{\max} = 118\,\text{GHz}$, and has an open-base collector-emitter breakdown voltage of more than 7 V. The 1-μm InP-HBT technology reviewed in [191] reaches $f_T = 180\,\text{GHz}$ and $f_{\max} = 200\,\text{GHz}$ at $V_{CE} = 1.5\,\text{V}$ and $I_C/A_E = 2.0\,\text{mA}/\mu\text{m}^2$ and has an open-base collector-emitter breakdown voltage of more than 2 V.

Another advantage of the InP-HBT technology specific to optical communication applications is that it permits the integration of long-wavelength optoelectronic devices. For example, a p-i-n photodetector sensitive in the 1.3- to 1.55-μm wavelength range may be integrated on the same chip by reusing the base-collector diode. A drawback of the InP technology is the present lack of large substrates.

Appendix E
Answers to the Problems

Chapter 2

2.1(a) $f = c/\lambda = (299.8\,\text{Mm/s})/(1.55\,\mu\text{m}) = 193.4\,\text{THz}$.

2.1(b) $\Delta f = c/\lambda^2 \cdot \Delta\lambda = (299.8\,\text{Mm/s})/(1.55\,\mu\text{m})^2 \cdot 0.1\,\text{nm} = 12.48\,\text{GHz}$.

2.2   The linear expression for $D(\lambda)$ is

$$D(\lambda) = 17\,\text{ps/(nm} \cdot \text{km)} \cdot \frac{\lambda - 1{,}300\,\text{nm}}{1{,}550\,\text{nm} - 1{,}300\,\text{nm}}.$$

Integrating $\partial\tau/\partial\lambda = D(\lambda) \cdot L$ (Eq. (2.2)) results in

$$\tau(\lambda) = 17\,\text{ps/(nm} \cdot \text{km)} \cdot \left(\frac{\lambda^2}{500\,\text{nm}} - 5.2 \cdot \lambda\right) \cdot L + \xi,$$

where ξ is an arbitrary constant. This is the quadratic relationship plotted in Fig. 2.3.

2.3 Convolving the Gaussian input pulse $x(t)$ with the impulse response $h(t)$ results in

$$y(t) = \int_{-\infty}^{\infty} h(t - t') \cdot x(t') \, dt'$$

$$= h(0) \cdot x(0) \cdot \int_{-\infty}^{\infty} \exp\left(-\frac{1}{2} \cdot \frac{(t - t')^2}{\sigma_T^2}\right) \cdot \exp\left(-\frac{1}{2} \cdot \frac{t'^2}{\sigma_{in}^2}\right) dt'$$

$$= y(0) \cdot \exp\left(-\frac{1}{2} \cdot \frac{t^2}{\sigma_{in}^2 + \sigma_T^2}\right).$$

It thus follows that $\sigma_{out} = \sqrt{\sigma_{in}^2 + \sigma_T^2}$, which, when multiplied by two, is equivalent to Eq. (2.7).

2.4 Calculating the Fourier transform of the impulse response $h(t)$ results in

$$H(f) = \int_{-\infty}^{\infty} h(t) \cdot \exp(-j\,2\pi f\,t) \, dt$$

$$= h(0) \cdot \int_{-\infty}^{\infty} \exp\left(-\frac{1}{2} \cdot \frac{t^2}{\sigma_T^2}\right) \cdot \exp(-j\,2\pi f\,t) \, dt$$

$$= H(0) \cdot \exp\left(-\frac{1}{2} \cdot \frac{(2\pi f)^2}{1/\sigma_T^2}\right),$$

which is equivalent to Eq. (2.9).

2.5 Inserting $f = B/2$ into Eq. (2.9) and comparing it with $0.794 \cdot H(0)$ for 1 dB of attenuation yields

$$\exp\left(-\frac{(\pi B)^2 (\Delta T/2)^2}{2}\right) \geq 0.794.$$

Solving for B gives $B \leq \sqrt{-8\ln(0.794)}/(\pi \Delta T)$, or approximately $B \leq 1/(2 \cdot \Delta T)$. This, in fact, is how the spreading limit given in Eq. (2.8) was derived in [46].

2.6 From Fig. 2.3, we see that for $D > 0$, shorter wavelengths propagate faster than longer wavelengths. A pulse with negative chirp has a longer wavelength during the leading edge (red shift) and a shorter wavelength (blue shift) during the trailing edge. Thus, the trailing edge will "catch up" with the leading edge, effectively compressing the pulse.

2.7(a) $24.3\,\text{dB}/(0.4\,\text{dB/km}) = 60.8\,\text{km}$.

2.7(b) $1/(2 \cdot 0.5\,\text{ps}/(\text{nm} \cdot \text{km}) \cdot 3\,\text{nm} \cdot 2.5\,\text{Gb/s}) = 133.3\,\text{km}$. The maximum transmission distance is 60.8 km, limited by attenuation.

2.8(a) 24.3 dB/(0.25 dB/km) = 97.2 km.

2.8(b) $1/(2 \cdot 17\,\text{ps/(nm} \cdot \text{km}) \cdot 3\,\text{nm} \cdot 2.5\,\text{Gb/s}) = 3.9\,\text{km}$. The maximum transmission distance is 3.9 km, limited by chromatic dispersion.

2.9(a) The dispersion-limited system of Problem 2.8.

2.9(b) The dispersion limit increases to 588 km. The system now is limited by attenuation to a distance of 97.2 km.

2.10 We don't have to worry about PMD. For the longest fiber of Problem 2.8, we have $\overline{\Delta T} = 2\,\text{ps}/\sqrt{\text{km}} \cdot \sqrt{97.2\,\text{km}} = 20\,\text{ps}$, which is significantly lower than $0.1/(2.5\,\text{Gb/s}) = 40\,\text{ps}$, thus the outage probability is extremely small.

Chapter 3

3.1 Optical attenuation, $40\,\text{km} \cdot 0.25\,\text{dB/km} = 10\,\text{dB}$; electrical attenuation, $2 \cdot 10\,\text{dB} = 20\,\text{dB}$.

3.2(a) The missing power comes from the voltage source used to reverse bias the photodetector.

3.2(b) Without a bias source, energy conservation requires $V_F \cdot \mathcal{R}P < P$, thus $V_F < 1/\mathcal{R}$.

3.3(a) $(\mathcal{R}P)^2 = 2q\mathcal{R}P \cdot BW_n$; thus $P = 2q/\mathcal{R} \cdot BW_n$. For $\eta = 1$, this is $P = 2hc/\lambda \cdot BW_n$.

3.3(b) $P/R_{ANT} = 4kT/R_{ANT} \cdot BW_n$; thus $P = 4kT \cdot BW_n$.

3.3(c) They become equal at the temperature $T = hc/(2\lambda k)$. This also is about the temperature at which the photodetector starts to fail because of excessive thermally-generated dark current.

3.4(a) The shot-noise current in the battery/resistor circuit is strongly "suppressed" and usually is not measurable. (However, the resistor R produces a thermal noise current, $\overline{i_n^2} = 4kT/R \cdot BW_n$, which is independent of the DC current.)

3.4(b) Even noise experts don't seem to agree on the explanation! But it seems that shot noise in its full strength, $\overline{i_n^2} = 2qI \cdot BW_n$, only occurs if the carriers cross from one electrode to another electrode, without "obstacles." This is the case to a good approximation in p-n junctions and vacuum tubes, but not in resistors.

3.5 The shot noise equation $\overline{i_n^2} = 2qI \cdot BW_n$ applies only to *randomly* arriving carriers. However, in the deterministic APD, each photon generates a group of M carriers with highly correlated arrival times. In fact, we could say that the current in the APD ($I_{APD} = MI_{PIN}$) consists of "coarse" carriers with the charge Mq. Substituting these quantities into the shot noise equation yields the correct result: $\overline{i_{n,APD}^2} = 2 \cdot (Mq) \cdot (MI_{PIN}) \cdot BW_n$.

3.6 The 6 terms have the following origins: (i) shot noise due to the signal current, (ii) shot noise due to the ASE power, (iii) shot noise due to the detector dark current, (iv) spontaneous-spontaneous beat noise, (v) signal-spontaneous beat noise, and (vi) shot-spontaneous beat noise.

3.7 The average signal power is $\overline{i_S^2} = \mathcal{R}^2 \cdot \overline{(P_S^2 - \overline{P_S})^2}$, which is equal to $\mathcal{R}^2 \overline{P_S}^2$ for a DC-balanced ideal NRZ signal with high extinction. The average noise power is $\overline{i_{n,ASE}^2} = \mathcal{R}^2 \cdot (2\overline{P_S} S_{ASE} + S_{ASE}^2 \cdot BW_O) \cdot BW_n$. Thus, Eqs (3.16) and (3.17) remain valid for a DC-balanced NRZ signal after the substitution $P_S \to \overline{P_S}$.

3.8(a) The noise figure according to the definition Eq. (3.18) is $F = \overline{i_{n,OA}^2}/[G^2 \cdot 2q^2 \lambda P/(hc) \cdot BW_n]$, where we used $I_{PIN} = q\lambda/(hc) \cdot P$ for the ideal p-i-n detector. Adding the shot noise term to Eq. (3.15) yields $\overline{i_{n,OA}^2} = (q\lambda)^2/(hc)^2 \cdot (2GPS_{ASE} + S_{ASE}^2 \cdot BW_O) \cdot BW_n + 2q^2 \lambda GP/(hc) \cdot BW_n$, where we used $P_S = GP$ and $\mathcal{R} = q\lambda/(hc)$ for the ideal p-i-n detector. Inserting the latter equation into the former one yields

$$F = \frac{\lambda}{hc} \cdot \left(\frac{S_{ASE}}{G} + \frac{S_{ASE}^2}{2 \cdot G^2 P} \cdot BW_O \right) + \frac{1}{G}.$$

3.8(b) For $S_{ASE} = 0$, we have $F = 1/G$, which is less than one for $G > 1$ (corresponding to a negative noise figure when expressed in dBs). Note that an ideal optical amplifier that amplifies the EM field before it is quantized in the p-i-n detector produces less shot noise than when the EM field first is quantized and then is amplified deterministically as assumed in the reference model for $F = 1$. But as we know, practical optical amplifiers always produce sufficient ASE noise to keep the noise figure at more than one.

3.9(a) Following the definition for the noise figure. The total output (shot) noise is $\overline{i_n^2} = 2q\mathcal{R}P/G \cdot BW_n$ where P is the optical input power. The output noise due to the source is $\overline{i_{n,S}^2} = 1/G^2 \cdot 2q\mathcal{R}P \cdot BW_n$. (To derive this expression, we assume that each photon from the source is "deterministically attenuated" in the fiber rather than randomly absorbed.) The ratio $\overline{i_n^2}/\overline{i_{n,S}^2}$ is the noise figure $F = G$.

3.9(b) Similar to Problem 3.9(a), the total output noise is $\overline{i_n^2} = F_2 \cdot G_2^2 \cdot 2q\mathcal{R}P/G_1 \cdot BW_n$ (Eq. (3.18) with $I_{PIN} = \mathcal{R}P/G_1$), the output noise due to the source is $\overline{i_{n,S}^2} = G_2^2/G_1^2 \cdot 2q\mathcal{R}P \cdot BW_n$, and thus $F = G_1 \cdot F_2$.

3.9(c) Similar to Problem 3.9(b), the total output noise is $\overline{i_n^2} = n \cdot F_2 \cdot G^2 \cdot 2q\mathcal{R}P/G \cdot BW_n$ (note that the input power of each segment is P and that the gain from each segment output to the system output is one; thus, the n noise contributions are equal and add up directly), the output noise due to the source is $\overline{i^2_{n.S}} = 2q\mathcal{R}P \cdot BW_n$, and thus $F = n \cdot G \cdot F_2$.

Chapter 4

4.1 If the input noise spectrum is approximately constant during the period $[t-\xi \ldots t]$, that is, $I^2_{n.PD}(f, [t-\xi \ldots t]) \approx I^2_{n.PD}(f, t)$, then we can rewrite Eq. (4.5) as $V^2_{n.PD}(f, t) = H(f) \cdot I^2_{n.PD}(f, t) \cdot \int_{-\infty}^{\infty} h(t') \cdot e^{j2\pi f t'} dt' = H(f) \cdot I^2_{n.PD}(f, t) \cdot H^*(f) = |H(f)|^2 \cdot I^2_{n.PD}(f, t)$. Thus, the approximation is valid for time values more than ξ after a bit transition.

4.2 The two unequal, unnormalized Gaussians can be written: $1/v^{rms}_{n.0} \cdot \text{Gauss}(v_O/v^{rms}_{n.0})$ and $1/v^{rms}_{n.1} \cdot \text{Gauss}(|v^{pp}_S - v_O|/v^{rms}_{n.1})$, where we assumed, without loss of generality, that the zero level is at 0. Equating these two distributions and solving for v_O yields the optimum decision threshold voltage V_{DTH}. Neglecting the different heights of the distributions, we find $V_{DTH} = v^{pp}_S \cdot v^{rms}_{n.0}/(v^{rms}_{n.0} + v^{rms}_{n.1})$. Integrating the two tails results in $BER = 1/2 \cdot 1/v^{rms}_{n.0} \cdot \int_{V_{DTH}}^{\infty} \text{Gauss}(v_O/v^{rms}_{n.0}) \, dv_O + 1/2 \cdot 1/v^{rms}_{n.1} \cdot \int_{v^{pp}_S - V_{DTH}}^{\infty} \text{Gauss}(v_O/v^{rms}_{n.1}) \, dv_O$ or in normalized coordinates $BER = 1/2 \cdot \int_Q^{\infty} \text{Gauss}(x) \, dx + 1/2 \cdot \int_Q^{\infty} \text{Gauss}(x) \, dx = \int_Q^{\infty} \text{Gauss}(x) \, dx$, where $Q = v^{pp}_S/(v^{rms}_{n.0} + v^{rms}_{n.1})$.

4.3 With $\xi = v^{rms}_{n.0}/v^{rms}_{n.1}$, we can rewrite Eq. (4.10) as $Q^2 = (v^{pp}_S)^2/[(\xi + 1)^2 \cdot \overline{v^2_{n.1}}]$ and Eq. (4.11) as $SNR = (v^{pp}_S)^2/[2 \cdot (\xi^2 + 1) \cdot \overline{v^2_{n.1}}]$. Thus, Eq. (4.12) can be generalized to

$$SNR = \frac{(\xi + 1)^2}{2(\xi^2 + 1)} \cdot Q^2.$$

4.4(a) If we normalize the noise power, $\overline{v^2_n}$, to 1, the swing of the RZ signal, v^{pp}_S, must be $2Q$ to achieve the specified BER. The time-averaged mean-free signal power, and thus the SNR, of that signal is $SNR = (1 - \xi/2) \cdot (-\xi Q)^2 + \xi/2 \cdot [(2-\xi)Q]^2 = (2-\xi)\xi \cdot Q^2$.

4.4(b) For a 50%-RZ signal and $BER = 10^{-12}$, $SNR = 0.75 \cdot 7.035^2 = 37.1$ (15.7 dB).

4.4(c) When sampling at the maximum eye opening, the sampled SNR does not depend on ξ and always is $SNR = Q^2$.

4.5(a) If we normalize the noise power to 1, the swing of the finite slope NRZ signal must be $2Q$ to achieve the specified BER. The time-averaged mean-free signal power, and thus the SNR, of that signal is $SNR = [\xi/6 + (1 - \xi) + \xi/6] \cdot Q^2 = (1 - 2/3 \cdot \xi) \cdot Q^2$.

4.5(b) For a 0.3-UI rise/fall-time NRZ signal and $BER = 10^{-12}$, $SNR = 0.8 \cdot 7.035^2 = 39.6$ (16.0 dB).

4.5(c) When sampling at the maximum eye opening, the sampled SNR does not depend on ξ and always is $SNR = Q^2$.

4.6(a) If we normalize the noise power to 1, the levels of the PAM-4 signal must all be separated by approximately $2Q$ to achieve the specified BER. The time-averaged, mean-free signal power, and thus the SNR, of that signal is $SNR = 1/4 \cdot (-3Q)^2 + 1/4 \cdot (-Q)^2 + 1/4 \cdot Q^2 + 1/4 \cdot (3Q)^2 = 5 \cdot Q^2$.

4.6(b) For a PAM-4 signal and $BER = 10^{-12}$, $SNR = 5 \cdot 7.035^2 = 247.5$ (23.9 dB).

4.7(a) E_b is the *signal power* times the information bit period $1/(r \cdot B)$. N_0 is the *noise power* divided by the noise bandwidth BW_n of the linear channel. Hence

$$\frac{E_b}{N_0} = SNR \cdot \frac{BW_n}{r \cdot B}.$$

4.7(b) When the noise bandwidth BW_n is equal to the information bit rate (a.k.a. the system bit rate), $r \cdot B$.

4.8 Given the average optical power $\overline{P}_S$ and the extinction ratio ER, the power for zeros and ones are $P_0 = 2\overline{P}_S/(ER+1)$ and $P_1 = 2\overline{P}_S \cdot ER/(ER+1)$, respectively; thus, $i_S^{pp} = \mathcal{R}(P_1 - P_0) = 2\mathcal{R}\overline{P}_S \cdot (ER-1)/(ER+1)$. Solving for $\overline{P}_S$ and inserting Eq. (4.19) for i_S^{pp} yields

$$\overline{P}_{sens} = \frac{ER+1}{ER-1} \cdot \frac{Q \cdot (i_{n,0}^{rms} + i_{n,1}^{rms})}{2\mathcal{R}}.$$

4.9 The rule is equivalent to Eq. (4.24) when written in the log domain and specialized for $BER = 10^{-12}$. The value $10 \log Q(BER = 10^{-12}) - 30 = -21.53$ [dBm] is the sensitivity of a receiver with $\mathcal{R} = 1$ A/W and $i_{n,\mathrm{amp}}^{rms} = 1\,\mu$A.

4.10 The noise power for the zeros is $\overline{i_{n,0}^2} = \overline{i_{n,\mathrm{amp}}^2}$, and with Eq. (3.15), the noise power for the ones is $\overline{i_{n,1}^2} = \overline{i_{n,\mathrm{amp}}^2} + 4\mathcal{R}^2 \overline{P}_S \cdot S_{ASE} \cdot BW_n$. With $OSNR = \overline{P}_S/(S_{ASE} \cdot BW_O)$, the noise for the ones can be rewritten in terms of OSNR as $\overline{i_{n,1}^2} = \overline{i_{n,\mathrm{amp}}^2} + 4\mathcal{R}^2 \overline{P}_S^2 \cdot BW_n/(OSNR \cdot BW_O)$. Inserting into Eq. (4.21), setting $\overline{P}_S = \overline{P}_{sens}$, and solving for $\overline{P}_{sens}$ yields

$$\overline{P}_{sens} = \frac{1}{1 - Q^2/OSNR \cdot BW_n/BW_O} \cdot \frac{Q \cdot i_{n,\mathrm{amp}}^{rms}}{\mathcal{R}}.$$

4.10(a) For $OSNR \to \infty$, the sensitivity becomes identical to $\overline{P}_{sens,PIN}$ in Eq. (4.24).

4.10(b) For a high received power, we need at least $OSNR = Q^2 \cdot BW_n/BW_O$ to meet the specified BER.

4.11(a) With Eq. (3.15), we find the noise power for the zeros as $\overline{i_{n.0}^2} = \overline{i_{n.\text{amp}}^2} + \mathcal{R}^2 \cdot S_{ASE}^2 \cdot BW_O \cdot BW_n$ and the noise power for the ones as $\overline{i_{n.1}^2} = \overline{i_{n.0}^2} + 4\mathcal{R}^2 G \overline{P_S} \cdot S_{ASE} \cdot BW_n$. With Eq. (4.21), the sensitivity is $\overline{P_{\text{sens.}OA}} = Q \cdot (i_{n.0}^{rms} + i_{n.1}^{rms})/(2G\mathcal{R})$. Setting $\overline{P_S} = \overline{P_{\text{sens.}OA}}$ and solving for $\overline{P_{\text{sens.}OA}}$ yields

$$\overline{P_{\text{sens.}OA}} = \frac{Q}{G\mathcal{R}} \cdot \sqrt{\overline{i_{n.\text{amp}}^2} + (\mathcal{R} \cdot S_{ASE})^2 \cdot BW_O \cdot BW_n} + \frac{Q^2 \cdot S_{ASE} \cdot BW_n}{G}.$$

4.11(b) Replacing S_{ASE} with $\eta \tilde{F} \cdot q \cdot G/\mathcal{R}$ (from Eqs. (3.23) and (3.2)) yields

$$\overline{P_{\text{sens.}OA}} = \frac{Q}{G\mathcal{R}} \cdot \sqrt{\overline{i_{n.\text{amp}}^2} + (\eta \tilde{F} \cdot q \cdot G)^2 \cdot BW_O \cdot BW_n} + \eta \tilde{F} \cdot \frac{Q^2 \cdot q \cdot BW_n}{\mathcal{R}},$$

which is a generalization of Eq. (4.29).

4.12 With Eq. (3.15) and $P_{ASE} = S_{ASE} \cdot BW_O$, we find the noise for the zeros as $i_{n.0}^{rms} = \mathcal{R} \cdot P_{ASE} \cdot \sqrt{BW_n/BW_O}$ and the noise for the ones as $i_{n.1}^{rms} = \mathcal{R}\sqrt{4\overline{P} \cdot P_{ASE} + P_{ASE}^2} \cdot \sqrt{BW_n/BW_O}$. The signal swing is $i_S^{pp} = 2\mathcal{R}\overline{P}$. With $Q = i_S^{pp}/(i_{n.0}^{rms} + i_{n.1}^{rms})$ from Eq. (4.19), we can generalize Eq. (4.32) to

$$Q = \frac{2 OSNR}{\sqrt{4 OSNR + 1} + 1} \cdot \sqrt{\frac{BW_O}{BW_n}}.$$

4.13 According to Eq. (4.31), the necessary transmit power is $\overline{P_{\text{out}}} = \eta nGF \cdot Q^2 \cdot q \cdot BW_n/\mathcal{R}$ when neglecting the amplifier noise. With Eq. (3.2), this power can be rewritten as $\overline{P_{\text{out}}} = nGF \cdot Q^2 \cdot hc/\lambda \cdot BW_n$. Solving Eq. (4.33) for $\overline{P_{\text{out}}}$ and inserting Eq. (4.32) for $OSNR$ also leads to $\overline{P_{\text{out}}} = nGF \cdot Q^2 \cdot hc/\lambda \cdot BW_n$.

4.14 The value $Q = 7.0$ for a $BER = 10^{-12}$ was calculated assuming a *Gaussian* noise distribution; however, the noise distribution produced by a p-i-n photodetector (without amplifier noise) is a *Poisson* distribution! For the Poisson distribution in Eq. (4.35), we find $Q = \bar{n}/\sigma(n) = M/\sqrt{M} = \sqrt{M}$, and with $BER = 1/2 \cdot e^{-M}$, we get $Q = \sqrt{-\ln(2 \cdot BER)}$ and all is well! Note that given a Poisson distribution, we need only $Q = 5.2$ for $BER = 10^{-12}$.

4.15 The total noise power is calculated as $\overline{i_n^2} = 1/H_0^2 \int |H(f)|^2 \cdot (\alpha_0 + \alpha_1 f + \alpha_2 f^2) \, df$. Expanding this expression and comparing it with $\overline{i_n^2} = \alpha_0 \cdot I_2 B + \alpha_1 \cdot I_f B^2 + \alpha_2 \cdot I_3 B^3$ reveals the relationship $I_f = 1/B^2 \cdot 1/H_0^2 \int |H(f)|^2 \cdot f \, df$.

4.16 The power penalty due to finite extinction is $PP = (ER + 1)/(ER - 1)$. This follows directly from the answer to Problem 4.8.

4.17(a) The smallest output value for a one occurs if it is preceded by a long sequence of zeros. This value corresponds to the step response $(0 \to 1)$ of the filter evaluated for $t = 1/B$. Similarly, the largest output value for a zero occurs if it is preceded by a long sequence of ones, which corresponds to the inverse step response $(1 \to 0)$ evaluated for $t = 1/B$. The difference between these two values is the worst-case output swing, and its inverse is the power penalty (assuming the full swing is normalized to one).

For a first-order low-pass filter, the step response is $1 - \exp(-2\pi \cdot BW_{3dB} \cdot t)$; the inverse step response is $\exp(-2\pi \cdot BW_{3dB} \cdot t)$. Thus, the power penalty is

$$PP = \frac{1}{1 - 2 \cdot \exp(-2\pi \cdot BW_{3dB}/B)}.$$

4.17(b) For a second-order Butterworth low-pass filter, the step response is [7] $1 - \sqrt{2} \cdot \exp(-\sqrt{2}\pi \cdot BW_{3dB} \cdot t) \cdot \sin(\sqrt{2}\pi \cdot BW_{3dB} \cdot t + \pi/4)$. Following the same procedure as in Problem 4.17(a), the power penalty is

$$PP = \frac{1}{1 - 2\sqrt{2} \cdot \exp(-\sqrt{2}\pi \cdot BW_{3dB}/B) \cdot \sin(\sqrt{2}\pi \cdot BW_{3dB}/B + \pi/4)}.$$

If the above equation yields $PP < 1$ as a result of over/undershoot, we set $PP = 1$.

4.17(c) The power penalty values are

Bandwidth BW_{3dB}/B	1st-Order Filter PP (dB)	2nd-Order Butterworth PP (dB)
1/3	1.23	2.97
2/3	0.13	0.00
4/3	0.00	0.01

4.18 Calculating the Fourier transform of the impulse response $h(t)$ results in $H(f) = \int_{-\infty}^{\infty} h(t) \cdot \exp(-j 2\pi f t) \, dt = \int_0^{1/B} \exp(-j 2\pi f t) \, dt = j/(2\pi f) \cdot [\exp(-j 2\pi f/B) - 1]$. The magnitude is

$$|H(f)| = \frac{1}{2\pi f} \sqrt{\left[\cos\left(\frac{2\pi f}{B}\right) - 1\right]^2 + \left[\sin\left(\frac{2\pi f}{B}\right)\right]^2} = \frac{\sin(\pi f/B)}{\pi f}.$$

The tangent of the phase is given by

$$\frac{\text{Im}[H(f)]}{\text{Re}[H(f)]} = \frac{\cos(2\pi f/B) - 1}{\sin(2\pi f/B)} = \tan\left(-\frac{\pi f}{B}\right).$$

Thus, $H(f)$ can be written as

$$H(f) = \frac{\sin(\pi f/B)}{\pi f} \cdot e^{-j\pi f/B}.$$

When normalized such that $|H(0)| = 1$, this expression is identical to Eq. (4.60).

4.19 The output signal equals the input signal delayed by $1/B$ plus c_1 times the undelayed input signal; thus, the frequency response is $H(f) = \exp(-j\,2\pi f/B) + c_1$. Its magnitude is

$$|H(f)| = \sqrt{\left[\cos\left(\frac{2\pi f}{B}\right) + c_1\right]^2 + \left[\sin\left(\frac{2\pi f}{B}\right)\right]^2},$$

and thus $|H(0)| = |1 + c_1|$ and $|H(B/2)| = |1 - c_1|$, corresponding to a high-pass response for $c_1 < 0$.

4.20 The output signal is $y(t) = A\{a_0 + X \cdot [\sin(\omega_1 \cdot t) + \sin(\omega_2 \cdot t)] + a_2 \cdot X^2 \cdot [\sin(\omega_1 \cdot t) + \sin(\omega_2 \cdot t)]^2\}$. After expanding and sorting with respect to frequencies,

$$\begin{aligned}
y(t) = &\; A \cdot [a_0 + a_2 \cdot X^2] \\
&+ A \cdot X \cdot [\sin(\omega_1 \cdot t) + \sin(\omega_2 \cdot t)] \\
&- A \cdot a_2/2 \cdot X^2 \cdot [\cos(2\omega_1 \cdot t) + \cos(2\omega_2 \cdot t)] \\
&+ A \cdot a_2 \cdot X^2 \cdot [\cos((\omega_1 - \omega_2) \cdot t) - \cos((\omega_1 + \omega_2) \cdot t)].
\end{aligned}$$

The first line represents the output offset, the second line represents the fundamental tones, the third line represents the second-order harmonic products, and the fourth line represents the second-order intermodulation products.

4.21(a) The jitter histogram consists of two Dirac pulses separated by ΔT: $1/2 \cdot [\delta(t_J - \Delta T/2) + \delta(t_J + \Delta T/2)]$ and the peak-to-peak jitter is $t_{DJ}^{pp} = \Delta T$.

4.21(b) With random jitter, the histogram consists of two Gaussians separated by ΔT: $1/2 \cdot \{\text{Gauss}[(t_J - \Delta T/2)/t_{RJ}^{rms}] + \text{Gauss}[(t_J + \Delta T/2)/t_{RJ}^{rms}]\}$, which looks like the histograms shown in Fig. 4.21.

4.22 The error probability as a function of the slice level is $BER = 1/2 \cdot [\int_{V_{DTH}}^{\infty} \text{Zero}(v_O)\,dv_O + \int_{-\infty}^{V_{DTH}} \text{One}(v_O)\,dv_O]$, which can be rewritten as $1/2 \cdot [1 - \int_{-\infty}^{V_{DTH}} \text{Zero}(v_O)\,dv_O + \int_{-\infty}^{V_{DTH}} \text{One}(v_O)\,dv_O]$. Taking the derivative and setting it to zero leads to $\partial BER/\partial V_{DTH} = 1/2 \cdot [-\text{Zero}(V_{DTH}) + \text{One}(V_{DTH})] = 0$, hence the optimum slice level is at the intersection point $\text{Zero}(V_{DTH}) = \text{One}(V_{DTH})$.

368 ANSWERS TO THE PROBLEMS

4.23 Given $BER = 10^{-4}$ and a frame size of 255 bytes results in an average of $M = 255 \cdot 8 \cdot 10^{-4} = 0.204$ bit errors per frame. The probability for 9 errors per frame (which is not correctable with RS(255,239), assuming each error is in a different byte) can be found with the Poisson distribution as $\exp(-M) \cdot M^9/9! = 1.38 \cdot 10^{-12}$. Neglecting the small possibility of more than 9 errors per frame, this number is the frame error rate at the output of the decoder. Converting the frame error rate back to the (payload) bit error rate yields $BER = 9/(239 \cdot 8) \cdot 1.38 \cdot 10^{-12} = 6.47 \cdot 10^{-15}$. A more precise analysis results in $5 \cdot 10^{-15}$ [53].

Chapter 5

5.1(a) The rms noise in the differential mode is $\sqrt{2}\,\text{mV} \approx 1.41\,\text{mV}$, and the rms noise in the common mode is $\sqrt{2}/2\,\text{mV} \approx 0.71\,\text{mV}$.

5.1(b) When reproducing the single-ended output noise, $i_{n.TIA}^{rms} = 1\,\text{mV}/0.5\,\text{k}\Omega = 2\,\mu\text{A}$. When reproducing the differential output noise, $i_{n.TIA}^{rms} = 1.41\,\text{mV}/1.0\,\text{k}\Omega = 1.41\,\mu\text{A}$.

5.2(a) The input overload current must be $i_{ovl}^{pp} > 2\mathcal{R}\overline{P}_{ovl} = 0.16\,\text{mA}$, and the input-referred rms noise current must be $i_{n.TIA}^{rms} < \mathcal{R}\overline{P}_{sens}/Q = 1.43\,\mu\text{A}$.

5.2(b) The averaged input-referred noise current density must be $I_{n.TIA} < 1.43\,\mu\text{A}/\sqrt{10\,\text{GHz}} = 14.3\,\text{pA}/\sqrt{\text{Hz}}$.

5.3(a) The transimpedance is $R_T = 25\,\Omega \cdot A = 2.5\,\text{k}\Omega$.

5.3(b) An amplifier with a 2-dB noise figure connected to a $50\,\Omega$ source has an input-referred noise power that is 2 dB (1.58×) stronger than that of the $50\,\Omega$ source alone. Thus, the input-referred rms noise current is $i_n^{rms} = \sqrt{4kT/50\,\Omega \cdot F \cdot BW_n} = 2.29\,\mu\text{A}$.

5.3(c) The sensitivity of the TIA receiver is better by $10\log(2.29/1.4) = 2.14\,\text{dB}$.

5.4(a) The transimpedance, input impedance, and output impedance expressions for $R_S \neq 0$ become

$$Z_T(s) = -R_T \cdot \frac{1}{1+s/\omega_p},$$

$$Z_I(s) = R_I \cdot \frac{1}{1+s/\omega_p},$$

$$Z_O(s) = R_O \cdot \frac{1+s/\omega_z}{1+s/\omega_p},$$

where

$$R_T = \frac{A}{A+1} \cdot (R_F - R_S/A), \quad R_I = \frac{R_F + R_S}{A+1}, \quad R_O = \frac{R_S}{A+1}$$

and
$$\omega_p = \frac{A+1}{(R_F + R_S)C_T}, \qquad \omega_z = \frac{1}{R_F C_T}.$$

Note that for $R_S = 0$, these results correspond to Eqs. (5.12) through (5.16).

5.4(b) The bandwidth of the transimpedance and input impedance is given by $\omega_p/(2\pi)$ corresponding to Eq. (5.15). The bandwidth of the output impedance is smaller and given by $\omega_z/(2\pi)$. For frequencies above $\omega_z/(2\pi)$, the output impedance becomes inductive.

5.5(a) Let the open-loop pole spacing be $\xi = R_F C_T / T_A$. From Eq. (5.22), we find that $Q = \sqrt{(A+1)\xi}/(\xi+1)$. For large values of ξ, we can simplify this expression to $Q \approx \sqrt{(A+1)/(\xi+2)}$; thus, the required pole spacing is

$$\xi \approx \frac{A+1}{Q^2} - 2.$$

5.5(b) For a Bessel response, we have $Q = 1/\sqrt{3}$, and thus $\xi \approx 3A + 1$.

5.5(c) For a critically damped response, we have $Q = 1/2$, and thus $\xi \approx 4A + 2$.

5.6 By simply plugging the numbers into Eq. (5.25), we find that at 2.5 Gb/s, $R_T \leq 7.63\,\text{k}\Omega$, at 10 Gb/s, $R_T \leq 477\,\Omega$, and at 40 Gb/s, $R_T \leq 29.8\,\Omega$. (As we know, in practice, higher transimpedance values can be achieved by adding a post amplifier.)

5.7(a) The transimpedance expression for $C_F \neq 0$ becomes

$$Z_T(s) = -R_T \cdot \frac{1}{1 + s/(\omega_0 Q) + s^2/\omega_0^2},$$

where

$$R_T = \frac{A}{A+1} \cdot R_F,$$

$$\omega_0 = \sqrt{\frac{A+1}{R_F(C_T + C_F) \cdot T_A}},$$

$$Q = \frac{\sqrt{(A+1) \cdot R_F(C_T + C_F) \cdot T_A}}{R_F[C_T + (A+1)C_F] + T_A}.$$

5.7(b) Setting $Q = 1/\sqrt{2}$ and solving for C_F (with $C_F \ll C_T$) yields

$$C_F \approx \sqrt{\frac{2C_T \cdot T_A}{(A+1)R_F} - \frac{T_A}{(A+1)R_F} - \frac{C_T}{A+1}}.$$

5.7(c) The amplifier time constant necessary for $Q = 1/\sqrt{2}$ is given by $T_A \approx R_F[C_T + (A+1)C_F]^2/[2(A+1)(C_T + C_F)]$, assuming $T_A \ll R_F(C_T + AC_F)$. Inserting this into the expression for ω_0 yields

$$BW_{3dB} = \frac{\omega_0}{2\pi} \approx \frac{1}{2\pi} \cdot \frac{\sqrt{2}(A+1)}{R_F[C_T + (A+1)C_F]}.$$

If we assume further that $C_T \ll AC_F$, the expression reduces to $BW_{3dB} \approx \sqrt{2}/(2\pi \cdot R_F C_F)$.

5.7(d) Combining the expressions for ω_0 and R_T, we find the transimpedance limit (for $Q \leq 1/\sqrt{2}$) to be

$$R_T \leq \frac{A \cdot f_A}{2\pi \cdot (C_T + C_F) \cdot BW_{3dB}^2},$$

which is lower than the limit in Eq. (5.25), but has the same form.

5.8(a) From Eqs. (5.37) and (5.39), ignoring the gate shot noise for a MOSFET TIA, we find the f^2-noise corner frequency

$$f_{c2} = \frac{1}{2\pi \cdot R_F C_T} \cdot \sqrt{\frac{g_m R_F}{\Gamma} + 1}.$$

5.8(b) With respect to R_F and C_T, the f^2-noise corner frequency is monotonically related to the TIA bandwidth $1/(2\pi \cdot R_F C_T) \cdot \sqrt{2A(A+1)}$. Thus, higher bit-rate TIAs generally have a higher f^2-noise corner frequency.

5.9 With Eqs. (5.37) and (5.38), ignoring the gate shot noise for a MOSFET TIA, we find the noise spectrum

$$I_{n.TIA}^2(f) = \frac{4kT}{R_F} + 4kT\Gamma \left[\frac{1}{g_m R_F^2} + \frac{f_c}{g_m R_F^2} \cdot 1/f \right.$$
$$\left. + \frac{(2\pi C_T)^2}{g_m} \cdot f^2 + \frac{(2\pi C_T)^2 f_c}{g_m} \cdot f \right].$$

5.10(a) With the simplifying assumptions and Eqs. (5.21) and (5.22), we find $\omega_0 \approx \sqrt{A/(R_F C_T T_A)}$ and $Q \approx \sqrt{AT_A/(R_F C_T)}$. Thus, varying R_F causes $\omega_0 \sim 1/\sqrt{R_F}$ and $Q \sim 1/\sqrt{R_F}$, that is, a decrease in R_F causes an increase in bandwidth and an increase in quality factor (peaking).

5.10(b) Varying A proportional to R_F causes both ω_0 and Q to remain constant.

5.10(c) Varying A and T_A proportional to $\sqrt{R_F}$ causes $\omega_0 \sim 1/\sqrt{R_F}$, whereas Q and $GBW = A/(2\pi T_A)$ remain constant.

5.11 From the transimpedance and the stage gains, we can estimate the value of the feedback resistor in the first stage: $R_F = 1/A_1 \cdot (A+1)/A \cdot R_T = 1\,\mathrm{k}\Omega$

(cf. Eq. (5.53)). The white-noise contribution of this $1\,\mathrm{k}\Omega$ resistor alone is about $4.07\,\mathrm{pA}/\sqrt{\mathrm{Hz}}$. The measured noise densities are suspiciously low!

5.12 The transimpedance of the idealized current-mode TIA is

$$Z_T(s) = -R_T \cdot \frac{1}{1 + s/\omega_p},$$

where

$$R_T = \frac{A}{A+1} \cdot R_F \quad \text{and} \quad \omega_p = \frac{A+1}{R_F C_L}.$$

Thus, the 3-dB bandwidth is $BW_{3\mathrm{dB}} = 1/(2\pi) \cdot (A+1)/(R_F C_L)$.

5.13(a) The transimpedance of the active-feedback TIA is

$$Z_T(s) = R_T \cdot \frac{1}{1 + s/(\omega_0 Q) + s^2/\omega_0^2},$$

where

$$R_T = \frac{1}{g_{mF}}, \quad \omega_0 = \sqrt{\frac{A \cdot g_{mF}}{C_T \cdot T_A}}, \quad Q = \sqrt{\frac{A \cdot g_{mF} \cdot T_A}{C_T}},$$

and $C_T = C_D + C_I$.

5.13(b) Setting $Q = 1/\sqrt{2}$ and solving for T_A yields $T_A = C_T/(2A \cdot g_{mF})$.

5.13(c) Inserting this time constant into the expression for ω_0 yields $BW_{3\mathrm{dB}} = 1/(2\pi) \cdot \sqrt{2} A \cdot g_{mF}/C_T$.

5.13(d) Combining the expressions for ω_0 and R_T, we find the same transimpedance limit as in Eq. (5.25).

5.14 The transimpedance of the TIA with bond wire inductance is

$$Z_T(s) = -R_T \cdot \frac{1}{1 + sR_I C_T + s^2 L_B C_D + s^3 R_I C_I L_B C_D},$$

where $R_T = A/(A+1) \cdot R_F$, $R_I = R_F/(A+1)$, and $C_T = C_D + C_I$. For the poles to assume the third-order Butterworth positions, $L_B = (R_I C_T)^2/(2 C_D)$, $C_I = 1/4 \cdot C_T$, and $C_D = 3/4 \cdot C_T$.

5.15(a) The (low-frequency) relationship between the detector current, i_I, and the single-ended TIA input voltage, v_{IP}, is $v_{IP} = V_{OCM} + (A+2)/(2A+2) \cdot R_F \cdot i_I$, where V_{OCM} is the output common-mode voltage and A is the differential gain of the feedback amplifier. Thus, the single-ended input resistance of the balanced TIA is

$$R_I = \frac{A+2}{2A+2} \cdot R_F.$$

372 ANSWERS TO THE PROBLEMS

5.15(b) For $i_{IP} = i_I$ and $i_{IN} = 0$, we find that the single-ended input resistance $R_I = \Delta v_{IP}/\Delta i_I$ can be written in terms of the differential and common-mode resistances as $R_I = R_{I.c} + 1/4 \cdot R_{I.d}$, which leads to the same result as in Problem 5.15(a).

Chapter 6

6.1 Multiplying the sensitivity of an ideal noise-limited receiver in Eq. (4.20) with the power penalty due to a finite MA gain and DEC sensitivity in Eq. (6.4) yields

$$\overline{P}_{sens} = \frac{Q \cdot i_{n.TIA}^{rms}}{\mathcal{R}} + \frac{V_{DTA}}{2\mathcal{R} \cdot A \cdot R_T}.$$

6.2(a) The "total output noise power" is $\int_{f_1}^{f_2} |A(f)|^2 \cdot V_{n.MA}^2(f)/4 \, df$; note that the *available* input-referred noise spectrum, $V_{n.MA}^2(f)/4$, is amplified by $|A(f)|^2$. The "fraction of the output noise power due to the thermal noise of the source resistance" is $\int_{f_1}^{f_2} |A(f)|^2 \cdot kTR_S \, df$; again, note that the *available* thermal noise spectrum, kTR_S, is amplified by $|A(f)|^2$. The "input SNR" is $\overline{v_S^2}$ to $1/A_0^2 \cdot \int_{f_1}^{f_2} |A(f)|^2 \cdot kTR_S \, df$. The "output SNR" is $A_0^2 \cdot \overline{v_S^2}$ to $\int_{f_1}^{f_2} |A(f)|^2 \cdot V_{n.MA}^2(f)/4 \, df$. For both definitions, we find that

$$F = \frac{\int_{f_1}^{f_2} |A(f)|^2 \cdot V_{n.MA}^2(f) \, df}{\int_{f_1}^{f_2} |A(f)|^2 \cdot 4kTR_S \, df}.$$

6.2(b) For a narrow bandwidth Δf located at frequency f, we can neglect the variation of A and $V_{n.MA}^2$ with frequency and write the integrals as products:

$$F(f) = \frac{|A(f)|^2 \cdot V_{n.MA}^2(f) \cdot \Delta f}{|A(f)|^2 \cdot 4kTR_S \cdot \Delta f} = \frac{V_{n.MA}^2(f)}{4kTR_S}.$$

6.2(c) For a wide bandwidth, the numerator integral can be rewritten in terms of the input-referred mean-square noise voltage (using Eq. (6.12)), and the denominator integral can be rewritten in terms of the noise bandwidth (using Eq. (4.44)):

$$F = \frac{A_0^2 \cdot \overline{v_{n.MA}^2}}{\int_0^\infty |A(f)|^2 \, df \cdot 4kTR_S} = \frac{\overline{v_{n.MA}^2}}{4kTR_S \cdot BW_n}.$$

6.3(a) The capacitor C causes the high-pass transfer function $H(s) = s2R_0C/(1+s2R_0C)$, and thus the low-frequency cutoff is $f_{LF} = 1/(4\pi R_0C)$.

6.3(b) Inserting the above result into Eq. (6.38) yields $PP = 1 + r/(2R_0 C \cdot B)$.

6.3(c) Solving for C yields $C \geq r/[(PP - 1) \cdot 2R_0 \cdot B]$. With $r = 72$ and $PP = 0.05$ dB, we find $C \geq 24.9$ nF for a 2.5-Gb/s system and $C \geq 6.2$ nF for a 10-Gb/s system.

6.4(a) The magnitude of the second-order Butterworth transfer function is $|H(f)| = \sqrt{1/(1 + (f/f_0)^4)}$. Setting $|H(f)|^n = 1/\sqrt{2}$ and solving for f reveals that the 3-dB bandwidth of n sections is $(2^{1/n} - 1)^{1/4} \cdot f_0$. Setting $n = 1$ yields the single-stage bandwidth f_0, and thus the bandwidth shrinkage is $(2^{1/n} - 1)^{1/4}$.

6.4(b) The stage gain is $A_S = A_{\text{tot}}^{1/n}$ and the stage bandwidth is $BW_S = (2^{1/n} - 1)^{-1/4} \cdot BW_{\text{tot}}$ from above. Thus, we have $GBW_S = A_{\text{tot}}^{1/n} \cdot BW_{\text{tot}} \cdot (2^{1/n} - 1)^{-1/4} = GBW_{\text{tot}} \cdot A_{\text{tot}}^{1/n-1} \cdot (2^{1/n} - 1)^{-1/4}$. The gain-bandwidth extension is $GBW_{\text{tot}}/GBW_S = A_{\text{tot}}^{1-1/n} \cdot (2^{1/n} - 1)^{1/4}$, in accordance with Eq. (6.46).

6.4(c) Differentiating $A_{\text{tot}}^{1-1/n} \cdot (2^{1/n} - 1)^{1/4}$ with respect to n and setting the result to zero reveals the optimum number of stages:

$$n_{\text{opt}} = \frac{\ln 2}{-\ln(1 - \frac{\ln 2}{4 \ln A_{\text{tot}}})}.$$

Using $\ln(1 + x) \approx x$ for $x \ll 1$, we can simplify this expression to $n_{\text{opt}} \approx 4 \ln A_{\text{tot}}$ for $A_{\text{tot}} \gg \sqrt[4]{2}$.

6.5(a) The input admittance is $Y(s) = (1 - A) \cdot sC$, and thus the effective input capacitance is $C_I = (1 - A) \cdot C$.

6.5(b) The input admittance is $Y(s) = (1 - A_0) \cdot [1 + sT/(1 - A_0)]/[1 + sT] \cdot sC$. At low frequencies, the effective input capacitance is $C_I = (1 - A_0) \cdot C$; at high frequencies, it is $C_I = C$.

6.6 The problem with the argument is that the definition of the differential current is used inconsistently (cf. Appendix B.2). The input capacitance is $C_{be}/2$ only if the differential base current is defined as $1/2 \cdot (i_{B.P} - i_{B.N})$; with the same definition for the differential collector current, the transconductance is $g_m/2 = 1/2 \cdot \Delta(i_{C.P} - i_{C.N})/\Delta(v_{BE.P} - v_{BE.N})$. Thus, no f_T doubling takes place.

6.7(a) The transfer function of the MOSFET stage with series feedback is

$$A(s) = -A_0 \cdot \frac{1 + s/\omega_z}{(1 + s/\omega_{p1})(1 + s/\omega_{p2})},$$

where

$$A_0 = \frac{g_m R_D}{g_m R_S + 1}, \qquad \omega_z = \frac{1}{R_S C_S},$$

$$\omega_{p1} = \frac{1}{R_D C_L}, \qquad \omega_{p2} = \frac{g_m R_S + 1}{R_S (C_S + C_{gs})}.$$

6.7(b) Setting $\omega_z = \omega_{p2}$ for a single-pole response leads to the condition $C_S = C_{gs}/(g_m R_S) \approx 1/(2\pi f_T \cdot R_S)$.

6.7(c) The input admittance of the MOSFET stage with series feedback is

$$Y(s) = s C_0 \cdot \frac{1 + s/\omega_z}{1 + s/\omega_{p2}},$$

where $C_0 = C_{gs}/(g_m R_S + 1)$; ω_z and ω_{p2} are the same as in Problem 6.7(a).

6.7(d) If $\omega_z = \omega_{p2}$ (same condition as for the single-pole response), the input admittance becomes purely capacitive with the value $C_I = C_{gs}/(g_m R_S + 1)$, in agreement with Eq. (6.52).

6.8(a) The input conductance of the MOSFET regulated cascode is

$$Y_I = 1/R_S + \frac{(|A_F| + 1) \cdot g_m + g_o}{g_o R_D + 1},$$

which for $R_D \ll 1/g_o$ and $(|A_F| + 1) \cdot g_m/g_o \gg 1$ can be simplified to $Y_I \approx 1/R_S + (|A_F| + 1) \cdot g_m$.

6.8(b) The output conductance of the MOSFET regulated cascode is

$$Y_O = 1/R_D + \frac{g_o}{[(|A_F| + 1) \cdot g_m + g_o] R_S + 1},$$

which for $(|A_F| + 1) \cdot g_m/g_o \gg 1$ and $(|A_F| + 1) \cdot g_m R_S \gg 1$ can be simplified to $Y_O \approx 1/R_D + g_o/[(|A_F| + 1) \cdot g_m R_S]$.

6.8(c) If we set $A_F = 0$ (DC bias applied to gate), the above expressions describe the simple cascode. Thus, the feedback mechanism in the regulated cascode reduces the input resistance from the FET by $|A_F| + 1$ and increases the output resistance from the FET by $|A_F| + 1$.

6.8(d) When including the base-emitter conductance g_m/β into our calculations, the approximate input conductance becomes $Y_I \approx 1/R_S + (1 + 1/\beta) \cdot (|A_F| + 1) \cdot g_m$, which only is a small modification from what we had before. However, the approximate output conductance changes more drastically to $Y_O \approx 1/R_D + g_o/\beta + g_o/[(|A_F| + 1) \cdot g_m R_S]$, which means that the output resistance can be boosted by at most β, no matter how large $|A_F|$ is made.

6.9(a) The bandwidth of the simple MOSFET stage is $BW = 1/(2\pi \cdot R_D C_L)$.

6.9(b) The bandwidth of the MOSFET stage with TIA load for $Q = 1/\sqrt{2}$ (Butterworth response) is given by Eq. (5.21) as $BW' = \sqrt{(|A_F|+1) \cdot BW_F/(2\pi \cdot R_F C_L)}$. Using $R_D = R_T = |A_F|/(|A_F|+1) \cdot R_F$ to make the gain of both stages equal, it follows that $BW' = \sqrt{|A_F| \cdot BW_F/(2\pi \cdot R_D C_L)}$.

6.9(c) Expressing BW' in terms of BW yields $BW' = \sqrt{|A_F| \cdot BW_F \cdot BW}$. Using $|A| \cdot BW = |A_F| \cdot BW_F$ to equate the GBW of the simple common-source stage with that of the feedback amplifier, it follows that $BW' = \sqrt{|A|} \cdot BW$, in agreement with Eq. (6.55).

6.10(a) The bandwidth of the two-stage MOSFET amplifier is $BW' = \sqrt{\sqrt{2}-1}/(2\pi \cdot R'_D C_L)$, where we have used the second term of Eq. (6.45) for the bandwidth shrinkage due to cascading two first-order stages. For the gain to be equal to the single-stage amplifier, we need $R'_D = R_D/\sqrt{|A|}$ (each sub-stage has gain $\sqrt{A}$), and thus we have $BW' = \sqrt{(\sqrt{2}-1) \cdot |A|}/(2\pi \cdot R_D C_L)$.

6.10(b) Expressing BW' in terms of BW yields $BW' = \sqrt{(\sqrt{2}-1) \cdot |A|} \cdot BW$, in agreement with Eq. (6.56). Thus, the bandwidth extension is $\sqrt{(\sqrt{2}-1) \cdot |A|}$.

6.11 The bandwidth of a MOSFET stage with active-feedback load for $Q = 1/\sqrt{2}$ (Butterworth response) is given by $BW' = \sqrt{|A_F| \cdot g_{mF} \cdot BW_F/(2\pi \cdot C_L)}$, which follows directly from the expression of ω_0 in the result to Problem 5.13(a). Using $R_D = R_T = 1/g_{mF}$ to make the gain of both stages equal, it follows that $BW' = \sqrt{|A_F| \cdot BW_F/(2\pi \cdot R_D C_L)}$. Because this bandwidth is the same as that obtained in Problem 6.9(b) for the shunt-feedback TIA load, the bandwidth extension is the same, too.

6.12(a) The differential admittance $Y(s) = 1/2 \cdot (I_p - I_n)/(V_p - V_n)$ is

$$Y(s) = -sC \cdot \frac{1 - s/\omega_z}{1 + s/\omega_p},$$

where

$$\omega_z = \frac{g_m}{C_{gs}}, \qquad \omega_p = \frac{g_m}{2C + C_{gs}}.$$

6.12(b) The bandwidth of the negative capacitance is $BW \approx g_m/[2\pi \cdot (2C + C_{gs})]$, as given by the pole of $Y(s)$. The negative capacitance at low frequencies is $-C$.

6.13(a) The transfer function of the MOSFET source follower is

$$A(s) = A_0 \cdot \frac{1 + s/\omega_z}{1 + s/\omega_p},$$

where

$$A_0 = \frac{g_m}{g_m + g_{mb} + g_o}, \quad \omega_z = \frac{g_m}{C_{gs}}, \quad \omega_p = \frac{g_m}{A_0 \cdot (C_L + C_{gs})}.$$

Assuming $\omega_z \gg \omega_p$, the buffer bandwidth is $BW_B = g_m/[2\pi A_0(C_L + C_{gs})]$.

6.13(b) The input admittance of the MOSFET source follower is

$$Y(s) = sC_0 \cdot \frac{1 + s/\omega_{z2}}{1 + s/\omega_p},$$

where

$$C_0 = (1 - A_0) \cdot C_{gs} + C_{gd},$$

$$\omega_{z2} = \frac{C_0}{C_\infty} \cdot \frac{g_m}{A_0 \cdot (C_L + C_{gs})} \quad \text{with} \quad C_\infty = \frac{C_{gs} \cdot C_L}{C_{gs} + C_L} + C_{gd},$$

and ω_p is the same as in Problem 6.13(a). For $\omega \ll \omega_{z2}$, the input capacitance is $C_I = C_0 = (1 - A_0) \cdot C_{gs} + C_{gd}$.

6.13(c) From Problem 6.13(a), we have $C_L = g_m/(2\pi \cdot BW_B \cdot A_0) - C_{gs}$. Expressed in terms of f_T: $C_L = f_T/BW_B \cdot (C_{gs} + C_{gd})/A_0 - C_{gs}$. With $C_I = (1 - A_0) \cdot C_{gs} + C_{gd}$ from Problem 6.13(b), we find the capacitance-transformation ratio

$$\kappa = \frac{C_{gs} + C_{gd}}{A_0[(1 - A_0)C_{gs} + C_{gd} + C_{I1}]} \cdot \left(\frac{f_T}{BW_B} - \frac{A_0(C_{gs} + C_{L0})}{C_{gs} + C_{gd}}\right).$$

Inserting the values leads to $\kappa = 2.32 \cdot (f_T/BW_B - 1.66)$, in accordance with [156].

6.14(a) The transfer function of the MOSFET common-source buffer is

$$A(s) = -A_0 \cdot \frac{1 - s/\omega_z}{1 + s/\omega_p},$$

where

$$A_0 = \frac{g_m}{1/R_D + g_o}, \quad \omega_z = \frac{g_m}{C_{gd}}, \quad \omega_p = \frac{g_m}{A_0 \cdot (C_L + C_{gd})}.$$

Assuming $\omega_z \gg \omega_p$, the buffer bandwidth is $BW_B = g_m/[2\pi A_0(C_L + C_{gd})]$.

6.14(b) The input admittance of the MOSFET common-source buffer is

$$Y(s) = sC_0 \cdot \frac{1 + s/\omega_{z2}}{1 + s/\omega_p},$$

where

$$C_0 = C_{gs} + (1 + A_0) \cdot C_{gd},$$

$$\omega_{z2} = \frac{C_0}{C_\infty} \cdot \frac{g_m}{A_0 \cdot (C_L + C_{gd})} \quad \text{with} \quad C_\infty = C_{gs} + \frac{C_{gd} \cdot C_L}{C_{gd} + C_L},$$

and ω_p is the same as in Problem 6.14(a). For $\omega \ll \omega_p$, the input capacitance is $C_I = C_0 = C_{gs} + (1 + A_0) \cdot C_{gd}$.

6.14(c) From Problem 6.14(a), we have $C_L = g_m/(2\pi \cdot BW_B \cdot A_0) - C_{gd}$. Expressed in terms of f_T: $C_L = f_T/BW_B \cdot (C_{gs} + C_{gd})/A_0 - C_{gd}$. With $C_I = C_{gs} + (1 + A_0) \cdot C_{gd}$ from Problem 6.14(b), we find the capacitance-transformation ratio

$$\kappa = \frac{C_{gs} + C_{gd}}{A_0[C_{gs} + (1 + A_0)C_{gd} + C_{I1}]} \cdot \left(\frac{f_T}{BW_B} - \frac{A_0(C_{gd} + C_{L0})}{C_{gs} + C_{gd}}\right).$$

Inserting the values leads to $\kappa = 0.93 \cdot (f_T/BW_B - 0.55)$, in accordance with [156]. In conclusion, the source-follower buffer is about twice as effective as a capacitance transformer compared with the common-source buffer.

6.15(a) The bandwidth of the amplifier is given by the stage bandwidth and the multi-stage bandwidth shrinkage: $BW_{\text{tot}} \approx 1/[2\pi R_{O0}(C_{O0} + C_{I0})] \cdot \sqrt{2^{1/n} - 1}$. Actually, the bandwidth is slightly larger than this because the last-stage pole is at the frequency $1/[2\pi \cdot R_{O0}/\xi_1 \cdot (\xi_1 \cdot C_{O0} + C_L)]$, which is higher than the frequency of the first $n - 1$ poles for $C_L < C_I$. The total power dissipation simply is $P_{\text{tot}} = n \cdot \xi_1 \cdot P_0$. Given the input capacitance $C_I = \xi_1 \cdot C_{I0}$, we find the scale parameter $\xi_1 = \xi_i = C_I/C_{I0}$. The input-referred mean-square noise voltage of the amplifier is given by the series $\overline{v_{n,\text{tot}}^2} = \overline{v_{n0}^2}/\xi_1 + \overline{v_{n0}^2}/(\xi_1 \cdot A_S^2) + \ldots = \sum_{i=0}^{n-1} 1/A_S^{2i} \cdot \overline{v_{n0}^2}/\xi_1 = (1 - 1/A_S^{2n})/(1 - 1/A_S^2) \cdot \overline{v_{n0}^2}/\xi_1$, where $A_S = g_{m0}R_0$ is the stage gain.

6.15(b) The optimum scale factor, κ, for which all n poles have the same frequency, follows from the boundary conditions $C_I = \xi_1 \cdot C_{I0}$ and $C_L = \xi_{n+1} \cdot C_{I0}$ as $\kappa = \sqrt[n]{C_I/C_L}$. The amplifier bandwidth now is $BW_{\text{tot}} = 1/[2\pi R_{O0}(C_{I0}/\kappa + C_{O0})] \cdot \sqrt{2^{1/n} - 1}$. The total power dissipation now is given by the series $P_{\text{tot}} = \xi_1 \cdot P_0 + \xi_1/\kappa \cdot P_0 + \ldots = \sum_{i=0}^{n-1} 1/\kappa^i \cdot \xi_1 \cdot P_0 = (1 - 1/\kappa^n)/(1 - 1/\kappa) \cdot \xi_1 \cdot P_0$. The input-referred noise power of the amplifier now is given by the series $\overline{v_{n,\text{tot}}^2} = \overline{v_{n0}^2}/\xi_1 + \overline{v_{n0}^2} \cdot \kappa/(\xi_1 \cdot A_S^2) + \ldots = \sum_{i=0}^{n-1} \kappa^i/A_S^{2i} \cdot \overline{v_{n0}^2}/\xi_1 = (1 - \kappa^n/A_S^{2n})/(1 - \kappa/A_S^2) \cdot \overline{v_{n0}^2}/\xi_1$.

6.15(c) The optimum scale factor is $\kappa = \sqrt[4]{16} = 2.0$. The improvement in bandwidth is

$$\frac{1 + C_{I0}/C_{O0}}{1 + (C_{I0}/C_{O0})/\kappa} = 1.429.$$

The savings in power are

$$\frac{1 - 1/\kappa^n}{(1 - 1/\kappa) \cdot n} = 0.469.$$

The increase in noise power is

$$\frac{1 - \kappa^n/A_S^{2n}}{1 - \kappa/A_S^2} \cdot \frac{1 - 1/A_S^2}{1 - 1/A_S^{2n}} = (1.106)^2.$$

6.16(a) The transfer function of the MOSFET stage with shunt peaking is

$$A(s) = A_0 \cdot \frac{1 + s/\omega_z}{1 + s/(Q\omega_0) + s^2/\omega_0^2},$$

where

$$A_0 = g_m R, \quad \omega_z = \frac{R}{L} = \frac{\omega_0}{Q}, \quad \omega_0 = \sqrt{\frac{1}{LC_L}}, \quad Q = \sqrt{\frac{L}{C_L}} \cdot \frac{1}{R}.$$

6.16(b) From the above pole expressions, we find that $L = Q^2 \cdot R^2 C_L$. Thus for $Q = 0.644$, the inductance value is $L = 0.415 \cdot R^2 C_L$, in accordance with Eq. (6.61).

6.17(a) The impedance of the MOSFET active inductor is

$$Z(s) = R_0 \cdot \frac{1 + s/\omega_z}{(1 + s/\omega_{p1})(1 + s/\omega_{p2})},$$

where

$$R_0 = \frac{1}{g_m + g_{mb} + g_o},$$

$$\omega_z = \frac{1}{R_G(C_{gs} + C_{gd})},$$

$$(\omega_{p1}^{-1} + \omega_{p2}^{-1})^{-1} = \frac{1}{R_0 C_{gs}[(g_{mb} + g_o)R_G + 1] + R_G C_{gd}},$$

$$\omega_{p1} \cdot \omega_{p2} = \frac{1}{R_0 R_G C_{gs} C_{gd}}.$$

If $\omega_{p2} \gg \omega_{p1}$, the dominant pole, ω_{p1}, is (conservatively) approximated by $(\omega_{p1}^{-1} + \omega_{p2}^{-1})^{-1}$, which is given above.

6.17(b) The impedance is inductive in the range ω_z to ω_{p1}. Assuming $\omega_{p2} \gg \omega_{p1}$, this frequency range is

$$\frac{1}{2\pi} \cdot \frac{1}{R_G(C_{gs} + C_{gd})} \quad \ldots \quad \frac{1}{2\pi} \cdot \frac{1}{R_0 C_{gs}[(g_{mb} + g_o)R_G + 1] + R_G C_{gd}}.$$

With suitable values for R_G, g_{mb}, and g_o, the upper frequency can be approximated as $f_T/2$.

6.17(c) For the impedance to be inductive at any frequency, $\omega_z < \omega_{p1}$ must hold. Using the above approximation for ω_{p1}, we find the condition $R_G > 1/g_m$.

6.17(d) For $\omega \ll \omega_{p1}, \omega_{p2}$: $Z(s) \approx R_0 + s R_0/\omega_z$. Thus, the value of the inductance is $L = R_0/\omega_z = R_G(C_{gs} + C_{gd})/(g_m + g_{mb} + g_o) \approx R_G/(2\pi f_T)$, in accordance with Eq. (6.63).

6.18 The integral evaluates to $\ln|Z_0| \cdot [\arcsin(1) - \arcsin(0)]$, and thus the inequality simplifies to $|Z_0| \leq 2/(2\pi \cdot BW \cdot C_\infty)$, in accordance with Eq. (6.64).

6.19(a) The transfer function of the MA with offset compensation is

$$A_{\text{tot}}(s) = T_0 \cdot \frac{s(1 + s/\omega_z)}{(1 + s/\omega_{p1})(1 + s/\omega_{p2})},$$

where

$$T_0 = \frac{AR_0C}{\xi AA_1 + 1}, \qquad \omega_z = \frac{1}{R_1C_1},$$

$$\omega_{p1} = \frac{1}{2R_0C}, \qquad \omega_{p2} = \frac{\xi AA_1 + 1}{R_1C_1}.$$

6.19(b) For $C \to \infty$, the LF cutoff is determined by the offset compensation loop and is given by ω_{p2}: $f_{LF} = (\xi AA_1 + 1)/(2\pi \cdot R_1 C_1)$. For $C_1 \to \infty$, the LF cutoff is determined by the AC coupling and is given by ω_{p1}: $f_{LF} = 1/(2\pi \cdot 2R_0 C)$.

Chapter 7

7.1(a) Given an average optical power $\overline{P}$ and an extinction ratio ER, the power for zeros and ones are $P_0 = 2\overline{P}/(ER+1)$ and $P_1 = 2\overline{P} \cdot ER/(ER+1)$, respectively. Assuming a signal-independent noise, we can write the proportion $Q \sim P_1 - P_0$. Inserting the power levels results in

$$Q \sim 2\overline{P} \cdot \frac{ER - 1}{ER + 1}.$$

The power penalty is the inverse of the multiplier of $\overline{P}$ normalized such that $PP(ER \to \infty) = 1$. Thus, we find that $PP = (ER+1)/(ER-1)$, in accordance with Eq. (7.5).

7.1(b) Assuming that the rms noise current is proportional to the square-root of the power level, we can write the proportion $Q \sim (P_1 - P_0)/(\sqrt{P_1} + \sqrt{P_0})$.

Inserting the power levels from Problem 7.1(a) and squaring the result yields

$$Q^2 \sim 2\overline{P} \cdot \frac{(ER-1)^2}{(ER+1) \cdot (\sqrt{ER}+1)^2}.$$

Again, the power penalty is the inverse of the multiplier of $\overline{P}$ normalized such that $PP(ER \to \infty) = 1$. Thus, we find that $PP = (ER+1) \cdot (\sqrt{ER}+1)^2/(ER-1)^2 = (\sqrt{ER}+1)/(\sqrt{ER}-1) \cdot (ER+1)/(ER-1)$, in accordance with Eq. (7.6).

7.2 In a photodetector, the bandgap energy must be *smaller* than the photon energy of the light it is designed to detect (to permit the generation of electron/hole pairs). In a laser, the bandgap energy must be *matched* to the photon energy of the light it is designed to emit (to permit stimulated emission).

7.3 Considering that (i) the phase change $\Phi(t)$ causes the frequency excursion $\Delta f = 1/(2\pi) \cdot \partial\Phi/\partial t$ and (ii) the equivalence $d/dt\{\ln P(t)\} = 1/P(t) \cdot dP(t)/dt$, Eq. (7.10) follows easily.

7.4 Given a DC-balanced NRZ signal with high extinction, the signal currents in the receiver for zeros and ones are 0 and $2\mathcal{R}\overline{P}$, respectively. The rms-noise currents in the presence of RIN noise for zeros and ones are $\sqrt{i_{n,\text{amp}}^2}$ and $\sqrt{i_{n,\text{amp}}^2 + RIN \cdot (2\mathcal{R}\overline{P})^2 \cdot BW_n}$, respectively (cf. Eq. (7.11)). Thus, we can write Q as

$$Q = \frac{2\mathcal{R}\overline{P}}{\sqrt{i_{n,\text{amp}}^2} + \sqrt{i_{n,\text{amp}}^2 + RIN \cdot (2\mathcal{R}\overline{P})^2 \cdot BW_n}}.$$

After rewriting this equation such that $\overline{P}$ occurs only once, we get $\mathcal{R}\overline{P} \cdot (Q^2 \cdot RIN \cdot BW_n - 1) + Q \cdot \sqrt{i_{n,\text{amp}}^2} = 0$. The power penalty is the inverse of the multiplier of $\overline{P}$ normalized such that $PP(RIN = 0) = 1$. Thus, we find that $PP = 1/(1 - Q^2 \cdot RIN \cdot BW_n)$, in accordance with Eq. (7.12).

7.5 In the case of destructive interference, the light at the combiner is radiated into a direction that does *not* couple into the output fiber.

7.6(a) The resulting signal has three levels: $0+0$, $\{0+1, 1+0\}$, and $1+1$.

7.6(b) The filter has the transfer function $H(f) = \exp(-j\,2\pi f/B) + 1$.

7.6(c) The spectrum of the NRZ (input) signal can be written as $j/(2\pi f) \cdot [\exp(-j\,2\pi f/B) - 1]$ (cf. Problem 4.18). After multiplying with the filter transfer function, $H(f)$, from above, we obtain $j/(2\pi f) \cdot [\exp(-j\,4\pi f/B) - 1]$. Note that the resulting duobinary spectrum has half the bandwidth of the original NRZ spectrum.

Chapter 8

8.1(a) The gate-source voltage of an FET as a function of the drain current is $V_{GS}(I_D) = \sqrt{L/W \cdot 2I_D/(\mu_n C'_{ox})} + V_{TH}$. Thus, the differential input voltage that switches I_M completely is $V_{GS}(I_M) - V_{GS}(0) = \sqrt{L/W \cdot 2I_M/(\mu_n C'_{ox})}$. This voltage is equal to the overdrive voltage of one of the FETs biased at I_M.

8.1(b) The base-emitter voltage of a BJT as a function of the collector current is $V_{BE}(I_C) = V_T \ln(I_C/I_{C0})$. Thus, the differential input voltage that switches $0.99 I_M$ to one output and $0.01 I_M$ to the other is $V_{BE}(0.99 I_M) - V_{BE}(0.01 I_M) = V_T(\ln 0.99 - \ln 0.01) = 4.6 V_T \approx 115\,\text{mV}$.

8.2 The rate of current change is $di/dt = (0.8 - 0.2) \cdot 50\,\text{mA}/30\,\text{ps} = 1\,\text{mA/ps}$ and the maximum permissible voltage drop across the inductance is $\Delta v = 5.2\,\text{V} - 2.5\,\text{V} - 1.5\,\text{V} = 1.2\,\text{V}$. The inductance thus has to be less than $L = \Delta v/(di/dt) = 1.2\,\text{nH}$, corresponding to about a 1.2-mm bond wire.

8.3(a) The laser current, I_L, can be found conveniently by calculating the contributions of each source (I'_M, I'_B, and V_{TH}) to I_L separately and then adding them:

$$I_L = \frac{I'_M \cdot R_0 + I'_B \cdot (2R_0 - R_{LD}) - V_{TH}}{2R_0}$$

$$= \frac{1}{2} \cdot I'_M + \left(1 - \frac{R_{LD}}{2R_0}\right) \cdot I'_B - \frac{V_{TH}}{2R_0}.$$

8.3(b) With the given values, the laser current is $I_L = 0.5 \cdot I'_M + 0.88 \cdot I'_B - 20\,\text{mA}$.

8.4 All three circuits are equivalent and implement the function $v_O = 2v_I + R \cdot i_O$, that is, a driver stage with output resistance R and unity gain when loaded with R. Circuit (a) is a direct implementation of the mathematical function, (b) corresponds to a current-steering output stage with passive back termination, and (c) corresponds to a current-steering output stage with active back termination.

8.5(a) The edge rate of the differential input signal is $(0.8 - 0.2) \cdot v_I^{pp}/t_R$, thus increasing the slice level by V_{OS} causes the ones to shrink on both sides by $t_R/(0.6 \cdot v_I^{pp}) \cdot V_{OS}$. Similarly the zeros will broaden on both sides by the same amount. Therefore, we have

$$t_{PWD} = \frac{1}{0.3} \cdot \frac{V_{OS}}{v_I^{pp}} \cdot t_R.$$

8.5(b) With the given values, $t_{PWD} = -0.1\,\text{UI} \ldots 0.1\,\text{UI}$ or, equivalently, the PWD adjustment range is $\pm 10\%$.

8.6(a) The differential output swing is $v_O^{pp} = 2R \cdot I_1$ and the differential switching voltage is $v_{I,\min}^{pp} = 2[V_{GS}(I_D = I_1) - V_{TH}]$ (cf. Problem 8.1). Using the quadratic model, we have $I_1 = \mu_n C'_{ox}/2 \cdot W/L \cdot (v_{I,\min}^{pp}/2)^2$, and thus $v_O^{pp} = R \cdot \mu_n C'_{ox} \cdot W/L \cdot (v_{I,\min}^{pp}/2)^2$. Solving for R and using the load capacitance $C = 2/3 \cdot C'_{ox} \cdot WL$ results in the time constant $RC = 8/3 \cdot L^2/\mu_n \cdot v_O^{pp}/(v_{I,\min}^{pp})^2$. Thus, the 20% to 80% rise/fall times are (cf. Eq. (8.15))

$$t_R = t_F = 3.7 \cdot \frac{L^2}{\mu_n} \cdot \frac{v_O^{pp}}{(v_{I,\min}^{pp})^2}.$$

8.6(b) For $v_O^{pp}/v_{I,\min}^{pp} = 1$, the rise/fall times become $3.7 \cdot L^2/\mu_n \cdot 1/v_O^{pp}$, thus a larger logic swing improves the speed. Note, however, that for submicron MOSFETs operated at large gate-source overdrive voltages, the square law no longer holds (cf. Appendix D). Thus, under these conditions, the relationship between speed and logic swing derived in this exercise is not accurate.

8.7 A small-signal AC analysis reveals the following. The laser current is the sum of the modulation and bias currents, $I_l = I_m + I_b$. The bias current is controlled by the APC loop as follows: $I_b = g_m \cdot (-A) \cdot R/(1+s \cdot RC) \cdot \xi \cdot I_l$. Solving for the transfer function of the transmitter, I_l/I_m, yields

$$\frac{1}{1 + \xi A g_m R} \cdot \frac{1 + s \cdot RC}{1 + s \cdot RC/(1 + \xi A g_m R)}.$$

Assuming that the loop gain, $\xi A g_m R$, is much larger than unity, the low-frequency cutoff is given by the pole of the above transfer function

$$f_{LF} = \frac{1}{2\pi} \cdot \frac{1 + \xi A g_m R}{RC},$$

and with Eq. (8.17), the power penalty follows as

$$PP \approx 1 + \frac{(1 + \xi A g_m R)r}{B \cdot RC}.$$

8.8(a) The power for zeros and ones are $P_0 = \xi \cdot (I_B - I_{TH})$ and $P_1 = \xi \cdot (I_B + I_M - I_{TH})$, respectively, as long as $I_B \geq I_{TH}$. The average power follows as $\overline{P} = \xi \cdot (I_B - I_{TH} + I_M/2)$ and is held constant by the APC, which means $I_B - I_{TH} = \overline{P}/\xi - I_M/2$. Thus, we have $P_0 = \overline{P} - \xi \cdot I_M/2$ and $P_1 = \overline{P} + \xi \cdot I_M/2$ and the ER becomes

$$ER = \frac{2\overline{P} + \xi \cdot I_M}{2\overline{P} - \xi \cdot I_M}.$$

8.8(b) For an infinite ER, we need $\xi \cdot I_M/(2\overline{P}) = 1.0$. When ξ degrades by 30%, the expression $\xi \cdot I_M/(2\overline{P})$ becomes 0.7; thus, $ER = (1+0.7)/(1-0.7) = 5.7\times$ (7.5 dB).

8.8(c) To achieve $ER = 6.6\times$ (8.2 dB), we need $\xi \cdot I_M/(2\overline{P}) = 0.737$. To permit a slope efficiency degradation of 10%, we need $\xi \cdot I_M/(2\overline{P}) = 0.737/0.9 = 0.819$, corresponding to $ER = 10.0\times$ (10.0 dB).

8.9 Given a 1-mW peak-to-peak optical output power, the modulation currents must be 12.5 mA, 14.3 mA, and 20.0 mA for the temperatures $-40°C$, $20°C$, and $80°C$, respectively. (i) The modulation current can be controlled as a function of the temperature using the above values and interpolations in between. Simpler, an average temperature coefficient of 0.0625 mA/°C could be used. (ii) The modulation current can be increased in proportion to the bias current increase: $\Delta I_M = 0.227 \cdot \Delta I_B$. (iii) A combination of the above two methods.

8.10 The output voltage of the mixer is $10 \cdot V_O \cdot \sin(\omega t) \cdot [1 \pm \xi \cdot \sin(\omega t)]$, which can be expanded into the series $\pm 5 \cdot V_O \cdot \xi + 10 \cdot V_O \cdot \sin(\omega t) \pm 5 \cdot V_O \cdot \xi \cdot \cos(2\omega t)$. After filtering out the components at ω and 2ω and amplifying the result by A, we obtain the bias voltage $V_B = \pm 5A \cdot V_O \cdot \xi$.

8.11 Without S_D and with $I_{REF}/2$, the integral in Eq. (8.18) evaluates to $1/C \cdot [I_{PD} \cdot n_1/B - I_{REF}/2 \cdot (n_0 + n_1)/B]$. With $(n_0 + n_1)/2 = n_1$ for 50% mark density, the above expression becomes identical to the second expression in Eq. (8.18).

Appendix F
Notation

Voltages and Currents.

- Constant voltages and currents (i.e., DC voltages and currents) are designated with uppercase letters and uppercase indices (e.g., V_{GS} or I_D).

- Total instantaneous voltages and currents are designated with lowercase letters and uppercase indices (e.g., v_{GS} or i_D).

- Small-signal voltages and currents in the time domain are designated with lowercase letters and lowercase indices (e.g., v_{gs} or i_d). These small-signal voltages and currents represent a small change in the total instantaneous value, and thus also are written as such with a Δ prefix (e.g., Δv_{GS} or Δi_D).

- Phasors for voltages and currents (complex quantities that describe the amplitude and phase of a small-signal sinusoid) are designated with uppercase letters and lowercase indices (e.g., V_{gs} or I_d).

- The peak-to-peak value (swing) of a voltage or current signal is designated with a lowercase letter and a *pp* superscript (e.g., v^{pp} or i^{pp}).

- The average value of a voltage or current signal is designated with a lowercase letter and a horizontal bar (e.g., $\bar{v}$ or $\bar{i}$).

Noise Quantities.

- Instantaneous noise voltages and currents in the time domain are designated with lowercase letters and the index n (e.g., v_n or i_n).

- Mean-square noise voltages and currents, measured in a given bandwidth, are designated with $\overline{v_n^2}$ and $\overline{i_n^2}$, respectively. These quantities also are referred to as *noise powers* for short. The mean indicates either the time average (for stationary noise) or the ensemble average (for nonstationary noise), as appropriate. These noise quantities can be calculated as the integral of an (output-referred) power spectral density over the given bandwidth. The integral of an input-referred power spectral density, however, does not lead to a meaningful input-referred noise power.

- Root-mean-square (rms) noise voltages and currents, measured in a given bandwidth, are designated with v_n^{rms} and i_n^{rms}, respectively. These quantities are the square roots of the mean-square noise quantities (see above).

- The power spectral densities for voltages and currents are designated with $V_n^2(f)$ and $I_n^2(f)$, respectively. These quantities also are referred to as *power spectra* for short. The power spectral densities used in this book all are *one sided*, and thus directly represent the noise power in a 1-Hz bandwidth at frequency f. (No matter if we say power spectral density or simply power spectrum, we always mean the power in a 1-Hz bandwidth.) Power spectral densities can be calculated as the Fourier transform of the autocorrelation function of the time domain signal.

- The root spectral densities for voltages and currents are designated with $V_n(f)$ and $I_n(f)$, respectively. These quantities are the square roots of the power spectral densities (see above) and represent the rms noise voltage or current in a 1-Hz bandwidth at frequency f.

Appendix G
Symbols

Latin Symbols

A	(i) Voltage gain; (ii) current gain
a	Fiber attenuation per unit length
A_0	Voltage gain at low frequencies
A_ν	Voltage gain of amplifier number ν
a_ν	Normalized power-series coefficient
A_C	Collector area
A_{cm}	Common-mode gain
A_d	Differential gain
A_{dd}	Positive power-supply gain
A_E	Emitter area ($= W_E L_E$)
A_F	Gain of feedback amplifier
A_S	Voltage gain per stage
A_{ss}	Negative power-supply gain
A_{tot}	Total voltage gain of a multistage amplifier
A_X	Gain to node x
B	(i) Bit rate; (ii) susceptance
BER	Bit-error rate (actually, bit-error probability)
BER_{in}	Bit-error rate before error correction

Symbol	Description
BER_{out}	Bit-error rate after error correction
BV	Breakdown voltage
BV_{CBO}	Open-emitter collector-base breakdown voltage
BV_{CEO}	Open-base collector-emitter breakdown voltage
BV_{CES}	Shorted-base collector-emitter breakdown voltage
BW	Bandwidth
BW_{3dB}	3-dB bandwidth
BW_ν	Bandwidth of block number ν
BW_B	Bandwidth of buffer
BW_D	Bandwidth of decision circuit
BW_F	Bandwidth of feedback amplifier
BW_n	Noise bandwidth (for white noise)
BW_{n2}	Noise bandwidth (for f^2 noise)
BW_O	Optical bandwidth
BW_S	Bandwidth of a stage
C	Capacitor or capacitance
c	Speed of light in vacuum, $c = 299.8 \cdot 10^6$ m/s
C_0	Capacitance at low frequencies
C_∞	Capacitance at high frequencies
c_ν	Filter coefficient number ν
C_B	Bridge capacitor
C_{bc}	Base-collector capacitance
C_{be}	Base-emitter capacitance
C_D	Capacitance of photodetector
C_{db}	Drain-bulk capacitance
C_E	Emitter capacitor
C_F	Feedback capacitor
C_{gd}	Gate-drain capacitance
C_{gs}	Gate-source capacitance
C_I	Input capacitance
C_{jc}	Collector junction capacitance
C_{je}	Emitter junction capacitance
C_L	Load capacitance
C_O	Output capacitance
C'_{ox}	Oxide capacitance per unit area
C_P	(i) Parasitic capacitance; (ii) bond-pad capacitance
C_{PD}	Internal capacitance of photodiode
C_{sb}	Source-bulk capacitance
C_T	Total capacitance at the TIA summing node ($= C_D + C_I$)
C_{TSI}	Capacitance of input transmission-line segment
C_{TSO}	Capacitance of output transmission-line segment
C_X	Capacitor at unused TIA input
$CMRR$	Common-mode rejection ratio
CNR	Carrier-to-noise ratio
CSO	Composite second-order distortion

CTB	Composite triple-beat distortion
D	Chromatic dispersion parameter
D_n	Diffusion constant for electrons
D_{PMD}	Polarization-mode dispersion parameter
E	Electric field
E_b	Energy per information bit
ER	Extinction ratio
F	(i) Noise figure; (ii) excess noise factor
$\tilde{F}$	Optical noise figure
f	(i) Frequency; (ii) function
Δf	Bandwidth
f_0	Resonance frequency of a second-order transfer function
f_ν	Frequency component number ν
f_A	Pole frequency of amplifier ($= 1/(2\pi \cdot T_A)$)
f_c	$1/f$-noise corner frequency of noise spectrum
f_{c2}	f^2-noise corner frequency of noise spectrum
f_{cutoff}	Cutoff frequency of transmission line
f_{LF}	Low-frequency cutoff
$f_{\max}$	Unity power-gain frequency of a transistor (maximum frequency of oscillation)
f_p	Frequency of pole
f_{SR}	Self-resonance frequency
f_T	Unity current-gain frequency of a transistor (transition frequency)
f_z	Frequency of zero
G	(i) Power gain; (ii) conductance
g_m	Transconductance of a transistor
g_{mb}	Bulk-input transconductance of a transistor
g_o	Output conductance of a transistor
$\text{Gauss}(x)$	Normalized Gaussian distribution: $1/\sqrt{2\pi} \cdot \exp(-x^2/2)$
GBW	Gain-bandwidth product
GBW_S	Gain-bandwidth product of a single stage
GBW_{tot}	Gain-bandwidth product of total amplifier
GC	Gain compression
$H(f)$	Transfer function of a linear system
h	Planck constant, $h = 6.626 \cdot 10^{-34}$ Js
$h(t)$	Impulse response of a linear system
H_0	Passband value of the transfer function $H(f)$
$H_{FRC}(f)$	Full raised-cosine spectrum
$H_{NRZ}(f)$	Spectrum of the ideal NRZ signal
$HD\nu$	νth-order harmonic distortion
i	Current
I_0	Reference current
I_1	First Personick integral
I_2	Second Personick integral
I_3	Third Personick integral

390 SYMBOLS

Symbol	Description
I_{AGC}	Automatic gain control current
i_{ANT}	Antenna current
I_{APD}	Avalanche photodetector current
I_B	(i) Base current; (ii) laser bias current
I_C	Collector current
I_{C0}	Collector current at $V_{BE} = 0$
I'_{C0}	Collector current density at $V_{BE} = 0$
i_{CM}	Common-mode current
I_D	(i) Drain current; (ii) differential current
I_{DK}	Dark current
I_E	Emitter current
I_f	Personick integral for f noise
I_G	Gate current
i_I	Input current
$i_{I.\text{incident}}$	Input current component due to incident wave
$i_{I.\text{outgoing}}$	Input current component due to outgoing wave
I_L	Laser current
$I_{L.\text{on}}$	Laser on current
i_{lin}^{pp}	Maximum input current for linear operation
I_M	Laser modulation current ($= i_L^{pp}$)
i_N	Current at the inverting terminal
i_n	Noise current
$i_{n.0}$	Noise current when receiving a zero
$i_{n.1}$	Noise current when receiving a one
$i_{n.\text{amp}}$	Input-referred noise current of the linear channel (amplifier)
$i_{n.APD}$	Noise current of an avalanche photodetector
$i_{n.APD.0}$	Noise current of an avalanche photodetector when receiving a zero
$i_{n.APD.1}$	Noise current of an avalanche photodetector when receiving a one
$i_{n.ASE}$	Noise current of a p-i-n photodetector due to amplified spontaneous emission
$i_{n.B}$	Noise current at the base of a BJT
$i_{n.C}$	Noise current at the collector of a BJT
$i_{n.D}$	Noise current at the drain of an FET (channel noise)
$i_{n.DK}$	Noise current due to dark current
$i_{n.\text{front}}$	Input-referred noise current of the TIA front-end
$i_{n.\text{front}.C}$	Input-referred noise current of the TIA front-end due to $i_{n.C}$
$i_{n.\text{front}.D}$	Input-referred noise current of the TIA front-end due to $i_{n.D}$
$i_{n.\text{front}.Rb}$	Input-referred noise current of the TIA front-end due to $i_{n.Rb}$
$i_{n.G}$	Noise current at the gate of an FET
$i_{n.MA}$	Input-referred (TIA input) noise current of the main amplifier
$i_{n.OA}$	Noise current of an optically preamplified p-i-n detector
$i_{n.OA.0}$	Noise current of an optically preamplified p-i-n detector when receiving a zero
$i_{n.OA.1}$	Noise current of an optically preamplified p-i-n detector when receiving a one

$i_{n.OA.S}$	Noise current of an optically preamplified p-i-n detector that is due to the quantum noise of the optical source
$i_{n.PD}$	Noise current of a photodetector
$i_{n.PIN}$	Noise current of a p-i-n photodetector
$i_{n.PIN,0}$	Noise current of a p-i-n photodetector when receiving a zero
$i_{n.PIN,1}$	Noise current of a p-i-n photodetector when receiving a one
$i_{n.Rb}$	Noise current of the intrinsic base resistance of a BJT
$i_{n.\text{res}}$	Noise current of a resistor
$i_{n.RIN}$	Noise current of a p-i-n photodetector due to laser RIN noise
$i_{n.TIA}$	Input-referred noise current of the TIA
I_{OA}	Current of an optically preamplifier p-i-n detector
I_{OS}	Offset current
i_{ovl}^{pp}	Input overload current
i_P	Current at the noninverting terminal
I_{PD}	Photodetector current
I_{PE}	Preemphasis current
I_{PIN}	p-i-n photodetector current
I_{REF}	Reference current
i_S	Signal current
I_{se}	Single-ended current
i_{sens}^{pp}	Electrical sensitivity of the TIA
I_{TH}	Laser threshold current
I_{TH0}	Laser threshold current at $T=0$
$IMDv$	vth-order intermodulation distortion
j	Imaginary unit, $j=\sqrt{-1}$
k	(i) Boltzmann constant, $k = 1.381 \cdot 10^{-23}\,\text{J/K} = 86.18\,\mu\text{eV/K}$; (ii) magnetic coupling factor
k_A	Ionization-coefficient ratio
L	(i) Length; (ii) inductor or inductance
L_B	Bond-wire inductance
L_E	Emitter length
L_I	Input inductor
L_O	Output inductor
L_{opt}	Optimum inductance
M	(i) Avalanche gain; (ii) mean of Poisson distribution
m	Modulation index
M_v	FET transistor number v
M_{opt}	Optimum avalanche gain
N	Number of bits
n	(i) Integer number; (ii) refractive index
N_0	Noise power spectral density
n_0	Number of zero bits in a burst
N_1	Atomic population for the ground state
n_1	Number of one bits in a burst
N_2	Atomic population for the excited state

n_{clad}	Refractive index of the fiber cladding
n_{cor}	Refractive index of the fiber core
N_{CSO}	Second-order beat count
N_{CTB}	Triple-beat count
n_{opt}	Optimum number of stages
$NECG$	Net electrical coding gain
$OSNR$	Optical signal-to-noise ratio
P	Power
$\overline{P}$	Average power
P_0	Optical power when transmitting a zero
P_1	Optical power when transmitting a one
P_{ASE}	Optical noise power due to amplified spontaneous emission
P_{in}	Optical input power (into modulator)
P_{out}	Optical output power (from transmitter)
$\overline{P}_{ovl}$	Optical overload power
P_S	Optical signal power
$\overline{P}_{sens}$	Optical receiver sensitivity
$\overline{P}_{sens0}$	Sensitivity reference value
$\overline{P}_{sens,APD}$	Sensitivity of receiver with an avalanche photodetector
$\overline{P}_{sens,OA}$	Sensitivity of receiver with an optically preamplified p-i-n detector
$\overline{P}_{sens,OAC}$	Sensitivity of receiver with a cascade of optical amplifiers
$\overline{P}_{sens,PIN}$	Sensitivity of receiver with a p-i-n photodetector
$\overline{P}_{sens,quant}$	Quantum limit of sensitivity
Poisson(n)	Poisson distribution: $\exp(-M) \cdot M^n/n!$
PP	Power penalty
$PSRR^+$	Positive power-supply rejection ratio
$PSRR^-$	Negative power-supply rejection ratio
Q	Quality factor of a second-order transfer function ($= 1/(2\zeta)$)
$\mathcal{Q}$	Personick Q
q	Electron charge, $q = 1.602 \cdot 10^{-19}$ C
Q_ν	Bipolar transistor number ν
$\mathcal{Q}_{in}$	Personick Q needed for BER_{out} with error correction
Q_n	Negative charge in an FET channel
$\mathcal{Q}_{out}$	Personick Q needed for BER_{out} without error correction
R	Resistor or resistance
ΔR	Deviation of resistance from R_0
$\mathcal{R}$	Responsivity
r	(i) Run length; (ii) code rate
R_0	(i) Characteristic impedance; (ii) resistance at low frequencies
R_{AGC}	Automatic gain control resistor
R_{ANT}	Antenna resistance
$\mathcal{R}_{APD}$	Responsivity of an avalanche photodetector
R_B	Base resistor
R_b	Intrinsic base resistance
R_C	Collector resistor

R_c	Common-mode resistance
R_D	(i) Drain resistor; (ii) dummy load resistor
R_d	Differential resistance
R_E	Emitter resistor
R_F	Feedback resistor
R_G	Gate resistor
R_g	Intrinsic gate resistance
R_I	Input resistance (resistor)
R_L	Load resistor
R_{LD}	Internal series resistance of laser diode
R_O	Output resistance (resistor)
R_P	(i) Parallel resistor; (ii) parasitic resistance
R_{PD}	Internal series resistance of photodiode
R_S	(i) Shunt resistor; (ii) series resistor; (iii) source resistor
R_T	(i) Transresistance ($= Z_T$, if Z_T is real); (ii) termination resistor
RIN	Relative intensity noise spectrum
S	(i) Switch; (ii) power spectral density; (iii) S parameter
S	S matrix
s	Complex frequency variable
S_0	S parameter at low frequencies
$S_{\mu\nu}$	S parameter from port ν to port μ
$S_{\mu\nu.se}$	Single-ended S parameter from port ν to port μ
$S_{mn\mu\nu}$	Mixed-mode S parameter from mode n and port ν to mode m and port μ
S_{ASE}	Power spectral density due to amplified spontaneous emission in both polarization modes ($= 2S'_{ASE}$)
S'_{ASE}	Power spectral density due to amplified spontaneous emission in a single polarization mode
S_D	Data switch
$\mathbf{S}_{dd}$	Submatrix of differential S parameters
S_R	Reset switch
SNR	Signal-to-noise ratio
T	(i) Temperature; (ii) bit interval ($= 1/B$)
ΔT	Pulse spreading ($= 2\sigma_T$)
t	Time
T_0	Characteristic temperature
T_A	Time constant of amplifier ($= 1/(2\pi \cdot f_A)$)
t_{DDJ}	Data-dependent jitter
t_{delay}	Delay time
t_{DJ}	Deterministic jitter
t_F	Fall time
T_{in}	Input pulse width ($= 2\sigma_{\text{in}}$)
t_J	Timing jitter
t_{JTOL}	Jitter tolerance
T_{out}	Output pulse width ($= 2\sigma_{\text{out}}$)

t_{PWD}	Pulse-width distortion
t_R	Rise time
t_{RJ}	Random jitter
t_S	Sampling instant
t_{TJ}	Total jitter
t_{TOD}	Turn-on delay
$TBD3$	Triple-beat (intermodulation) distortion
THD	Total harmonic distortion
v	Voltage
V_{AGC}	Automatic gain control voltage
V_{APD}	Avalanche photodetector bias voltage
V_B	Bias voltage for modulator
V_{BC}	Control voltage for the bias current/voltage of a laser/modulator
V_{BE}	Base-emitter voltage
V_{BI}	Bias voltage
$V_{BI\nu}$	Bias voltage number ν
V_{CC}	Positive power-supply voltage of a bipolar circuit
V_{CE}	Collector-emitter voltage
v_{CM}	Common-mode voltage
v_D	Differential voltage
V_{DD}	Positive power-supply voltage of an FET circuit
V_{DRIFT}	Voltage drift
V_{DS}	Drain-source voltage
V_{DTA}	Decision threshold ambiguity width or sensitivity of a decision circuit
V_{DTH}	Decision threshold voltage
V_E	Voltage of vertical eye opening
V_{EE}	Negative power-supply voltage of a bipolar circuit
v_{EOL}	End-of-life indicator voltage
V_{GS}	Gate-source voltage
v_I	Input voltage
$v_{I.\text{incident}}$	Input voltage component due to incident wave
$v_{I.\text{outgoing}}$	Input voltage component due to outgoing wave ($= v_{I.\text{reflected}}$ or $v_{I.\text{transmitted}}$)
$v_{I.ovl}^{pp}$	Maximum permissible input voltage swing
$v_{I.\text{reflected}}$	Input voltage component due to reflected wave
$v_{I.\text{transmitted}}$	Input voltage component due to transmitted wave
v_{ICM}	Common-mode input voltage
v_{IN}	Voltage at inverting input
v_{IP}	Voltage at noninverting input
V_L	Voltage drop across a laser
v_{lin}^{pp}	Maximum input voltage for linear operation
v_{LOS}	Loss-of-signal indicator voltage
V_M	Modulator voltage
v_{M1}	Voltage at the first input port of an MZ modulator
v_{M2}	Voltage at the second input port of an MZ modulator

V_{MC}	Control voltage for the modulation current of a laser
v_N	Voltage at the inverting terminal
v_n	(i) Noise voltage; (ii) electron velocity
$v_{n.0}$	Noise voltage when receiving a zero
$v_{n.1}$	Noise voltage when receiving a one
$v_{n.\mathrm{amp}}$	Noise voltage due to the linear channel (amplifier)
$v_{n.MA}$	Input-referred noise voltage of the MA
$v_{n.MA.\mathrm{av}}$	Available input-referred noise voltage of the MA
$v_{n.ON}$	Noise voltage at inverting output
$v_{n.OP}$	Noise voltage at noninverting output
$v_{n.PD}$	Noise voltage due to the photodetector
$v_{n.S}$	Noise voltage of the source resistance
v_O	Output voltage
$v_{O.\mathrm{incident}}$	Output voltage component due to incident wave
$v_{O.\mathrm{outgoing}}$	Output voltage component due to outgoing wave ($= v_{O.\mathrm{reflected}}$ or $v_{O.\mathrm{transmitted}}$)
$v_{O.ovl}^{pp}$	Maximum permissible output voltage swing
$v_{O.\mathrm{reflected}}$	Output voltage component due to reflected wave
$v_{O.\mathrm{transmitted}}$	Output voltage component due to transmitted wave
v_{OCM}	Common-mode output voltage
v_{ON}	Voltage at inverting output
v_{OP}	Voltage at noninverting output
V_{OS}	Offset voltage
$V_{OS.SA}$	Offset voltage for slice-level adjustment
V_{OSv}	Offset voltage of amplifier number v
V_{OSN}	Offset-control voltage at the inverting input
V_{OSP}	Offset-control voltage at the noninverting input
v_{ovl}^{pp}	Input overload voltage
v_P	Voltage at the noninverting terminal
v_{PE}	Preemphasis voltage
V_R	Voltage drop across resistor
V_{REF}	Reference voltage
V_S	Modulation voltage (or voltage swing) for modulator ($= v_M^{pp}$)
v_S	(i) Signal voltage; (ii) source voltage
V_{SA}	Slice-level adjustment voltage
V_{SC}	Control voltage for voltage swing of a modulator
V_{se}	Single-ended voltage
v_{sens}^{pp}	Sensitivity of the MA
V_{SS}	Negative power-supply voltage of an FET circuit
V_{SW}	Switching voltage
V_T	Thermal voltage, $V_T = 25.256\,\mathrm{mV}$ @ 20°C
V_{TH}	Threshold voltage
V_{TODC}	Turn-on delay compensation voltage
v_X	Voltage at node x
V_π	Switching voltage of Mach-Zehnder modulator

W	Width
W_E	Emitter width
X	Amplitude of the input signal x
x	Input signal to filter or amplifier
x_B	Base thickness
Y	Admittance
y	Output signal from filter or amplifier
Y_D	Admittance of photodetector
Y_I	Input admittance
Y_O	Output admittance
Z	Impedance
Z_0	Impedance at low frequencies
Z_c	Common-mode impedance
Z_{cd}	Mode-conversion impedance
Z_d	Differential impedance
Z_{dc}	Mode-conversion impedance
Z_I	Input impedance
Z_O	Output impedance
Z_{se}	Single-ended impedance
Z_T	Transimpedance
Z_{TL}	Impedance of transmission line
Z_{TLI}	Impedance of input transmission line
Z_{TLO}	Impedance of output transmission line
Z_{TS}	Impedance of transmission-line segment
Z_{TSI}	Impedance of input transmission-line segment
Z_{TSO}	Impedance of output transmission-line segment

Greek Symbols

α	Laser chirp parameter
α_v	Noise parameter for f^v-noise
β	Current gain of a bipolar transistor
Γ	Channel-noise factor of FET
Γ_{in}	Input reflection coefficient ($= S_{11}$)
Γ_{out}	Output reflection coefficient ($= S_{22}$)
γ	Current spreading from emitter to collector
δ	(i) Relative offset error in decision threshold; (ii) relative ISI in signal
ζ	Damping factor ($= 1/(2Q)$)
η	Quantum efficiency
κ	(i) Capacitance transformation ratio; (ii) scale factor
λ	Wavelength
$\Delta\lambda$	Spectral linewidth
$\Delta\lambda_S$	Spectral linewidth of the unmodulated source
μ_n	Electron mobility

ξ	Generic parameter used in exercises
σ	Standard deviation or rms value
σ_{in}	Input pulse rms width
σ_{out}	Output pulse rms width
σ_T	RMS impulse spread
τ	(i) Group delay; (ii) transit time
$\Delta\tau$	Group delay variation
$\Delta\tau_{AM}$	Delay variation due to AM modulation
τ_c	Carrier lifetime
τ_F	Transit time of a bipolar transistor
Φ	Phase
ω	Angular frequency
ω_0	Pole angular frequency of a second-order transfer function
ω_p	Angular frequency of a (single) pole
ω_T	Unity current-gain angular frequency of a transistor
ω_z	Angular frequency of a (single) zero

Special Symbols

$\approx$	Approximately equal to
$\sim$	Proportional to ($y \sim x$ means $y = \text{const} \cdot x$)
...	(i) Additional terms, etc.; (ii) range ($1\ldots2\,\text{k}\Omega$ means in the range of $1\,\text{k}\Omega$ to $2\,\text{k}\Omega$)
$\rightarrow$	(i) Good moment for solving the indicated problem; (ii) changes to, goes to

Appendix H
Acronyms

A/D	Analog-to-Digital converter
ABC	Automatic Bias Control
AC	Alternating Current
AGC	Automatic Gain Control
AM	Amplitude Modulation
APC	Automatic Power Control
APD	Avalanche Photodetector or Avalanche Photodiode
APON	ATM Passive Optical Network
ASE	Amplified Spontaneous Emission
ASK	Amplitude-Shift Keying
ATC	Adaptive Threshold Control
ATM	Asynchronous Transfer Mode
AWG	Arrayed Waveguide Grating
BCH	Bose-Chaudhuri-Hocquenghem code
BER	Bit-Error Rate
BERT	Bit-Error Rate Test set
BGA	Ball Grid Array
BiCMOS	BJT + CMOS
BJT	Bipolar Junction Transistor
BPON	Broadband Passive Optical Network

BW	Bandwidth
CATV	Community-Antenna Television
CDR	Clock and Data Recovery
CID	Consecutive Identical Digits
CK	Clock
CML	Current-Mode Logic
CMOS	Complementary MOS
CMRR	Common-Mode Rejection Ratio
CMU	Clock Multiplication Unit
CNR	Carrier-to-Noise Ratio
CO	Central Office
CPM	Cross-Phase Modulation
CRZ	Chirped Return-to-Zero
CS-RZ	Carrier-Suppressed Return-to-Zero
CSO	Composite Second-Order distortion
CTB	Composite Triple-Beat distortion
CW	Continuous Wave
D/A	Digital-to-Analog converter
DBR	Distributed Bragg Reflector (laser)
DC	Direct Current
DCF	Dispersion Compensating Fiber
DDA	Differential Difference Amplifier
DDJ	Data-Dependent Jitter
DEC	Decision Circuit
DFB	Distributed Feedback (laser)
DFE	Decision-Feedback Equalizer
DGD	Differential Group Delay
DHBT	Double Heterojunction Bipolar Transistor
DJ	Deterministic Jitter
DML	Directly Modulated Laser
DMT	Discrete MultiTone modulation
DMUX	Demultiplexer
DOP	Degree Of Polarization
DPSK	Differential Phase-Shift Keying
DQE	Differential Quantum Efficiency
DSF	Dispersion-Shifted Fiber
DSL	Digital Subscriber Line
DSP	Digital Signal Processor
DTAW	Decision Threshold Ambiguity Width
DUT	Device Under Test
DWDM	Dense Wavelength Division Multiplexing
E/O	Electrical to Optical converter
EA	Electroabsorption
EAM	Electroabsorption Modulator
ECL	Emitter-Coupled Logic

EDFA	Erbium-Doped Fiber Amplifier
EFM	Ethernet in the First Mile
EM	Electromagnetic
EML	Electroabsorption Modulated Laser
EOL	End Of Life
EPON	Ethernet Passive Optical Network
ER	Extinction Ratio
ESCON	Enterprise Systems CONnection
ESD	Electrostatic Discharge
FDDI	Fiber Distributed Data Interface
FEC	Forward Error Correction
FET	Field-Effect Transistor
FFE	Feed-Forward Equalizer
FIR	Finite Impulse Response filter
FIT	Failures In Time (failures per 10^9 hours)
FITL	Fiber In The Loop
FKE	Franz-Keldysh Effect
FM	Frequency Modulation
FOPA	Fiber Optical Parametric Amplifier
FP	Fabry-Perot (laser)
FSAN	Full Service Access Network
FSK	Frequency-Shift Keying
FTTC	Fiber To The Curb
FTTH	Fiber To The Home
FTTP	Fiber To The Premise
FWHM	Full Width at Half Maximum
FWM	Four-Wave Mixing
GaAs	Gallium-Arsenide
GbE	Gigabit Ethernet
GBW	Gain-Bandwidth product
GPON	Gigabit-capable Passive Optical Network
GRIN	Graded Index
GVD	Group-Velocity Dispersion
HBT	Heterojunction Bipolar Transistor
HD	Harmonic Distortion
HEMT	High Electron-Mobility Transistor
HFC	Hybrid Fiber-Coax
HFET	Heterostructure Field-Effect Transistor
HPF	High-Pass Filter
I/V	Current vs. Voltage
IC	Integrated Circuit
IEEE	Institute of Electrical and Electronics Engineers
IIP3	Input-referred 3rd-order Intercept Point
IM/DD	Intensity Modulation with Direct Detection
IMD	Intermodulation Distortion

InGaAs	Indium-Gallium-Arsenide
InGaAsP	Indium-Gallium-Arsenide-Phosphide
InP	Indium-Phosphide
IP	Internet Protocol
	Intellectual Property
ISDN	Integrated Services Digital Network
ISI	Intersymbol Interference
JFET	Junction Field-Effect Transistor
JTOL	Jitter Tolerance
L/I	Light vs. Current
LA	Limiting Amplifier
LAN	Local-Area Network
LD	Laser Diode
LDPC	Low-Density Parity-Check code
LED	Light-Emitting Diode
LF	Low Frequency
$LiNbO_3$	Lithium Niobate
LMS	Least Mean Square
LOS	Loss of Signal
LPF	Low-Pass Filter
MA	Main Amplifier
MAC	Medium Access Control
MAN	Metropolitan-Area Network
MES	Metal-Semiconductor
MESFET	MES + FET
MLM	Multiple-Longitudinal Mode (laser)
MMF	MultiMode Fiber
MMIC	Monolithic Microwave IC
MODFET	Modulation-Doped Field-Effect Transistor
MOS	Metal-Oxide-Semiconductor
MOSFET	MOS + FET
MPN	Mode-Partition Noise
MQW	Multiple Quantum Well
MSM	Metal-Semiconductor-Metal (photodetector)
MSR	Mode-Suppression Ratio
MTTF	Mean-Time To Failure
MUX	Multiplexer
MZ	Mach-Zehnder
MZM	Mach-Zehnder Modulator
NA	Numerical Aperture
Nd	Neodymium
NECG	Net Electrical Coding Gain
NF	Noise Figure
NIC	Negative Impedance Converter
NRZ	Non-Return-to-Zero

NRZ1	Non-Return-to-Zero change-on-Ones
NWA	Network Analyzer
NZ-DSF	NonZero Dispersion-Shifted Fiber
O/E	Optical to Electrical converter
OA	Optical Amplifier
OADM	Optical Add-Drop Multiplexer
OC	Optical Carrier
OEIC	Optoelectronic Integrated Circuit
OFDM	Orthogonal Frequency Division Multiplexing
OLT	Optical Line Termination
OMA	Optical Modulation Amplitude
ONU	Optical Network Unit
OOK	On-Off Keying
OPA	Optical Parametric Amplifier
OSNR	Optical Signal-to-Noise Ratio
OTDM	Optical Time-Division Multiplexing
OXC	Optical cross Connect
P2MP	Point-to-Multipoint network
P2P	Point-to-Point connection
PAM	Pulse Amplitude Modulation
PAR	Peak-to-Average Ratio
PD	Photodetector or Photodiode
PHEMT	Pseudomorphic High Electron-Mobility Transistor
PHFET	Pseudomorphic Heterostructure Field-Effect Transistor
PIC	Photonic Integrated Circuit
PJ	Periodic Jitter
PLL	Phase-Locked Loop
PM	Phase Modulation
PMD	Polarization-Mode Dispersion
	Physical Medium Dependent (Ethernet layer)
PMF	Polarization-Maintaining Fiber
POF	Plastic Optical Fiber
PON	Passive Optical Network
POTS	Plain Old Telephone Service
PP	Power Penalty
	Peak to Peak
PRBS	PseudoRandom Bit Sequence
PSK	Phase-Shift Keying
PSP	Principal State of Polarization
PSRR	Power-Supply Rejection Ratio
PSTN	Public Switched Telephone Network
PWC	Pulse-Width Control
PWD	Pulse-Width Distortion
QAM	Quadrature Amplitude Modulation
QCSE	Quantum-Confined Stark Effect

RF	Radio Frequency
RFC	Radio-Frequency Choke
RIN	Relative Intensity Noise
RJ	Random Jitter
RMS	Root Mean Square
RN	Remote Node
ROSA	Receiver Optical Sub-Assembly
RS	Reed-Solomon code
RX	Receiver
RZ	Return-to-Zero
SBS	Stimulated Brillouin Scattering
SCFL	Source-Coupled FET Logic
SCM	SubCarrier Multiplexing
SDH	Synchronous Digital Hierarchy
SDM	Space Division Multiplexing
SDV	Switched Digital Video
SFF	Small Form-Factor module
SFP	Small Form-factor Pluggable module
Si	Silicon
SiGe	Silicon-Germanium
SiO_2	Silicon Oxide
SLM	Single-Longitudinal Mode (laser)
SMF	Single-Mode Fiber
SNR	Signal-to-Noise Ratio
SOA	Semiconductor Optical Amplifier
SONET	Synchronous Optical Network
SOP	State Of Polarization
SPICE	Simulation Program with Integrated Circuit Emphasis
SPM	Self-Phase Modulation
SRS	Stimulated Raman Scattering
STM	Synchronous Transport Module
STS	Synchronous Transport Signal
TAS	Transadmittance Stage
TCM	Time Compression Multiplexing
TCP	Transmission Control Protocol
TDD	Time Division Duplexing
TDM	Time Division Multiplexing
TDMA	Time Division Multiple Access
TEC	Thermoelectric Cooler
TEGFET	Two-dimensional Electron-Gas Field-Effect Transistor
THD	Total Harmonic Distortion
Ti	Titanium
TIA	Transimpedance Amplifier
TIS	Transimpedance Stage
TJ	Total Jitter

TOD	Turn-On Delay
TOSA	Transmitter Optical Sub-Assembly
TQFP	Thin Quad Flat Pack
TTL	Transistor-Transistor Logic
TV	Television
TX	Transmitter
TZA	Transimpedance Amplifier
UI	Unit Interval
VCSEL	Vertical-Cavity Surface-Emitting Laser
VGA	Variable-Gain Amplifier
VLSI	Very Large-Scale Integration
VSB	Vestigial Sideband
WAN	Wide-Area Network
WDM	Wavelength Division Multiplexing
XOR	Exclusive Or
XPM	Cross-Phase Modulation
YAG	Yttrium-Aluminum-Garnet ($Y_3Al_5O_{12}$)

References

1. Agere Systems. Using the lithium niobate modulator: electro-optical and mechanical connections. Agere Systems, Technical Note, April 1998.

2. Agere Systems. Low-cost, high-voltage APD bias circuit with temperature compensation. Agere Systems, Application Note, January 1999.

3. Agere Systems. Electroabsorptive modulated laser (EML): setup and optimization. Agere Systems, Technical Note, May 2000.

4. Agere Systems. Relationship between chirp and voltage in Agere Systems' Mach-Zehnder lithium niobate modulators. Agere Systems, Technical Note, March 2002.

5. Govind P. Agrawal. *Fiber-Optic Communication Systems*. John Wiley & Sons, New York, 2nd edition, 1997.

6. Stephen B. Alexander. *Optical Communication Receiver Design*. SPIE Press, copublished with IEE, Bellingham, Washington, 1997.

7. Phillip E. Allen and Douglas R. Holberg. *CMOS Analog Circuit Design*. Holt, Rinehart and Winston, New York, 1987.

8. Behnam Analui and Ali Hajimiri. Bandwidth enhancement for transimpedance amplifiers. *IEEE J. Solid-State Circuits*, SC-39(8):1263–1270, August 2004.

9. Kamran Azadet, Erich F. Haratsch, Helen Kim, Fadi Saibi, Jeffrey H. Saunders, Michael Shaffer, Leilei Song, and Meng-Lin Yu. Equalization and FEC techniques for optical transceivers. *IEEE J. Solid-State Circuits*, SC-37(3):317–327, March 2002.

10. Y. Baeyens, R. Pullela, J. P. Mattia, H.-S. Tsai, and Y.-K. Chen. A 74-GHz bandwidth InAlAs/InGaAs-InP HBT distributed amplifier with 13-dB gain. *IEEE Microwave Guided Wave Lett.*, 9(11):461–463, November 1999.

11. M. J. Bennett. Dispersion characteristics of monomode optical-fiber systems. *IEE Proceedings, Pt. H*, 130(5):309–314, August 1983.

12. Andrew J. Blanksby and Chris J. Howland. A 690-mW 1-Gb/s 1024-b, rate-1/2 low-density parity-check code decoder. *IEEE J. Solid-State Circuits*, SC-37(3):404–412, March 2002.

13. Henry A. Blauvelt. Predistorter for linearization of electronic and optical signals. U.S. Patent No 5,252,930, October 1993.

14. Henry A. Blauvelt, Israel Ury, David B. Huff, and Howard L. Loboda. Broadband optical receiver with passive tuning network. U.S. Patent No 5,179,461, January 1993.

15. David E. Bockelman and William R. Eisenstadt. Combined differential and common-mode scattering parameters: theory and simulation. *IEEE Trans. on Microwave Theory and Techniques*, MTT-43(7):1530–1539, July 1995.

16. Hendrik W. Bode. *Network Analysis and Feedback Amplifier Design*. D. Van Nostrand Company, New York, 1945.

17. Simona Brigati, Paolo Colombara, Lucio D'Ascoli, Umberto Gatti, Tibor Kerekes, and Piero Malcovati. A SiGe BiCMOS burst-mode 155-Mb/s receiver for PON. *IEEE J. Solid-State Circuits*, SC-37(7):887–894, July 2002.

18. Aaron Buchwald. Multi gigabit-per-second serial data links, March 2001. Lecture Notes, MEAD Microelectronics.

19. Aaron Buchwald and Ken Martin. *Integrated Fiber-Optic Receivers*. Kluwer Academic Publishers, Boston, 1995.

20. H. Bülow, F. Buchali, W. Baumert, R. Ballentin, and T. Wehren. PMD mitigation at 10Gb/s using linear and nonlinear integrated electronic equaliser circuits. *Electronics Letters*, 36(2):163–164, January 2000.

21. Klaas Bult and Govert Geelen. A fast-settling CMOS op amp with 90 dB dc-gain and 116 MHz unity-gain frequency. In *ISSCC Dig. Tech. Papers*, pages 108–109, February 1990.

22. E. M. Cherry and D. E. Hooper. The design of wide-band transistor feedback amplifiers. *Proceedings IEE*, 110(2):375–389, February 1963.

23. Walter Ciciora, James Farmer, and David Large. *Modern Cable Television Technology: Video, Voice, and Data Communications.* Morgan Kaufmann, San Francisco, 1999.

24. Alexandru A. Ciubotaru and Javier Sánchez García. An integrated direct-coupled 10-Gb/s driver for common-cathode VCSELs. *IEEE J. Solid-State Circuits,* SC-39(3):426–433, March 2004.

25. Donald Estreich. Basic building blocks. In Ravender Goyal, editor, *High-Frequency Analog Integrated Circuit Design,* pages 127–169. John Wiley & Sons, Inc., New York, 1995.

26. Donald Estreich. Wideband amplifiers. In Ravender Goyal, editor, *High-Frequency Analog Integrated Circuit Design,* pages 170–240. John Wiley & Sons, Inc., New York, 1995.

27. Donald B. Estreich. A monolithic wide-band GaAs IC amplifier. *IEEE J. Solid-State Circuits,* SC-17(6):1166–1173, December 1982.

28. Dennis L. Feucht. *Handbook of Analog Circuit Design.* Academic Press, San Diego, 1990.

29. Daniel A. Fishman and B. Scott Jackson. Transmitter and receiver design for amplified lightwave systems. In Ivan P. Kaminow and Thomas L. Koch, editors, *Optical Fiber Telecommunications IIIB,* pages 69–114. Academic Press, San Diego, 1997.

30. FSAN. Full service access network. http://www.fsanweb.org.

31. Sherif Galal and Behzad Razavi. 10-Gb/s limiting amplifier and laser/modulator driver in 0.18-μm CMOS technology. *IEEE J. Solid-State Circuits,* SC-38(12):2138–2146, December 2003.

32. Sherif Galal and Behzad Razavi. Broadband ESD protection circuits in CMOS technology. *IEEE J. Solid-State Circuits,* SC-38(12):2334–2340, December 2003.

33. Richard D. Gitlin, Jeremiah F. Hayes, and Stephen B. Weinstein. *Data Communications Principles.* Plenum Press, New York, 1992.

34. Paul R. Gray and Robert G. Meyer. *Analysis and Design of Analog Integrated Circuits.* John Wiley & Sons, New York, 1977.

35. Alan B. Grebene. *Bipolar and MOS Analog Integrated Circuit Design.* John Wiley & Sons, New York, 1984.

36. Michael Green. Broadband data signals and circuits, April 2002. Lecture Notes, MEAD Microelectronics.

37. Yuriy M. Greshishchev. Front-end circuits for optical communications, February 2001. ISSCC'2001 Tutorial.

38. Yuriy M. Greshishchev and Peter Schvan. A 60-dB gain, 55-dB dynamic range, 10-Gb/s broad-band SiGe HBT limiting amplifier. *IEEE J. Solid-State Circuits*, SC-34(12):1914–1920, December 1999.

39. Yuriy M. Greshishchev, Peter Schvan, Jonathan L. Showell, Mu-Liang Xu, Jugnu J. Ojha, and Jonathan E. Rogers. A fully integrated SiGe receiver IC for 10Gb/s data rate. In *ISSCC Dig. Tech. Papers*, pages 52–53, February 2000.

40. Edward Harstead and Pieter H. van Heyningen. Optical access networks. In Ivan P. Kaminow and Tingye Li, editors, *Optical Fiber Telecommunications IVB*, pages 438–513. Academic Press, San Diego, 2002.

41. D. Hassin and R. Vahldieck. Feedforward linearization of analog modulated laser diodes: theoretical analysis and experimental verification. *IEEE Trans. on Microwave Theory and Techniques*, MTT-41(12):2376–2382, December 1993.

42. Simon Haykin. *Communication Systems*. John Wiley & Sons, New York, 4th edition, 2001.

43. Jeff Hecht. *City of Light: The Story of Fiber Optics*. Oxford University Press, New York, 1999.

44. Fred Heismann, Steven K. Korotky, and John J. Veselka. Lithium niobate integrated optics: selected contemporary devices and system applications. In Ivan P. Kaminow and Thomas L. Koch, editors, *Optical Fiber Telecommunications IIIB*, pages 377–462. Academic Press, San Diego, 1997.

45. Lindor Henrickson, David Shen, Uno Nellore, Alan Ellis, Joong Oh, Hui Wang, Giovanni Capriglione, Ali Atesoglu, Alice Yang, Peter Wu, Syed Quadri, and David Crosbie. Low-power fully integrated 10-Gb/s SONET/SDH transceiver in 0.13-μm CMOS. *IEEE J. Solid-State Circuits*, SC-38(10):1595–1601, October 2003.

46. Paul S. Henry, R. A. Linke, and A. H. Gnauck. Introduction to lightwave systems. In Stewart E. Miller and Ivan P. Kaminow, editors, *Optical Fiber Telecommunications II*, pages 781–831. Academic Press, San Diego, 1988.

47. Timothy H. Hu and Paul R. Gray. A monolithic 480Mb/s parallel AGC/decision/clock-recovery circuit in 1.2-μm CMOS. *IEEE J. Solid-State Circuits*, SC-28(12):1314–1320, December 1993.

48. IEEE. Ethernet in the first mile, task force IEEE 802.3ah, 2001. http://www.ieee802.org/3/efm/.

49. Mark Ingels, Geert Van der Plas, Jan Crols, and Michel Steyaert. A CMOS 18THzΩ 240Mb/s transimpedance amplifier and 155Mb/s LED-driver for low

cost optical fiber links. *IEEE J. Solid-State Circuits*, SC-29(12):1552–1559, December 1994.

50. Noboru Ishihara, Makoto Nakamura, Yukio Akazawe, Naoto Uchida, and Yhuji Akahori. 3.3V, 50Mb/s CMOS transceiver for optical burst-mode communication. In *ISSCC Dig. Tech. Papers*, pages 244–245, 1997.

51. ITU-T. Digital line systems based on the synchronous digital hierarchy for use on optical fibre cables, recommendation G.958. International Telecommunication Union, Geneva, Switzerland, November 1994.

52. ITU-T. Broadband optical access systems based on passive optical networks (PON), recommendation G.983.1. International Telecommunication Union, Geneva, Switzerland, October 1998.

53. ITU-T. Forward error correction for submarine systems, recommendation G.975. International Telecommunication Union, Geneva, Switzerland, October 2000.

54. Takanori Iwai, Kenji Sato, and Ko-ichi Suto. Signal distortion and noise in AM-SCM transmission systems employing the feedforward linearized MQW-EA external modulator. *IEEE J. Lightwave Technology*, LT-13(8):1606–1612, August 1995.

55. Renuka P. Jindal. Gigahertz-band high-gain low-noise AGC amplifiers in fine-line NMOS. *IEEE J. Solid-State Circuits*, SC-22(4):512–521, August 1987.

56. Renuka P. Jindal. Silicon MOS amplifier operation in the integrate and dump mode for gigahertz band lightwave communication systems. *IEEE J. Lightwave Technology*, LT-8(7):1023–1026, July 1990.

57. David Johns and Ken Martin. *Analog Integrated Circuit Design*. John Wiley & Sons, New York, 1997.

58. Ivan P. Kaminow and Thomas L. Koch. *Optical Fiber Telecommunications IIIA*. Academic Press, San Diego, 1997.

59. Ivan P. Kaminow and Thomas L. Koch. *Optical Fiber Telecommunications IIIB*. Academic Press, San Diego, 1997.

60. Ivan P. Kaminow and Tingye Li. *Optical Fiber Telecommunications IVA: Components*. Academic Press, San Diego, 2002.

61. Ivan P. Kaminow and Tingye Li. *Optical Fiber Telecommunications IVB: Systems and Impairments*. Academic Press, San Diego, 2002.

62. Bryon L. Kasper. Receiver design. In Stewart E. Miller and Ivan P. Kaminow, editors, *Optical Fiber Telecommunications II*, pages 689–722. Academic Press, San Diego, 1988.

63. Bryon L. Kasper, Alfred R. McCormick, Charles A. Burrus Jr., and J. R. Talman. An optical-feedback transimpedance receiver for high sensitivity and wide dynamic range at low bit rates. *IEEE J. Lightwave Technology*, LT-6(2):329–338, February 1988.

64. Bryon L. Kasper, Osamu Mizuhara, and Young-Kai Chen. High bit-rate receivers, transmitters, and electronics. In Ivan P. Kaminow and Tingye Li, editors, *Optical Fiber Telecommunications IVA*, pages 784–851. Academic Press, San Diego, 2002.

65. Sanjay Kasturia and Jack H. Winters. Techniques for high-speed implementation of nonlinear cancellation. *IEEE J. Select. Areas Commun.*, SAC-9(5):711–717, June 1991.

66. Nicolas Kauffmann, Sylvain Blayac, Miloud Abboun, Philippe André, Frédéric Aniel, Muriel Riet, Jean-Louis Benchimol, Jean Godin, and Agnieszka Konczykowska. InP HBT driver circuit optimization for high-speed ETDM transmission. *IEEE J. Solid-State Circuits*, SC-36(4):639–647, April 2001.

67. M. Kawi, H. Watanabe, T. Ohtsuka, and K. Yamaguchi. Smart optical receiver with automatic decision threshold setting and retiming phase alignment. *IEEE J. Lightwave Technology*, LT-7(11):1634–1640, November 1989.

68. Haideh Khorramabadi, Liang D. Tzeng, and Maurice J. Tarsia. A 1.06Gb/s, −31dBm to 0dBm BiCMOS optical preamplifier featuring adaptive transimpedance. In *ISSCC Dig. Tech. Papers*, pages 54–55, February 1995.

69. Ulrich Killat, editor. *Access to B-ISDN via PONs: ATM Communication in Practice*. John Wiley and B. G. Teubner, Chichester, England, 1996.

70. Helen Kim and Jonathan Bauman. A 12GHz, 30dB modular BiCMOS limiting amplifier for 10Gb/s SONET receiver. In *ISSCC Dig. Tech. Papers*, pages 160–161, February 2000.

71. Helen H. Kim, S. Chandrasekhar, Charles A. Burrus, and Jon Bauman. A Si BiCMOS transimpedance amplifier for 10Gb/s SONET receiver. *IEEE J. Solid-State Circuits*, SC-36(5):769–776, May 2001.

72. Bendik Kleveland, Carlos H. Diaz, Dieter Vook, Liam Madden, Thomas H. Lee, and S. Simon Wong. Exploiting CMOS reverse interconnect scaling in multigigahertz amplifier and oscillator design. *IEEE J. Solid-State Circuits*, SC-36(10):1480–1488, October 2001.

73. Thomas L. Koch. Laser sources for amplified and WDM lightwave systems. In Ivan P. Kaminow and Thomas L. Koch, editors, *Optical Fiber Telecommunications IIIB*, pages 115–162. Academic Press, San Diego, 1997.

74. Herwig Kogelnik, Robert M. Jopson, and Lynn E. Nelson. Polarization-mode dispersion. In Ivan P. Kaminow and Tingye Li, editors, *Optical Fiber Telecommunications IVB*, pages 725–861. Academic Press, San Diego, 2002.

75. John D. Kraus. *Antennas*. McGraw Hill, New York, 2nd edition, 1988.

76. P. I. Kuindersma, M. W. Snikkers, G. P. J. M. Cuypers, J. J. M. Binsma, E. Jansen, A. van Geelen, and T. van Dongen. Universality of the chirp-parameter of bulk active electro absorption modulators. European Conference on Optical Communication (ECOC), Madrid, Spain, 1998.

77. Manfred Lang, Zhi-Gong Wang, Zhihao Lao, Michael Schlechtweg, Andreas Thiede, Michaela Rieger-Motzer, Martin Sedler, Wolfgang Bronner, Gudrun Kaufel, Klaus Köhler, Axel Hülsmann, and Brian Raynor. 20-40Gb/s, 0.2-μm GaAs HEMT chip set for optical data receiver. *IEEE J. Solid-State Circuits*, SC-32(9):1384–1393, September 1997.

78. Zhihao Lao, Manfred Berroth, Volker Hurm, Andreas Thiede, Roland Bosch, Peter Hofman, Alex Hülsmann, Canute Moglestue, and Klaus Köhler. 25Gb/s AGC amplifier, 22GHz transimpedance amplifier and 27.7GHz limiting amplifier ICs using AlGaAs/GaAs-HEMTs. In *ISSCC Dig. Tech. Papers*, pages 356–357, February 1997.

79. Zhihao Lao, Andreas Thiede, Ulrich Nowotny, Hariolf Lienhart, Volker Hurm, Michael Schlechtweg, Jochen Hornung, Wolfgang Bronner, Klaus Köhler, Alex Hülsmann, Brian Raynor, and Theo Jakobus. 40-Gb/s high-power modulator driver IC for lightwave communication systems. *IEEE J. Solid-State Circuits*, SC-33(10):1520–1526, October 1998.

80. Lawrence E. Larson, Chia-Shing Chou, and Michael J. Delaney. An ultrahigh-speed GaAs MESFET operational amplifier. *IEEE J. Solid-State Circuits*, SC-24(6):1523–1528, December 1989.

81. Edward A. Lee and David G. Messerschmitt. *Digital Communication*. Kluwer Academic Publishers, Boston, 2nd edition, 1994.

82. Thomas H. Lee. *The Design of CMOS Radio-Frequency Integrated Circuits*. Cambridge University Press, Cambridge, U.K., 1998.

83. Max Ming-Kang Liu. *Principles and Applications of Optical Communications*. Irwin, McGraw-Hill, Chicago, 1996.

84. Y. H. Lo, P. Grabbe, M. Z. Iqbal, R. Bhat, J. L. Gimlett, J. C. Young, P. S. D. Lin, A. S. Gozdz, M. A. Koza, and T. P. Lee. Multigigabit/s 1.5 μm λ/4-shifted DFB OEIC transmitter and its use in transmission experiments. *IEEE Photonics Technology Letters*, 2(9):673–674, September 1990.

85. John R. Long and Miles A. Copeland. The modeling, characterization, and design of monolithic inductors for silicon RF IC's. *IEEE J. Solid-State Circuits*, SC-32(3):357–369, March 1997.

86. Toru Masuda, Ken-ichi Ohhata, Fumihiko Arakawa, Nobuhiro Shiramizu, Eiji Ohue, Katsuya Oda, Reiko Hayami, Masamitchi Tanabe, Hiromi Shimamoto, Masao Kondo, Takashi Harada, and Katsuyoshi Washio. 45GHz

transimpedance, 32dB limiting amplifier, and 40Gb/s 1:4 high-sensitivity demultiplexer with decision circuit using SiGe HBTs for 40Gb/s optical receiver. In *ISSCC Dig. Tech. Papers*, pages 60–61, February 2000.

87. Toru Masuda, Ken-ichi Ohhata, Eiji Ohue, Katsuya Oda, Masamitchi Tanabe, Hiromi Shimamoto, T. Onai, and Katsuyoshi Washio. 40Gb/s analog IC chipset for optical receiver using SiGe HBTs. In *ISSCC Dig. Tech. Papers*, pages 314–315, February 1998.

88. Maxim Integrated Products. Maintaining the extinction ratio of optical transmitters using k-factor control. Maxim Application Note HFAN-2.2.1, June 2002.

89. Mounir Meghelli, Michel Bouché, and Agnieszka Konczykowska. High power and high speed InP DHBT driver IC's for laser modulation. *IEEE J. Solid-State Circuits*, SC-33(9):1411–1416, September 1998.

90. Pablo V. Mena, Sung-Mo Kang, and Thomas A. DeTemple. Rate-equation-based laser models with a single solution regime. *IEEE J. Lightwave Technology*, LT-15(4):717–730, April 1997.

91. Robert G. Meyer and William D. Mack. A wideband low-noise variable-gain BiCMOS transimpedance amplifier. *IEEE J. Solid-State Circuits*, SC-29(6):701–706, June 1994.

92. Robert G. Meyer and William D. Mack. Monolithic AGC loop for a 160Mb/s transimpedance amplifier. *IEEE J. Solid-State Circuits*, SC-31(9):1331–1335, September 1996.

93. Stewart E. Miller and Ivan P. Kaminow. *Optical Fiber Telecommunications II*. Academic Press, San Diego, 1988.

94. Miyo Miyashita, Naohito Yoshida, Yoshiki Kojima, Toshiaki Kitano, Norio Higashisaka, Junichi Nakagawa, Tadashi Takagi, and Mutsuyuki Otsubo. An AlGaAs/InGaAs pseudomorphic HEMT modulator driver IC with low power dissipation for 10-Gb/s optical transmission systems. *IEEE Trans. on Microwave Theory and Techniques*, MTT-45(7):1058–1064, July 1997.

95. Sunderarajan S. Mohan, Maria del Mar Hershenson, Stephen P. Boyd, and Thomas H. Lee. Bandwidth extension in CMOS with optimized on-chip inductors. *IEEE J. Solid-State Circuits*, SC-35(3):346–355, March 2000.

96. Mehran Mokhtari, Thomas Swahn, Robert H. Walden, William E. Stanchina, Michael Kardos, Tarja Juhola, Gerd Schuppener, Hannu Tenhunen, and Thomas Lewin. InP-HBT chip-set for 40-Gb/s fiber optical communication systems operational at 3V. *IEEE J. Solid-State Circuits*, SC-32(9):1371–1383, September 1997.

97. M. Möller, T. F. Meister, R. Schmid, J. Rupeter, M. Rest, A. Schöpflin, and H.-M. Rein. SiGe retiming high-gain power MUX for direct driving an EAM up to 50Gb/s. *Electronics Letters*, 34(18):1782–1784, September 1998.

98. M. Möller, H.-M. Rein, and H. Wernz. 13Gb/s Si-bipolar AGC amplifier IC with high gain and wide dynamic range for optical-fiber receivers. *IEEE J. Solid-State Circuits*, SC-29(7):815–822, July 1994.

99. J. J. Morikuni, A. Dharchoudhury, Y. Leblebici, and S. M. Kang. Improvements to the standard theory for photoreceiver noise. *IEEE J. Lightwave Technology*, LT-12(7):1174–1184, July 1994.

100. Th. Mosch and P. Solina. Burst mode communication. In Ulrich Killat, editor, *Access to B-ISDN via PONs: ATM Communication in Practice*, pages 157–175. John Wiley and B. G. Teubner, Chichester, England, 1996.

101. Richard S. Muller and Theodore I. Kamins. *Device Electronics for Integrated Circuits*. John Wiley & Sons, New York, 1977.

102. J. Müllrich, T. F. Meister, M. Rest, W. Bogner, A. Schöpflin, and H.-M. Rein. 40Gb/s transimpedance amplifier in SiGe bipolar technology for receiver in optical-fibre TDM links. *Electronics Letters*, 34(5):452–453, March 1998.

103. Jens Müllrich, Herbert Thurner, Ernst Müllner, Joseph F. Jensen, William E. Stanchina, M. Kardos, and Hans-Martin Rein. High-gain transimpedance amplifier in InP-based HBT technology for receiver in 40-Gb/s optical-fiber TDM links. *IEEE J. Solid-State Circuits*, SC-35(9):1260–1265, September 2000.

104. Makoto Nakamura, Noboru Ishihara, and Yukio Akazawa. A 156-Mb/s CMOS optical receiver for burst-mode transmission. *IEEE J. Solid-State Circuits*, SC-33(8):1179–1187, August 1998.

105. Makoto Nakamura, Noboru Ishihara, Yukio Akazawa, and Hideaki Kimura. An instantaneous response CMOS optical receiver IC with wide dynamic range and extremely high sensitivity using feed-forward auto-bias adjustment. *IEEE J. Solid-State Circuits*, SC-30(9):991–997, September 1995.

106. Moshe Nazarathy, Josef Berger, Anthony J. Ley, Israel M. Levi, and Yishai Kagan. Progress in externally modulated AM CATV transmission systems. *IEEE J. Lightwave Technology*, LT-11(1):82–105, January 1993.

107. NCITS. Fibre channel: methodologies for jitter specification 2, T11.2 / Project 1316-DT / Rev 0.0. National Committee for Information Technology Standardization, April 2000. http://www.t11.org.

108. Michael Neuhäuser, Hans-Martin Rein, and Horst Wernz. Low-noise, high-gain Si-bipolar preamplifiers for 10Gb/s optical-fiber links – design and realization. *IEEE J. Solid-State Circuits*, SC-31(1):24–29, January 1996.

109. Kwok K. Ng. *Complete Guide to Semiconductor Devices*. John Wiley & Sons, New York, 2002.

110. Kwok K. Ng, Michel R. Frei, and Clifford A. King. Reevaluation of the $f_t BV_{ceo}$ limit of Si bipolar transistors. *IEEE Trans. Electron Devices*, ED-45:1854–1855, August 1998.

111. Ernst H. Nordholt. *The Design of High-Performance Negative-Feedback Amplifiers*. Elsevier, Amsterdam, The Netherlands, 1983.

112. Yong-Hun Oh, Sang-Gug Lee, and H. H. Park. A 2.5Gb/s CMOS transimpedance amplifier using novel active inductor load. In *Digest of European Solid-State Circuits Conference*, Villach, Austria, September 2001.

113. Kenichi Ohhata, Toru Masuda, Kazuo Imai, Ryoji Takeyari, and Katsuyoshi Washio. A wide-dynamic-range, high-transimpedance Si bipolar preamplifier IC for 10-Gb/s optical fiber links. *IEEE J. Solid-State Circuits*, SC-34(1):18–24, January 1999.

114. Kenichi Ohhata, Toru Masuda, Eiji Ohue, and Katsuyoshi Washio. Design of a 32.7-GHz bandwidth AGC amplifier IC with wide dynamic range implemented in SiGe HBT. *IEEE J. Solid-State Circuits*, SC-34(9):1290–1297, September 1999.

115. Christian Ølgaard. A laser control chip combining power regulator and a 622-MBit/s modulator. *IEEE J. Solid-State Circuits*, SC-29(8):947–951, August 1994.

116. N. A. Olsson. Lightwave systems with optical amplifiers. *IEEE J. Lightwave Technology*, LT-7(7):1071–1082, July 1989.

117. Yusuke Ota. High speed non-biased semiconductor laser diode driver for high speed digital communication. U.S. Patent No 6,018,538, January 2000.

118. Yusuke Ota and Robert G. Swartz. Burst-mode compatible optical receiver with a large dynamic range. *IEEE J. Lightwave Technology*, LT-8(12):1897–1903, December 1990.

119. Yusuke Ota, Robert G. Swartz, Vance D. Archer III, Steven K. Korotky, Mihai Banu, and Alfred E. Dunlop. High-speed, burst-mode, packet-capable optical receiver and instantaneous clock recovery for optical bus operation. *IEEE J. Lightwave Technology*, LT-12(2):325–331, February 1994.

120. Yusuke Ota, Robert G. Swartz, John S. Schafer, Mihai M. Banu, Alfred E. Dunlop, Wilhelm C. Fischer, and Thaddeus. J. Gabara. Low cost, low power digital optical receiver module for 50Mb/s passive optical network. *Int. J. High Speed Electronics Systems*, 7(4):471–489, 1996.

121. Taiichi Otsuji, Koichi Murata, Koichi Narahara, Kimikazu Sano, Eiichi Sano, and Kimiyoshi Yamasaki. 20-40-Gbit/s-class GaAs MESFET digital ICs for

future optical fiber communications systems. In Keh-Chung Wang, editor, *High-Speed Circuits for Lightwave Communications*, pages 87–123. World Scientific, Singapore, 1999.

122. Taiichi Otsuji, Mikio Yoneyama, Koichi Murata, and Eiichi Sano. A super-dynamic flip-flop circuit for broad-band applications up to 24Gb/s utilizing production-level 0.2-μm GaAs MESFET's. *IEEE J. Solid-State Circuits*, SC-32(9):1357–1362, September 1997.

123. Patrick K. D. Pai and Asad A. Abidi. A 40-mW 55Mb/s CMOS equalizer for use in magnetic storage read channels. *IEEE J. Solid-State Circuits*, SC-29(4):489–499, April 1994.

124. Sung-Min Park, Jaeseo Lee, and Hoi-Jun Yoo. 1-Gb/s 80-dBΩ fully differential CMOS transimpedance amplifier in multichip on oxide technology for optical interconnects. *IEEE J. Solid-State Circuits*, SC-39(6):971–974, June 2004.

125. Sung-Min Park and Hoi-Jun Yoo. 1.25-Gb/s regulated cascode CMOS transimpedance amplifier for Gigabit Ethernet applications. *IEEE J. Solid-State Circuits*, SC-39(1):112–121, January 2004.

126. Kenneth Pedrotti. High speed circuits for lightwave communications. In Keh-Chung Wang, editor, *High-Speed Circuits for Lightwave Communications*, pages 1–34. World Scientific, Singapore, 1999.

127. S. D. Personick. Receiver design for digital fiber optic communication systems. *Bell Syst. Tech. J.*, 52(6):843–886, July-August 1973.

128. Anders K. Petersen, Kürşad Kiziloğlu, Ty Yoon, Freddie Williams Jr., and Martin R. Sandor. Front-end CMOS chipset for 10Gb/s communication. In *IEEE Radio Frequency Integrated Circuits Symposium*, pages 93–96, Seattle, 2002.

129. Khoman Phang and David A. Johns. A CMOS optical preamplifier for wireless infrared communications. *IEEE Trans. Circuits Syst. – II*, CASII-46(7):852–859, July 1999.

130. Khoman Phang and David A. Johns. A 1V 1mW CMOS front-end with on-chip dynamic gate biasing for 75Mb/s optical receiver. In *ISSCC Dig. Tech. Papers*, pages 218–219, February 2001.

131. Philips Semiconductors. A hybrid wideband amplifier module for digital CATV networks with the BGD902. Philips Semiconductors, Application Note AN98109, February 1999.

132. Mary R. Phillips and Thomas E. Darcie. Lightwave analog video transmission. In Ivan P. Kaminow and Thomas L. Koch, editors, *Optical Fiber Telecommunications IIIA*, pages 523–559. Academic Press, San Diego, 1997.

133. Wolfgang Pöhlmann. A silicon-bipolar amplifier for 10 Gbit/s with 45 dB gain. *IEEE J. Solid-State Circuits*, SC-29(5):551–556, May 1994.

134. David M. Pozar. *Microwave Engineering*. Addison-Wesley Publishing, Reading, MA, 1990.

135. A. J. Price and K. D. Pedrotti. Optical transmitters. In Jerry D. Gibson, editor, *The Communications Handbook*, pages 774–788, Boca Raton, 1997. CRC Press.

136. Rajiv Ramaswami and Kumar N. Sivarajan. *Optical Networks: A Practical Perspective*. Morgan Kaufmann Publishers, San Francisco, 1998.

137. Hans Ransijn. Receiver and transmitter IC design, May 2001. CICC'2001 Ed. Session 3-2.

138. Hans Ransijn, Gregory Salvador, Dwight D. Daugherty, and Kenneth D. Gaynor. A 10-Gb/s laser/modulator driver IC with dual-mode actively matched output buffer. *IEEE J. Solid-State Circuits*, SC-36(9):1314–1320, September 2001.

139. Behzad Razavi. A 1.5V 900MHz downconversion mixer. In *ISSCC Dig. Tech. Papers*, pages 48–49, February 1996.

140. Behzad Razavi. A 622Mb/s, 4.5pA/$\sqrt{\text{Hz}}$ CMOS transimpedance amplifier. In *ISSCC Dig. Tech. Papers*, pages 162–163, February 2000.

141. Behzad Razavi. *Design of Integrated Circuits for Optical Communications*. McGraw-Hill, New York, 2003.

142. Behzad Razavi, Ran-Hong Yan, and Kwing F. Lee. Impact of distributed gate resistance on the performance of MOS devices. *IEEE Trans. Circuits Syst. – I*, CASI-41(11):750–754, November 1994.

143. Reinhard Reimann and Hans-Martin Rein. Bipolar high-gain limiting amplifier IC for optical-fiber receivers operating up to 4Gbit/s. *IEEE J. Solid-State Circuits*, SC-22(4):504–511, August 1987.

144. Reinhard Reimann and Hans-Martin Rein. A single-chip bipolar AGC amplifier with large dynamic range for optical-fiber receivers operating up to 3Gbit/s. *IEEE J. Solid-State Circuits*, SC-24(6):1744–1748, December 1989.

145. H.-M. Rein. Si and SiGe bipolar ICs for 10 to 40Gb/s optical-fiber TDM links. In Keh-Chung Wang, editor, *High-Speed Circuits for Lightwave Communications*, pages 35–71. World Scientific, Singapore, 1999.

146. H.-M. Rein and M. Möller. Design considerations for very-high-speed Si-bipolar IC's operating up to 50Gb/s. *IEEE J. Solid-State Circuits*, SC-31(8):1076–1090, August 1996.

147. H.-M. Rein, R. Schmid, P. Wenger, T. Smith, T. Herzog, and R. Lachner. A versatile Si-bipolar driver circuit with high output voltage swing for external and direct laser modulation in 10Gb/s optical-fiber links. *IEEE J. Solid-State Circuits*, SC-29(9):1014–1021, September 1994.

148. Hans-Martin Rein. Multi-gigabit-per-second silicon bipolar IC's for future optical-fiber transmission systems. *IEEE J. Solid-State Circuits*, SC-23(3):664–675, June 1988.

149. Hans-Martin Rein. Design of high-speed Si/SiGe bipolar ICs for optical-fiber systems with data rates up to 40Gb/s, March 2001. Lecture Notes, MEAD Microelectronics.

150. Mario Reinhold, Claus Dorschky, Eduard Rose, Rajasekhar Pullela, Peter Mayer, Frank Kunz, Yves Baeyens, Thomas Link, and John-Paul Mattia. A fully integrated 40-Gb/s clock and data recovery IC with 1:4 DEMUX in SiGe technology. *IEEE J. Solid-State Circuits*, SC-36(12):1937–1945, December 2001.

151. Matthias Rickelt and Hans-Martin Rein. A novel transistor model for simulating avalanche-breakdown effects in Si bipolar circuits. *IEEE J. Solid-State Circuits*, SC-37(9):1184–1197, September 2002.

152. Cathleen Rooman, Daniël Coppée, and Maarten Kuijk. Asynchronous 250-Mb/s optical receivers with integrated detector in standard CMOS technology for optocoupler applications. *IEEE J. Solid-State Circuits*, SC-35(7):953–958, July 2000.

153. K. Runge, P. J. Zampardi, R. L. Pierson, R. Yu, P. B. Thomas, S. M. Beccue, and K. C. Wang. AlGaAs/GaAs HBT circuits for optical TDM communications. In Keh-Chung Wang, editor, *High-Speed Circuits for Lightwave Communications*, pages 161–191. World Scientific, Singapore, 1999.

154. Klaus Runge, Detlef Daniel, R. D. Standley, James L. Gimlett, Randall B. Nubling, Richard L. Pierson, Steve M. Beccue, Keh-Chung Wang, Neng-Haung Sheng, Mau-Chung F. Chang, Dong Ming Chen, and Peter M. Asbeck. AlGaAs/GaAs HBT IC's for high-speed lightwave transmission systems. *IEEE J. Solid-State Circuits*, SC-27(10):1332–1341, October 1992.

155. Eduard Säckinger. Theory and monolithic CMOS integration of a differential difference amplifier. In W. Fichtner, W. Guggenbühl, H. Melchior, and G. S. Moschytz, editors, *Series in Microelectronics*. Hartung-Gorre Verlag, Konstanz, Germany, 1989.

156. Eduard Säckinger and Wilhelm C. Fischer. A 3-GHz, 32-dB CMOS limiting amplifier for SONET OC-48 receivers. *IEEE J. Solid-State Circuits*, SC-35(12):1884–1888, December 2000.

157. Eduard Säckinger, Josef Goette, and Walter Guggenbühl. A general relationship between amplifier parameters, and its application to PSRR improvement. *IEEE Trans. Circuits Syst.*, CAS-38(10):1173–1181, October 1991.

158. Eduard Säckinger and Walter Guggenbühl. A versatile building block: the CMOS differential difference amplifier. *IEEE J. Solid-State Circuits*, SC-22(2):287–294, April 1987.

159. Eduard Säckinger and Walter Guggenbühl. A high-swing, high-impedance MOS cascode circuit. *IEEE J. Solid-State Circuits*, SC-25(1):289–298, February 1990.

160. Eduard Säckinger and Yusuke Ota. Burst-mode laser techniques. U.S. Patent No 6,229,830, May 2001.

161. Eduard Säckinger and Yusuke Ota. Burst-mode laser techniques. U.S. Patent No 6,219,165, April 2001.

162. Eduard Säckinger, Yusuke Ota, Thaddeus J. Gabara, and Wilhelm C. Fischer. A 15-mW, 155-Mb/s CMOS burst-mode laser driver with automatic power control and end-of-life detection. *IEEE J. Solid-State Circuits*, SC-35(2):269–275, February 2000.

163. E. Sano, K. Sano, T. Otsuij, K. Kurishima, and S. Yamahata. Ultra-high speed, low power monolithic photoreceiver using InP/InGaAs double heterojunction bipolar transistors. *Electronics Letters*, 33(12):1047–1048, June 1997.

164. Jafar Savoj and Behzad Razavi. A CMOS interface circuit for detection of 1.2Gb/s RZ data. In *ISSCC Dig. Tech. Papers*, pages 278–279, February 1999.

165. Norman Scheinberg, Robert J. Bayruns, and Timothy M. Laverick. Monolithic GaAs transimpedance amplifiers for fiber-optic receivers. *IEEE J. Solid-State Circuits*, SC-26(12):1834–1839, December 1991.

166. R. Schmid, T. F. Meister, M. Rest, and H.-M. Rein. 40Gb/s EAM driver IC in SiGe bipolar technology. *Electronics Letters*, 34(11):1095–1097, May 1998.

167. R. Schmid, T. F. Meister, M. Rest, and H.-M. Rein. SiGe driver circuit with high output amplitude operating up to 23Gb/s. *IEEE J. Solid-State Circuits*, SC-34(6):886–891, June 1999.

168. John M. Senior. *Optical Fiber Communications: Principles and Practice*. Prentice Hall, Hertfordshire, England, 1985.

169. T. M. Shen and Govind P. Agrawal. Pulse-shape effects on frequency chirping in single-frequency semiconductor lasers under current modulation. *IEEE J. Lightwave Technology*, LT-4(5):497–503, May 1986.

170. M. Sherif and P. A. Davies. Decision-point steering in optical fibre communication systems: theory. *IEE Proceedings, Pt. J*, 136(3):169–176, June 1989.

171. Jiann-Chyi Shieh, Jun Cao, and Cheng-Chung Shih. CMOS 125MHz fiber/TP media converter with auto offset cancellation post amplifier and pre-emphasis LED driver. In *ISSCC Dig. Tech. Papers*, pages 312–313, February 2000.

172. Hisao Shigematsu, Masaru Sato, Tatsuya Hirose, and Yuu Watanabe. A 54-GHz distributed amplifier with 6-V_{pp} output for a 40-Gb/s LiNbO$_3$ modulator driver. *IEEE J. Solid-State Circuits*, SC-37(9):1100–1105, September 2002.

173. Hisao Shigematsu, Masaru Sato, Toshihide Suzuki, Tsuyoshi Takahashi, Kenji Imanishi, Naoki Hara, Hiroaki Ohnishi, and Yuu Watanabe. A 49-GHz preamplifier with a transimpedance gain of 52dBΩ using InP HEMTs. *IEEE J. Solid-State Circuits*, SC-36(9):1309–1313, September 2001.

174. P. W. Shumate. Lightwave transmitters. In Stewart E. Miller and Ivan P. Kaminow, editors, *Optical Fiber Telecommunications II*, pages 723–757. Academic Press, San Diego, 1988.

175. Stefanos Sidiropoulos and Mark Horowitz. A 700-Mb/s/pin CMOS signaling interface using current integrating receivers. *IEEE J. Solid-State Circuits*, SC-32(5):681–690, May 1997.

176. M. K. Simon. Nonlinear analysis of an absolute value type of an early-late gate bit synchronizer. *IEEE Trans. Communication Technology*, COM-18(5):589–596, October 1970.

177. R. G. Smith and S. D. Personick. Receiver design for optical fiber communication systems. In H. Kressel, editor, *Topics in Applied Physics Vol. 39: Semiconductor Devices for Optical Communication*. Springer Verlag, Berlin, Germany, 1982.

178. Masaaki Soda, Hiroshi Tezuka, Fumihiko Sato, Takasuke Hashimoto, Satoshi Nakamura, Toru Tatsumi, Tetsuyuki Suzaki, and Tsutomu Tashiro. Si-analog IC's for 20Gb/s optical receiver. *IEEE J. Solid-State Circuits*, SC-29(12):1577–1582, December 1994.

179. Bang-Sup Song and David C. Soo. NRZ timing recovery technique for band-limited channels. *IEEE J. Solid-State Circuits*, SC-32(4):514–520, April 1997.

180. Michiel S. J. Steyaert, Wim Dehaene, Jan Craninckx, Máirtín Walsh, and Peter Real. A CMOS rectifier-integrator for amplitude detection in hard disk servo loops. *IEEE J. Solid-State Circuits*, SC-30(7):743–751, July 1995.

181. Yasuyuki Suzuki, Hidenori Shimawaki, Yasushi Amamiya, Nobuo Nagano, Hitoshi Yano, and Kazuhiko Honjo. A 40-Gb/s preamplifier using AlGaAs/InGaAs HBT's with regrown base contacts. *IEEE J. Solid-State Circuits*, SC-34(2):143–147, February 1999.

182. Yasuyuki Suzuki, Tetsuyuki Suzaki, Yumi Ogawa, Sadao Fujita, Wendy Liu, and Akihiko Okamoto. Pseudomorphic 2DEG FET IC's for 10-Gb/s optical

communication systems with external optical modulation. *IEEE J. Solid-State Circuits*, SC-27(10):1342–1346, October 1992.

183. S. M. Sze. *Physics of Semiconductor Devices*. John Wiley & Sons, New York, 2nd edition, 1981.

184. S. M. Sze, editor. *Modern Semiconductor Device Physics*. John Wiley & Sons, New York, 1998.

185. Kiyoto Takahata, Yoshifumi Muramoto, Hideki Fukano, Kazutoshi Kato, Atsuo Kozen, Shunji Kimura, Yuhki Imai, Yutaka Miyamoto, Osaake Nakajima, and Yutaka Matsuoka. Ultrafast monolithic receiver OEIC composed of multimode waveguide p-i-n photodiode and HEMT distributed amplifier. *IEEE J. Select. Topics Quantum Electron.*, 6(1):31–37, January 2000.

186. Akira Tanabe, Masaaki Soda, Yasushi Nakahara, Takao Tamura, Kazuyoshi Yoshida, and Akio Furukawa. A single-chip 2.4-Gb/s CMOS optical receiver IC with low substrate cross-talk preamplifier. *IEEE J. Solid-State Circuits*, SC-33(12):2148–2153, December 1998.

187. Akira Tanabe, Masayuki Soda, Yasushi Nakahara, Akio Furukawa, Takao Tamura, and Kazuyoshi Yoshida. A single chip 2.4Gb/s CMOS optical receiver IC with low substrate crosstalk preamplifier. In *ISSCC Dig. Tech. Papers*, pages 304–305, February 1998.

188. Telcordia Technologies. SONET transport systems: common criteria, GR-253-CORE, Issue 3. Telcordia Technologies (formerly Bellcore), Piscataway, NJ, September 2000.

189. Rodney S. Tucker. High-speed modulation of semiconductor lasers. *IEEE J. Lightwave Technology*, LT-3(6):1180–1192, December 1985.

190. Tongtod Vanisri and Chris Toumazou. Integrated high frequency low-noise current-mode optical transimpedance preamplifiers: theory and practice. *IEEE J. Solid-State Circuits*, SC-30(6):677–685, June 1995.

191. S. P. Voinigescu, D. S. McPherson, F. Pera, S. Szilagyi, M. Tazlauanu, and H. Tran. A comparison of silicon and III-V technology performance and building block implementations for 10 and 40 Gb/s optical networking ics. *Int. J. High Speed Electronics Systems*, 13(1), March 2003.

192. Sorin P. Voinigescu, Timothy O. Dickson, Rudy Beerkens, and Paul Westergaard. A comparison of Si CMOS, SiGe BiCMOS, and InP HBT technologies for high-speed and millimeter-wave ICs. Si Monolithic Integrated Circuits in RF Systems, Atlanta, September 2004.

193. O. Wada, T. Hamaguchi, S. Miura, M. Makiuchi, K. Nagai, H. Horimatsu, and T. Sakurai. AlGaAs/GaAs p-i-n photodiode/preamplifier monolithic photoreceiver integrated on semi-insulating GaAs substrate. *Appl. Phys. Lett.*, 46(10):981–983, May 1985.

194. Robert H. Walden. A review of recent progress in InP-based optoelectronic integrated circuit receiver front-ends. In Keh-Chung Wang, editor, *High-Speed Circuits for Lightwave Communications*, pages 319–330. World Scientific, Singapore, 1999.

195. Richard C. Walker, Kuo-Chiang Hsieh, Thomas A. Knotts, and Chu-Sun Yen. A 10Gb/s Si-bipolar TX/RX chipset for computer data transmission. In *ISSCC Dig. Tech. Papers*, pages 302–303, February 1998.

196. Zhi-Gong Wang, Manfred Berroth, Ulrich Nowotny, Manfred Ludwig, Peter Hofmann, Alex Hülsmann, Klaus Köhler, Brian Raynor, and Joachim Schneider. Integrated laser-diode voltage driver for 20-Gb/s optical systems using 0.3-μm gate length quantum-well HEMT's. *IEEE J. Solid-State Circuits*, SC-28(7):829–834, July 1993.

197. Keh-Chung Wang, editor. *High-Speed Circuits for Lightwave Communications*. World Scientific, Singapore, 1999.

198. A. X. Widmer and P. A. Franaszek. A DC-balanced, partitioned-block, 8B/10B transmission code. *IBM J. Res. Develop.*, 27(5):440–451, September 1983.

199. Brett Wilson and Jason D. Drew. Novel transimpedance amplifier formulation exhibiting gain-bandwidth independence. In *IEEE International Symposium on Circuits and Systems Proceedings*, pages 169–172, 1997.

200. Jack H. Winters and Richard D. Gitlin. Electrical signal processing techniques in long-haul fiber-optic systems. *IEEE Trans. on Communications*, COM-38(9):1439–1453, September 1990.

201. Thomas T. Y. Wong. *Fundamentals of Distributed Amplification*. Artech House, Boston, 1993.

202. Thomas Y. K. Wong, Al P. Freundorfer, Bruce C. Beggs, and John Sitch. A 10Gb/s AlGaAs/GaAs HBT high power fully differential limiting distributed amplifier for III-V Mach-Zehnder modulator. *IEEE J. Solid-State Circuits*, SC-31(10):1388–1393, October 1996.

203. T. K. Woodward and A. V. Krishnamoorthy. 1Gb/s CMOS photoreceiver with integrated detector operating at 850 nm. *Electronics Letters*, 34(12):1252–1253, June 1998.

204. Shinji Yamashita, Satoshi Ide, Kazuyuki Mori, Atsushi Hayakawa, Norio Ueno, and Kazuhiro Tanaka. Novel cell-AGC technique for burst-mode CMOS preamplifier with wide dynamic range and high sensitivity for ATM-PON system. *IEEE J. Solid-State Circuits*, SC-37(7):881–886, July 2002.

205. James Daniel Yoder. Optical receiver preamplifier dynamic range enhancing circuit and method. U.S. Patent No 5,734,300, March 1998.

206. K. Yonenaga, S. Kuwano, S. Norimatsu, and N. Shibata. Optical duobinary transmission system with no receiver sensitivity degradation. *Electronics Letters*, 31(4):302–304, February 1995.

207. John L. Zyskind, Jonathan A. Nagel, and Howard D. Kidorf. Erbium-doped fiber amplifiers for optical communications. In Ivan P. Kaminow and Thomas L. Koch, editors, *Optical Fiber Telecommunications IIIB*, pages 13–68. Academic Press, San Diego, 1997.

Index

10-Gigabit Ethernet, 6, 237, 319
1000Base-LX, 6
1000Base-SX, 6
100Base-FX, 6, 304
100Base-TX, 6, 304
2-dimensional electron-gas field-effect transistor, 350
3R receiver, 1
4B5B code, 6
64B66B code, 6
8B10B code, 5, 53, 174

A

ABC, *see* automatic bias control
Absolute bandwidth, 82
Absolute jitter, 94
Absorption region, 31
AC coupling
  and low-frequency cutoff, 205
  of burst-mode amplifier, 212
  of laser, 270
  of source follower, 192
Active back termination, 275, 296–297, 309
Active-feedback load, 213
Active-feedback TIA, 135, 157, 189, 222, 230
Active inductor, 192, 195, 223, 232, 277, 281
Active regime, 351, 353
Active star network, 7
Adaptive equalizer, 84

Adaptive threshold control, 142
Adaptive transimpedance, 108, 130–131, 132, 146, 151, 157, 161
  burst mode, 141
AGC, *see* automatic gain control
Aging monitor, 285
AM, *see* amplitude modulation
AM-to-PM conversion, 175
AM-VSB, *see* amplitude modulation with vestigial sideband
Amplified spontaneous emission, 35
Amplitude detector, 207, 210–211
Amplitude modulation, 175
Amplitude modulation with vestigial sideband, 4, 52, 86, 143, 290
Amplitude-shift keying, 65
Analog laser driver, 290
Analog modulator driver, 290
Analog receiver, 143
Analog transmitter, 290
APC, *see* automatic power control
APD, *see* avalanche photodetector
APON, *see* asynchronous transfer mode PON
ASE, *see* amplified spontaneous emission
ASK, *see* amplitude-shift keying
Asynchronous transfer mode, 6, 8
Asynchronous transfer mode PON, 8
ATC, *see* adaptive threshold control
ATM, *see* asynchronous transfer mode
ATM-PON, *see* asynchronous transfer mode PON

425

AT&T Bell Laboratories, 247
Automatic bias control, 263, 286, 293, 302, 311
Automatic gain control, 159, 207
  at the detector, 32
  at the optical preamplifier, 36
Automatic gain control amplifier, 1, 82, 100, 154, 159–162, 164, 170–171, 205, 207, 214–221, 224, 226
  burst mode, 290
Automatic gain control range, 162
Automatic power control, 243, 260, 282, 284–285, 304, 310–311
  burst mode, 287–288, 290
  dual loop, 284
  single loop, 283
Available input-referred noise voltage, 166–167, 170
Avalanche breakdown, 353
Avalanche gain, 31–33, 61
Avalanche noise, 32, 34
Avalanche photodetector, 25, 30–31, 31–34, 36, 38, 42, 45, 51, 57–59, 61, 64, 71–72, 95, 130, 206, 256
Averaged input-referred noise current density, 110, 128, 156

B

Back termination, 273–275, 295, 298–299, 303
  active, 275, 296, 309
  passive, 273, 275, 297, 309
Back-up battery, 263, 307
Bandwidth
  absolute, 82
  for RZ signals, 80
  noise, 68, 126
  Nyquist, 81
  of automatic power control loop, 283
  of avalanche photodetector, 34
  of buffer, 191
  of decision circuit, 68
  of distributed amplifier, 201
  of fiber, 13, 21
  of high-impedance front-end, 113
  of jitter, 93
  of laser modulation, 245
  of LED modulation, 245
  of low-impedance front-end, 113
  of main amplifier, 164
  of monitor photodiode, 284
  of multistage amplifier, 178
  of offset compensation loop, 205
  of optical preamplifier, 34
  of p-i-n photodiode, 28
  of receiver, 73
  of stage with buffer, 191
  of stage with inductive interstage network, 198
  of stage with inductive load, 197
  of stage with shunt peaking, 193
  of stage with TIA load, 188
  of the TIA's feedback amplifier, 117
  of transimpedance amplifier, 111, 114
  optical, 35, 234
Bandwidth allocation, 76
Bandwidth extension, 188–189, 193, 198–199, 201
Bandwidth shrinkage, 178
Base pushout, 180, 353
Base resistance, 181, 353
Baseline wander, 5, 173
Bathtub curve, 92, 318
BCH, see Bose-Chaudhuri-Hocquenghem code
BER, see bit-error rate
BERT, see bit-error rate test set
BERT scan, 92, 318
Bessel response, 68, 116
Bias current
  of BJT for maximum speed, 180
  of BJT for minimum noise, 128
  of laser, 245, 260
Bias current range (of laser driver), 259
Bias T, 251, 271, 286, 302
Bias voltage
  of avalanche photodetector, 31
  of electroabsorption modulator, 248, 262
  of modulator, 262
  of MZ modulator, 251, 262
  of p-i-n photodiode, 29
Bias voltage range (of modulator driver), 261
BiCMOS technology, 148, 355
Bipolar junction transistor, 115, 145, 213, 294, 351
Birefringence, 17
Bit error, 47
Bit-error rate, 49, 318
  and E_b/N_0, 53
  and dynamic range, 56
  and forward error correction, 98
  and jitter, 95
  and OSNR, 63
  and Personick Q, 49
  and sensitivity, 55
  and SNR, 52
Bit-error rate floor, 61, 247
Bit-error rate plots, 59
Bit-error rate test set, 92, 318
BJT, see bipolar junction transistor
Block coding, 5
Bode-Fano limit, 337
Bode network theorem, 196, 232
Bode plot, 194
Bond-wire inductor, 136, 195, 277
Bose-Chaudhuri-Hocquenghem code, 97
Bounded uncorrelated jitter, 91
BPON, see broadband PON

INDEX   **427**

Broadband PON, 8, 212
Broadband technique, 179
  and f_T boosting, 182
  and buffer stage, 190
  and capacitive peaking, 185
  and cascode, <u>186</u>
  and distributed amplifier, <u>200</u>
  and emitter peaking, <u>185</u>
  and fast transistor, 179
  and inductive interstage network, 197
  and inductive load, 196
  and inductive peaking, 193, 196
  and negative capacitance, 189
  and series feedback, <u>184</u>
  and shunt feedback, 187
  and shunt peaking, <u>193</u>
  and source peaking, 185
  and stage scaling, 190
  and TIA load, 187
Buffer layer, <u>348</u>
Buffer stage, 191, 216, 277
Burst, <u>6</u>, 8, 141, 212
Burst mode, <u>6</u>
Burst-mode laser driver, 287, 303, 307
Burst-mode main amplifier, 142, 212
Burst-mode receiver, 142, 212
Burst-mode signal, 212
Burst-mode transimpedance amplifier, 141
Burst-mode transmission, 6, <u>8</u>
Butterworth response, 68, 73, <u>116</u>
BW, *see* bandwidth

C

Capacitance transformer, <u>191</u>
Carrier-suppressed return-to-zero, 4, <u>252</u>
Carrier-to-noise ratio, <u>52</u>, 144, 290
Cascode, 134, <u>186</u>, 277
  regulated, 150, <u>187</u>
CATV, *see* community-antenna television
CDR, *see* clock and data recovery circuit
Central office, <u>7</u>–8
Channel capacity theorem, <u>97</u>
Channel layer, <u>348</u>
Channel-length modulation, <u>345</u>
Characteristic impedance, 197, 200, 202, 273, 298, <u>330</u>
Charge pump, 223
Chatter, <u>143</u>
Cherry-Hooper stage, 187, <u>215</u>–218, 277, 298
Chirp, 22, <u>234</u>, 245–248, 250, 252–253, 264, 291
Chirp parameter, <u>235</u>, 245, 248, 252, 257, 262
Chirped return-to-zero, 4, <u>252</u>
Chromatic dispersion, 13, <u>15</u>–16, 18–22, 253, 292
CID, *see* consecutive identical digits
Clad fiber, <u>14</u>
Clock and data recovery circuit, <u>1</u>, 113, 159

Clock multiplication unit, <u>1</u>, 267
Closed-loop frequency response, 115, 118, 132, 151
CML, *see* current-mode logic
CMOS, *see* complementary metal-oxide-semiconductor
CMRR, *see* common-mode rejection ratio
CMU, *see* clock multiplication unit
CNR, *see* carrier-to-noise ratio
CO, *see* central office
Code rate, <u>53</u>, 98
Coding gain, <u>98</u>
Coherent detection, <u>19</u>
Color grading, <u>314</u>
Common-base input stage, <u>133</u>
Common-gate input stage, <u>133</u>, 150
Common-mode current, <u>324</u>
Common-mode rejection ratio, <u>327</u>
Common-mode voltage, <u>322</u>
Community-antenna television, <u>4</u>, 86, 143, 290
Complementary metal-oxide-semiconductor, 145, 213, 294, <u>346</u>
Compliance voltage, <u>261</u>
Composite second order, <u>90</u>, 143, 243, 290
Composite triple beat, <u>90</u>, 143, 243, 290
Confocal waveguide, <u>12</u>
Consecutive identical digits, 173
Continuous mode, <u>6</u>
Continuous-mode main amplifier, 212
Continuous-mode transmission, 6
Continuous wave, <u>3</u>, 233
Cooled laser, 237, <u>240</u>, 307
Copackaging
  of receiver, <u>136</u>
  of transmitter, <u>272</u>, 307
Corning Glass Works, <u>12</u>
Cost function, <u>84</u>
CPM, *see* cross-phase modulation
Cross-phase modulation, <u>18</u>
CRZ, *see* chirped return-to-zero
CS-RZ, *see* carrier-suppressed return-to-zero
CSO, *see* composite second order
CTB, *see* composite triple beat
Current-mode logic, <u>268</u>, 280
Current-mode technique, <u>135</u>
Current-mode transimpedance amplifier, 134
Current steering, <u>268</u>
Current switching, <u>303</u>
Cutoff frequency, <u>197</u>–198, 200–202, 337
CW, *see* continuous wave
Cycle-to-cycle jitter, <u>94</u>

D

Dark current, <u>30</u>, 34, 38, 72
Data-dependent jitter, <u>90</u>, 111, 164, 173
Data retiming, <u>280</u>

DBR, see distributed Bragg reflector laser
DC balance, 4, 6, 51, 142, 212, 288
DC wander, 5
DCF, see dispersion compensating fiber
DDA, see differential difference amplifier
DDJ, see data-dependent jitter
DEC, see decision circuit
Decision circuit, 47, 49, 54, 68, 77, 79, 82–84, 90, 93, 95, 98–99, 162, 207
Decision circuit sensitivity, 162
Decision-feedback equalizer, 82
Decision point, 90
Decision-point steering, 95
Decision threshold ambiguity width, 162
Decision threshold control, 95
Decision threshold offset, 70
Demultiplexer, 1
Dense wavelength division multiplexing, 14, 18, 175, 235, 240
Depletion-mode MESFET, 347
Detector model, 45
Deterministic jitter, 91, 95, 266, 317
DFB, see distributed-feedback laser
DFE, see decision-feedback equalizer
DGD, see differential group delay
DHBT, see double heterojunction bipolar transistor
Differential current switch, 268
Differential difference amplifier, 204–206
Differential group delay, 17
Differential-mode current, 324
Differential-mode voltage, 322
Differential quantum efficiency, 242
Differential TIA, 125, 127, 138, 140, 146, 148–149
Direct detection, 19
Direct modulation, 233, 291
Discrete multitone, 4
Dispersion
 chromatic, 15, 19, 253
 group-velocity, 15
 modal, 14
 polarization-mode, 17, 253
Dispersion compensating fiber, 16, 254
Dispersion compensation, 16
Dispersion parameter, 15, 254
Dispersion penalty, 235, 253
Dispersion-shifted fiber, 16, 254
Distributed amplifier, 200, 203, 220, 277, 301
Distributed Bragg reflector laser, 240
Distributed-feedback laser, 237, 240, 253
Dithering, 85
DJ, see deterministic jitter
DMT, see discrete multitone
DMUX, see demultiplexer
Donor layer, 348
Double heterojunction bipolar transistor, 358
Double reflection, 273, 275, 295–296, 298, 302

Downstream direction, 8
DQE, see differential quantum efficiency
Drift
 of baseline, 5, 173, 282
 of Mach-Zehnder modulator, 251, 286
 of peak detector, 288
Drift field, 25–26, 31, 355
Drift transistor, 357
Driving point impedance, 106
Droop, 210
DSF, see dispersion-shifted fiber
DTAW, see decision threshold ambiguity width
Dual-drive MZM, 250–252, 302
Dual-gate FET, 208
Dual-loop APC, 284
Duty-cycle distortion jitter, 91, 266
DWDM, see dense wavelength division multiplexing
Dynamic flip-flop, 281
Dynamic range
 of amplifier, 88
 of main amplifier, 169
 of receiver, 56, 61
 of transimpedance amplifier, 108, 130, 132, 140
 of VGA stage, 207

E

E_b/N_0, 52, 54, 99
EAM, see electroabsorption modulator
EAM driver, 262, 294–295, 299, 301, 305, 307
Early effect, 352
ECL, see emitter-coupled logic
EDFA, see erbium-doped fiber amplifier
Edge-emitting laser, 240
Electrical receiver sensitivity, 54
Electroabsorption modulated laser, 247
Electroabsorption modulator, 247
Electrooptic effect, 250
Emitter-coupled logic, 280, 354
Emitter peaking, 185
Emitter pole, 185
EML, see electroabsorption modulated laser
End-of-life detection, 285, 290
Enhancement-mode MESFET, 347
EOL, see end-of-life detection
EPON, see Ethernet PON
Equalizer, 15–16, 18, 78, 82, 113, 161
 adaptive, 84
 decision-feedback, 82
 feed-forward, 82
 implementation issues of, 85
 postcursor, 83
 precursor, 84
Equivalent noise current source, 109
Equivalent noise voltage source, 165
ER, see extinction ratio

INDEX   **429**

Erbium-doped fiber, 34
Erbium-doped fiber amplifier, 13, <u>34</u>, 36–37, 40
Error propagation, <u>83</u>
Ethernet, 6, 153, 304
Ethernet PON, 8, 212
Even mode, <u>322</u>
Excess bandwidth, <u>81</u>
Excess noise factor, <u>33</u>
External modulation, <u>233</u>, 291
Extinction ratio, 38, 55, 102–103, <u>236</u>–237, 245, 283, 287, 303, 310
Eye diagram, 73, 78, 90, 264–267, <u>313</u>
Eye-diagram mask margin, <u>267</u>
Eye-diagram mask test, <u>267</u>
Eye margin, <u>318</u>
Eye monitor, <u>84</u>, 96
Eye opening, <u>267</u>
Eye pattern, <u>313</u>

F

f_{max}, *see* maximum frequency of oscillation
f_T, *see* transition frequency
f_T-doubler, 183
Fabry-Perot laser, <u>237</u>, 254
Fall time, 262, <u>264</u>
Fast Ethernet, 3, 6, 304
FDDI, *see* fiber distributed data interface
FEC, *see* forward error correction
Feed-forward equalizer, <u>82</u>
Feeder fiber, 8
FET, *see* field-effect transistor
FFE, *see* feed-forward equalizer
Fiber attenuation, <u>11</u>, 253
Fiber bandwidth, <u>13</u>, 21
Fiber Channel, 6, 174
Fiber distributed data interface, 3, 6–7
Fiber loss, <u>11</u>, 253
Fiber optical parametric amplifier, 13
Fiber-to-the-curb system, <u>8</u>
Fiber-to-the-home system, <u>8</u>, 307
Field-effect transistor, 343
Finite impulse response, <u>82</u>
FIR, *see* finite impulse response
FKE, *see* Franz-Keldysh effect
Flip-chip technology, 154, 273, 307
Flip-flop, 263, 280–281, 296, 305
FOPA, *see* fiber optical parametric amplifier
Forward error correction, 64, <u>97</u>, 145
Forward transmission coefficient, <u>332</u>
Four-wave mixing, <u>18</u>
FP, *see* Fabry-Perot laser
Franz-Keldysh effect, <u>248</u>
Frequency-shift keying, <u>65</u>
FSK, *see* frequency-shift keying
FTTC, *see* fiber-to-the-curb system
FTTH, *see* fiber-to-the-home system

Full-wave rectifier, 210–211
FWM, *see* four-wave mixing

G

GaAs, *see* gallium-arsenide technology
Gain
  and Miller effect, 186
  of avalanche photodetector, 31
  of Cherry-Hooper stage, 216
  of coding (FEC), <u>98</u>
  of main amplifier, <u>161</u>
  of multistage amplifier, 177
  of offset compensation loop, 204
  of optical amplifier, 35
  of predriver, 276
  of stage for optimum GBW extension, 178
  of stage with series feedback, 185
  of stage with TIA load, 187
  of the TIA's feedback amplifier, 114
  of two port, <u>332</u>
Gain-bandwidth extension, <u>176</u>
Gain-bandwidth product, <u>176</u>
Gain compression, 31, <u>87</u>, 162, 170
Gain control, 207
Gallium-arsenide technology, 146, 153, 213–214, 226, 294–296, 301, 305, 347, 350, 357
Gas lens, <u>12</u>
Gauge transformation, <u>327</u>
Gaussian distribution, 50, 61, 65, 91, 95, 204
Gaussian noise, 49, 54, 74, 97, 317
Gaussian pulse, 20
Gaussian spectrum, 20
GbE, *see* gigabit Ethernet
GBW, *see* gain-bandwidth product
Gigabit Ethernet, 6–7, 174, 285
Gilbert cell, 209, <u>217</u>
Golden PLL, <u>93</u>
Graded-index multimode fiber, <u>14</u>
Gradient descent, <u>85</u>
GRIN, *see* graded-index multi-mode fiber
Gross coding gain, <u>98</u>
Group-delay variation
  of main amplifier, <u>164</u>
  of transimpedance amplifier, <u>111</u>
Group-velocity dispersion, <u>15</u>
GVD, *see* group-velocity dispersion

H

Half-rate clock, 282, 296
Half section, <u>202</u>, 337
Hard-decision decoder, <u>99</u>
Harmonic distortion, <u>87</u>–89, 108, 170
HBT, *see* heterojunction bipolar transistor
HD, *see* harmonic distortion
Headroom, 118, 181, 187, 192, 209, 223, 272, 298
HEMT, *see* high electron-mobility transistor

H

Heterojunction bipolar transistor, 145, 213, 294, 355
Heterostructure field-effect transistor, 145, 213, 294, 348
HFC, *see* hybrid fiber-coax
HFET, *see* heterostructure field-effect transistor
High electron-mobility transistor, 350
High-impedance front-end, 112
Horizontal eye closure, 317
Horizontal eye margin, 318–319
Horizontal eye opening, 76, 86, 265–266, 316
Hybrid fiber-coax, 4, 86, 143, 290

I

I/V curve (of laser), 241
IIP3, *see* input-referred 3rd-order intercept point
IMD, *see* intermodulation distortion
Impedance matching, 144, 203, 248, 273, 276, 291, 299, 302, 333
Incident wave, 330
Indium-phosphide technology, 153, 220, 226, 299, 305, 351, 357
Inductive interstage network, 198, 276–277
Inductive load, 196, 276–277
Information capacity theorem, 97
InP, *see* indium-phosphide technology
Input buffer, 215, 218, 295, 297–300
Input dynamic range, 169
Input offset voltage, 171
Input overload current
  of receiver, 56
  of transimpedance amplifier, 108, 114, 130, 151
Input overload voltage, 170, 224
Input-referred 3rd-order intercept point, 89
Input-referred noise current, 109, 121
Input-referred noise current spectrum, 109, 122, 126
Input-referred noise voltage, 165
Input-referred noise voltage spectrum, 165
Input-referred rms noise current, 110, 126–127, 129
Input-referred rms noise voltage, 165, 170
Input reflection coefficient, 331
Input return loss, 331, 336–338
Integrate and dump, 79, 154, 288
Interferometer, 250
Interleaver, 97
Intermodulation distortion, 88, 90, 108, 170
Internet, 9
Interstage buffer, 191, 216, 277
Intersymbol interference, 73–74, 78, 80–82, 91, 313
Inverse scaling, 192, 223, 277
Ioffe Physical Institute, 247
Ionization-coefficient ratio, 33
ISI, *see* intersymbol interference

ISI canceler, 82
Isolation, 187, 332

J

Jitter, 90, 107–108, 170, 175, 212, 261, 280, 313
  absolute, 94
  bounded uncorrelated, 91
  cycle-to-cycle, 94
  data-dependent, 90, 111, 164, 173
  deterministic, 91, 95, 266, 317
  duty-cycle distortion, 91, 266
  period, 94
  periodic, 91
  random, 90–91, 95, 266, 317
  sampling, 76–77, 79, 93
  total, 92, 266–267
  wideband, 93
Jitter analyzer, 93
Jitter bandwidth, 93, 267
Jitter generation, 93, 265–266, 269
Jitter tolerance, 94–95
Jitter transfer, 93, 265
Johnson limit, 354

K

Kirk effect, 180, 353

L

L/I curve, 242–243
LA, *see* limiting amplifier
Laser
  continuous wave, 233
  cooled, 237, 240, 307
  distributed Bragg reflector, 240
  distributed-feedback, 237, 240, 253
  Fabry-Perot, 237, 254
  semiconductor, 237, 241, 245, 247
  uncooled, 237, 239, 261, 307
  vertical-cavity surface-emitting, 237, 240–242, 271
  yttrium-aluminum-garnet, 291
Laser diode, 3, 237, 269–271, 290
Laser driver, 1, 259, 270, 273, 291, 296–298, 303, 305
  analog, 290
  and automatic power control, 282
  and back termination, 273
  and current-steering output stage, 268
  and data retiming, 280
  and end-of-life detection, 285
  and predriver, 276
  and pulse-width control, 279
  burst-mode, 287
  specifications, 259
Laser rate equation, 243
Lateral BJT, 352

LD, *see* laser diode
LDPC, *see* low-density parity-check code
Least-mean-square algorithm, 85
LED, *see* light-emitting diode
LED driver, 304
Light-emitting diode, 237, 240, 254, 304
Limiting amplifier, 1, 47, 159–161, 164, 170–171, 175, 206, 212, 215, 222–224
Limits in optical communication systems, 253
  due to attenuation, 255
  due to chromatic dispersion, 253
  due to polarization mode dispersion, 255
$LiNbO_3$, *see* lithium-niobate modulator
Line code, 4, 173
Linear channel, 46–48, 53, 78, 82, 86, 100
Linear regime, 131–132, 208–209, 221, 261, 344
Linearization technique, 292
Linewidth enhancement factor, 235
Lithium-niobate modulator, 250
LMS, *see* least-mean-square algorithm
Long-haul transmission, 3, 7, 11–12, 15, 19, 62, 64, 77, 82, 109, 237
Longitudinal mode, 322
LOS, *see* loss of signal
Loss
  due to substrate, 357
  of fiber, 11, 253
  of interstage network, 198
  of MZ modulator, 252
  of transmission line, 202, 220, 302
Loss of signal, 211
Low-density parity-check code, 99
Low-frequency cutoff, 173–174, 192, 205, 282
Low-impedance front-end, 112–113, 143
Low-noise amplifier, 169

M

M-derived half section, 202, 337
MA, *see* main amplifier
Mach-Zehnder modulator, 247, 250, 286
Main amplifier, 1, 45, 107, 159
  and automatic gain control, 207
  and broadband techniques, 179
  and loss of signal detection, 211
  and multistage architecture, 176
  and offset compensation, 203
  burst-mode, 212
  specifications, 161
Mark density, 5, 284, 289
Mark-density compensation, 285, 289
Mask hit, 267
Matched filter, 53, 77–78, 81
Maximum frequency of oscillation, 179–181, 346, 353
Maximum input current for linear operation, 108, 151
Maximum input voltage for linear operation, 170, 208, 224
Mean-time to failure, 247, 285
MES, *see* metal-semiconductor
MESFET, *see* metal-semiconductor field-effect transistor
Metal-oxide semiconductor field-effect transistor, 343
Metal-oxide-semiconductor, 346
Metal-semiconductor, 347
Metal-semiconductor field-effect transistor, 145, 213, 294, 347
Metal-semiconductor-metal photodetector, 153, 351
Metallic mode, 322
Miller capacitance, 186–190, 201, 216, 220, 271, 277
Miller effect, 183–184, 189, 205
Mixed-mode S parameter, 339
MLM, *see* multiple-longitudinal mode
MMF, *see* multimode fiber
Modal dispersion, 14
Mode-partition noise, 247
Mode-suppression ratio, 247
Mode voltage, 322–323
MODFET, *see* modulation-doped field-effect transistor
Modulation bandwidth, 245
Modulation current, 260, 269–270, 272, 278, 284, 298–299, 303
Modulation current range, 259–260
Modulation-doped field-effect transistor, 350
Modulation index, 144, 243, 293
Modulation voltage, 262, 269, 271
Modulation voltage range, 261, 263
Modulator, 3, 233, 247
  dual-drive MZ, 250
  electroabsorption, 247
  lithium-niobate, 250
  Mach-Zehnder, 247, 250, 286
  single-drive MZ, 251
  tandem, 233
Modulator driver, 1, 259, 261, 291, 295–298, 300–301, 303, 305
  analog, 290
  and automatic bias control, 286
  and back termination, 273
  and current-steering output stage, 268
  and data retiming, 280
  and predriver, 276
  and pulse-width control, 279
  specifications, 259
Monitor photodiode, 243, 260, 282, 284–286, 288, 307
MOS, *see* metal-oxide-semiconductor

MOSFET, *see* metal-oxide-semiconductor field-effect transistor
MPN, *see* mode-partition noise
MQW, *see* multiple quantum well
MSM, *see* metal-semiconductor-metal photodetector
MTTF, *see* mean-time to failure
Multimode fiber, 14, 240
Multiple-longitudinal mode, 239
Multiple quantum well, 239
Multiplexer, 1, 85, 282, 299
Multiplication factor, 31, 354
Multiplication region, 31
Multistage amplifier, 118, 150, 168, 176–178, 188, 192–193, 199, 207
MUX, *see* multiplexer
MZM, *see* Mach-Zehnder modulator
MZM driver, 262, 272, 301

N

National Association of Broadcasters, 52
NECG, *see* net electrical coding gain
Negative capacitance, 189–190, 230, 276–277, 302
Negative impedance converter, 190, 230, 276, 302
Net electrical coding gain, 99
Network analyzer, 330
Neutralization, 190
NF, *see* noise figure
NIC, *see* negative impedance converter
Noise
  and bit-error rate, 49
  of analog receiver, 144
  of analog transmitter, 290
  of avalanche photodetector, 32
  of laser, 246
  of main amplifier, 165
  of optical amplifier, 36
  of p-i-n photodetector, 29
  of receiver, 47
  of transimpedance amplifier, 108, 121
Noise bandwidth, 29, 53–54, 58–59, 68, 78, 80, 126
Noise equivalent bandwidth, 68
Noise figure, 37–40, 43, 59, 62–63, 165–167, 168, 170, 220, 228
Noise matching, 125, 137
Noise optimization (TIA), 121, 127
Non-return-to-zero, 3, 21, 49, 51, 53, 73, 76, 78, 80, 107, 233–234, 264, 313, 322
Non-return-to-zero change-on-ones, 3
Nonlinearity, 86
  and composite second-order distortion, 90
  and composite triple-beat distortion, 90
  and gain compression, 87
  and harmonic distortion, 87
  and intermodulation distortion, 88

of decision circuit, 47
of decision-feedback equalizer, 82
of external modulator, 292
of fiber, 18
of intensity modulation with direct detection, 19
of laser, 274, 290
of limiting amplifier, 160
of MZ modulator, 293
of transmitter, 292–293
Nonzero dispersion-shifted fiber, 18, 254
NRZ, *see* non-return-to-zero
NRZ1, *see* non-return-to-zero change-on-ones
NWA, *see* network analyzer
Nyquist bandwidth, 81
Nyquist pulse, 81
NZ-DSF, *see* nonzero dispersion-shifted fiber

O

OC, *see* optical carrier
Odd mode, 322
OEIC, *see* optoelectronic integrated circuit
OFDM, *see* othogonal frequency division multiplexing
Offset compensation, 203, 205–206
Offset control, 131, 140–141, 149, 151, 212
Offset voltage
  for pulse-width control, 279
  for slice-level adjustment, 206
  of main amplifier, 171, 203
  of modulator, 262
  of transimpedance amplifier, 140, 142
Ohmic regime, 344
OMA, *see* optical modulation amplitude
On-chip termination, 295, 297–298, 300, 333
On-off keying, 3, 48, 65
One-port interstage network, 197
OOK, *see* on-off keying
Open collector, 273
Open-collector output stage, 297–298
Open drain, 273
Open-drain output stage, 295
Open-loop frequency response, 115, 117–118, 151, 205, 283
Open-loop pole, 117, 131–134, 141, 156
Optical amplifier, 35, 37, 40, 62, 95, 109, 132
  erbium-doped fiber amplifier, 13, 34
  fiber optical parametric amplifier, 13
  Raman amplifier, 13, 40
  semiconductor optical amplifier, 34, 239
Optical carrier (SONET)
  OC-12, 151
  OC-192, 7, 56, 153
  OC-3, 151
  OC-48, 56, 93, 108, 151, 267
  OC-768, 153
Optical duobinary, 252, 262

INDEX **433**

Optical feedforward linearization, 292
Optical fiber, 11, 13–14
Optical filter, 35–36
Optical in-line amplifier, 13, 62, 256
Optical isolator, 35
Optical modulation amplitude, 237
Optical noise figure, 39
Optical overload power, 56, 61, 131
Optical receiver sensitivity, 55, 57–58, 62, 131, 139
Optical signal-to-noise ratio, 37, 54, 63–64, 111, 132
Optically preamplified p-i-n detector, 25, 34, 36, 38–39, 46, 51, 57–59, 62, 64–65, 130, 256
Optimum APD gain, 33, 61
Optimum number of stages, 177–178
Optimum receiver bandwidth, 73
Optimum receiver response, 76–78
Optimum slice level, 95–96, 206
Optoelectronic integrated circuit
  for receiver, 153
  for transmitter, 307
Orthogonal frequency division multiplexing, 4
OSNR, *see* optical signal-to-noise ratio
Outgoing wave, 330
Output buffer, 132, 146, 150, 215, 218, 275
Output reflection coefficient, 332
Output return loss, 332
Output voltage range (of laser driver), 261
Overload limit, 56, 61, 131
Overmodulation, 262

P

P-i-n FET, 153
P-i-n photodetector, 25–26, 28–30, 38, 45, 57–58, 129, 143–144, 153, 246, 256
P-i-n photodiode, 25
P2MP, *see* point-to-multipoint network
P2P, *see* point-to-point connection
Package parasitics, 28, 136, 242, 271, 307, 336–337
PAM, *see* pulse amplitude modulation
Passive back termination, 273, 275, 297, 309
Passive optical network, 8–9, 287, 303
Passive star network, 7
PD, *see* photodetector *and* photodiode
Peak detector, 141–142, 210, 288
Period jitter, 94
Periodic jitter, 91
Personick integral, 69, 126
Personick Q, 50
Phantom zero, 118
Phase-locked loop, 93–94, 314
Phase modulation, 175, 252
Phase noise, 93, 246
Phase-shift keying, 65

PHEMT, *see* pseudomorphic high electron-mobility transistor
PHFET, *see* pseudomorphic heterostructure field-effect transistor
Photodetector, 1, 25
Photodiode, 28, 153, 243, 282, 284, 289
Pilot tone, 284, 286, 293
PJ, *see* periodic jitter
Plastic optical fiber, 12
PLL, *see* phase-locked loop
PM, *see* phase modulation
PMD, *see* polarization-mode dispersion
PMF, *see* polarization-maintaining fiber
POF, *see* plastic optical fiber
Point-to-multipoint network, 7–8
Point-to-point connection, 7
Poisson distribution, 57, 65
Polarization maintaining fiber, 18
Polarization-mode dispersion, 17, 82, 253, 255
Polarization-mode dispersion parameter, 17
PON, *see* passive optical network
Post amplifier, 1, 159
  and transimpedance amplifier, 107, 132, 151
Postcursor equalizer, 83
Postcursor ISI, 82
Power dissipation
  and active termination, 297
  and back termination, 274
  and TIA load, 188
  and transmission line, 273
  of burst-mode laser driver, 303
  of laser/modulator driver, 263
  of multistage amplifier, 178
Power MUX, 299
Power penalty, 70, 161
  and APC bandwidth, 282
  and decision threshold offset, 70
  and detector dark current, 72
  and extinction ratio, 236
  and fiber alignment, 15
  and intersymbol interference, 74
  and MA gain, 162
  and MA low-frequency cutoff, 173
  and MA noise, 168
  and MA offset, 171
  and MA sensitivity, 170
  and polarization-mode dispersion, 17
  and pulse spreading, 21, 235
  and relative intensity noise, 246
Power splitter/combiner, 8
Power-supply rejection ratio, 327
Power wave, 330
PP, *see* power penalty
PRBS, *see* pseudorandom bit sequence
Precursor equalizer, 84
Precursor ISI, 83

Predistortion linearization, 292–293
Predriver, 263, 271, 276–279, 302
Preemphasis, 304
Primary dark current, 34
Principle state of polarization, 17
Pseudo bit-error rate, 96
Pseudomorphic heterostructure field-effect transistor, 350
Pseudomorphic high electron-mobility transistor, 350
Pseudorandom bit sequence, 5, 173, 284, 314, 316–317
PSK, *see* phase-shift keying
PSP, *see* principal state of polarization
PSRR, *see* power-supply rejection ratio
Pulse amplitude modulation, 4, 102
Pulse spreading, 16, 19–22, 235, 253, 255
Pulse-width control, 265, 279
Pulse-width distortion, 107–108, 142, 170–171, 173, 212, 261, 265–266, 269, 279–280, 288
Push-pull topology, 291
PWC, *see* pulse-width control
PWD, *see* pulse-width distortion

Q

QAM, *see* quadrature amplitude modulation
QCSE, *see* quantum-confined Stark effect
Quadrature amplitude modulation, 4, 52, 86
Quadrature point, 262, 287, 293
Quantum-confined Stark effect, 248
Quantum efficiency, 25, 27, 37, 56, 243
Quantum limit, 65, 130
Quantum well, 239

R

Raised-cosine filtering, 69, 78, 80–82
Raman amplifier, 13, 40
Random jitter, 90–91, 95, 266, 317
Re-amplification, 1
Re-shaping, 1
Re-timing, 1
Receiver bandwidth, 73–76, 81, 111, 164
Receiver model, 45, 47
Rectangular filter, 77–80
Rectifier, 210
Reed-Solomon code, 97
Reference receiver, 268
Reflected wave, 330
Regenerator, 62, 95, 266–267, 280
Regulated cascode, 150, 187, 229
Relative intensity noise, 144, 246, 290–291
Relaxation oscillation, 244–246, 267
Remote node, 8
Responsivity, 27, 31, 35–36, 55, 57, 108, 144, 243
Return-to-zero, 3–4, 56, 80–81, 233

Return-to-zero differential phase-shift keying, 4, 252, 262
Reverse transmission coefficient, 332
RF choke, 221, 242, 269–270, 290, 299, 309
RFC, *see* RF choke
RIN, *see* relative intensity noise
Ring network, 7
Rise time, 262, 264
Rise-time budget, 264
RJ, *see* random jitter
RN, *see* remote node
Root-locus plot, 194
RS, *see* Reed-Solomon code
RS(255 239), 97
Run length, 4, 173–174, 283
RZ, *see* return-to-zero
RZ-DPSK, *see* return-to-zero differential phase-shift keying

S

S parameter, 162, 329
 and matching, 333
 differential, 339
 single ended, 329
Sampling jitter, 76–77, 79, 93
Sampling offset, 79
Saturated regime
 of bipolar junction transistor, 115, 261, 353
 of field-effect transistor, 281, 345
Saturation current (of photodiode), 30
SBS, *see* stimulated Brillouin scattering
SCFL, *see* source-coupled FET logic
SCM, *see* subcarrier multiplexing
Scrambling, 5, 173
SDH, *see* synchronous digital hierarchy
SDM, *see* space division multiplexing
Self-phase modulation, 18, 22, 175
Semi-insulating substrate, 348, 357
Semiconductor laser, 237, 241, 245, 247
Semiconductor optical amplifier, 34, 239
Sensitivity, 54
 and attenuation limit, 255
 and bit rate, 128
 and dynamic range, 56
 and extinction ratio, 237
 and MA noise, 165
 and optical signal-to-noise ratio, 63
 and power penalty, 70
 and quantum limit, 65
 and receiver bandwidth, 73
 and reference bit-error rate, 56
 and TIA noise optimization, 126
 electrical, 54
 of APD receiver, 57–58
 of decision circuit, 162
 of main amplifier, 170

of p-i-n receiver, 57–58
of receiver with optical preamplifier, 57–58
of transimpedance amplifier, 110
optical, 55
Sensitivity limit, 47, 56, 61, 131, 163, 170
Series feedback, 179, 184, 209, 215–216, 229, 277, 291
Series peaking, 199, 277, 303
Shot noise, 29, 32–33, 36–37, 42, 58–59, 72, 121–122, 124, 127, 144–145, 246
Shunt feedback, 113, 136, 139, 179, 277, 291
Shunt-feedback TIA, 107, 112–113, 114–115, 117, 119, 121, 130, 132–135, 138, 154, 187, 189
Shunt peaking, 178, 193, 195–196, 199, 277–278
Si, *see* silicon technology
Sidegating, 348
SiGe, *see* silicon-germanium technology
Signal-spontaneous beat noise, 36, 39
Signal-spontaneous beat noise limited noise figure, 39
Signal-to-noise ratio, 37, 50–51, 51–52, 54, 97, 111, 132, 144, 166, 246–247
Silica glass, 11
Silicon-germanium technology, 148, 153, 215, 218–219, 226, 305, 357
Silicon technology, 346, 354
Single-drive MZM, 251, 271
Single-longitudinal mode, 240
Single-loop APC, 283–284, 310–311
Single-mode fiber, 14–16, 20–21, 235, 253–254, 256
Slice-level adjust, 95, 206
Slice-level error, 171
Slice-level steering, 95, 161, 171
SLM, *see* single-longitudinal mode
Slope efficiency, 242–243, 283–284, 290, 304
Slow-start circuit, 286
SMF, *see* single-mode fiber
Smith chart, 334
Smooth-start circuit, 286
SNR, *see* signal-to-noise ratio
SNR per bit, 52
SOA, *see* semiconductor optical amplifier
Soft-decision decoder, 99–100, 161
Soliton, 22
SONET, *see* synchronous optical network
Source-coupled FET logic, 280
Space division multiplexing, 8
Spectrum analyzer, 93
Spiral inductor, 148, 195–196, 199, 277, 281, 338, 357
SPM, *see* self-phase modulation
Spontaneous emission, 35, 240, 242, 244, 246
Spontaneous-spontaneous beat noise, 36–39
Spot noise figure, 167
SRS, *see* stimulated Raman scattering

Stage scaling, 192
Static flip-flop, 281
Stimulated Brillouin scattering, 18
Stimulated emission, 34, 237, 240, 244, 246
Stimulated Raman scattering, 18, 40, 175
Subcarrier multiplexing, 4, 52
Switching curve, 248, 251–252, 262, 279, 286–287, 293
Switching voltage, 248, 251–252, 262
Synchronous digital hierarchy, 5, 7, 173, 236, 285
Synchronous optical network, 5, 7, 56, 93, 97, 108, 154, 173–174, 211, 226, 236, 266–267, 280, 285
Synthetic inductor, 195

T

T-coil network, 199, 223, 276–277, 302, 338
Tandem modulator, 233
TAS, *see* transadmittance stage
TCM, *see* time compression multiplexing
TDD, *see* time division duplexing
TDM, *see* time division multiplexing
TDMA, *see* time division multiple access
TEC, *see* thermoelectric cooler
TEGFET, *see* 2-dimensional electron-gas field-effect transistor
Temperature dependence
of avalanche photodetector, 32
of laser, 240, 243
Terminal current, 324–325
Terminal voltage, 322–325
THD, *see* total harmonic distortion
Thermoelectric cooler, 240, 263, 307
Threshold current, 241–242, 243, 245, 261, 283–285, 290
Threshold voltage, 49, 51, 71, 77, 181, 211, 223, 344
TIA, *see* transimpedance amplifier
Time compression multiplexing, 8
Time division duplexing, 8
Time division multiple access, 8
Time division multiplexing, 8
Time interval analyzer, 93
Timing jitter, 265
TIS, *see* transimpedance stage
TJ, *see* total jitter
TOD, *see* turn-on delay
Total harmonic distortion, 88
Total input-referred noise current, 110, 126
Total jitter, 92, 266–267
Transadmittance stage, 216
Transceiver, 3–4, 7, 98, 263, 307, 321
Transfer impedance, 106
Transform limited pulse, 234, 253, 255
Transformer, 144, 291

436 INDEX

Transimpedance, 76, 106–108, 110–114, 116, 163, 168, 172
Transimpedance amplifier, 1, 25, 45, 64, 105, 112–113, 159, 187
 active-feedback, 135
 adaptive, 130
 analog, 143
 and noise optimization, 121
 as load, 187
 burst-mode, 141
 current-mode, 134
 differential, 137
 shunt-feedback, 113
 specifications, 105
 with common-base/gate input stage, 133
 with inductive input coupling, 136
 with post amplifier, 132
Transimpedance limit, 117–118, 121, 129, 134, 151
Transimpedance stage, 186, 216
Transistor-transistor logic, 354
Transit time, 28, 179, 182, 344–345, 352–353, 355
Transition density, 5
Transition frequency, 179, 181, 345, 353
Transmission line, 132, 154, 159, 202, 220, 242, 248, 250, 252, 269, 272–273, 276, 295–298, 307, 329–331
 artificial, 137, 196, 198, 200–201, 301, 337
Transmitted wave, 330
Transponder, 3
Transresistance, 106, 187
Transversal mode, 322
Traveling-wave amplifier, 200
Traveling-wave photodetector, 26
Triode regime, 344
Triple beat, 89
TTL, *see* transistor-transistor logic
Turbo code, 99
Turn-on delay, 244–245, 265, 279, 288, 303
Turn-on delay compensation, 288, 303
Turn-on delay jitter, 244–245, 266, 288, 303
Two-port interstage network, 197
Typical values, 10
TZA, *see* transimpedance amplifier

U

UI, *see* unit interval
Uncooled laser, 237, 239, 261, 307
Undermodulation, 262
Unilateralization, 187

Unit interval, 264
Upstream direction, 8

V

Variable-gain amplifier, 160, 207
Variable-gain stage, 207
VCSEL, *see* vertical-cavity surface-emitting laser
Vertical BJT, 352
Vertical-cavity surface-emitting laser, 237, 240–242, 271
Vertical eye closure, 75, 264, 316–317
Vertical eye margin, 318–319
Vertical eye opening, 74, 84, 316
VGA, *see* variable-gain amplifier
Viterbi decoder, 77, 82
Voltage swing
 at DEC input, 162
 at driver input, 259
 at EAM driver output, 262
 at flip-flop output, 281
 at high-impedance front-end, 113
 at MA input, 170
 at modulator driver output, 262, 286, 299, 302, 305
 at MZ modulator, 251–252
 at MZM driver output, 262, 272
 at predriver output, 276, 278
 at TIA input, 114, 139
 at TIA output, 115
 of differential circuit, 138, 322

W

Waterfall curve, 98–99
Waveguide photodetector, 26
Wavelength division multiplexing, 8, 64, 175, 240
Wavelength division multiplexing PON, 9
WDM, *see* wavelength division multiplexing
WDM-PON, *see* wavelength division multiplexing PON
Weight perturbation, 85
Wideband jitter, 93
Wideband noise figure, 167

X

XOR gate, 96
XPM, *see* cross-phase modulation

Y

YAG, *see* yttrium-aluminum-garnet laser
Yttrium-aluminum-garnet laser, 291

TK
7871.
.B74
S23

2005